Wavefront Shaping for Biomedical Imaging

Learn about the theory, techniques, and applications of wavefront shaping in biomedical imaging using this unique text. With authoritative contributions from researchers who are defining the field, cutting-edge theory is combined with real-world practical examples, experimental data, and the latest research trends to provide the first book-level treatment of the subject. It is suitable for both background reading and use in a course, with coverage of essential topics such as adaptive optical microscopy, deep tissue microscopy, time reversal and optical phase conjugation, and tomography. The latest images from the forefront of biomedical imaging are included, and full-color versions are available in the eBook version. Researchers, practitioners, and graduate students in optics, biophotonics, biomedical engineering, and biology who use biomedical imaging tools and are looking to advance their knowledge of the subject will find this an indispensable resource.

Joel Kubby was Professor in the Jack Baskin School of Engineering at the University of California at Santa Cruz. He also authored *A Guide to Hands-On MEMS Design and Prototyping* (Cambridge, 2011).

Sylvain Gigan is Professor at Sorbonne Université, Paris and Group Leader in the Laboratoire Kastler-Brossel, Paris.

Meng Cui is Assistant Professor of Electrical and Computer Engineering and Biology at Purdue University.

Advances in Microscopy and Microanalysis

Microscopic visualization techniques range from atomic imaging to visualization of living cells at near nanometer spatial resolution, and advances in the field are fueled by developments in computation, image detection devices, labeling, and sample preparation strategies. Microscopy has proven to be one of the most attractive and progressive research tools available to the scientific community, and remains at the forefront of research in many disciplines, from nanotechnology to live cell molecular imaging. This series reflects the diverse role of microscopy, defining it as any method of imaging objects of micrometer scale or less, and includes both introductory texts and highly technical and focused monographs for researchers and practitioners in materials and the life sciences

Series Editors
Patricia Calarco, *University of California, San Francisco*
Michael Isaacson, *University of California, Santa Cruz*

Series Advisors
Bridget Carragher, *The Scripps Research Institute*
Wah Chiu, *Baylor College of Medicine*
Christian Colliex, *Université Paris Sud*
Ulrich Dahmen, *Lawrence Berkeley National Laboratory*
Mark Ellisman, *University of California, San Diego*
Peter Ingram, *Duke University Medical Center*
J. Richard McIntosh, *University of Colorado*
Giulio Pozzi, *University of Bologna*
John C. H. Spence, *Arizona State University*
Elmar Zeitler, *Fritz-Haber Institute*

Books in Series

Published
Heide Schatten, *Scanning Electron Microscopy for the Life Sciences*
Frances Ross, *Liquid Cell Electron Microscopy*
Joel Kubby, Sylvain Gigan, and Meng Cui, *Wavefront Shaping for Biomedical Imaging*

Forthcoming
Michael Isaacson, *Microscopic Nanocharacterization of Materials*
Chris Jacobsen, *X-Ray Microscopy*
Thomas F. Kelly, Simon Ringer, and Brian Gorman, *Atomic-Scale Tomography*
Eric Lifshin, *The Scanning Electron Microscope*

Wavefront Shaping for Biomedical Imaging

Edited by

JOEL KUBBY
Formerly of University of California, Santa Cruz

SYLVAIN GIGAN
Sorbonne Université and Laboratoire Kastler-Brossel

MENG CUI
Purdue University

CAMBRIDGE
UNIVERSITY PRESS

CAMBRIDGE
UNIVERSITY PRESS

University Printing House, Cambridge CB2 8BS, United Kingdom

One Liberty Plaza, 20th Floor, New York, NY 10006, USA

477 Williamstown Road, Port Melbourne, VIC 3207, Australia

314–321, 3rd Floor, Plot 3, Splendor Forum, Jasola District Centre, New Delhi – 110025, India

79 Anson Road, #06–04/06, Singapore 079906

Cambridge University Press is part of the University of Cambridge.

It furthers the University's mission by disseminating knowledge in the pursuit of
education, learning, and research at the highest international levels of excellence.

www.cambridge.org
Information on this title: www.cambridge.org/9781107124127
DOI: 10.1017/9781316403938

© Cambridge University Press 2019

First published 2019

Printed in the United Kingdom by TJ International Ltd, Padstow Cornwall

A catalogue record for this publication is available from the British Library.

Library of Congress Cataloging-in-Publication Data
Names: Kubby, Joel A., editor. | Gigan, Sylvain, editor. | Cui, Meng, editor.
Title: Wavefront shaping for biomedical imaging / edited by Joel Kubby,
Sylvain Gigan, Meng Cui.
Other titles: Advances in microscopy and microanalysis.
Description: Cambridge, United Kingdom ; New York, NY : Cambridge University
Press, 2019. | Series: Advances in microscopy and microanalysis | Includes
bibliographical references and index.
Identifiers: LCCN 2018057968 | ISBN 9781107124127 (hardback : alk. paper)
Subjects: | MESH: Optical Imaging | Microscopy – methods | Light | Image
Interpretation, Computer-Assisted – methods
Classification: LCC RE79.I42 | NLM WN 195 | DDC 616.07/545–dc23
LC record available at https://lccn.loc.gov/2018057968

ISBN 978-1-107-12412-7 Hardback

In memory of Joel

Contents

Color plates can be found between pages 204 and 205

Contributors

Débora M. Andrade
University of Oxford, UK

Jacopo Antonello
University of Oxford, UK

Alexandre Aubry
PSL Research University, France

Oscar Azucena
Quanergy, CA, USA

Amaury Badon
Boston University, MA, USA

A. Claude Boccara
PSL Research University, France

Martin J. Booth
University of Oxford, UK

Stephen A. Boppart
University of Illinois at Urbana-Champaign, IL, USA

Emmanuel Bossy
Université Grenoble Alpes, France

Antonio M. Caravaca-Aguirre
Université Grenoble Alpes, France

Wonshik Choi
Korea University, Korea

Youngwoon Choi
Korea University, Korea

Meng Cui
Purdue University, IN, USA

Kishan Dholakia
University of St. Andrews, UK

Mathias Fink
PSL Research University, France

Sylvain Gigan
Sorbonne Université and Laboratoire Kastler-Brossel, France

Roarke Horstmeyer
Duke University, NC, USA

Na Ji
University of California, Berkeley, CA, USA

Benjamin Judkewitz
Humboldt University of Berlin, Germany

Sungsam Kang
Korea University, Korea

Joel Kubby
Formerly of University of California, Santa Cruz, CA, USA

Puxiang Lai
Hong Kong Polytechnic University, China

Yuan-Zhi Liu
University of Illinois at Urbana-Champaign, IL, USA

Yang Lu
Harvard Medical School, MA, USA

Cheng Ma
Tsinghua University, China

Hari P. Paudel
Harvard Medical School, MA, USA

Rafael Piestun
University of Colorado at Boulder, CO, USA

Cristina Rodriguez
Janelia Research Campus, Howard Hughes Medical Institute, VA, USA

Alexander Rohrbach
University of Freiburg, Germany

Nathan D. Shemonski
University of Illinois at Urbana-Champaign, IL, USA

Fredrick A. South
University of Illinois at Urbana-Champaign, IL, USA

Xiaodong Tao
University of California, Santa Cruz, CA, USA

Ivo M. Vellekoop
University of Twente, The Netherlands

Tom Vettenburg
University of Dundee, UK

Lihong V. Wang
California Institute of Technology, CA, USA

Peng Xiao
Zhongshan Ophthalmic Center, Sun Yat-sen University, China

Xiao Xu
Washington University in St. Louis, MO, USA

Preface

Wavefront shaping has greatly expanded the capability of optical microscopy and measurements in biological systems. It is the next step beyond the use of adaptive optics for biological imaging and it is about a decade behind in biological application based on the citation counts for these two search terms in the National Library of Medicine. A pivotal publication in wavefront shaping was an article by Vellekoop and Mosk,[1] which is referenced in more than half of the chapters in this volume and has been referenced more than 700 times over the last decade. Wavefront shaping can be used to focus light through or even inside scattering media. Even though light scatters in a complex manner, because the scattering process is linear and deterministic, by shaping the incident wavefront faster than changes in the media, light can be focused deep within scattering materials. Recent breakthroughs in measuring and controlling high-order optical wavefronts have led to many important applications, including deep tissue microscopy with improved imaging quality and depth, focusing light through scattering media, optical phase conjugation, structured illumination, and optical tomography.

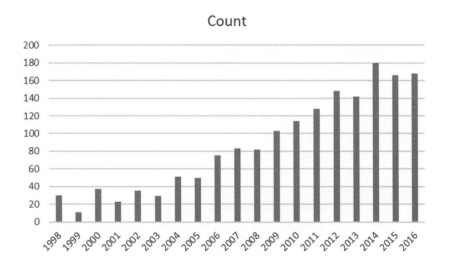

Count

1 I. M. Vellekoop and A. P. Mosk, "Focusing coherent light through opaque strongly scattering media," *Opt. Lett.,* vol. 32, pp. 2309–2311, 2007.

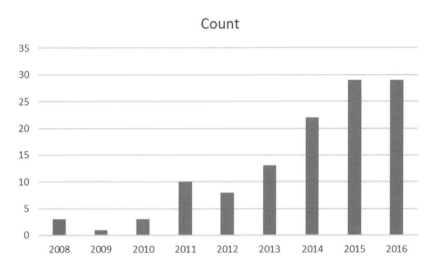

Figure 0.1 Number of publications listed in PubMed (National Library of Medicine) found in a keyword search for "adaptive optics."

Number of publications listed in PubMed (National Library of Medicine) found in a keyword search for "wavefront shaping."

This book includes contributions from leading experts in a variety of research fields that employ innovative wavefront control technologies for biomedical applications, with the latest developments from the forefront of biomedical imaging. Each of the contributors is a leader in the field, and most are blazing new trails in their respective subject areas, so the terminology and symbols used are just now developing. A "Notations" section provides uniformity in notation throughout the book. This book is aimed at graduate students, researchers, and practitioners in the field of biomedical imaging, particularly those in the field of neurobiology. It is suitable for use in a graduate-level course for students who have a background in imaging optics.

We start with two imaging approaches in Part I, "Adaptive Optical Microscopy for Biological Imaging." The first approach uses indirect, image-based wavefront sensing. Here the entire image is used to make an iterative AO correction, similar to the approach used for finding the correct prescription for eyeglasses. A trial aberration, such as astigmatism, is tested, and the improvement of the image measured is related to a merit function, such as overall image intensity. The second approach uses direct, guide star–based wavefront sensing. Here the aberration is measured in the neighborhood of the guide star called the *isoplanatic patch*. The isoplanatic patch is a region around the guide star where the image can be well compensated, which can be smaller than the size of the image. The improvement is measured by the Strehl ratio, the ratio of the maximum intensity of the actual point spread function to a perfect point spread function. Both approaches use deformable mirrors for low-order wavefront correction of refractive image aberrations, similar to the use of adaptive optics in astronomy and vision science. As the imaging depth goes deeper, scattering of light becomes more significant than refractive image aberrations, and less ballistic light is available to form an image for adaptive optical correction.

Higher-order wavefront shaping is required to compensate for scattering. Higher-order wavefront shapers, such as spatial light modulators, are used, as described in Part II, "Deep Tissue Microscopy." The first approach, called Iterative Multi-Photon Adaptive Compensation Technique (IMPACT), makes use of iterative feedback and non-linearity to focus light onto the brightest target. The second approach uses iterative zonal methods for deep tissue imaging. Rather than measuring the wavefront segments in parallel, like Shack–Hartmann wavefront sensing, the wavefront segments are measured serially by segmenting the optical pupil. The shift in the image of a guide star is measured to correct for local phase gradients that cause light rays to miss the focal spot, and the fluorescent intensity of the focal spot is used to obtain constructive interference.

As the amount of scattering increases, two different approaches are discussed for focusing light in, and through, turbid media in Parts III and IV. The first approach, discussed in Part III, "Focusing Light through Turbid Media Using the Scattering Matrix," measures the scattering matrix, which connects the input optical modes with the optical output optical modes for light propagating through a turbid medium. The matrix is first measured and then is used to compute inputs corresponding to the desired output. The scattering matrix can be measured in either transmission, using a CCD camera behind the sample or using photoacoustic imaging instead of a CCD camera to measure the optical intensity inside the sample, or in reflection, using time-domain measurements of the time-resolved reflection matrix.

The second approach, discussed in Part IV, "Focusing Light through Turbid Media Using Feedback Optimization," uses feedback optimization. In feedback optimization, a guide star is used for optimizing the focus on a point-by-point basis. The optimization problem is to find the incident electric field that maximizes transmission into a desired output mode. The feedback signal can be fluorescence, with either linear (one-photon) or nonlinear (multiphoton) excitation, from an implanted guide star or a fluorescently labeled structure of interest. A benefit of nonlinear excitation is that the optimization algorithm causes the light to converge in time and space to a single diffraction-limited focus. The optimization needs to be fast enough to keep up with changes that occur in the sample. The initial demonstration of feedback optimization used a liquid crystal spatial light modulator ($\sim$100 Hz frame rate) on static samples, but later work used micro-electro-mechanical spatial light modulators ($\sim$10 kHz frame rate) to optimize the focus in dynamically changing samples with millisecond decorrelation times. The use of feedback-based optimization of a guide star for scanned imaging will require on the order of 10^{12} measurements of the transmission matrix elements. To make this practical in real-time imaging will require making use of correlations between the transmission matrix elements, as discussed in the final chapter in Part IV.

The use of iterative time reversal, where light or ultrasound traveling through an aberrating medium is focused using a time reversal mirror, is discussed in Part V, "Time Reversal, Optical Phase Conjugation." The first approach uses the decomposition of the time reversal operator (DORT), while the second approach uses time-reversed ultrasonically encoded light to form a focus in scattering media. Here diffuse light is tagged with ultrasonic modulation and then selectively detected for time reversal. Essentially, the ultrasonic focus acts as a guide star for the focal spot.

In addition to correcting refractive aberrations and the loss of light due to scattering in conventional forms of microscopy, wavefront shaping is being used in light sheet microscopy in the illumination path for optical sectioning with improved contrast and less sample exposure. Instead of using Gaussian beams, Bessel and Airy beams have been used to improve performance, as described in Part VI, "Shaped Beams for Light Sheet Microscopy." Bessel beams can self-reconstruct in inhomogeneous media, maintaining a thin plane of illumination with improved axial resolution, increased imaging depth, and fewer artifacts. Airy beams have also been used for propagation-invariant light sheet illumination and for optical manipulation. In addition to wavefront shaping along the illumination path for the light sheet, adaptive optics are being used to maintain image resolution in the detection path.

Wavefront shaping has also been used in optical coherence tomography (OCT) in biological imaging, with many applications in vision science. In Part VII, "Tomography," the benefits of wavefront shaping in optical tomography, in both hardware and computational implementations, are discussed. This application has led to improvements in the resolution and signal-to-noise ratio for retinal imaging. Using computational rather than hardware-based adaptive optics, the optical system can be simplified, leading to decreased cost. The ability for postprocessing in the computational approach also allows for more flexible aberration sensing and correction but is limited to interference-based imaging.

Finally, we wish to thank all of the contributors for sharing their knowledge.

Joel Kubby
Sylvain Gigan
Meng Cui

Notation

c_{US} speed of ultrasound
D pupil diameter
g anisotropy factor
ℓ_{s} scattering mean free path
ℓ_{tr} transport mean free path
ℓ_{a} $= 1/\mu_a$, absorption length
R $= r(.)$, reflection matrix
S Strehl ratio
T $= (t_{mn})$, transmission matrix or $T(x, y)$ or $T(k_x, k_y) \ldots$
η enhancement

Part I

Adaptive Optical Microscopy for Biological Imaging

1 Adaptive Optical Microscopy Using Image-Based Wavefront Sensing

Jacopo Antonello, Débora M. Andrade, and Martin J. Booth

1.1 Introduction

The most common approach to adaptive optics (AO), as originally employed in astronomical telescopes, has been to use a wavefront sensor to measure aberrations directly. In situations where such sensing provides reliable measurement, this is clearly the ideal method (see Chapter 2), but this approach has limitations, and particularly so in the context of microscopy. In order to understand this, one should consider the constraints the use of a wavefront sensor places on the nature of the optical configuration. A wavefront is only well defined in particular situations, for example, when light is emitted by a small or distant, pointlike object, such as a star for a telescope or a minuscule bead in a microscope. In these situations, a wavefront sensor provides a clear and reliable measurement, and this phenomenon has been used to great effect, as explained in Chapter 2. However, not all sources of light have these necessary properties. For example, a large, luminous object comprises an arrangement of individual emitters, each of which produces its associated wavefront. In this case, a wavefront sensor would respond to all of the light impinging upon it, thus giving potentially ambiguous measurements. In an extreme case, such as where light is emitted throughout the volume of the specimen, the sensor would be swamped with light and thus be unable to provide sensible aberration measurement. For this reason, in microscopy, direct wavefront sensing has been effective where pointlike sources have been employed, either through the introduction of fluorescent beads [1, 2] or using localized fluorescent markers [3] and nonlinear excited guide stars [4, 5, 6].

What happens if such sources cannot be employed? The use of beads or additional fluorescence markers in specimens is often not practical; nonlinear guide stars can only be produced in certain microscopes – particularly two-photon fluorescence microscopes – and require specialized and expensive laser sources. Moreover, all of these options are only relevant to fluorescence microscopes, but there are many other microscope modalities for which aberration correction might be required. A related challenge is that, if point sources cannot be used, then the source is by definition an extended object. How can one be certain that the aberration measured corresponds to that induced in the path to or from the focus, rather than to or from another region, perhaps out of focus or displaced laterally? A key observation is realizing that a microscope already has a method for determining the light that comes from the focal plane, as that is the light that forms the useful part of the image. The image itself can therefore be used indirectly as the source for wavefront measurement, as long as the information contained within the image can be interpreted correctly.

The general approach to wavefront sensorless adaptive optics (or "sensorless AO" for short) relies upon the optimization of a measurement that is known to be related to the aberration content. For example, in an adaptive laser focusing system, one might maximize the intensity at the center of the focused beam. In image-based AO systems, which are a subcategory of sensorless AO systems, an appropriately chosen image property is optimized. In certain microscopes (e.g., confocal or multiphoton microscopes), the total image intensity (sum of all pixel values) is an appropriate optimization metric [7, 8, 9, 10, 11, 12, 13, 14, 15], as it exhibits a maximum value when no aberration is present. If the aberration in the system is nonzero due to refractive index variations in the specimen, then the value of the metric will be lower than its optimal value. The goal of the sensorless AO routine would be to use the adaptive element to maximize the metric by minimizing the total aberration in the system.

This concept can be illustrated through analogy with a user refocusing a microscope. The refocusing operation through translation of the specimen is optically equivalent to the application of a spherical phase aberration. The user has a good intrinsic appreciation of what constitutes an in-focus image, namely, that the features are sharp and the image shows the greatest amount of detail and contrast. If the object is not in focus, then the image appears blurred. The user changes the focus of the microscope until the quality of the image is maximized. Although this operation is performed subjectively, it is equivalent to a search algorithm, where the input variable (focal position) is changed until the quality metric (image sharpness or contrast, for example) is maximized.

A sensorless AO system is an extension of this concept, where the input variables correspond to all of the different aberration modes (e.g., spherical, coma, astigmatism) that can be corrected by the system and the image quality metric is defined mathematically, rather than subjectively. For practical operation, the algorithm used to find the optimal correction should be efficient, so that the number of measurements (and hence the time taken) is as small as possible.

There are other advantages to sensorless AO in microscopes. As the measurements are derived from images, the detected aberrations are certain to correspond to those that have affected the image formation. This is not necessarily the case when a wavefront sensor is employed, as different aberrations could affect the imaging and sensing optical paths, due to differences in their optical properties or to relative misalignment between the two paths. These errors in measuring the aberrations that affect the final image would not arise in the sensorless AO arrangement. In many practical systems, this can be an important consideration, as instrument aberrations can be appreciable. This is illustrated in Figure 1.1.

In principle, these sensorless AO schemes are broadly applicable across all microscopes, as there are relatively simple hardware and software requirements. On the hardware side, an adaptive element must be introduced into the microscope; on the software side, the only requirements are that a suitable combination of image quality metric and optimization procedure can be defined, in order to control the adaptive element. However, the wide range of microscope modalities means that each implementation

Deformable mirror flat

Microscope aberrations
corrected

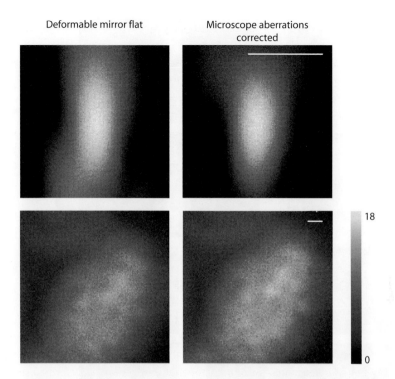

18

0

Figure 1.1 Correcting instrument aberrations with adaptive optics. (top) Axial sections of a confocal microscope PSF before (left) and after (right) correction of instrument aberrations. (bottom) Confocal image of a chloroplast located 30 μm deep in intact plant tissue where instrument aberrations were not corrected (left) and where instrument aberrations were corrected (right). Even though specimen-induced aberrations were not yet compensated, there was a measurable improvement in the image after correction. Scale bars are 500 nm. Images are raw data.

will be different. Examples of a range of such implementations are shown in Figure 1.2. In this chapter, we elaborate on the principles behind the indirect, sensorless approach to wavefront measurement and outline the application of these schemes to a range of microscopes.

1.2   Simple Sensorless Adaptive Optics System

In order to explain the principles behind model-based AO systems, we use a simple optical system consisting of a focusing lens, an adaptive element, and a pinhole detector. While this system is not an imaging system – and hence cannot be used for image-based AO – it provides a useful basis for outlining the principles of sensorless aberration correction. The system, as illustrated in Figure 1.3, consists of an input beam whose wavefronts contain phase aberration $\Psi(r,\theta)$, which passes through an adaptive element that adds an aberration $\Phi(r,\theta)$. The beam is then focused by the lens onto a pinhole aperture placed before a photodetector. The pinhole diameter is chosen to be slightly

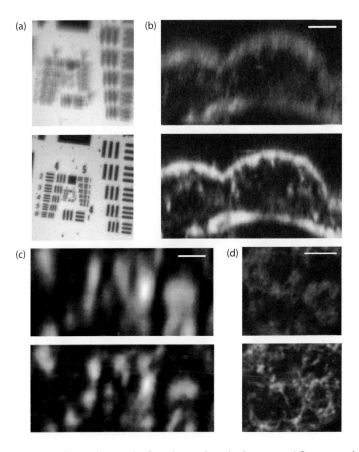

Figure 1.2 Correction results from image-based microscope AO systems, in each case showing images before and after aberration compensation. (**a**) Transmission microscope image of a US Air Force test chart. Reprinted with permission from [16], *Optics Express*. (**b**) Axial sections from third harmonic microscopy of mouse embryos; scale bar: 5 μm. Reprinted with permission from [12], *Optics Express*. (**c**) Axial sections from an AO 3D stimulated emission-depletion microscope (the precorrection image has been increased in brightness to highlight detail); scale bar: 1 μm. Reprinted with permission from [17], *Optics Express*. (**d**) Two-photon fluorescence images of neurons in a *Drosophila melanogaster* brain; scale bar: 5 μm. Images reproduced from references with the permission of the Optical Society (OSA).

$\Psi(r, \theta)$ $\Phi(r, \theta)$ P

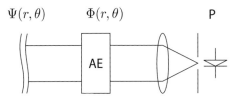

Figure 1.3 Simple optical system to illustrate the principles of sensorless adaptive optics. An input beam with phase aberration $\Psi(r, \theta)$ passes through an adaptive element (AE), which adds a phase aberration $\Phi(r, \theta)$, and is focused onto a photodetector covered by a pinhole (P).

smaller than the diffraction-limited spot, so that it samples the on-axis focal intensity. If the residual aberration $\Phi(r,\theta) + \Psi(r,\theta) = 0$, most of the light will pass through the pinhole and fall onto the photodetector. On the contrary, if the residual aberration is nonzero, a larger part of the light will be blocked by the pinhole, resulting in a lower signal being measured by the photodetector. In this simple situation, one can use the signal measured by the photodetector as a metric to apply aberration correction. In practice, to obtain aberration correction, one must sequentially apply different aberrations with the adaptive element and measure the corresponding value of the metric until this is maximized.

Frequently in AO, it is helpful to express aberrations using a modal expansion. The most common modal expansion uses Zernike polynomials [18] to describe aberrations, so that $\Phi(r,\theta)$ is expanded as follows:

$$\Phi(r,\theta) = \sum_{i=1}^{\infty} u_i \mathcal{Z}_i(r,\theta), \tag{1.1}$$

where u_i is the coefficient of the ith Zernike polynomial, $\mathcal{Z}_i$. Each coefficient u_i can be related to a geometrical aberration [19] such as defocus or astigmatism. In most practical situations, it is possible to represent significant aberrations with a small number of modes, so that the summation in Eq. (1.1) need only be taken over N terms. It is also sensible to neglect the first (piston) mode and assume $u_1 = 0$, as the piston mode constitutes the average value of the phase $\Phi(r,\theta)$, and it has no effect on the photodetector signal.

We consider the simple case of correcting 0.5 radians of defocus aberration $\mathcal{Z}_4$, so that we have $\Psi(r,\theta) = x_4 \mathcal{Z}_4(r,\theta)$, where $x_4 = 0.5$. A plot of the photodetector signal, which we use as our optimization metric m, is shown in Figure 1.4. It can be seen that m is a function of u_4, which is the amount of defocus that we introduce with the adaptive element. For $u_4 = -0.5$, the residual aberration $x_4 + u_4$ is zero, and the metric assumes the maximum value.

It is possible to treat the system in Figure 1.3 as a black box, where one can control u_4 and measure the corresponding value of the metric, but otherwise one has no further knowledge of the underlying system. In this scenario, one could apply a general optimization algorithm to maximize the metric. For example, in Figures 1.5 and 1.6, we applied, respectively, a genetic algorithm [20] and the Nelder–Mead algorithm [21]. The figures show the sequence of measurements required in each case to locate the maximum of the curve and the residual aberration at each stage of the process.

On the other hand, one can create a model for the system in Figure 1.3 that can be used to reduce the number of measurements that are necessary to maximize the metric. For example, when considering a vanishingly small pinhole, the metric signal measured with the photodetector corresponds to the intensity of the point-spread function measured at the origin of the image plane, I_0. From a classical result [22], it is known that for small aberrations, the Strehl ratio S, defined as the ratio between the aberrated and unaberrated focal intensities, is given by

$$S = \frac{I_0(x_4 + u_4)}{I_0(0)} \approx 1 - (x_4 + u_4)^2. \tag{1.2}$$

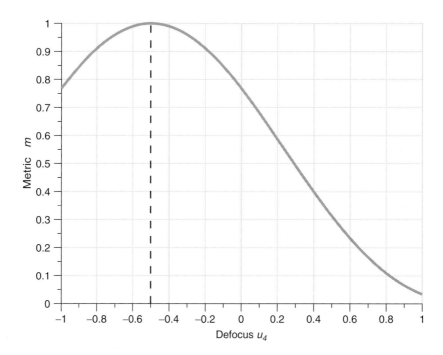

Figure 1.4 Plot of the optimization metric m (photodetector signal) as a function of applied correction aberration u_4. The metric has been normalized to its maximum value.

This suggests that one can model the metric in Figure 1.4 with a parabola, i.e., $m \approx I_0(0)(-u_4^2 + \alpha u_4 + \beta)$, where α and β are the coefficients of the parabola. By taking two measurements of the metric and solving a curve-fitting problem, one can estimate the coefficients α and β, which allows to determine the optimal value of u_4 that maximizes the metric. An example of applying this principle is depicted in Figure 1.7. As long as the initial aberration is in the region where the metric curve is approximately parabolic, the estimation of the aberration will be correct. It can be seen that this optimization method is more efficient than the previous black-box approach, where many more measurements were required.

If more aberration modes are present, as considered in Eq. (1.1), this procedure can be extended to include maximization of each mode. In this particular case, the mathematical principle is the same, but the parabola is replaced by a multivariate quadratic polynomial. An example comparison between the approach explained above and the Nelder–Mead algorithm is shown in Figure 1.8, where it can be seen that the former method is more efficient.

1.3   Image-Based Adaptive Optics Systems

An image-based AO system operates on many of the same sensorless AO principles outlined in the previous section. The main difference is that a single measurement now

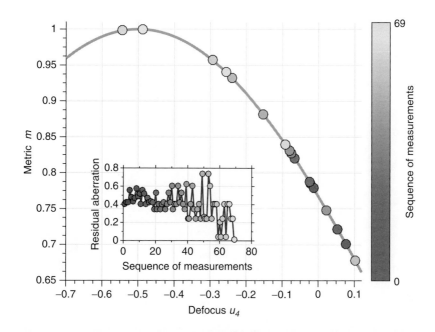

Figure 1.5 Optimization of the metric m using a genetic algorithm (GA). A GA is a search algorithm that tries to maximize a fitness function by simulating the process of natural selection. For the problem of aberration correction, the metric m is used as the fitness function. The population consists of a set of candidate values for u_4. For each member of the population, the GA measures the corresponding value of the metric m. The algorithm iteratively generates a new population by stochastically combining the members of the current population that exhibit a high fitness. This process results in a sequence of measurements of the metric m. Some values from such a sequence are plotted along the metric curve using a color map to indicate the order in which the measurements were acquired. The inset plot reports the residual aberration $|x_4 + u_4|$ for each measurement in the sequence.

involves acquiring a 2D image with a camera or recording a 2D image or a 3D volume when using a scanning microscope. The measurement obtained in this way is converted into a number using an image quality metric. One then optimizes the metric, as described in the previous section, using an appropriate optimization algorithm. In this section, we discuss additional aspects that should be considered when designing image-based AO systems.

1.3.1   Image Quality Metrics

Image quality is inherently a subjective measure that depends upon the type of microscope, the specimen, and the application. However, it is generally possible to define mathematically a metric that reflects the necessary properties that a user would attribute to a good image, i.e., an image obtained with zero aberration. For example, in a conventional wide-field microscope, an aberration-free image would have the sharpest features, so a measure of image sharpness or high spatial frequency content would be appropriate;

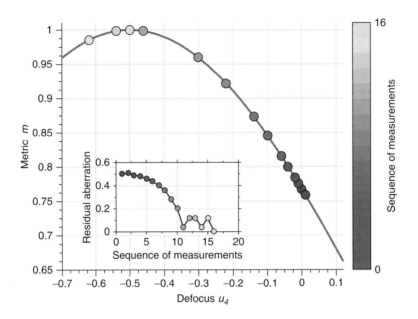

Figure 1.6 Optimization of the metric m using the Nelder–Mead (NM) algorithm. The NM is a search algorithm that repeatedly applies geometrical operations, such as reflection and expansion, to a simplex until it converges to the optimal value that maximizes m. In this particular case, with only one variable, the simplex consists of a line segment.

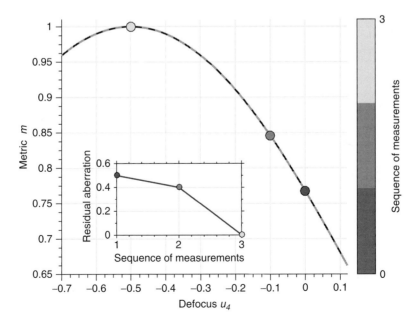

Figure 1.7 Optimization of the metric m by fitting a parabola through the sequence of measurements. The parabola is shown in the black dashed line and closely approximates the actual metric function.

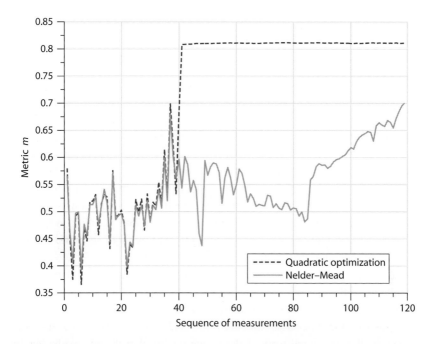

Figure 1.8 Example comparing one application of multivariate quadratic optimization algorithm [23] and the Nelder–Mead algorithm [21] using the same initial aberration for 37 aberration modes.

in a multiphoton microscope, the image brightness is strongly reduced by aberrations, so in this case, a metric based upon the sum of all pixel values (i.e., the total image intensity) would be relevant. In fact, multiple definitions of the image quality may be satisfactory for a given microscope type, which reflects the inherent subjectivity in assigning a numerical value to the quality of an image.

A suitable metric should satisfy some essential mathematical properties. In particular, it should ideally exhibit a global extremum (maximum or minimum) at which the residual aberration $x + u$ is zero by definition. The metric should not exhibit multiple local extrema close to the global extremum, as in such a case, apparently optimal solutions corresponding to nonzero residual aberration might be found. In this situation, the aberration correction problem would be mathematically ill posed and, in practice, prone to failure. As a final property, an ideal metric would resemble a convex or concave smooth function for a large range of values of the residual aberration, so that a simple optimization algorithm can be used to apply the aberration correction. Once we have defined the metric function, we can use knowledge of its properties and its dependence upon aberrations to design an estimation algorithm that provides efficient estimation of the required correction.

1.3.2 Effects of Specimen Structure

It is important to consider the intertwined effects of the specimen structure and aberrations on image formation. In many practical cases, an image can be expressed by a convolution between a function representing the specimen structure, $f(\mathbf{p})$, and the point spread function (PSF) of the microscope, $h(\mathbf{p})$, where the vector $\mathbf{p}$ represents spatial position in three dimensions. In an incoherent imaging system, the image would therefore be expressed as

$$I(\mathbf{p}) = f(\mathbf{p}) \otimes h(\mathbf{p}) = \iiint f(\mathbf{p}' - \mathbf{p})h(\mathbf{p}')\,d\mathbf{p}'. \tag{1.3}$$

As a specific example, in a fluorescence microscope, f would represent the distribution of fluorescent markers, and h would be the intensity PSF of the microscope. It should be clear at this stage that the image I depends both on aberrations, via the PSF h, and on the specimen structure f. Ideally, an image-based AO scheme should be insensitive to specimen structure, providing the same aberration correction irrespective of the form of f. It is therefore important to take into account possible variation in specimen structure when designing the AO scheme. An illustrative example of such effects, comparing two different specimen structures, is found in Figure 1.9.

In certain situations, the specimen structure may have a particularly simple form. For example, if the specimen is a "guide star," as discussed in Chapter 2, one may replace f with the pointlike Dirac distribution, so that one is effectively measuring the PSF of the microscope. In such a case, the functional form of $I(\mathbf{p})$ is well known, and consequently, the functional form of the image quality metric can also be obtained. On the contrary, when f is given by a more involved function, or when one considers variations between specimens with significantly different structure, it is more challenging

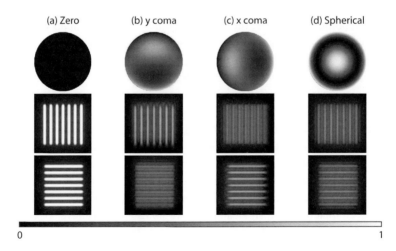

0   1

Figure 1.9 Illustration showing how different aberrations affect the image quality when considering two specimen structures, each consisting of parallel lines in different orientations. Column (**a**) shows the images of the specimens when no aberration is present; column (**b**) shows the effect of y coma, column (**c**) the effect of x coma, and column (**d**) the effect of spherical aberration. The first row shows plots of the phase aberration function. Two specimens consisting of vertical lines (second row) and horizontal lines (third row) are shown using the same color map. Spherical aberration has a similar effect on the image quality of both specimens. On the contrary, y coma and x coma affect the image quality of the two specimens in different ways.

to predict the functional form of the image quality metric. This variability in the functional form affects the number of measurements that must be taken to unambiguously determine the correction aberration.

1.3.3 Effects of Different Aberration Modes

When one investigates the effects of aberrations on the image formation process of a microscope, it becomes apparent that different aberration modes affect images to differing degrees – the same magnitude of one mode can severely degrade the image, whereas another can have little or no effect. An obvious example of a mode that has no effect in most microscopes is piston (constant phase offset). Tip, tilt, and defocus modes, which collectively give rise to three-dimensional image displacements, could also be considered to have no effect on image quality, as they merely cause a translation in image position. It is usually sensible to remove such modes from the aberration correction procedure [14, 24], as they could produce undesirable image shifts, and applying them with the adaptive element may reduce the stroke that is available for correcting the other aberration modes that affect the image quality.

Some modes may be more dominant in a particular microscope due to the nature of the specimen – spherical aberration from a mismatch in refractive index is perhaps the most common example. In other cases, the image formation process of the microscope can enhance sensitivity to certain modes. This was revealed in structured illumination microscopy [25], where certain modes that affected the gridlike illumination pattern had greater effect than other modes. It was also seen that specimen-induced astigmatism could severely affect localization microscopes, which relied upon the same astigmatism mode to retrieve the axial position of emitters [26]. An example of the effect of incrementally correcting different aberrations modes is found in Figure 1.10.

1.3.4 Choice of Optimization Algorithm

The effectiveness of an image-based AO scheme depends greatly on the choice of optimization algorithm, alongside the modal representation (Section 1.2) and the definition of the image quality metric (Section 1.3.1). Image-based AO systems can in principle be built around any optimization algorithm, but considerable differences in performance can be observed between different approaches. We classify these algorithms into two different categories. The first class, which we refer to as model-free algorithms, comprises general optimization routines that were not specifically designed to tackle aberration correction problems. These algorithms make no assumption about the functional form of the image quality metric. The second class we denote as model-based algorithms. These methods assume that the image quality metric has a certain functional form, which constitutes the model.

Many different heuristic algorithms for model-free optimization have been developed [27]. Examples of such algorithms that have been applied in microscopy and imaging are hill-climbing algorithms [28, 29, 9, 30], evolutionary algorithms [7, 31, 9, 10, 11, 32], conjugate gradient methods [33], and stochastic parallel gradient descent

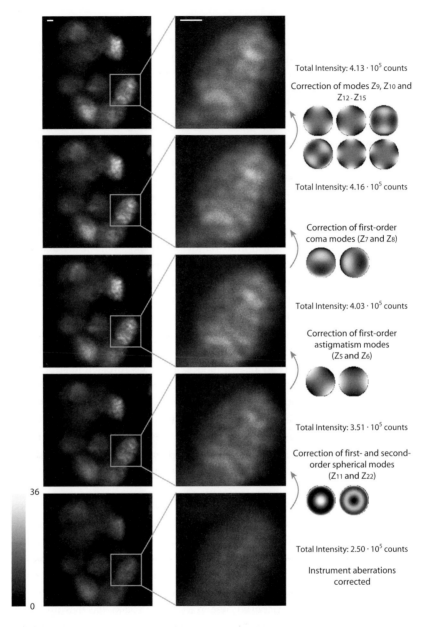

Figure 1.10 Chloroplasts in intact plant tissue imaged by confocal microscopy of endogenous fluorescence, 30 μm deep into the specimen. (bottom) When only instrument aberrations are corrected, imaging is still significantly precluded by specimen-induced aberrations. Image quality is gradually increased as the most severe aberrations are corrected, in this case, spherical aberrations, astigmatism, and coma. Typically, the improvement in image quality is very prominent for the first aberration modes corrected and stabilizes after a certain number of modes are corrected. The number of modes necessary to reach such a plateau depends on the specimen structure and possibly on the field of view. After aberration correction, the stacks of thylakoids within the chloroplasts, structures whose diameter ranges from approximately 300 to 500 nm, are clearly visible. Scale bars are 1 μm. The same linear color scale applies to all images. Images are raw data.

methods [34]. Such algorithms are designed to solve problems where very little background knowledge is available and where, contrary to the discussion in Section 1.3.1, the metric may have any functional form. For example, evolutionary algorithms, such as genetic algorithms, are typically used to tackle combinatorial problems, where differentiation cannot be applied, or strongly multimodal optimization problems [20]. As a result, these methods require a large number of measurements of the metric [35, 23, 36] and still may not converge to the global optimum [21]. In a sensorless AO system, this requires taking a large number of images before the aberration correction can be applied, which leads to long image acquisition times and increased specimen exposure. Overall, this means that model-free algorithms may be far from ideal in practical scenarios where one desires rapid measurement and low exposure.

In most microscopes, the effects of aberrations on the imaging process can be modeled using theory or measured through empirical observations. As such, an imaging model is usually available and can be used to inform the design of an efficient image-based AO scheme that should outperform any model-free approach. Image-based AO schemes built around such models have been implemented in a range of microscopes, including confocal [8], two-photon [37, 13, 14], harmonic-generation [12, 38, 24], wide-field [16], structured-illumination [25], stimulated emission depletion [39, 40, 17] (STED), and single-molecule localization microscopes [26]. In each case, an appropriate optimization metric had been defined that exhibited the desired properties outlined earlier. In several cases, such as confocal or multiphoton methods, the total image intensity was found to be appropriate, whereas other microscopes required more complex metric definitions, such as in the STED microscope, where a metric combining image brightness and sharpness provided the required properties.

Zernike polynomials often provide a convenient basis to concisely represent aberrations, as they include the three image displacement modes (tip, tilt, and defocus) and modes such as astigmatism, coma, and spherical aberration, which are frequently encountered in microscopy. Nevertheless, better performance might be achieved through use of modes derived from the mirror deformation functions [41]. The reason for this is that deformable mirrors can produce aberrations that are restricted by the mirror construction, for example, the arrangement of the actuators, and therefore can generate each Zernike mode with a varying degree of accuracy [42, 24]. It is, however, possible to derive [41] a set of mirror modes that exhibit the same orthogonality properties of Zernike modes but more closely represent the range of aberrations that the deformable mirror can generate. Generally, aberration correction has been applied considering around 10 to 20 low-order aberration modes, as this provided a sensible trade-off for most applications between the required capability of the active element and the expected benefit of the aberration correction [43]. See also Figure 1.10.

The optimization metric function has in different applications been approximated by quadratic [44, 35, 45, 16, 25, 13, 38, 46, 47, 14], Gaussian [25, 38], or Lorentzian functions [16, 13, 14]. Such functions provide an accurate model of the metric for aberrations within a certain range, which depends upon the type of microscope and the definition of the metric. In some cases, one may use multiple image quality metrics, for example, modeled by two different Gaussian functions. These can be used to handle aberrations of different magnitudes, for example, by first correcting larger aberrations with a coarse accuracy and then correcting smaller aberrations with a finer accuracy [16].

The minimum number of measurements required to be able to compute a solution to the fitting problem depends on the number of free parameters of the function used to model the image quality metric. For instance, when considering Figure 1.7, the parabola has two unknown coefficients. The number of unknowns can reflect the variation that the metric exhibits for different specimen structures. For example, in sensorless adaptive optics systems where a final image is not formed [35, 48, 46, 23], the functional form of the metric is not affected by any specimen structure. In this case, correction of N aberration modes has been shown using as few as $N + 1$ measurements of the metric. On the contrary, for fully fledged imaging systems, such as [16, 25, 13, 38, 14, 15], aberration correction of N modes has been shown using at least $2N + 1$ measurements, as exemplified in Figure 1.11. In this case, the additional measurements are necessary to account for the interaction of the specimen structure with the PSF, as outlined in

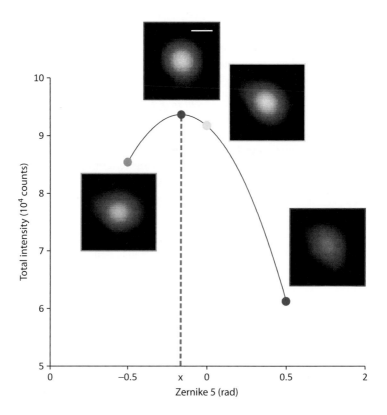

Figure 1.11 Basic principle of image-based adaptive optics. Here we exemplify the aberration correction procedure for the first-order astigmatism mode (Zernike 5) in a confocal microscope. By computing the total intensity of *x-y* images of a gold bead for symmetrical biases (±0.5 rad rms) and for no bias, and fitting the plot with a parabola, the optimal bias may be determined as the coordinate of the peak of the parabola. This method is called the "$2N + 1$ algorithm" [25] because it requires a minimum of 2 times the number of modes to be corrected (symmetrical biases) plus 1 (0 bias) measurements. Scale bar is 200 nm and images are raw data.

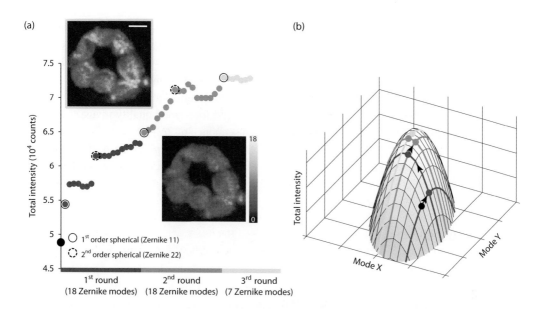

(a)

(b)

Figure 1.12 Convergence of optimization algorithm in sensorless adaptive optics. Here we show the variation in total intensity of an image (the adopted metric) throughout the optimization procedure. (**a**) Chloroplasts located 30 μm deep in a broad bean plant leaf were imaged by confocal microscopy, and the improvement in the image quality metric can be seen as a function of individual Zernike modes corrected. In total, 18 Zernike modes were used, and the same correction procedure was repeated. The first- and second-order spherical modes were highlighted to exemplify the significant influence of these modes when imaging deep into tissue with prominent refractive index mismatch. The scale bar is 3 μm and images are raw data. (**b**) Schematic illustrating why repetition of the aberration correction procedure may further improve image quality, for the simple case where only two aberration modes are present. The metric's dependence on the two orthogonal modes may be approximated by a paraboloid. If the main axes of such paraboloid are not aligned with the x and y axes [25], the convergence of the described optimization algorithm requires more than one round of correction [15]. In both (**a**) and (**b**), the black dot represents the initial state of the system. The first, second, and third rounds of corrections are indicated with the color map. More than one round of correction is required, as the axes of the paraboloid are not aligned with the coordinate axes. However, another set of modes can usually be chosen for which the axes are aligned [25], so that more efficient correction can be implemented. The image before the correction is reported in the inset with the color bar. The image after the correction is reported using the same color scale in the inset with the light border.

Eq. (1.3), which may result in detectable spatial variations in the functional form of the metric [24]. We further remark that collecting extra measurements in addition to the minimum necessary number improves the accuracy of the aberration correction [14]. Ultimately, factors such as the specific structure of the specimen, image noise, and the required level of correction will dictate how many measurements are required, especially if a mathematically optimum set of modes cannot be used, e.g., due to the limitations of the deformable mirror. Figure 1.12 demonstrates such a correction procedure in a confocal microscope using Zernike modes.

It should be noted that the speed of image-based aberration correction is limited only by the rate at which the necessary images can be acquired. As the metric is calculated from a whole image, useful measurements can be taken using images with far lower exposure levels that is usually required for the final scientific image. This means that the measurement images can be acquired much more quickly with fewer pixels and lower illumination levels than usually required for a good quality image. For example, measurement of $N = 10$ modes through a sequence of $2N + 1 = 21$ image acquisitions could be readily implemented in less than a second in a typical confocal microscope. As aberrations are predominantly static over an image period, the image-based AO method implemented in this way would be highly practical.

1.4   Conclusion

We have outlined in this chapter the principles behind adaptive optical microscopy through image-based aberration measurement. This approach has broad application across microscopy, although different schemes are required for each type of microscope. On the hardware side, it requires only the addition of an adaptive element into an existing microscope design. Once set up, image-based AO schemes are straightforward to operate and do not require complex computation. An image-based AO scheme requires specification of an image quality metric, a modal aberration expansion, and an optimization algorithm. While there are a wide range of model-free methods that can be used to optimize the metric, it is generally the case that model-based algorithms provide more efficient aberration correction, requiring a smaller number of images and hence less time and lower specimen exposure.

A drawback of current model-based algorithms is that they provide good correction only over a limited range of aberration amplitudes. However, this range can be extended through use of schemes that are sensitive over different aberration ranges. There is scope for further development of such schemes, for example, through the use of other optimization metrics and advanced optimization algorithms. Further advances could be made where spatial variations in aberrations are present. Current methods measure, in effect, an average of the aberrations across the image field, whereas it is known that significant variations may occur within this region [15, 49]. Image-based AO could be extended to measure these variations and be combined with multiconjugate AO correction to provide more accurate aberration compensation [50, 51].

In conclusion, we note that there are similarities between the concepts of sensorless AO explained in this chapter and the methods of scattering compensation described in Chapter 5 and subsequent chapters. In both cases, an image quality metric, such as total intensity, is optimized using a suitable algorithm. The main difference between these methods is the regime in which they operate: AO is concerned with compensation of low-order aberrations arising from the nonuniform refractive index distribution throughout relatively transparent specimens, whereas the methods of Chapter 5 and subsequent chapters are applicable to opaque, strongly scattering media, where multiple scattering events cause significant changes in amplitude and phase. Furthermore, in AO, the expected performance of the corrected system would be equivalent to a Strehl ratio approaching the ideal value of 1; compensation of scattering, on the other hand, results in much lower Strehl ratios, although the improvement ratio after correction may be considerably higher than for AO systems.

1.5 References

[1] O. Azucena, J. Crest, J. Cao, W. Sullivan, P. Kner, D. Gavel, D. Dillon, S. Olivier, and J. Kubby, "Wavefront aberration measurements and corrections through thick tissue using fluorescent microsphere reference beacons," Opt. Express **18**, 17521–17532 (2010).

[2] X. Tao, Z. Dean, C. Chien, O. Azucena, D. Bodington, and J. Kubby, "Shack–Hartmann wavefront sensing using interferometric focusing of light onto guide-stars," Opt. Express **21**, 31282–31292 (2013).

[3] X. Tao, O. Azucena, M. Fu, Y. Zuo, D. C. Chen, and J. Kubby, "Adaptive optics microscopy with direct wavefront sensing using fluorescent protein guide stars," Opt. Lett. **36**, 3389–3391 (2011).

[4] X. Tao, A. Norton, M. Kissel, O. Azucena, and J. Kubby, "Adaptive optical two-photon microscopy using autofluorescent guide stars," Opt. Lett. **38**, 5075–5078 (2013).

[5] R. Aviles-Espinosa, J. Andilla, R. Porcar-Guezenec, O. E. Olarte, M. Nieto, X. Levecq, D. Artigas, and P. Loza-Alvarez, "Measurement and correction of in vivo sample aberrations employing a nonlinear guide-star in two-photon excited fluorescence microscopy," Biomed. Opt. Express **2**, 3135–3149 (2011).

[6] K. Wang, D. E. Milkie, A. Saxena, P. Engerer, T. Misgeld, M. E. Bronner, J. Mumm, and E. Betzig, "Rapid adaptive optical recovery of optimal resolution over large volumes," Nat. Methods **11**, 625–628 (2014).

[7] O. Albert, L. Sherman, G. Mourou, T. B. Norris, and G. Vdovin, "Smart microscope: an adaptive optics learning system for aberration correction in multiphoton confocal microscopy," Opt. Lett. **25**, 52–54 (2000).

[8] M. J. Booth, M. A. A. Neil, and T. Wilson, "New modal wave-front sensor: application to adaptive confocal fluorescence microscopy and two-photon excitation fluorescence microscopy," J. Opt. Soc. Am. A **19**, 2112–2120 (2002).

[9] A. J. Wright, D. Burns, B. A. Patterson, S. P. Poland, G. J. Valentine, and J. M. Girkin, "Exploration of the optimisation algorithms used in the implementation of adaptive optics in confocal and multiphoton microscopy," Microsc. Res. Techniq. **67**, 36–44 (2005).

[10] S. P. Poland, A. J. Wright, and J. M. Girkin, "Evaluation of fitness parameters used in an iterative approach to aberration correction in optical sectioning microscopy," Appl. Opt. **47**, 731–736 (2008).

[11] W. Lubeigt, S. P. Poland, G. J. Valentine, A. J. Wright, J. M. Girkin, and D. Burns, "Search-based active optic systems for aberration correction in time-independent applications," Appl. Opt. **49**, 307–314 (2010).

[12] A. Jesacher, A. Thayil, K. Grieve, D. Débarre, T. Watanabe, T. Wilson, S. Srinivas, and M. Booth, "Adaptive harmonic generation microscopy of mammalian embryos," Opt. Lett. **34**, 3154–3156 (2009).

[13] D. Débarre, E. J. Botcherby, T. Watanabe, S. Srinivas, M. J. Booth, and T. Wilson, "Image-based adaptive optics for two-photon microscopy," Opt. Lett. **34**, 2495–2497 (2009).

[14] A. Facomprez, E. Beaurepaire, and D. Débarre, "Accuracy of correction in modal sensorless adaptive optics," Opt. Express **20**, 2598–2612 (2012).

[15] J. Zeng, P. Mahou, M.-C. Schanne-Klein, E. Beaurepaire, and D. Débarre, "3D resolved mapping of optical aberrations in thick tissues," Biomed. Opt. Express **3**, 1898–1913 (2012).

[16] D. Débarre, M. J. Booth, and T. Wilson, "Image based adaptive optics through optimisation of low spatial frequencies," Opt. Express **15**, 8176–8190 (2007).

[17] B. R. Patton, D. Burke, D. Owald, T. J. Gould, J. Bewersdorf, and M. J. Booth, "Three-dimensional STED microscopy of aberrating tissue using dual adaptive optics," Opt. Express **24**, 8862–8876 (2016).

[18] F. Zernike, "Beugungstheorie des Schneidenverfahrens und seiner verbesserten Form, der Phasenkontrastmethode," Physica **1**, 689–704 (1934).

[19] V. N. Mahajan, "Aberrated point-spread functions for rotationally symmetric aberrations," Appl. Opt. **22**, 3035–3041 (1983).

[20] X. Yu and M. Gen, *Introduction to evolutionary algorithms* (Springer Science & Business Media, 2010).

[21] J. C. Lagarias, J. A. Reeds, M. H. Wright, and P. E. Wright, "Convergence properties of the Nelder–Mead simplex method in low dimensions," SIAM J. Optimiz. **9**, 112–147 (1998).

[22] M. Born and E. Wolf, *Principles of Optics*, 7th ed. (Cambridge University Press, 1999).

[23] J. Antonello, M. Verhaegen, R. Fraanje, T. van Werkhoven, H. C. Gerritsen, and C. U. Keller, "Semidefinite programming for model-based sensorless adaptive optics," J. Opt. Soc. Am. A **29**, 2428–2438 (2012).

[24] J. Antonello, T. van Werkhoven, M. Verhaegen, H. H. Truong, C. U. Keller, and H. C. Gerritsen, "Optimization-based wavefront sensorless adaptive optics for multiphoton microscopy," J. Opt. Soc. Am. A **31**, 1337–1347 (2014).

[25] D. Débarre, E. J. Botcherby, M. J. Booth, and T. Wilson, "Adaptive optics for structured illumination microscopy," Opt. Express **16**, 9290–9305 (2008).

[26] D. Burke, B. Patton, F. Huang, J. Bewersdorf, and M. J. Booth, "Adaptive optics correction of specimen-induced aberrations in single-molecule switching microscopy," Optica **2**, 177–185 (2015).

[27] Z. Michalewicz and D. B. Fogel, *How to solve it: modern heuristics* (Springer Science & Business Media, 2013).

[28] G. Vdovin, "Optimization-based operation of micromachined deformable mirrors," Proc. SPIE **3353**, 902–909 (1998).

[29] P. Marsh, D. Burns, and J. Girkin, "Practical implementation of adaptive optics in multiphoton microscopy," Opt. Express **11**, 1123–1130 (2003).

[30] L. P. Murray, J. C. Dainty, and E. Daly, "Wavefront correction through image sharpness maximisation," Proc. SPIE **5823**, 40–47 (2005).

[31] L. Sherman, J. Y. Ye, O. Albert, and T. B. Norris, "Adaptive correction of depth-induced aberrations in multiphoton scanning microscopy using a deformable mirror," J. Microsc. **206**, 65–71 (2002).

[32] K. F. Tehrani, J. Xu, Y. Zhang, P. Shen, and P. Kner, "Adaptive optics stochastic optical reconstruction microscopy (AO-STORM) using a genetic algorithm," Opt. Express **23**, 13677–13692 (2015).

[33] J. R. Fienup and J. J. Miller, "Aberration correction by maximizing generalized sharpness metrics," J. Opt. Soc. Am. A **20**, 609–620 (2003).

[34] M. A. Vorontsov, "Decoupled stochastic parallel gradient descent optimization for adaptive optics: integrated approach for wave-front sensor information fusion," J. Opt. Soc. Am. A **19**, 356–368 (2002).

[35] M. J. Booth, "Wave front sensor-less adaptive optics: a model-based approach using sphere packings," Opt. Express **14**, 1339–1352 (2006).

[36] A. J. Wright, S. P. Poland, J. M. Girkin, C. W. Freudiger, C. L. Evans, and X. S. Xie, "Adaptive optics for enhanced signal in CARS microscopy," Opt. Express **15**, 18209–18219 (2007).

[37] M. A. A. Neil, R. Juškaitis, M. J. Booth, T. Wilson, T. Tanaka, and S. Kawata, "Adaptive aberration correction in a two-photon microscope," J. Microsc. **200**, 105–108 (2000).

[38] N. Olivier, D. Débarre, and E. Beaurepaire, "Dynamic aberration correction for multiharmonic microscopy," Opt. Lett. **34**, 3145–3147 (2009).

[39] T. J. Gould, D. Burke, J. Bewersdorf, and M. J. Booth, "Adaptive optics enables 3D STED microscopy in aberrating specimens," Opt. Express **20**, 20998–21009 (2012).

[40] T. J. Gould, E. B. Kromann, D. Burke, M. J. Booth, and J. Bewersdorf, "Auto-aligning stimulated emission depletion microscope using adaptive optics," Opt. Lett. **38**, 1860–1862 (2013).

[41] M. Booth, T. Wilson, H.-B. Sun, T. Ota, and S. Kawata, "Methods for the characterization of deformable membrane mirrors," Appl. Opt. **44**, 5131–5139 (2005).

[42] G. Vdovin, O. Soloviev, A. Samokhin, and M. Loktev, "Correction of low order aberrations using continuous deformable mirrors," Opt. Express **16**, 2859–2866 (2008).

[43] M. Schwertner, M. Booth, and T. Wilson, "Characterizing specimen induced aberrations for high NA adaptive optical microscopy," Opt. Express **12**, 6540–6552 (2004).

[44] M. J. Booth, "Wavefront sensorless adaptive optics, modal wavefront sensing, and sphere packings," Proc. SPIE **5553**, 150–158 (2004).

[45] M. J. Booth, "Wavefront sensorless adaptive optics for large aberrations," Opt. Lett. **32**, 5–7 (2007).

[46] H. Linhai and C. Rao, "Wavefront sensorless adaptive optics: a general model-based approach," Opt. Express **19**, 371–379 (2011).

[47] D. Débarre, A. Facomprez, and E. Beaurepaire, "Assessing correction accuracy in image-based adaptive optics," Proc. SPIE **8253**, 82530F (2012).

[48] H. Song, R. Fraanje, G. Schitter, H. Kroese, G. Vdovin, and M. Verhaegen, "Model-based aberration correction in a closed-loop wavefront-sensor-less adaptive optics system," Opt. Express **18**, 24070–24084 (2010).

[49] M. J. Booth, "Adaptive optical microscopy: the ongoing quest for a perfect image," Light Sci. Appl. **3**, e165 (2014).

[50] R. D. Simmonds and M. J. Booth, "Modelling of multi-conjugate adaptive optics for spatially variant aberrations in microscopy," J. Opt. **15**, 094010 (2013).

[51] A. von Diezmann, M. Y. Lee, M. D. Lew, and W. E. Moerner, "Correcting field-dependent aberrations with nanoscale accuracy in three-dimensional single-molecule localization microscopy," Optica **2**, 985–993 (2015).

2 Adaptive Optical Microscopy Using Guide Star–Based Direct Wavefront Sensing

Xiaodong Tao, Oscar Azucena, and Joel Kubby

2.1 Introduction

In this chapter we will review adaptive optics (AO) in biological imaging using direct wavefront measurement. Here light from a point source in the specimen is used to measure the wavefront with a detector such as a Shack–Hartmann wavefront sensor, similar to the approach that is used in astronomy. The benefit of direct wavefront measurement relative to the sensorless methods discussed in Chapter 1 is that the wavefront can be measured quickly in one step. Typically sensorless methods are iterative, requiring a number of measurements. Taking multiple measurements can take more time and may expose the sample to more light which can lead to photobleaching. Another benefit is that some indirect methods use optimization of a merit function such as image sharpness or image intensity. In direct wavefront sensing the wavefront aberration is directly measured and corrected rather than optimized. As we shall discuss in this chapter, a common metric for direct wavefront measurement and correction is the Strehl ratio which is defined as the ratio of the on-axis beam intensity to the diffraction-limited beam intensity. The Strehl ratio indicates how close the imaging is to the diffraction limit. The objective is getting as close as possible to the diffraction limit. The price that is paid for direct wavefront sensing is system complexity and cost for the addition of the wavefront sensor. In addition it is necessary to provide a point source reference for measuring the wavefront. In this chapter we will discuss how this has been done in different forms of microscopy (wide-field, confocal and multiphoton) after a brief introduction to the general field of wavefront measurement in astronomy and biological imaging.

2.2 Background

2.2.1 Wavefront Sensing and Correction in Astronomy

AO has been used for applications in astronomy [1, 2, 3] vision science [4, 5], and microscopy [6, 7, 8]. In astronomy, AO has been used to overcome the image aberrations caused by the Earth's atmosphere. Light from a distant star, which can be considered a point source since it is so far away, travels through the vacuum of space as a spherical wave. When the wavefront enters Earth's atmosphere, which is initially nearly planar for the small section that is collected by the telescope, the wavefront is distorted due to dynamic changes in the index of refraction of the atmosphere caused by winds and

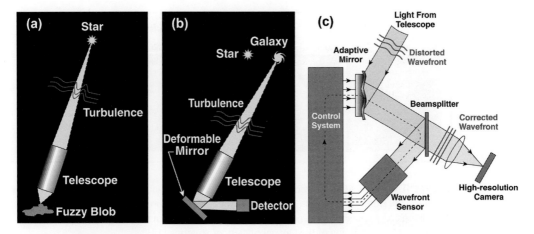

Figure 2.1 AO in astronomy. (**a**) The image of a star appears as a fuzzy blob in a telescope due to the wavefront aberrations caused by turbulence in the atmosphere. (**b**) If a deformable mirror is used to correct the image of the star to make it back into a point of light, a nearby galaxy with more complex structure will also be corrected. (**c**) The wavefront distortions can be measured with a wavefront sensor. The wavefront distortions measured using the guide star are fed back through a control system to deform the adaptive mirror to a shape that is the opposite of the distorted wavefront, correcting the reflected wavefront of the "*guide star*" back into a plane wave. Credit: Claire Max, Center for Adaptive Optics [11].

temperature fluctuations. These fluctuations in the index of refraction cause changes in the velocity of the wavefront, so that some portions travel faster than others, leading to the distorted wavefront shown in Figure 2.1. These dynamic distortions are what cause stars to appear to twinkle. When the star is imaged in a telescope, it appears as a fuzzy blob rather than a point of light, as shown in Figure 2.1a. By measuring the wavefront distortions from the star using a wavefront sensor [9, 10], the opposite shape of the wavefront distortion can be applied to an adaptive mirror to correct the reflected image, as shown in Figures 2.1b and 2.1c. When a star is used as a reference point source for making wavefront corrections, it is called a "guide star." If light from a nearby galaxy travels through the same part of the atmosphere, the guide star can be used to correct the image of the galaxy, as shown in Figure 2.1b.

Examples of the use of AO in astronomy and vision science are shown in Figure 2.2. An image of the planet Neptune is shown in Figure 2.2a with conventional optics and corrected with AO [12]. Images of clouds on the planet can be seen in the corrected image. An uncorrected and corrected image of a living human retina is shown in Figure 2.2b [13]. The mosaic of the individual rods and cones can be seen in the corrected image. The goal for the use of AO in both astronomy and vision science is to obtain diffraction-limited imaging given by the Rayleigh criterion [14]:

$$\sin(\theta) = 1.22\frac{\lambda}{D} \tag{2.1}$$

Where θ is the angular resolution, λ is the wavelength of light that is being imaged and D is the diameter of the aperture that is used for imaging. A commonly used metric used

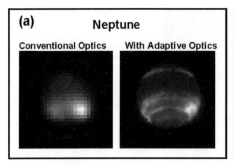
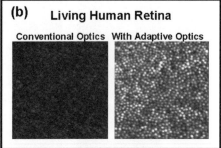

Figure 2.2 The use of AO to correct image aberrations. (**a**) Astronomy: Neptune observed in the near-infrared (1.65 microns) with and without AO. © AAS. Reproduced with permission from [12]. (**b**) Vision science: Imaging of individual rods and cones in the living human retina. Reproduced by permission of the Optical Society of America [13].

for characterizing how close the image is to the diffraction limit is the "*Strehl ratio*," which is defined as the ratio of the measured on-axis beam intensity to the diffraction-limited intensity. An equation for the Strehl ratio S is [15]:

$$S = e^{\left(-\left(\frac{2\pi\sigma}{\lambda}\right)^2\right)}$$
(2.2)

where σ is the root-mean-square deviation of the wavefront and λ is the wavelength. An unaberrated optical system would have a Strehl ratio of 1, but typically an optical system with a Strehl ratio greater than 0.8 is considered diffraction-limited.

Very often there are no guide stars located sufficiently close to the galaxy of interest in astronomical imaging such that light from the guide star travels through the same part of the atmosphere as the light from the galaxy, so an "*artificial guide star*" is created where it is desired by exciting fluorescence in the layer of sodium atoms in the Earth's mesosphere at an altitude of approximately 100 km. A powerful laser (sodium D2a line at 598 nm) is projected upward from the dome of the telescope (after checking to make sure there are no airplanes in the vicinity that would be blinded by the light), as shown in Figure 2.3.

2.2.2   Wavefront Sensing and Correction in Biology

In biological imaging, there very often are no pointlike "natural guide stars" that can be used for making wavefront measurements [6], so they need to be created. An example of a guide star that has been used previously for making wavefront measurements to characterize the "point spread function" (PSF) of a microscope is a fluorescent bead [18, 19]. This technique does not require the use of a subresolution object to obtain the three-dimensional microscope PSF, so that a bright fluorescent bead that is larger than the diffraction limit of the microscope can be used. Because the wavefront measurement using a Shack–Hartmann wavefront sensor (SHWS) is determined by the motion of

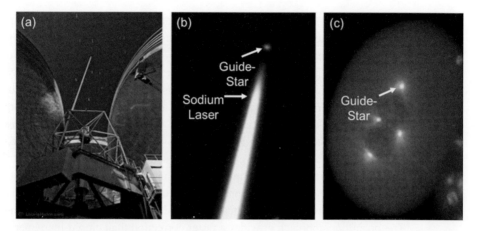

Figure 2.3 Artificial guide stars in astronomy and biology. (**a**) A sodium laser is projected up from the dome of the telescope. Credit: Keck II, photography by Laurie Hatch (www.lauriehatch.com). (**b**) Where the sodium laser intersects a layer of sodium atoms in the Earth's mesosphere, at an altitude of approximately 100 km above the Earth's surface, the laser excites fluorescence in the sodium layer creating a small point of light that can be used as a guide star. Photography by Chris Dainty, reproduced with permission from the Royal Astronomical Society [16]. (**c**) In biology, fluorescently labeled features such as centrosomes can be used as guide stars for wavefront measurements in microscopy. © Kim and Roy. Journal of Cell Biology. Reproduced with permission [17].

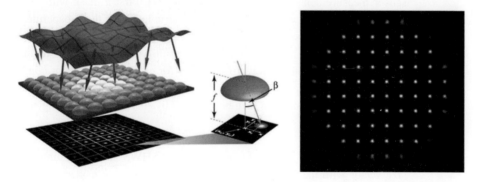

Figure 2.4 (Left) Shack–Hartmann wavefront sensor. An aberrated wavefront is incident on an array of lenslets, forming an array of images of the guide star on a charge-coupled device (CCD) camera placed at the focal length f behind the lenslet array. If there is a local slope to the wavefront in a subaperture, as indicated by the arrows, the spot will be shifted by an angle β. By measuring the shifts in the centroids of the spots, the local slope of the wavefront can be determined. (Credit: Laser Lab Göttingen/Germany). (Right) Hartmann spots from a fluorescent bead.

spot centroids, the object size is not limited by the diffraction limit of the objective. An example of a SHWS is shown in Figure 2.4 (left).

The aberrated wavefront can be regenerated from the displacement of the "Hartmann spots" [20]. Some typical Hartmann spots from a fluorescent bead are shown in

Figure 2.4 (right). The spot displacement in each subaperture is directly proportional to the product of the mean slope and the focal length of the lenslet. The image is processed to obtain the location of the Hartmann spots by using a cross-correlation centroiding algorithm [21]. The slope measurements are then processed to obtain the wavefront using a Fast Fourier Transform (FFT) reconstruction algorithm [22]. If the opposite shape of the aberrated wavefront is placed on an adaptive mirror, such as shown in Figure 2.1c, the reflected image is the corrected wavefront. The wavefront sensor and the adaptive mirror can be used in a closed-loop system as shown in Figure 2.1c, where the wavefront is measured and an error signal is sent to modify the shape of the mirror to flatten the wavefront (i.e., to bring the Hartmann spots into the center of each subaperture on the CCD camera), or in open-loop by characterizing the mirror so that a precise shape that is the opposite of the measured wavefront can be directly placed on the mirror. The open-loop approach provides a faster correction than closed-loop operation where an iterative approach to wavefront measurement and correction is used, but it requires a well-calibrated mirror [23, 24].

There are various ways of estimating a wavefront from the Hartmann slopes [25, 26]. Two essential pieces of information are needed: (1) the phase difference (the slope measurement times the subaperture size) from each subaperture and, (2) the geometrical layout of the subapertures. The wavefront can then be calculated by relating the slope measurement to the phases at the edge of the subaperture in the correct geometrical order. A method for directly obtaining the deformable mirror commands from wavefront sensor measurements is described by Tyson [27]. First a mask with the subapertures must be created; this will generate the geometric layout of the subapertures in the aperture. The next step is to measure and record the response of all the subaperture slope changes while actuating each actuator. The results will be a set of linear equations which shows the response of the wavefront sensor for each of the actuator commands that is called the "poke" matrix (also called the actuator influence matrix). The deformable mirror (DM) commands can then be obtained by solving the following equation:

$$s = Av \tag{2.3}$$

where s is an n size vector obtained from the SHWS slope measurements, v is an m size vector with the DM actuator commands, and A is an $n \times m$ size poke matrix. In the linear approximation, Equation 2.3 can be pseudo-inverted to obtain an estimate of the DM command matrix. Note that some DMs are nonlinear devices so that applying a large change in voltage to an actuator will not necessarily result in the same change in shape every time, but the matrix given in Equation 2.3 performs well in a closed-loop system since only very small voltage changes occur in each feedback cycle thus reducing the nonlinear effects. There are various methods for inverting the matrix A including singular value decomposition (SVD). The advantage of using SVD is that the mode space can be directly calculated. The noisier modes, and all the null space modes by default, can then be removed by setting a threshold on the singular value space [26, 28].

To use this approach for making wavefront measurements in wide-field microscopy, we have injected fluorescent bead "guide stars" into the samples that are then used

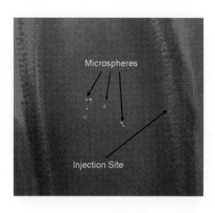
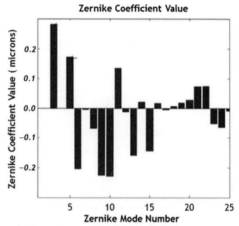

Figure 2.5 (Left) Fluorescent microspheres injected into a fruit fly embryo for use as fluorescent guide stars in wide-field microscopy. Combination of a differential interference contrast image (DIC) and a confocal image of injected microspheres in fruit fly embryo 40 μm below the surface of the embryo. Reproduced with permission from [29], SPIE. (Right) Aberrations that were measured for a 1 μm fluorescent bead at a depth of 45 μm below the surface of a Drosophila embryo. Reproduced with permission from [31], OSA.

for making SHWS measurements (29,303–3). An example of fluorescent beads (microspheres) that have been injected into a fruit fly embryo is shown in Figure 2.5 (left). Each of these beads can be used as an artificial guide star for making a wavefront measurement. It is also possible to use a fluorescent bead that is excited at a different wavelength than the labeling that is used for imaging the sample to minimize photobleaching of the sample, so long as the correction at the guide star wavelength works for the correction of the emission light from the sample labeling [33, 34]. This will depend on the dispersion of the tissue. So far tissue dispersion has not been an issue.

Figure 2.6 shows the design of the system used to measure the wavefront aberration introduced by a Drosophila embryo. Two different objectives 20× and 40× (Melles Griot, Rochester, NY) were used with numerical apertures of 0.4 and 0.65 respectively. L1 and L2 are 65 mm focal length lenses that image the aperture of the objective (plane P1) onto the Shack–Hartmann wavefront sensor (plane P2). The field stop between L1 and L2 blocks the light coming from other microspheres in the field of view, allowing only the light from one microsphere to pass. The field stop could be moved further down the optical system since it is only needed by the Shack–Hartmann sensor. The minimum size of the field stop was set to 1 mm in image space to prevent excessive spatial filtering of the wavefront [35]. The large distance between L1 and the aperture (P1) allows for the excitation laser (HeNe $\lambda = 633$ nm) to be placed in this area. The laser is directed to the optical path via the 45° beam splitter BS1 (Semrock, Rochester, NY). An emission filter was also added after L2 to reduce the effect of scattered laser excitation light by the embryo and to allow for the Shack–Hartmann wavefront sensor to only see the fluorescent emission light. By using a 90/10 beam splitter the microsphere

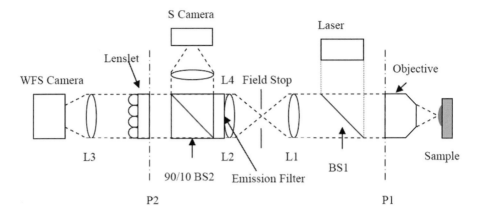

Figure 2.6 Microscope set up with a Shack–Hartmann wavefront sensor. Beam splitter BS1 allows the laser light to be focused onto the sample. Beam splitter BS2 allows for both the science camera and the WFS to simultaneously see the fluorescent microsphere. Reproduced with permission from [29], SPIE.

can be simultaneously imaged by the Shack–Hartmann sensor and the science camera (S Camera). Immediately following the beam splitter is the wavefront sensor which is composed of a lenslet array (AOA Inc., Cambridge, MA) and a cooled CCD camera (Roper Scientific, Acton, NJ). The lenslet array has 1936 (44 × 44) lenses, each with a focal length of 24 mm and a diameter d_{LA} of 328 μm. The lens L3 de-magnifies the pupil by a factor 2 so that it can fit into the cooled camera.

The aberrations that have been measured for a fluorescent bead at a depth of 45 μm below the surface of a Drosophila embryo are shown in Figure 2.5 (right). The guide star reference source used to measure the wavefront was one of the fluorescent microspheres shown in Figure 2.5 (left). The aberrations are expressed in terms of Zernike polynomials, a sequence of orthogonal polynomials defined on a unit disk [36]. The Zernike mode number corresponds to different polynomials (0-piston, 1-tip, 2-tilt, 3-defocus, 4-astigmatism 0°, 5-astigmatism 45°, 6-X coma, 7-Y coma, 8-third-order spherical) with names that are indicative of the type of aberration. In general, as the order increases, the spatial frequency of the polynomial increases. As can be seen in Figure 2.5 (right), the aberrations tend to decrease with increasing order [37]. Here astigmatism and other spherical aberrations dominate the wavefront error. This is mainly due to the index mismatches in the optical path as well as the curved body of the embryo, which mostly introduced lower-order aberrations. The optical aberrations due to the cover slip and air-glass interface, including tip, tilt, and focus, have been removed using a reference image with a guide star but without a biological specimen. The Zernike mode decomposition for a variety of other biological specimens has been measured and follows the same trend of decreasing aberration at higher order [37].

One of the challenges in designing a SHWS is the amount of light the reference source can provide. Fluorescent microspheres are composed of fluorescent dye. Since the amount of light emitted is proportional to the radius cubed, smaller beads provide less light. The size of the beads should be smaller than the diffraction limit of one subaperture of the Shack–Hartmann wavefront sensor. Note that this is larger than the diffraction limit of the microscope aperture by the ratio D(size of the aperture)/d_{LA}(lenslet array diameter). Since the diffraction limit of the microscope is inversely proportional to the numerical aperture (NA) smaller beads are needed for higher numerical aperture systems. Fortunately the light gathered by the objective also increases with increasing NA (light gathering power $\sim NA^2$). Thus, increasing the wavefront sampling by a factor of 4 increases the size of the microsphere radius by a factor of 2, and the amount of light emitted by a factor of 8. The only way to determine if a microsphere, or any fluorescent source, will work is to image it with a SHWS using an objective, as shown in Figure 2.6. In order to increase the speed of the AO correction loop the bead size should be maximized.

The size of the bead d_{bead} should be smaller than the diffraction limit of the wavefront sensor when imaged through the microscope objective:

$$d_{bead} < d_{diffraction\ limit} = 2.44 \frac{\lambda}{2NA_{ob}} \frac{D_o}{d_{LA}} = d_{DLO} \times N_{D/d} \qquad (2.4)$$

Where λ is the wavelength at which the fluorescent beads are emitting, NA_{ob} is the Numerical Aperture of the objective, D_o is the limiting aperture of the objective, and d_{LA} is the diameter of the lenslets in the array. This could also be represented as the diffraction limit of the objective d_{DLO} times the number of subapertures across the limiting pupil. Using this technique we can measure the aberration introduced by a biological sample by injecting a fluorescent bead into the sample.

2.3   Techniques

2.3.1   AO Wide-Field Microscopy

To make both wavefront measurements and corrections, an AO system was added to the back port of an Olympus IX71 inverted microscope (Olympus Microscope, Center Valley, PA) as shown in Figure 2.7. The AO system was designed around an Olympus 60× oil immersion objective with a numerical aperture of 1.42 and a working distance of 0.15 mm. Lenses L1 and L2 have 180 mm and 85 mm focal lengths, respectively, and are used to image the back pupil of the 60× objective onto the deformable mirror (DM) (Boston Micromachines, Boston, MA). The DM has 140 actuators on a square array with a pitch of 400 μm, a stroke of 3.5 μm and a 4.4 mm aperture. L3 and L4 are 275 mm and 225 mm focal length lenses, respectively, and are used to reimage the back pupil of the objective onto the Shack–Hartmann Wavefront Sensor (SHWS). The system has an illumination arm that couples laser light in through a dichroic mirror D. A confocal illuminator (not shown) is used for excitation of the guide star reference beads. This confocal illuminator allows the illumination of a single guide star fluorescent bead to create a single spot. The beam splitter (BS) lets 90% of the emitted light coming from the guide star go to the SHWS for wavefront measurement and 10% for imaging in the

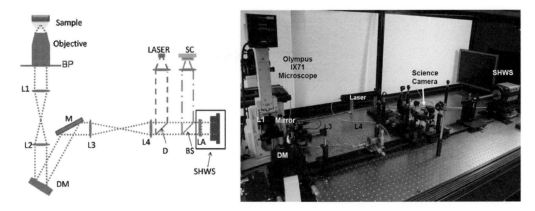

Figure 2.7 AO wide-field microscope. (Left) Schematic diagram of the optical system. The back pupil (BP) plane of an objective lens is imaged onto a deformable mirror (DM) by the relay telescope formed by lenses L1 & L2. After reflection from the DM, the light is reflected off of a fold mirror (M) and is imaged onto a Shack–Hartmann wavefront sensor (SHWS) through the relay telescope formed by lenses L3 & L4. The SHWS comprises a lenslet array (LA) and a CCD camera. A beam splitter BS directs some of the light onto a second CCD camera that captures the image (science camera SC). A laser illuminates the sample to cause the guide star to fluoresce. Laser excitation light is coupled in through a dichroic mirror D. (Right) Implementation of the AO wide-field microscope. Adapted with permission from [32] SPIE and [33], OSA.

Science Camera (SC). The SHWS is composed of a 44×44 element lenslet array (AOA Inc., Cambridge, MA) and a cooled CCD camera (Roper Scientific, Acton, NJ).

2.3.1.1   Results

Demonstration of wavefront correction is shown in Figure 2.8, where each panel shows the results of an additional correction step taken 10 ms apart. Each correction was done using the light coming from a single bead to directly measure the wavefront. The measurement was then fed back to the deformable mirror by using a proportional gain of 0.4 which was the highest possible gain for this sample before the onset of oscillations (Lyapunov stability criteria) [38]. In AO DM correction usually requires a gradual change in shape to account for the nonlinearity of the wavefront sensor and DM. This results mainly from the nonlinear effects of the DM and secondly, usually much smaller, the nonlinear effects of the SHWS. The nonlinear effects of the DM come from the nonlinear dependence of the electrostatic actuation force on the applied voltage and plate separation for a parallel plate actuator and the nonlinear restoring force from stretching of both the mechanical spring layer as well as the mirror surface [39]. Figure 2.8a shows the original point spread function (PSF) of the microsphere taken with the science camera before correction. Figure 2.8b shows the result of correcting for 40% of the measured wavefront error in Figure 2.8a. These steps were repeated until there was no additional significant reduction in wavefront error (i.e., less than 7 nm). Figure 2.8e demonstrates the results of correcting the wavefront after 4 steps in the AO loop. Each image has been normalized to its own maximum to clearly show the details of the PSF.

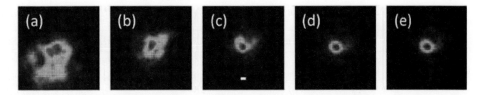

Figure 2.8 AO microscope loop correction steps. (**a**) The original point spread function (PSF) of the microsphere before correction taken with the science camera. (**b**) The result of correcting for 40% of the measured wavefront error in (**a**). These steps were repeated until there was no additional significant reduction in wavefront error (i.e., less than 7 nm). (**e**) The results of correcting the wavefront after 4 steps in the AO loop. The length of the white bar in (**c**) is equal to the diffraction limit of the 40× (0.75 NA) objective lens, 0.45 μm. The bead was located 100 μm beneath the surface of the embryo. Reproduced with permission from [31], OSA.

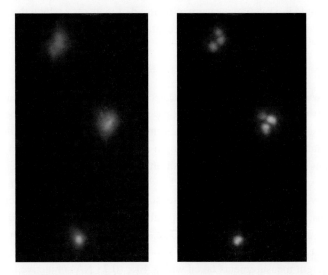

Figure 2.9 Images of 1 μm fluorescent beads through 20-μm-thick Drosophila embryo tissue with (right) and without (left) adaptive optical correction. Reproduced with permission from [33], OSA.

The bar in Figure 2.8c is approximately equal to the diffraction limit of the 40× objective, 0.45 μm. The improvement in the Strehl ratio was approximately 10×. As can be seen, the original PSF in Figure 2.8 does not look like a point, but does after wavefront correction in Figure 2.8e. This improvement is important for imaging features at the diffraction limit of the microscope, and for use in deconvolution software that assumes an ideal PSF.

The benefit of adaptive optical correction can be seen when small fluorescent beads (1 μm green fluorescent beads (Invitrogen, Carlsbad, CA)) are spaced closely so that the individual beads cannot be resolved, as shown in Figure 2.9 (left). After AO correction, the individual beads can be resolved in Figure 2.9 (right). This figure is similar to images

Table 2.1 Isoplanatic angle measurements for the 40× magnification, 0.75 NA objective lens [31]

#	d [μm]	Angle [arcmin]	RMS(1)[rads]	RMS(2)[rads]	RMS(12−)[rads]
1	14	10.7	1.60	1.90	0.73
2	18	13.8	0.98	1.24	0.77
3	25	19.1	1.69	1.10	1.30
Mean	**19 ± 5.57**	**14.5 ± 4.25**	**1.42 ± 0.39**	**1.41 ± 0.43**	**0.93 ± 0.32**

of closely spaced stars in astronomy [40]. Without AO the stars are not resolved, but can be resolved with the use of AO.

2.3.1.2   Isoplanatic Angle

The isoplanatic angle is a relative measure of the field of view over which the AO system can correct the wavefront aberration to an acceptable level ($\sim$1 rad^2 residual wavefront error):

$$\sigma_\theta^2 = \langle (\varphi\,(X,0) - \varphi\,(X,\theta_0))^2 \rangle = 1\,rad^2 \tag{2.5}$$

where φ is the wavefront in radians, X is a vector representing the two-dimensional coordinates, θ_0 is the isoplanatic angle, and σ_θ^2 is the mean-square error between the measured and observed wavefront. We can determine the isoplanatic half-width by multiplying the isoplanatic angle by the focal length of the objective. In order to determine the isoplanatic angle we took wavefront measurements from two microspheres separated by a distance d. A microsphere was excited by shining a laser on it. Each microsphere was excited individually. Each wavefront sensor measurement was collected over a period of 500 ms, much longer than the typical AO loop bandwidth ($\sim$50 ms). This insures that there is little noise in the data. The standard deviation for each individual wavefront was measured to be better than 1% of the wavelength at 647 nm. Table 2.1 shows three different measurements taken with a 40× (0.75 NA) objective lens.

The first measurement shows that the wavefront error for the bead located at the center of the field of view RMS(1) is 1.60 radians, the wavefront error for the bead located 14 μm from the center RMS(2) is 1.90 radians, and the wavefront error between the two measurements RMS(12−) is 0.73 radians. Taking the average of three measurements shows the isoplanatic half-width is 19 ± 5.6 μm. This results show that a reference microsphere together with an AO system can help to improve the quality of the images taken, not just at the location of the microsphere but also within a circle 10 microns in radius from the location of the microsphere.

An emerging field is tomography AO, where multiple light sources together with multiple SHWSs are used. The information from each wavefront sensor is then processed using a reconstructor to acquire a tomographic image of the changes in the index of refraction in the optical path [1]. One of the advantages of using tomography AO is that it can provide information on the depth dependence of variations in the index of refraction in the tissue which allows for the AO system to correct for the wavefront aberrations only in the optical path. This technique can also extend the isoplanatic angle by correcting wavefront aberrations that are common to a larger field of view. By depositing multiple fluorescent beads into the biological sample and using multiple wavefront sensors we can also apply the tomographic techniques that have been developed for astronomical AO [41].

Multiconjugate AO uses multiple wavefront correctors that are placed conjugate to multiple layers of wavefront aberration, rather than the more typical approach in biological imaging where adaptive optical correction is accomplished with a single deformable mirror DM in a plane conjugated to the pupil of the microscope objective lens that is used for imaging. Using multiconjugate AO, the isoplantic patch can be widened. This approach has been implemented in astronomy [42, 43, 44, 45], vision science [46, 47], and is now being investigated for use in biological imaging [48, 49, 50]. In astronomy there are relatively well defined layers in the atmosphere that can be conjugated to wavefront correctors (e.g., ground layer, high altitude) [51]. In vision science one plane was conjugate to the pupil of the eye while another plane was conjugate to the retina [46, 47]. In biological imaging the aberrations are in general spread throughout a volume, although for some samples the aberrations may be concentrated within a particular plane [52].

In scanning laser microscopy, AO correction over large volumes was accomplished using a laser-induced guide star and direct wavefront sensing. They are able to demonstrate adaptive correction of complex optical aberrations at high numerical aperture (NA) with a 14 ms update rate. This enables recovery of diffraction-limited imaging over large volumes (>240 μm per side) [53].

2.3.2   AO Confocal Microscopy

A similar approach for measuring wavefront aberrations with a Shack–Hartmann wavefront using guide star reference beacons can also be implemented in confocal microscopy [54, 55]. A benefit of scanned laser microscopy is that a similar approach to the laser guide stars that are used in astronomy can be implemented in biological imaging without the need for injection of fluorescent beads. Biological samples are typically labeled with fluorescent markers which can be illuminated by the scanned laser beam to generate a small point of fluorescent light for use as a guide star. As described above, in astronomy the atomic sodium layer in the Earth's mesosphere can be used to generate an artificial guide star, as shown in Figure 2.3. The small point of fluorescent light that is excited where a sodium laser intersects the sodium layer can be used as a reference for making wavefront measurements with a wavefront sensor. Even though the sodium layer is extended around the Earth, only the small spot where the laser penetrates the sodium layer is excited, resulting in a small spot. In biological samples, the intersection of a focused laser beam with a fluorescently labeled structure can result in a similar small spot of light for wavefront measurements. The only requirements are that light from other focal planes are rejected and that the resulting spot of light is smaller than the diffraction limit of the wavefront sensor, which is considerably larger than the diffraction limit of the microscope imaging system as shown in Equation 2.4. Another approach uses backscattered light rather than fluorescence; however this double-pass configuration leads to lower sensitivity to odd-symmetry aberration modes [56].

An example of a scanning laser confocal microscope with AO is shown in Figure 2.10. A solid state laser ($\lambda = 515$ nm) provides the excitation light for both fluorescence imaging and wavefront sensing. The light is fed into an objective lens (60× water objective, NA 1.2, Olympus) and scanned on the sample in a raster pattern with two galvo scanners (6215H, Cambridge Technology). The emission light from yellow fluorescent protein (YFP) is divided by a beam splitter (BS). Half of the light is collected by a photomultiplier tube (PMT, H422–50, Hamamatsu). The other half of the light is used for wavefront sensing. To eliminate the intensity loss from the division of light by the BS, a switchable mirror can be included instead of the BS to maximize the signal into the PMT or the wavefront sensor. The AO system includes a deformable mirror (DM, 140 actuators, Boston Micromachines) and a Shack–Hartmann wavefront sensor (SHWS). The SHWS is composed of a 44 × 44 element lenslet array with a lenslet diameter of 400 μm and focal length of 24 mm (AOA Inc., Cambridge, MA) and an electron multiplying (EM) CCD camera (Cascade II, Photometrics). In order to minimize the amount of out-of-focus light that enters the SHWS, irises I1 and I2 are placed in the light path. These irises also block stray light from the DM, scanner, and lenses. The iris acts as a low-pass spatial filter. However, the higher-order wavefront will only give a small contribution to the overall aberration [37]. For mouse brain tissue, the first 14 Zernike modes (third-order) contribute the most aberration. The wavefront error with an iris can be estimated [18, 35]. For a 50 μm thick mouse brain tissue sample, the RMS wavefront error with and without an iris is only 0.0549λ ($\lambda = 527$ nm) when the size of the iris is set to the diameter of the point spread function (PSF) with 80% of the encircled energy (150 μm for I1). This setting will block 53% of the light from a plane within 1.5 μm of the focus. A cross-correlation centroiding algorithm and a fast Fourier transform reconstruction algorithm were implemented to obtain the wavefront [21, 22]. To make an accurate measurement, the diameter of the guide star should be smaller than the diffraction limit of the wavefront sensor, defined in Equation 2.4. In our current setup, the diffraction limit for the wavefront sensor is equal to 5.64 μm. Because the fluorescent light from a given point is proportional to the light intensity illuminating that point, the size of the guide star is limited by the illumination PSF, similar to the case in astronomy for a laser guide star illuminating the sodium layer in the mesosphere. The PSF can be calculated from the wavefront measurement [18]. For yellow fluorescent protein (YFP) at a depth of 70 μm, the diameter of the PSF with 80% of the encircled energy is 1.4 μm, which is small enough to be used as a guide star.

2.3.2.1   Results

The AO confocal microscope has been used to image mouse brain samples with cell bodies and dendrites labeled with YFP to study neural plasticity, and Drosophila to study early development deep within the embryo. For the studies of mouse brain samples, both labeled dendrites and cell bodies have been used as guide stars for wavefront measurements, as shown in Figure 2.11 (right) [57, 58]. A fixed brain tissue slice from a YFP-H line transgenic mouse was prepared. YFP is labeled on the cell bodies and

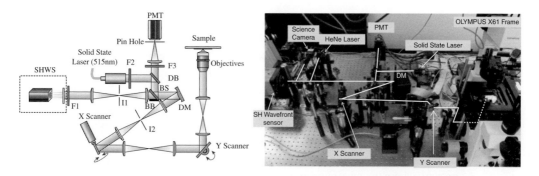

Figure 2.10 AO confocal microscope. (Left) Schematic diagram of the confocal optical system. The emission light from yellow fluorescent protein (YFP) is divided by a beam splitter (BS). Half of the light is collected by a photomultiplier tube (PMT, H422–50, Hamamatsu). The other half of the light is used for wavefront sensing. The AO system includes a deformable mirror (DM) and a Shack–Hartmann wavefront sensor (SHWS). F1, F2, and F3 are filters for the SHWS, the Solid State Laser, and the PMT, respectively. BB is a beam blocker. (Right) Layout of the optical system. Reproduced with permission of [55], SPIE.

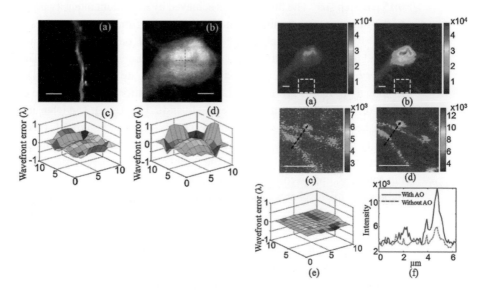

Figure 2.11 (Left) Wavefront measurements and corrections using YFP labeled structures as guide stars. The excitation light source is focused on a dendrite in (**a**) and on a cell body in (**b**). The scale bars are 5 μm. The crossed lines indicate the location of the focal point. The wavefront measurements are shown for a dendrite in (**c**) and a cell body in (**d**) (x, y scale in subapertures). (Right) Confocal fluorescence imaging of mouse brain tissue. The maximum intensity projection image (**a**) before and (**b**) after correction. The dashed boxes indicate enlarged images (**c**) before and (**d**) after correction. (**e**) Wavefront error after correction (x, y scale in subapertures). (**f**) Intensity profiles along the dashed lines in (**c**) and (**d**). The scale bar is 5 μm. Reproduced with permission from [57], OSA. A black and white version of this figure appears in some formats. For the color version, please refer to the plate section.

protrusions of the neurons. Sample brain coronal sections (100 μm) were mounted with antifade reagents (Invitrogen). The spherical aberration induced by the cover plate was compensated by adjustment of a correction collar on the objective lens. The approximate structure of the YFP is initially identified in the confocal image without wavefront correction. The laser beam is then steered to the region of the YFP. The YFP distribution in the background may affect the accuracy of the wavefront measurement. Similar challenges are found for nonuniformities of the sodium layer in Earth's mesosphere in astronomical laser guide stars. Nonetheless, for the mouse brain tissue studied here, the YFP in the neuron cell body with a diameter of 20 μm provides a good uniform background, where the maximum emission is from the focal plane. For other small structures, a refocusing operation is performed to achieve the best focal plane by maximizing the intensity of the Hartmann spots on the wavefront sensor. The emission light from the focal plane then makes the greatest contribution to the wavefront measurement. The DM corrects the aberration in a closed loop using the direct slope algorithm [31]. The corrected confocal image of the isoplanatic region around the YFP guide star is then captured with the optimal shape on the DM.

Confocal images are collected by scanning along the Z-axis with a 3 μm range and a 0.15 μm step size. The final image is achieved by maximum intensity projection applied to the images. After turning on the DM, the wavefront error converges after approximately 10 iterations, which takes ∼0.30 s. The YFP on the cell body is used as a guide star and located at a depth of 70 μm, as shown in Figure 2.11b (left). The RMS wavefront errors without correction and with correction are 0.35λ and 0.034λ, respectively, as shown in Figures 2.11c and 2.11d (left). The confocal images before correction and after correction are shown in Figures 2.11a and 2.11b (right). The intensity profile along the dashed lines across a dendrite and a spine is shown in Figure 2.11f (right). The peak intensity increases by 3×. The image of the dendrite and spines is much clearer after correction with improved contrast. The Strehl ratio is improved from 0.29 to 0.96, a significant 3.3× improvement. To validate the proposed method, mouse brain tissues with fluorescent microspheres (1 μm diameter, Bangs Laboratories, Inc.) are prepared. The wavefront error is corrected using the fluorescent protein first. The wavefront is measured again using a nearby fluorescent microsphere, which is located 3 μm from the fluorescent protein. Twenty measurements are made. The average RMS wavefront error on the fluorescent protein and microsphere are 0.0352λ and 0.0348λ, respectively, with variances of $5.9 \times 10^{-7}\lambda^2$ and $1.9 \times 10^{-7}\lambda^2$, respectively. The average RMS difference between these two methods is only 0.0004λ.

The AO confocal microscope has also been used for dynamic live imaging of the Drosophila embryo [59]. Here spatially and temporally dependent optical aberrations induced by the inhomogeneous refractive index of the live Drosophila samples limit the resolution for live dynamic imaging. We used green fluorescent protein (GFP) guide stars, with GFP-tagged centrosome proteins Polo and Cnn, for live imaging. The results demonstrate the ability to correct the aberrations and achieve near diffraction-limited images of medial sections of large Drosophila embryos. GFP-polo labeled centrosomes can be observed clearly after correction but cannot be observed before correction.

The samples, upstream activation sequence (UAS) EGFP-Cnn; Nanos-Gal4 flies [60] and GFP-Polo flies where GFP is cloned into the 5′ coding region of the Polo gene [61], were reared on standard cornmeal and molasses media. Embryos were collected for one hour on grape juice agar then aged at room temperature for 30–60 minutes. Dechorionation was done by hand using double-sided tape and the embryos were adhered to coverslips. Embryos were covered in Halocarbon oil (Sigma) to allow oxygen permeation and inhibit desiccation [62].

In order to find a suitable guide star, a guide star searching algorithm was developed to localize the desired guide star automatically during the confocal imaging process. The algorithm first initializes the guide star size R_s, the threshold value T for image thresh-holding and the maximum star number N_{max}. The number of detected guide stars depends on these settings. The noise in the image is first removed using Gaussian filters. The location S_{max} of the global maximum of the image is determined, which is used as the first guide star. The next guide star is located by assigning zero intensity to the area of the previous guide star and searching for the maximum of the modified image. The searching loop stops when the predefined maximum number of guide stars N_{max} is achieved or the intensity of the star is less than the threshold value T. In searching for the best guide star, three criteria are available: the brightest guide star, the star closest to the center of the image and the brightest star in a defined area. To achieve an accurate wavefront measurement, the size of the guide stars should be small enough to provide a diffraction-limited image on the wavefront sensor as defined in Equation 2.5. For our setup, $\lambda = 0.509$ μm; $NA = 1.1$; $D_0 = 4 \times 10^3$ μm; $d_l = 400$ μm. So $d_{diffraction\ limit}$ is equal to 5.64 μm. For laser scanning fluorescent microscopy, the emission light from a given point on the FP labeled structure is proportional to the light intensity illuminating that point. If the size of the FP labeled structure is smaller than the illumination PSF, the size of the guide star is equal to the size of the real structure. If the size of the FP labeled structure is larger than the PSF, then the size of the guide star depends on the illumination PSF. In this case, it is like the laser guide star used in astronomy that is excited in the sodium layer. The only part of the sodium layer that fluoresces is where it is illuminated by the laser. The illumination PSF is defined as [63]:

$$h\left(x_2, y_2\right) = \frac{i}{\lambda} \iint_\Sigma P\left(x_1, y_1\right) e^{-ik\varphi(x_1, y_1)} \frac{e^{-ik(r-R)}}{rR} \cos\left(n, r\right) dS \qquad (2.6)$$

where (x_2, y_2) are the coordinates in the focusing plane. (x_1, y_1) are the coordinates in the pupil plane. $P(x_1, y_1)$ and $\varphi(x_1, x_2)$ are the light field in amplitude and phase, respectively. k is the wave number. n is the unit normal of the pupil plane. r is the unit vector from (x_1, y_1) to (x_2, y_2). R is the distance from the pupil plane to (x_2, y_2). dS is the area element on the pupil plane. λ is the wavelength of the illumination light. The size of the guide star can be determined by the area of the PSF with 80% encircled energy. Owing to the high numerical aperture of the objective lens, the guide star size for the diffraction limit of the wavefront sensor is larger than that of the microscope system in most cases. Due to the short exposure time (~50 ms) for wavefront measurement, the theoretical maximum speed of the guide star motion in the lateral plane can be as high as 49 μm/s at a depth of 90 μm.

The light from out-of-focus planes adds noise to the wavefront measurement. A spatial filter (SF) can be placed at the focusing plane between the relay lenses in front of the wavefront sensor. However, the SF also removes high-order wavefront from the sample. Fortunately for most biological tissues, higher-order Zernike modes give only a small contribution to the overall aberration [64]. For the Drosophila embryo, the first 14 Zernike modes (without piston, tip, and tilt) give the major contributions [33]. The size of the pinhole can be determined from the band-limit of the wavefront measurement, which depends on the number of the Zernike modes to be measured [4, 35, 65]. Fourteen lenslets are the minimum number for a reliable measurement of the aberrations up to the first 14 Zernike modes. The measured wavefront is bandwidth limited at $d_{sub}/2$ because of aliasing, where d_{sub} is the width of the subaperture [31]. A spatial filter with a width of λ/d_{sub} can attenuate the high spatial frequency content above $d_{sub}/2$. With an aperture of 4 mm, d_{sub} for 14 lenslets is 0.85 mm. The angular size of the spatial filter is 6×10^{-4} rad for $\lambda = 509$ nm, which corresponds to a pinhole size of 150 μm.

The ability for making a wavefront measurement using a fluorescent protein centrosome guide star (FPCGS) was tested for measurement of spatially dependent wavefront aberrations in the Drosophila embryo with EGFP-Cnn labeled centrosomes at four different positions as shown in Figures 2.12a–d. In this experiment, we are looking at the outer edge of an ovoid-shaped embryo and therefore a small fraction of the distance is through cytoplasm and the other fraction is through the mounting medium. The illumination PSF and corresponding wavefront error show its high dependence on the sample orientation and imaging location. To analyze the specific aberrations, the wavefront measurements can be decomposed into different Zernike polynomials as shown in Figure 2.12e. In contrast to mouse brain tissue, where spherical aberrations are the dominant aberrations, the curved edge of the Drosophila embryo induces a large amount of astigmatism (Zernike modes 5 and 6) aberrations because of its cylindrical shape. The signs of these modes change according to the location. Those measurements also verified the necessity to correct these spatially dependent aberrations using AO. The results of these corrections are shown in confocal microscopy images without and with corrections that are captured at a depth of 83 μm below the coverslip, as shown in Figures 2.12f and 2.12g. *The GFP-polo labeled centrosomes can be observed clearly after correction but cannot be observed before correction.* This illustrates how the use of AO is critical for imaging these features. The size of the PSF decreases from 1.7 μm, before correction, to 0.21 μm, after correction. The Strehl ratio calculated based on the PSF shows an increase from 3.3×10^{-3} to 0.7.

The penetration depth of the AO microscope for live imaging of a Drosophila embryo was tested by performing AO correction during Z scanning from the top surface to a depth of 100 μm below the coverslip with a 1 μm z-step size. The AO correction is performed at each z-step. The purpose of the guide star searching algorithm is to search and calculate the location of EGFP-Cnn at each depth, acting as a potential guide star, in a cycle 13 embryo. Figures 2.13a and 2.13b (left) show the maximum intensity projection (MIP) produced from a scan series without and with correction, respectively. The GFP at the edge of embryo at different depths can be observed. Before correction, the EGFP-Cnn labeled centrosomes can only be observed down to 60 μm in depth. After

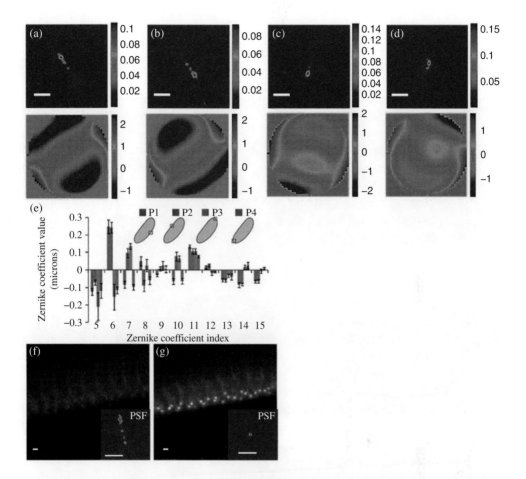

Figure 2.12 Wavefront measurement and correction. (**a–d**) The averaged point spread function (PSF) and wavefront errors over 6 measurements using EGFP-Cnn labeled centrosomes of a cycle 14 Drosophila embryo at four different locations (P1, P2, P3, and P4) at a depth of 60 μm. (**e**) The averaged coefficient value of the first 15 Zernike polynomial modes at these four locations. The error bar is the standard deviation for six measurements. (**f–g**) The images and PSF without and with correction for a cycle 14 Drosophila embryo with GFP-polo at a depth of 83 μm. Scale bars, 2 μm. Reproduced with permission from [59], OSA. A black and white version of this figure appears in some formats. For the color version, please refer to the plate section.

correction, they can be observed below a depth of 80 μm. Using the 3D view function in ImageJ [66] with a resampling factor of 2, the 3D images of the Drosophila embryo show the imaging depth increases from 60 μm to 95 μm, with more than a 50% increase in imaging depth as shown in Figures 2.13c and 2.13d (left). Figures 2.13e and 2.13f (left) show the enlarged images without and with correction at depths of 60 μm and 90 μm, respectively.

Figure 2.13 (right) is a deeper (200 μm) AO confocal image of the zebrafish brain. The zebrafish is more transparent than the fly embryo making it possible to image deeper into the specimen before scattering becomes the limiting factor for the imaging depth.

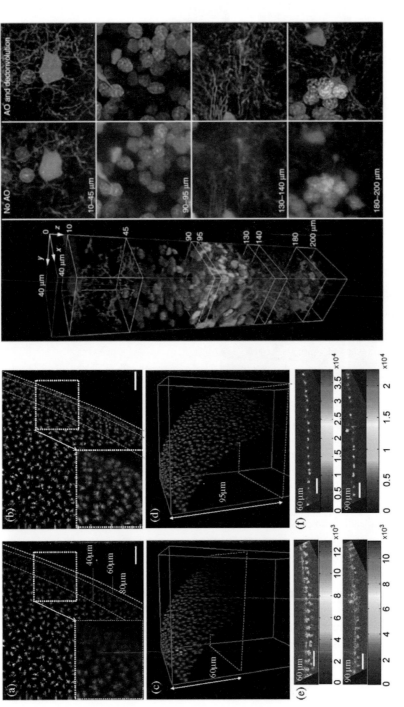

Figure 2.13 (Left) Comparison of the three-dimensional imaging without (left) and with (right) correction for imaging of cycle 13 fly embryos with EGFP-Cnn label. The maximum intensity projection of the scan series from the top surface to 100 µm without (**a**) and with (**b**) AO. The 3D reconstructions without (**c**) and with (**d**) AO. The confocal images without (**e**) and with (**f**) AO at the depths of 60 µm and 90 µm. The color maps are scaled to show the image data over its full range. Scale bar, 10 µm. Reproduced with permission from [59], OSA. (**Right**) Two-color confocal imaging with AO provided by a de-scanned two-photon guide star deep in the living zebrafish brain. 3D volume rendering (left) of oligodendrocytes (magenta) and neuronal nuclei (green) from the optic tectum through the midbrain. MIP before (center) and after (right) AO correction across four subvolumes spanning depths indicated by yellow boxes (left) demonstrate the recovery of diffraction-limited resolution throughout the 200-µm-deep imaging volume. Adapted by permission from Macmillan Publishers Ltd: Methods [53], copyright 2014. A black and white version of this figure appears in some formats. For the color version, please refer to the plate section.

This work used a two-photon guide star and direct wavefront sensing to demonstrate adaptive correction of complex optical aberrations up to the 45th Zernike mode [53].

For the fly embryo the wavefront error is measured at each depth before correction. After correction, the wavefront is measured again, directly from the sample with an updated correction by the DM. PSFs are then calculated from the wavefront measurements. Figures 2.14a and 2.14b show the wavefront measurement and estimated PSF at a depth of 90 μm below the coverslip without and with correction, respectively. The RMS wavefront errors at different depths with and without correction are shown in Figure 2.14c. Without correction, the RMS wavefront error reaches approximately 0.8λ when the imaging depth is 90 μm. The Zernike coefficient values without AO with the change of depth are shown in Figure 2.14d. Below a depth of 50 μm, the astigmatism and coma aberrations (Zernike index 5, 6, 7, and 8) begin to decrease with the increase of trefoil $\times$ aberration (index 9). Those contribute a RMS wavefront error of around $0.8\ \lambda$ from 50 μm to 90 μm. However the increases in the high-order aberration (the third order) generates a larger PSF with an increase in depth. The decrease of the Strehl ratio in Figure 2.14e shows the degradation of the optical performance with the imaging depth. After correction, even at a depth of 90 μm, the system can still achieve a Strehl ratio of 0.6 with an RMS wavefront error of 0.1λ. Aside from improving the penetration depth, AO also improves the optical resolution. Although the EGFP-Cnn labeled centrosomes can be observed at a depth of 60 μm without AO, the resolution is still poor because of the aberrations as shown in Figures 2.14e and 2.14f. Before correction, the size of the PSF is 1.67 μm at a depth of 60 μm. After correction, it decreases to 0.22 μm as shown in Figure 2.14f. At a depth of 90 μm, it shows a significant improvement of the PSF by a factor of nine.

Enabled by fast wavefront measurement and correction ability, AO microscopy using fluorescent protein guide stars can be used for time-lapse 4D imaging. We recorded EGFP-Cnn labeled centrosomes of an early Drosophila embryo for 4D imaging at a depth of 80 μm with an image size of 512×512 pixels and a time resolution of 30 s for five consecutive focal planes (1 plane/μm) over 20 minutes. At the beginning of each time period, the wavefront error is corrected at the third focal plane. In every time interval, images with and without correction are collected sequentially to compare results. Wavefront errors are measured directly from the sample before and after correction. The image sequence with a frame rate of 30 seconds was achieved by the maximum intensity projection in each time period. A single frame without and with correction is shown in Figure 2.15a. It shows a significant improvement in contrast and resolution. Without correction, the measured wavefront shows a dynamic change during the imaging time. The variation of the wavefront can be seen from the coefficient value change of different Zernike modes as shown in Figure 2.15b. The short-term data with 20 minutes imaging time are dominated by measurement noise from the wavefront sensor due to the low level of photon emission. The enlarged PSF, approximately 1.8 μm, makes it impossible to obtain high-resolution images. After correction, the coefficient value for Zernike modes are all below 0.05 μm. The Strehl ratio increases to approximately 0.6 as shown in Figure 2.15c. The AO compensates those dynamic aberrations and produces a near perfect PSF with a spot size of 0.22 μm as shown in Figure 2.15d.

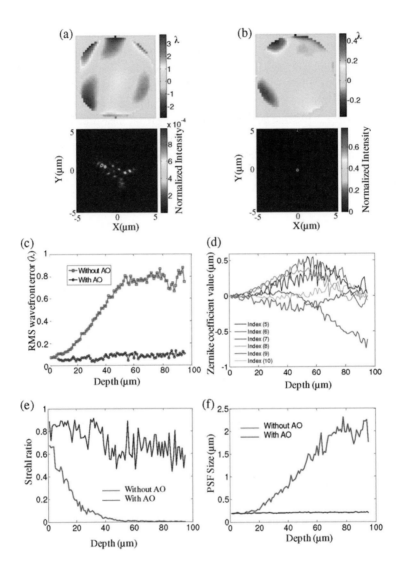

Figure 2.14 Comparison of the wavefront measurements and the PSFs without and with AO for different depths. The wavefront measurements and PSF without (**a**) and with (**b**) AO at the depth of 90 μm. The RMS wavefront errors change with the depth (**c**). The red and blue lines indicate the measurement without and with AO respectively. The Zernike coefficient values without AO with the change of depth (**d**). The Strehl ratio (**e**) and PSF (**f**) size change for different depths. The red and blue lines indicate without and with AO, respectively. (λ = 509 nm). Reproduced with permission from [59], OSA. A black and white version of this figure appears in some formats. For the color version, please refer to the plate section.

These results show that AO microscopy with wavefront sensing using a fluorescent protein guide star can correct for the dynamic aberrations induced by the ovoid shape of the Drosophila embryo at high resolution for live imaging. Compared to the use of a fluorescent microsphere as a reference guide star (28–33), using fluorescent protein

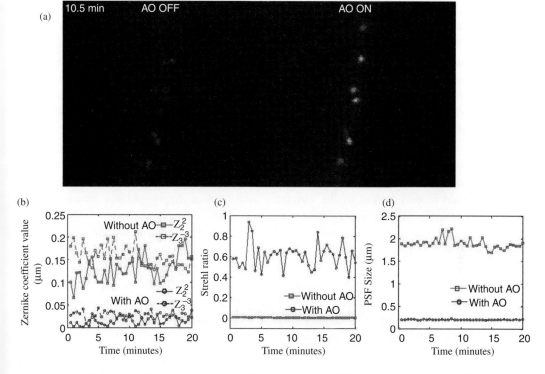

Figure 2.15 4D imaging of cycle 13 fly embryos with EGFP-Cnn label at a depth of 80 μm. A single frame without and with correction of a video movie (**a**). The coefficient value changes for Zernike modes Z_2^2 (astigmatism x, dashed line) and Z_3^{-3} (trefoil y, solid line) with and without AO during 20 minutes (**b**). The Strehl ratio change with (blue) and without (red) AO during 20 minutes (**c**). PSF size change with (blue) and without (red) AO during 20 minutes (**d**). Reproduced with permission from [59], OSA.

labeled subcellular structures as a noninvasive method simplifies the tissue preparation process and avoids the potential side effects of injecting fluorescent beads in live imaging. The guide star could be the same fluorescent protein used for imaging or a specially designed protein for use only as a guide star. A fluorescent protein with a long excitation wavelength such as td-Tomato with high quantum efficiency yield, less photobleaching and more stability will further extend the correction depth and provide a more robust wavefront correction for live imaging [67]. Using fluorescent proteins with excitation wavelengths different from the imaging fluorescent proteins can avoid the need to switch the light path between the wavefront sensing and imaging paths. Of particular interest is the use of a centrosome as a guide star. We can always find one or two centrosomes per cell in all animal cells depending on their position within the cell cycle. It broadens the applications of the wavefront sensing method for different kinds of cells and sample types. Moreover, the size of the centrosome is approximately 1 μm, which is particularly suitable as a guide star for the SHWS. The centrosomes are usually located close to the nucleus. The separation of centrosomes in two cells is often large enough in three-dimensional space for wavefront measurement with less background noise and less influence from the neighboring guide stars.

In comparison with the image-based AO method, the wavefront sensing method used here requires less time for wavefront measurement, which is particularly suitable to correct the dynamic aberrations induced in live specimens [68]. The total wavefront correction time is 600 ms which includes a 50 ms exposure time for the wavefront measurement, a 50 ms DM control time and a 500 ms flipper mirror switching time. The last one can be minimized to less than 1 ms by using a fast steering mirror, or eliminated using a beam splitter. To further minimize the DM control time, open-loop control to update the DM using the wavefront measurement and an accurate DM model can be applied after the calibration of the DM [23, 24, 39]. In this case, the correction speeds would be only limited by the wavefront measurement. Applying Field Programmable Gate Arrays (FPGA) in wavefront measurement will be beneficial for time-critical applications. For tissues with small isoplanatic patches, the guide star searching algorithm can find the optimal local guide star in each patch. A larger field of view with correction can be provided by stitching those patches together or by using conjugate AO. Finally, the application of FPCGS also simplifies the design of the hardware and software. Due to sharing a similar concept of wavefront sensing based on a laser guide star in astronomy and vision science, the knowledge of AO application in those fields has facilitated its application in microscopy.

2.3.3 AO Two-Photon Microscopy

The ability to image intact tissues and living animals with high resolution and depth penetration makes two-photon microscopy an invaluable tool for studying structure and function of cellular constituents within scattering tissue. Compared with the visible light used in confocal microscopy, the near-infrared light used in two-photon microscopy experiences less scattering in biological tissue [69]. The light detection is more efficient since both ballistic and diffuse emission light is collected [70]. However the penetration depth is still limited by scattering and optical aberration [71]. The optical aberration is caused by the variation in refractive index from the inhomogeneous optical properties of the tissues and the refractive index mismatch between the tissue and mounting medium. Most AO two-photon microscopes (AO-TPM) are based on indirect methods of aberration measurement which utilize processing of the final image [72–75]. The extended exposure time for indirect methods can cause photobleaching and may limit the bandwidth for live imaging. To increase the wavefront correction speed, wavefront measurement is a promising method for dynamic *in vivo* imaging applications. Coherence-gated wavefront sensing (CGWS) has been used for measuring the wavefront [76–79]. This approach is based on backscattered light rather than fluorescent light. This approach requires a complicated interferometric arrangement, but it does eliminate the need for a Shack–Hartmann wavefront sensor. An alternative method for adaptive correction in two-photon microscopy used reflected light and confocal imaging for depth selection [80]. This approach simplifies the optics in the wavefront detection light path relative to interferometry and reduces the potential of inducing additional aberrations. Researchers utilized a Shack–Hartmann wavefront sensor that allowed the wavefront to be measured by acquiring a single image, improving the bandwidth of the AO feedback loop. Another approach uses pupil segmentation, however this approach is too slow for dynamic live imaging of events that change on timescales of seconds [81–83]. This approach also used small fluorescent beads as references for measuring image shifts. Improvements to this approach using multiplexed aberration measurement have increased the speed [84] and accuracy for discontinuous wavefronts [85].

2.3.3.1 Nonlinear Guide Star

Direct wavefront measurement with a Shack–Hartmann wavefront sensor and a nonlinear guide star using two-photon excitation has been demonstrated for correcting aberrations in two-photon microscopy [86, 87]. An advantage of a two-photon guide star is that is light is confined to just the focal region by the nonlinear excitation process, eliminating the need for a pinhole to eliminate out-of-focus fluorescence that occurs for a one-photon guide star. However, the long 800 ms integration time for making wavefront measurements is too slow for dynamic *in vivo* imaging. A faster approach (30 ms) uses autofluorescent guide stars [88]. As a label-free method, no special sample preparation is required. Autofluorescence from retinal lipofuscin by one-photon excitation has been used as a linear guide star to measure the wavefront of the human eye in vision science [89]. Autofluorescence can also be used in dynamic *in vivo* biological microscopy by combining it with two-photon excitation, where only the fluorophores at the focus plane are excited for generation of a guide star with minimal background noise. Here the intrinsic fluorophores are illuminated by a near-infrared ultrashort pulsed laser with an appropriate wavelength, which is often different from the excitation wavelength of common fluorescent proteins such as green fluorescent protein (GFP) and red fluorescent protein (RFP) [90]. This induces less photobleaching to the labeled fluorescent proteins during wavefront sensing. This method is particularly effective when the fluorescence from fluorescent proteins is too weak for wavefront sensing. After wavefront measurement, the measured wavefront error can be compensated by a deformable mirror using open-loop control [91]. High-speed and high-performance wavefront measurement and correction are the critical advantages of wavefront measurement over other sensorless adaptive optical systems. This method is used for live imaging of a Drosophila embryo labeled with (GFP) and (RFP), enabling measurements of aberrations in the middle of a Drosophila embryo [88].

The layout of the AO-TPM is shown in Figure 2.16. Two-photon excitation is generated by a tunable (680–1080 nm) mode-locked Ti:Sapphire laser (140 fs, 80 MHz, Chameleon Ultra II, Coherent). The intensity of the laser is modulated by an electro-optic modulator (model 350–80LA, Conoptics Inc.). A $60\times$ water immersion objective with a numerical aperture of 1.2 was used (Olympus Microscope, Center Valley, PA) for imaging. In order to correct the wavefront, a deformable mirror (DM) (Boston Micromachines) with 140 actuators and 3.5 μm of stroke is placed in the optical path, where it is conjugate to the exit pupil of the objective, the X and Y scanners and the wavefront sensor. The optical system includes three telescope relay subsystems. Lenses L1 and

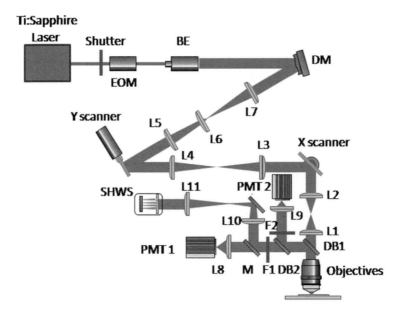

Figure 2.16 Schematic of adaptive optical two-photon microscope. L, lens. F, filter. M, mirror flipper. BE, beam expender. EOM, electro-optic modulator. DM, deformable mirror. DB, dichroic beam splitters. SHWS, Shack–Hartmann wavefront sensor. PMT, photomultiplier tube. Reproduced with permission from [88], OSA.

L2 image the exit pupil of the objective lens onto the X scanner. They de-magnify the pupil from 7.2 mm to 4 mm. Lenses L3 and L4 relay the X scanner conjugate to the Y scanner. This design minimizes the movement of the scanning beam at the exit pupil of the objective lens and the emission light at the DM, which is important for accurate wavefront measurement and correction. Lenses L2 and L4 also serve as scanning lenses. The current design is optimized for an optical scanning angle of 4.4 degree, which provides a field of view of 128 μm on the sample with a 60× objective. Lenses L5, L6, and L7 image the pupil of the Y scanner onto the DM. For nondescanned detection, two photomultiplier tubes (PMT) (H7422–20 and H7421–40, Hamamatsu) and lenses L8, L9 are located immediately after the objective lens, which collects the emitted light. The emission light is separated by single edge dichroic beam splitters (FF705-Di01, Semrock) for dual-color imaging. During wavefront measurement, two-photon images are captured first. A guide star searching algorithm detects the best guide star based on the intensity and location from the image [59]. Then two galvo scanners (6215H, Cambridge Technology) steer the beam to the best location in the sample to generate a guide star. The Shack–Hartmann wavefront sensor (SHWS) with a 44 × 44 element lenslet array (AOA Inc., Cambridge, MA) and an electron multiplying CCD camera (Photometrics) utilizes the emission light from the sample for wavefront measurement. Because the SHWS and DM are located on the emission and excitation paths, respectively, open-loop DM control is applied. This configuration will eliminate additional dichroic mirrors and switchable mounts for closed-loop configuration, where the SHWS should be located behind the DM. It is also possible to locate the SHWS in a de-scanned position behind the scanners to enable averaging of the wavefront measurements [53].

In this system, a mathematical model for an accurate open-loop control of the deformable mirror is applied. To achieve open-loop control of the DM, an accurate model and calibration procedure are required [23, 24, 39]. In our system, we apply mathematical modeling of the mirror surface based on the thin plate equation as follows:

$$\nabla^4 \omega(r) = f_P(r)/D \qquad (2.7)$$

where $f_P(r)$ is the plate force. ω is the displacement of the plate. D is the flexural rigidity of the plate. The plate force is the sum of spring forces $f_S(\omega)$ and electrostatic forces $f_E(\upsilon, \omega)$, where υ denotes voltage applied on the actuator. $f_S(\omega)$ and $f_E(\upsilon, \omega)$ can be calculated from displacement measurement during calibration. Accurate look-up tables of these two parameters for different υ and ω are generated. During the system operation, the desired wavefront for compensation of the tissue-induced aberration is measured by the SHWS. Then the plate force $f_P(r)$ is calculated using Equation 2.7 and $f_S(\omega)$ is retrieved from look-up tables based on the desired displacement ω of the mirror surface. Finally the desired voltage υ on the DM is estimated from the loop-up table $f_E(\upsilon, \omega)$. A Zygo interferometer was used to measure the surface displacement during the calibration. The root-mean-square (RMS) error of the open-loop control is around 17 nm for a DM displacement of 500 nm.

2.3.3.2 Results

To make an accurate wavefront measurement using the SHWS, the guide star should be smaller than the diffraction limit of the wavefront sensor. Also the number of photons coming from the guide star should be high enough to provide the required signal-to-noise ratio for the wavefront measurement. Thanks to the nonlinear excitation, the fluorescent emission is localized to a small area, which is often small enough to use as a guide star [86]. However, the intensity of the guide star depends on the distribution of the fluorescently labeled structure. An example shown in Figure 2.17 is the two-photon image of a Drosophila embryo labeled with GFP and RFP excited by a laser with different wavelengths at a depth of 51 μm. The laser power at the sample is 17.5 mW. At 1000 nm excitation, most of the RFP labeled nuclei structures can be observed near the membrane. In the middle of the embryo, because the yolk induces a large amount of aberration and scattering, the intensity of the nuclei structures becomes much dimmer. At 920 nm excitation, GFP labeled centrosomes are also seen to be distributed near the membrane as shown in Figure 2.17b. It is extremely difficult to use this fluorescence as a guide star in the middle of the embryo. Unfortunately, the mitosis process early in the development of the embryo (first nine cycles) occurs at the middle of the embryo, where the first cycles lasts for 10 min and begins once the egg has been laid [92]. High-resolution and high-speed imaging of this process are critical for this study. In the Drosophila embryo, a major source of autofluorescence is from yolk granules and the vitelline membrane [93] as shown in Figure 2.17c. Although the yolk does not contain any RFP or GFP, autofluorescence arises from endogenous fluorophores with emission spectra similar to that of nicotinamide adenine dinucleotide (NADH) [94]. At

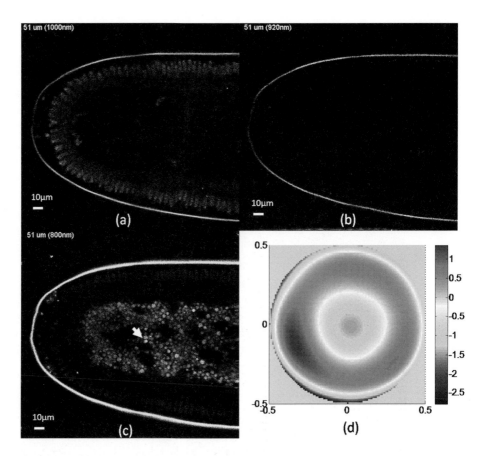

Figure 2.17 Two-photon images of a Drosophila embryo at a depth of 51 μm with excitation wavelengths of 1000 nm (**a**), 920 nm (**b**), and 800 nm (**c**). The wavefront (**d**) at the middle of embryo is measured by a SHWS in 30 ms using an autofluorescent guide star. During wavefront sensing, the laser illuminates the intrinsic fluorophores indicated by an arrow shown in (**c**). The scale bars are 10 μm. Reproduced with permission from [88], OSA. A black and white version of this figure appears in some formats. For the color version, please refer to the plate section.

800 nm excitation wavelength, the fluorescence from the yolk is even brighter than the RFP. Because autofluorescence is a natural emission from biological structures, it exists even before the first mitosis cycle. This special feature makes it particularly suitable as a guide star for imaging the embryo at the early stages. Figure 2.17d shows the wavefront measurement at a depth of 51 μm using autofluorescence for the guide star. During wavefront measurement, the laser is parked on the fluorophores indicated by the arrow. The laser power at the sample is 30 mW. The exposure time for a single measurement is 30 ms. The RMS error of the wavefront is 0.387 μm.

The two-photon images of yolk autofluorescence at a depth of 51 μm before and after correction are shown in Figures 2.18a and 2.18b. Each image is the maximum intensity projection of the three consecutive focal planes (1 plane/μm), which takes 6 seconds to achieve the whole stack. Because of the fast motion of the yolk during this imaging time, there is a small variation in structures shown in these two images. The

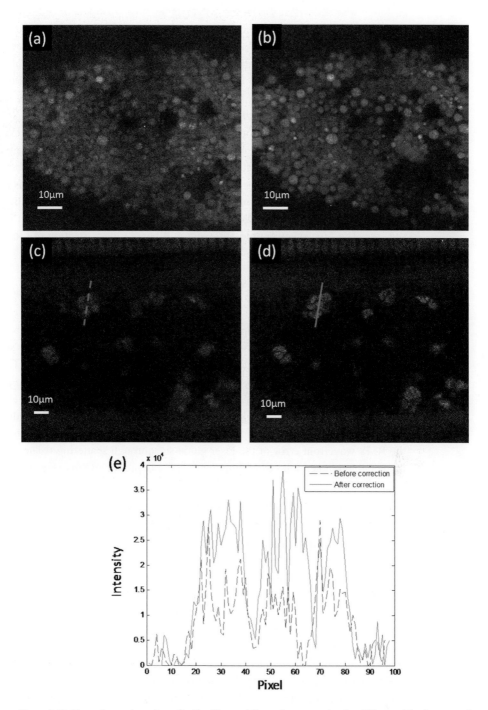

Figure 2.18 Two-photon imaging of a live Drosophila embryo at a depth of 51 μm. The images of yolk autofluorescence before (**a**) and after correction (**b**). The RFP labeled nuclei structure before (**c**) and after (**d**) correction. The intensity profiles (**e**) along lines in (**c**) and (**d**). Upper curve after correction, lower curve before correction. The scale bars are 10 μm. Reproduced with permission from [88], OSA.

excitation wavelength is set at 800 nm for maximizing the yolk autofluorescence. The structures of the endogenous fluorophores in yolk are much clearer after correction, and the noise is reduced dramatically. A two-photon image of RFP labeled nuclei structure without correction is shown in Figure 2.18c. Compared with the yolk autofluorescence, less RFP labeled structures can be selected for use as the guide star and the intensity of the fluorescence is much lower. Figure 2.18d shows the two-photon image after correction. The image after correction is much brighter than before correction. The intensity profile along the dashed lines across nuclei is shown in Figure 2.18e. The peak intensity increased by 2×. The fast direct wavefront measurement is particularly suitable to correct dynamic aberrations induced by live specimens. To validate this advantage, time-lapse wavefront aberrations were measured at a depth of 50 μm with a time resolution of 10 seconds. The RMS wavefront changed by around 0.1 λ in the first 10 seconds. The fast wavefront measurement will benefit the correction of these dynamic aberrations.

2.3.4   Imaging Deeper

Since Shack-Hartman wavefront sensing depends on ballistic light, it will be limited in depth by scattering. This approach can be extended to tissues that strongly scatter visible light by exploiting the reduced scattering of near-infrared guide stars. This method enables *in vivo* two-photon morphological and functional imaging as deep as 700 μm inside the mouse brain [67]. Another approach uses feedback-based wavefront shaping to focus light onto a guide star through scattering tissue [95]. With feedback-based wavefront shaping for focusing light onto the guide star the signal-to-noise ratio and the RMS wavefront error of the laser guide star through scattering tissue can be more than doubled, potentially extending the imaging depth for AO microscopy, as shown in Figure 2.19.

2.4   Conclusions and Future Directions

In this chapter we have reviewed the use of direct wavefront sensing for AO in biological imaging including wide-field, confocal, and multiphoton microscopy. Adaptive optical microscopy using direct wavefront sensing has been used to increase the resolution and signal intensity. Some of the benefits for this approach are faster correction for live imaging, lower-light exposure and less photobleaching since the wavefront sensing is accomplished in a single measurement rather than iterative measurements required for sensorless approaches (96–98). An additional benefit for using direct wavefront sensing is that the wavefront aberration is directly measured and corrected rather than optimized. The objective is getting as close as possible to the diffraction limit of the optical system, as measured by the Strehl ratio, rather than optimization of the overall image brightness or sharpness. The cost for these benefits is increased system complexity due to the requirements for the formation of a guide star in wide-field and confocal imaging,

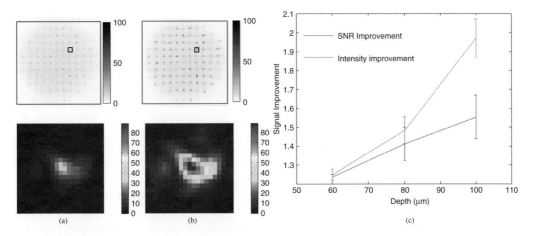

Figure 2.19 Wavefront measurement results using feedback-based wavefront shaping to focus light onto guide stars through scattering tissue. Complete images from the Shack–Hartmann wavefront sensor (grayscale inverted for clarity) and images of a single spot indicated by the white square before (**a**) and after (**b**) wavefront shaping. (**c**) The SNR improvement of wavefront measurement and intensity improvement for various depths. The error bars represent the standard deviation of the mean. Reproduced with permission from [95], OSA. A black and white version of this figure appears in some formats. For the color version, please refer to the plate section.

and the need for a wavefront sensing system such as a Shack–Hartmann wavefront sensor or an interferometer. In multiphoton microscopy a point guide star is automatically formed at the focus. Future work will most likely combine AO with higher-order wavefront shaping as imaging proceeds deeper into the sample where both refraction and scattering limit the imaging resolution, intensity, and depth.

Acknowledgments

We would like to acknowledge the research groups that contributed to results in this chapter including Profs. Yi Zuo and William Sullivan, MCD Biology, UC Santa Cruz, Prof. John Sedat, Department of Biochemistry and Biophysics in the School of Medicine, UC San Francisco, Dr. Donald Gavel and Daren Dillon, UCO/Lick Observatory, UC Santa Cruz, Scot Olivier and Diana Chen, Lawrence Livermore National Laboratory.

Funding

This material is based upon work supported by the National Science Foundation under Grant Numbers 852742, 1353461, & 1429810 and the California Institute for Regenerative Medicine (CIRM) under Grant Numbers RT1–01095 & CL1–00506–1.1. The results presented herein were obtained at the W. M. Keck Center for Adaptive

Optical Microscopy (CfAOM) at University of California Santa Cruz. The CfAOM was made possible by the generous financial support of the W. M. Keck Foundation. This material is also based upon work supported by the UC Office of the President for the UC Work Group for Adaptive Optics in Biological Imaging, by the Multicampus Research Programs and Initiatives (MRPI), Grant #MR-15–327968.

2.5   References

[1] John W. Hardy, *Adaptive Optics for Astronomical Telescopes*, Oxford Series in Optical and Imaging Sciences (Oxford University Press, 1998), vol. 16.

[2] Robert K. Tyson, *Introduction to Adaptive Optics* (SPIE Tutorial Texts in Optical Engineering, 2000), vol. 41.

[3] Robert Tyson, *Adaptive Optics Engineering Handbook*, Optical Science and Engineering (CRC Press, 1999).

[4] Jason Porter, Hope Queener, Julianna Lin, and Karen Thorn, *Adaptive Optics for Vision Science: Principles, Practices, Design and Applications*, Wiley Series in Microwave and Optical Engineering (John Wiley, 2006).

[5] David P. Biss, Daniel Sumorok, Stephen A. Burns, Robert H. Webb, Yaopeng Zhou, Thomas G. Bifano, Daniel Côté, Israel Veilleux, Parisa Zamiri, and Charles P. Lin, In vivo *fluorescent imaging of the mouse retina using adaptive optics*, Opt. Lett. **32** (2007), 659–661.

[6] Martin Booth, *Adaptive optics in microscopy*, Philos. Trans. R. Soc. London Ser. A **365** (2007), 2829–2843.

[7] Joel A. Kubby (ed.), *Adaptive Optics for Biological Imaging* (CRC Press/Taylor & Francis, 2013).

[8] Martin J. Booth, *Adaptive optical microscopy: the ongoing quest for a perfect image*, Light: Sci. Appl. **3** (2014), e165.

[9] Ben C. Platt and Roland Shack, *History and principles of Shack–Hartmann wavefront sensing*, J. Refract. Surg. **17** (2001), S573–S577.

[10] Robert K. Tyson (ed.), *Adaptive Optics Engineering Handbook* (Marcel Dekker, 2000).

[11] www.ucolick.org/~max/289/ (Last accessed on July 6, 2015).

[12] C. E. Max, B. A. Macintosh, S. G. Gibbard, D. T. Gavel, H. G. Roe, I. de Pater, A. M. Ghez, D. S. Acton, O. Lai, P. Stomski, and P. L. Wizinowich, *Cloud structures on Neptune observed with Keck telescope adaptive optics*, Astron. J. **125** (2003), 364–375.

[13] Y. Zhang, S. Poonja, and A. Roorda, *MEMS-based adaptive optics scanning laser ophthalmoscopy*, Opt. Lett. **31** (2006), 1268–1270.

[14] Max Born and Emil Wolf, *Principles of Optics* (Cambridge University Press, 1999), p. 461.

[15] Virendra N. Mahajan, *Strehl ratio for primary aberrations in terms of their aberration variance*, J. Opt. Soc. Am. **73** (1983), 860–861.

[16] L. Michaille, A. D. Cañas, J. C. Dainty, J. Maxwell, T. Gregory, J. C. Quartel, F. C. Reavell, R. W. Wilson, and N. J. Wooder, *A laser beacon for monitoring the mesospheric sodium layer at La Palma*, MNRAS **318**(1) (2000), 139–144.

[17] Dae Young Kim and Richard Roy, *Cell cycle regulators control centrosome elimination during oogenesis in Caenorhabditis elegans*, J. Cell Biol. **174** (2006), 751–757.

[18] J. L. Beverage, R. V. Shack, and M. R. Descour, *Measurement of the three-dimensional microscope point spread function using a Shack–Hartmann wavefront sensor*, J. Microsc. **205** (2002), 61–75.

[19] Marcus Reicherter, Witold Gorski, Tobias Haist, and Wolfgang Osten, *Dynamic correction of aberrations in microscopic imaging systems using an artificial point source*, Proc. SPIE **5462** (2004), 68.

[20] W. H. Southwell, *Wave-front estimation from wave-front slope measurements*, J. Opt. Soc. Am. **70** (1980), 998–1006.

[21] S. Thomas, T. Fusco, A. Tokovinin, M. Nicolle, V. Michau, and G. Rousset, *Comparison of centroid computation algorithms in a Shack–Hartmann sensor*, MNRAS **371** (2006), 323–336.

[22] Lisa A. Poyneer, D. T. Gavel, and J. M. Brase, *Fast wave-front reconstruction in large adaptive optics systems with use of the Fourier transform*, J. Opt. Soc. Am. **A 19** (2003), 2100–2111.

[23] Alioune Diouf, Andrew P. Legendre, Jason B. Stewart, Thomas G. Bifano, and Yang Lu, *Open-loop shape control for continuous microelectromechanical system deformable mirror*, Appl. Opt. **49** (2010), G148–154.

[24] Katie Morzinski, Luke C. Johnson, Donald T. Gavel, Bryant Grigsby, Daren Dillon, Marc Reinig, and Bruce A. Macintosh, *Performance of MEMS-based visible-light adaptive optics at Lick Observatory: closed- and open-loop control*, Proc. SPIE **7736** (2010), 77361O.

[25] J. Liang, D. R. Williams, and D. T. Miller, *Supernormal vision and high-resolution retinal imaging through adaptive optics*, J. Opt. Soc. Am. **A 14** (1997), 2884–2892.

[26] D. Gavel, *Suppressing anomalous localized waffle behavior in least squares wavefront reconstructor*, Proc. SPIE **4839** (2003), 972–980.

[27] R. K. Tyson, *Principles of Adaptive Optics*, 2nd edition (Academic Press, 1998).

[28] Oscar A. Azucena, *Adaptive Optics Wide-Field Microscopy Using Direct Wavefront Sensing* (ProQuest, UMI Dissertation Publishing, 2012).

[29] Oscar Azucena, Joel Kubby, Justin Crest, Jian Cao, William Sullivan, Peter Kner, Donald Gavel, Daren Dillon, and Scot Olivier, *Implementation of a Shack–Hartmann wavefront sensor for the measurement of embryo-induced aberrations using fluorescent microscopy*, Proc. SPIE **7209** (2009), 720906.

[30] Oscar Azucena, Justin Crest, Jian Cao, William Sullivan, Peter Kner, Don Gavel, Daren Dillon, Scot Olivier, and Joel Kubby, *Implementation of adaptive optics in fluorescent microscopy using wavefront sensing and correction*, Proc. SPIE **7595** (2010), 75950I.

[31] Oscar Azucena, Justin Crest, Jian Cao, William Sullivan, Peter Kner, Donald Gavel, Daren Dillon, Scot Olivier, and Joel Kubby, *Wavefront aberration measurements and corrections through thick tissue using fluorescent microsphere reference beacons*, Opt. Express **18** (2010), 17521–17532.

[32] Oscar Azucena, Xiaodong Tao, Justin Crest, Shaila Kotadia, William Sullivan, Donald Gavel, Marc Reinig, and Joel Kubby, *Adaptive optics wide-field microscope corrections using a MEMS DM and Shack–Hartmann wavefront sensor*, Proc. SPIE **7931**(2011), 79310J.

[33] Oscar Azucena, Justin Crest, Shaila Kotadia, William Sullivan, Xiaodong Tao, Marc Reinig, Donald Gavel, Scot Olivier, and Joel Kubby, *Adaptive optics wide-field microscopy using direct wavefront sensing*, Opt. Lett. **36** (2011), 825–827.

[34] P. Vermeulen, E. Muro, T. Pons, V. Loriette, and A. Fragola, *Adaptive optics for fluorescence wide-field microscopy using spectrally independent guide star and markers*, J. Biomed. Opt. **16** (2011), 076019.

[35] Michael Shaw, Kevin O'Holleran, and Carl Paterson, *Investigation of the confocal wavefront sensor and its application to biological microscopy*, Opt. Express **21** (2013), 19353–19362.

[36] Eric P. Goodwin and James C. Wyant, *Field Guide to Interferometric Optical Testing* (SPIE Press, 2006).

[37] M. Schwertner, M. J. Booth, and T. Wilson, *Characterizing specimen induced aberrations for high NA adaptive optical microscopy*, Opt. Express **12** (2004), 6540–6552.

[38] A. M. Lyapunov, *Stability of Motion* (Academic Press, 1966).

[39] C. R. Vogel and Q. Yang, *Modeling, simulation, and open-loop control of a continuous facesheet MEMS deformable mirror*, J. Opt. Soc. Am. A **23** (2006), 1074–1081.

[40] http://ao.jpl.nasa.gov/Palao/Results/dec98/HR2499.html

[41] Fabrice Vidal, Eric Gendron, and Gérard Rousset, *Tomography approach for multi-object adaptive optics*, J. Opt. Soc. Am. **A 27** (2010), A253–264.

[42] J. M. Beckers, *Increasing the size of the isoplanatic patch within multiconjugate adaptive optics*, Proceedings of European Southern Observatory Conference and Workshop on Very Large Telescopes and Their Instrumentation (European Southern Observatory, 1988), 693–703.

[43] D. C. Johnston and B. M. Welsh, *Analysis of multiconjugate adaptive optics*, J. Opt. Soc. Am. **A 11** (1994), 394–408.

[44] R. Ragazzoni, E. Marchetti, and G. Vatente, *Adaptive-optics corrections available for the whole sky*, Nature **403** (2000), 54–56.

[45] Tokovinin, M. Le Louarn, and M. Sarazin, *Isoplanatism in a multiconjugate adaptive optics system*, J. Opt. Soc. Am. **A 17** (2000), 1819–1827.

[46] J. Thaung, P. Knutsson, Z. Popovic, and M. Owner-Petersen, *Dual-conjugate adaptive optics for wide-field high-resolution retinal imaging*, Opt. Express **17** (2009), 4454–4467.

[47] Z. Popovic, P. Knutsson, J. Thaung, M. Owner-Petersen, and J. Sjostrand, *Noninvasive imaging of human foveal capillary network using dual-conjugate adaptive optics*, Invest. Ophthalmol. Visual Sci. **52** (2011), 2649–2655.

[48] Richard D. Simmonds and Martin J. Booth, *Modelling of multi-conjugate adaptive optics for spatially variant aberrations in microscopy*, J. Opt. **15** (2013), 4010.

[49] Tsai-wei Wu and Meng Cui, *Numerical study of multi-conjugate large area wavefront correction for deep tissue microscopy*, Opt. Express **23** (2015), 7463–7470.

[50] Jerome Mertz, Hari Paudel, and Thomas G. Bifano, *Field of view advantage of conjugate adaptive optics in microscopy applications*, Appl. Opt. **54** (2015), 3498–3506.

[51] J. Vernin and F. Roddier, *Experimental determination of two-dimensional spatiotemporal power spectra of stellar light scintillation: evidence for a multilayer structure of the air turbulence in the upper troposphere*, J. Opt. Soc. Am. **63** (1973), 270–273.

[52] J. H. Park, W. Sun, and M. Cui, *High-resolution* in vivo *imaging of mouse brain through the intact skull*, Proc. Natl. Acad. Sci. U.S.A. **112** (2015), 9236–9241.

[53] K. Wang, D. E. Milkie, A. Saxena, P. Engerer, T. Misgeld, M. E. Bronner, J. Mumm, and E. Betzig, *Rapid adaptive optical recovery of optimal resolution over large volumes*, Nat. Methods **11** (2014), 625–628.

[54] Xiaodong Tao, Bautista Fernandez, Oscar Azucena, Min Fu, Denise Garcia, Yi Zuo, Diana C. Chen, and Joel Kubby, *Adaptive optics confocal microscopy using direct wavefront sensing*, Opt. Lett. **36** (2011), 1062–1064.

[55] Xiaodong Tao, Bautista Fernandez, Diana C. Chen, Oscar Azucena, Min Fu, Yi Zuo, and Joel Kubby, *Adaptive optics confocal fluorescence microscopy with direct wavefront sensing for brain tissue imaging*, Proc. SPIE **7931** (2011), 79310L.

[56] Saad A. Rahman and Martin J. Booth, *Direct wavefront sensing in adaptive optical microscopy using backscattered light*, Appl. Opt. **52** (2013), 5523–5532.

[57] Xiaodong Tao, Oscar Azucena, Min Fu, Yi Zuo, Diana C. Chen, and Joel Kubby, *Adaptive optics microscopy with direct wavefront sensing using fluorescent protein guide stars*, Opt. Lett. **36** (2011), 3389–3391.

[58] Xiaodong Tao, Oscar Azucena, Min Fu, Yi Zuo, Diana C. Chen, and Joel Kubby, *Adaptive optics confocal microscopy using fluorescent protein guide-stars for brain tissue imaging*, Proc. SPIE **8253** (2012), 82530M.

[59] Xiaodong Tao, Justin Crest, Shaila Kotadia, Oscar Azucena, Diana C. Chen, William Sullivan, and Joel Kubby, *Live imaging using adaptive optics with fluorescent protein guide-stars*, Opt. Express **20** (2012), 15969–15982.

[60] J. Zhang and T. L. Megraw, *Proper recruitment of gamma-tubulin and D-TACC/Msps to embryonic Drosophila centrosomes requires Centrosomin Motif 1*, Mol. Biol. Cell **18** (2007), 4037–4049.

[61] T. Moutinho-Santos, P. Sampaio, I. Amorim, M. Costa, and C. E. Sunkel, *In vivo localisation of the mitotic POLO kinase shows a highly dynamic association with the mitotic apparatus during early embryogenesis in Drosophila*, Biol. Cell **91** (1999), 585–596.

[62] W. F. Rothwell and W. Sullivan, *Fluorescent analysis of drosophila embryos, Drosophila Protocols*, W. Sullivan, M. Ashburner, and R. S. Hawley, eds. (Cold Spring Harbor Laboratory Press, 2000), 141–157.

[63] M. Gu, *Advanced Optical Imaging Theory* (Springer, 1999).

[64] M. Schwertner, M. J. Booth, M. A. Neil, and T. Wilson, *Measurement of specimen-induced aberrations of biological samples using phase stepping interferometry*, J. Microsc. **213** (2004), 11–19.

[65] M. A. R. Jewel, V. Akondi, and B. Vohnsen, *3-D Analysis of Pinhole Size Optimization for a Confocal Signal-Based Wavefront Sensor*, Frontiers in Optics 2014, OSA Technical Digest (online) (Optical Society of America, 2014), JW3A.40.

[66] C. A. Schneider, W. S. Rasband, and K. W. Eliceiri, *NIH Image to ImageJ: 25 years of image analysis*, Nat. Methods **9** (2012), 671–675.

[67] Kai Wang, Wenzhi Sun, Christopher T. Richie, Brandon K. Harvey, Eric Betzig, and Na Ji, *Direct wavefront sensing for high-resolution* in vivo *imaging in scattering tissue*, Nat. Comm. **6** (2015), 7276.

[68] Jonathan M. Taylor, Christopher D. Saunter, Cyril Bourgenot, John M. Girkin, and Gordon D. Love, *Realtime wavefront sensing in a SPIM microscope, and active aberration tracking*, Proc. SPIE **9335** (2015), 93350A.

[69] Demirhan Kobat, Michael E. Durst, Nozomi Nishimura, Angela W. Wong, Chris B. Schaffer, and Chris Xu, *Deep tissue multiphoton microscopy using longer wavelength excitation*, Opt. Express **17** (2009), 13354–13364.

[70] Helmchen and W. Denk, *Deep tissue two-photon microscopy*, Nat. Methods **2** (2005), 932–940.

[71] J. M. Girkin, S. Poland, and A. J. Wright, *Adaptive optics for deeper imaging of biological samples*, Curr. Opin. Biotechnol. **20** (2009), 106–110.

[72] O. Albert, L. Sherman, G. Mourou, T. B. Norris, and G. Vdovin, *Smart microscope: an adaptive optics learning system for aberration correction in multiphoton confocal microscopy*, Opt. Lett. 25 (2000), 52–54.

[73] P. Marsh, D. Burns, and J. Girkin, *Practical implementation of adaptive optics in multiphoton microscopy*, Opt. Express **11** (2003), 1123–1130.

[74] D. Debarre, E. J. Botcherby, T. Watanabe, S. Srinivas, M. J. Booth, and T. Wilson, *Image-based adaptive optics for two-photon microscopy*, Opt. Lett. **34** (2009), 2495–2497.

[75] N. Ji, D. E. Milkie, and E. Betzig, *Characterization and adaptive optical correction of aberrations during* in vivo *imaging in the mouse cortex*, Nat. Methods **7** (2010), 141–147.

[76] Marcus Feierabend, Markus Rückel, and Winfried Denk, *Coherence-gated wave-front sensing in strongly scattering samples*, Opt. Lett. **29** (2004), 2255–2257.

[77] M. Rueckel, J. A. Mack-Bucher, and W. Denk, *Adaptive wavefront correction in two-photon microscopy using coherence-gated wavefront sensing*, Proc. Natl. Acad. Sci. U.S.A. **103** (2006), 17137–17142.

[78] Markus Rueckel and Winfried Denk, *Properties of coherence-gated wavefront sensing*, J. Opt. Soc. Am. A **24** (2007), 3517–3529.

[79] T. I. M. van Werkhoven, J. Antonello, H. H. Truong, M. Verhaegen, H. C. Gerritsen, and C. U. Keller, *Snapshot coherence-gated direct wavefront sensing for multi-photon microscopy*, Opt. Express **22** (2014), 9715–9733.

[80] J. W. Cha, J. Ballesta, and P. T. C. So, *Shack–Hartmann wavefront-sensor-based adaptive optics system for multiphoton microscopy*, J. Biomed. Opt. **15** (2010), 046022.

[81] N. Ji, D. E. Milkie, and E. Betzig, *Adaptive optics via pupil segmentation for high-resolution imaging in biological tissues*, Nat. Methods **7** (2010), 141–147.

[82] D. E. Milkie, E. Betzig, and N. Ji, *Pupil-segmentation-based adaptive optical microscopy with full-pupil illumination*, Opt. Lett. **36** (2011), 4206–4208.

[83] N. Ji, T. R. Sato, and E. Betzig, *Characterization and adaptive optical correction of aberrations during* in vivo *imaging in the mouse cortex*, Proc. Natl. Acad. Sci. U.S.A. **109** (2012), 22–27.

[84] C. Wang, R. Liu, D. E. Milkie, W. Sun, Z. Tan, A. Kerlin, T. Chen, D. S. Kim, and N. Ji, *Multiplexed aberration measurement for deep tissue imaging* in vivo, Nat. Methods **11** (2014), 1037–1040.

[85] R. Liu, D. E. Milkie, A. Kerlin, B. Maclennan, and N. Ji, *Direct phase measurement in zonal wavefront reconstruction using multidither coherent optical adaptive technique*, Opt. Express **22** (2014), 1619–1628.

[86] R. Aviles-Espinosa, J. Andilla, R. Porcar-Guezenec, O. E. Olarte, M. Nieto, X. Levecq, D. Artigas, and P. Loza-Alvarez, *Measurement and correction of* in vivo *sample aberrations employing a nonlinear guide-star in two-photon excited fluorescence microscopy*, Biomed. Opt. Express **2** (2011), 3135–3149.

[87] Rodrigo Aviles-Espinosa, Jordi Andilla, Rafael Porcar-Guezenec, Xavier Levecq, David Artigas, and Pablo Loza-Alvarez, *Depth aberrations characterization in linear and nonlinear microscopy schemes using a Shack–Hartmann wavefront sensor*, Proc. SPIE **8227** (2012), 82271D.

[88] Xiaodong Tao, Andrew Norton, Matthew Kissel, Oscar Azucena, and Joel Kubby, *Adaptive optical two-photon microscopy using autofluorescent guide-stars*, Opt. Lett. **38** (2013), 5075–5078.

[89] L. D. S. Haro and J. C. Dainty, *Single-pass measurements of the wave-front aberrations of the human eye by use of retinal lipofuscin autofluorescence*, Opt. Lett. **24** (1999), 61–63.

[90] R. Zipfel, R. M. Williams, R. Christie, A. Y. Nikitin, B. T. Hyman, and W. W. Webb, *Live tissue intrinsic emission microscopy using multiphoton-excited native fluorescence and second harmonic generation*, Proc. Natl. Acad. Sci. U.S.A. **100** (2003), 7075–7080.

[91] K. M. Morzinski, K. B. W. Harpsøe, D. T. Gavel, and S. M. Ammons, *The open loop control of MEMS: modeling and experimental results*, Proc. SPIE **6467** (2007), 64670G.

[92] M. Ashburner, K. G. Golic, and R. S. Hawley, *Drosophila: A Laboratory Handbook*, 2nd edition (Cold Spring Harbor Laboratory Press, 2005).

[93] M. Mavrakis, R. Rikhy, M. Lilly, and J. Lippincott-Schwartz, *Fluorescence imaging tech-niques for studying Drosophila embryo development*, Curr. Protoc. Cell Biol. **4** (2008), 18.

[94] J. P. Ogilvie, D. Debarre, X. Solinas, J.-L. Martin, E. Beaurepaire, and M. Joffre, *Use of coherent control for selective two-photon fluorescence microscopy in live organisms*, Opt. Express **14** (2006), 759–766.

[95] Xiaodong Tao, Ziah Dean, Christopher Chien, Oscar Azucena, Dare Bodington, and Joel Kubby, *Shack–Hartmann wavefront sensing using interferometric focusing of light onto guide-stars*, Opt. Express **21** (2013), 31282–31292.

[96] Ignacio Izeddin, Mohamed El Beheiry, Jordi Andilla, Daniel Ciepielewski, Xavier Darzacq, and Maxime Dahan, *PSF shaping using adaptive optics for three-dimensional single-molecule super-resolution imaging and tracking*, Opt. Express **20** (2012), 4957–4967.

[97] Christopher D. Saunter, Cyril Bourgenot, John M. Girkin, and Gordon D. Love, *Closed loop adaptive optics with a laser guide-star for biological light microscopy*, Proc. SPIE **8253** (2012), 82530J.

[98] C. Bourgenot, C. D. Saunter, G. D. Love, and J. M. Girkin, *Comparison of closed loop and sensorless adaptive optics in widefield optical microscopy*, J. Eur. Opt. Soc. Rap. Public. **8** (2013), 13027.

Part II

Deep Tissue Microscopy

3 Deep Tissue Fluorescence Microscopy

Meng Cui

3.1 Introduction

Technology development often drives discoveries in biological sciences. In the past few decades, the rapid progress of optical microscopy has led to explosive progress in many research fields [1]. Compared to other imaging modalities, optical imaging offers several advantages. First, optical imaging is noninvasive, which allows repeated measurement on the same cell, tissue culture and even behaving animals [2]. Second, optical imaging provides high spatial resolution. Typical resolution can reach a few hundred nanometers, which can resolve subcellular features. The frontiers of superresolution imaging methods can further improve the resolution by one or two orders of magnitude [3]. Third, optical imaging can provide high temporal resolutions. Volumetric imaging at tens or even hundreds of Hz is achievable in behaving animals [4]. For imaging highly transparent tissues, kHz frame rate has been demonstrated for imaging voltage signals [5]. Fourth, optical imaging can enjoy a variety of contrast mechanisms. In particular, fluorescence optical imaging can achieve single-molecule sensitivity. Combined with the rapid development of various fluorescence based functional indicators, optical imaging can directly reveal the molecular dynamics and signals in complex biological systems [6].

A major limitation of optical imaging is the superficial imaging depth. The most commonly used wide-field imaging [7] can achieve tens of microns imaging depth. Beyond that, the out-of-focus signal can overwhelm the true signal from the focal plane. The photon shot noise from the out of the focus signal can severely reduce the achievable signal-to-noise ratio (SNR). After all, SNR is perhaps the most important factor that determines whether a measurement approach is feasible for practical applications [8]. A way to reduce the out-of-focus signal is to employ confocal excitation and detection, which works very well in transparent and thin samples and achieves slightly improved spatial resolutions [9]. Confocal microscopes are widely used by biologists worldwide. In the past decade, scientists have revived plan illumination microscopy or light sheet microscopy [10, 11], in which a thin sheet of light illuminates the sample from the side while a second objective collects the signal in the perpendicular direction. Compared to confocal, light sheet imaging avoids the illumination above and below the focal plane and therefore reduces photobleaching and phototoxicity, which has been employed to image embryo development, cellular dynamics, and calcium signal of neurons. However, its performance in scattering tissue is similar or worse than that of conventional confocal microscopy. As many interesting dynamics happen far below the tissue surface, deep tissue imaging methods are greatly desired in many applications. Currently, the method of choice is laser scanning two-photon fluorescence microscopy (TPM) [12].

TPM was originally developed for reducing the photobleaching and phototoxicity in cellular imaging applications [13]. A few years later, the research community started to realize and appreciate the major advantages of TPM for *in vivo* imaging in scattering tissue. In particular, the *in vivo* imaging of mammalian brain plasticity and the calcium imaging of neuronal activity triggered a wide excitement and acceptance of TPM in neuroscience community and other research field such as immunology [14]. Several factors contribute to the huge success of TPM. First, TPM uses near-infrared (NIR) laser for excitation, which suffers much lower scattering and absorption from the tissue than the visible light used in one-photon excitation. Second, two-photon excitation is a nonlinear process. As a result, the signal is highly confined in the focus. Out-of-focus light and scattered light are typically too weak to excite any fluorescent signal, which reduces photobleaching and phototoxicity and lowers the imaging background. Third, as the excitation is 3D confined, we can use all the collectible fluorescence emission to represent the signal from the focus, which allows us to use both scattered fluorescence emission and ballistic fluorescence emission. This feature makes a huge difference for imaging in scattering tissue. If confocal detection were used, the fluorescence signal collection efficiency will decay exponentially as a function of depth. But with TPM, the signal collection efficiency has only a moderate change over a large imaging depth. It is worth noting that the forward propagating signal is typically lost in confocal microscopy. However, in scattering tissue, some of the forward propagating signal can in fact be collected in TPM due to scattering. All of these factors contribute to the deep tissue imaging capability of TPM.

Despite the great success of TPM, the imaging depth is still insufficient in many applications, in which the biological events of interest take place far below the surface. For example, TPM can only image the top $\sim$200 μm of the millimeter scale mouse popliteal lymph nodes [15]. The internal dynamics are inaccessible. In neuroscience, high quality imaging is typically performed for the top $\sim$500 μm of the neocortex of the mouse brain. It remains challenging to access 500–1000 μm depth in the cortex, not to mention the hippocampus below the cortex. Therefore, there is great need to improve the performance of TPM at large depth. It is worth noting that improving imaging depth is not merely to push for greater depth at the cost of other factors such as SNR, photodamage, photobleaching, background, and speed. All of these factors still need to be maintained at a level that is acceptable for practical applications.

3.2   Background

Many labs have realized this depth limitation of optical imaging and developed a variety of solutions. Generally, these efforts can be categorized to two directions. One is to change the tissue and the other is to change the light. With NIR wavelength, scattering is the major factor that limits the imaging depth, resulting from the inhomogeneous distribution of optical refractive index. Therefore, refractive index matching has been explored to change the scattering property of tissue (tissue clearing). The recent development of tissue clearing methods [16] has achieved great success, which has been applied to image the entire mouse brain. However, these tissue clearing methods have been limited to fixed tissues as the clearing processes can chemically change the tissue. For *in vivo* imaging applications, the measurement has to be noninvasive (no change to the physiological conditions). Therefore, the natural solution is to change the light. An important property of light scattering is that it is wavelength dependent. At longer wavelength, the refractive index variation in tissue is reduced. Moreover, an even longer optical path length difference is needed to generate the same level of wavefront distortion. Three-photon excited fluorescence microscopy has been developed [17] to use longer excitation wavelength to gain imaging depth. Another parameter of light we can control is the optical wavefront. After all, the decreased Strehl ratio is the result of accumulated optical wavefront distortion. Controlling the optical wavefront can in fact cancel the wavefront distortion and restore the Strehl ratio. Wavefront control is a rather general approach, which can work with any imaging techniques including TPM and three-photon methods. Overall, the collective control of optical wave (both wavelength and wavefront) is expected to extend the working depth of optical imaging. Here we will focus on the latest development of optical wavefront control techniques for microscopy applications.

Using wavefront control to compensate wavefront distortion in imaging has a long history. In the 1950s, the idea of adaptive optics (AO) was developed in astronomy to compensate the air turbulence induced wavefront distortions [18]. Currently, every large diameter ground based telescope is equipped with AO. The deformable mirror used in AO has a rather limited number of degrees of freedom and therefore can only handle moderate level of wavefront distortions. In the 1960s, the pioneers of laser holography at the University of Michigan in Ann Arbor used holographic recording to control a huge number of spatial modes [19]. Through experiments, they showed that one can focus or form an image through highly scattering media. In essence, this experiment showed that no matter how scattering the medium is there is always a solution to focus light through it, which generalized the idea of AO to a broad range of media. Overall, aberration or random scattering is about linear wave propagation, which is a time reversible process. Given a point source, if we can measure the distorted wavefront through the media (air turbulence or a glass of milk), we can always generate a perfect focus using a reversed (compensated) wavefront. This is also the principle behind many optical phase-conjugation experiments in the 1970s and 1980s [20, 21]. The concept of AO and holography (either static [19] or real time [21]) based time reversal involves measuring the optical wavefront and then reversing it. In the 1970s, a different approach was explored to compensate wavefront distortion without direct wavefront measurement. The method used optical phase modulation and a feedback loop to ensure that every beamlet constructively interfered at the target location. This coherent optical interferometry based approach is named Coherent Optical Adaptive Technique (COAT) [22]. The basic concept is illustrated in Figure 3.1a. Suppose there are two beams of coherent light, one of which is controlled by a phase modulator. After the two beams

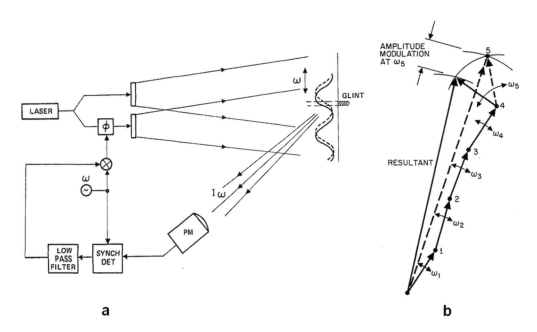

Figure 3.1 (**a**) The concept of the Coherent Optical Adaptive Technique. (**b**) The concept of multidither measurement. Images are from *Appl. Opt*, 13(2) (1974), 291–300.

propagate through some distance, the beams will diverge and overlap. The interference at the overlapped region appears as sinusoidal fringes. In COAT, the goal is to focus light through air turbulence onto a target that reflects light. If we use the phase modulator to change the phase of one beam, the interference fringe at the target plane will move. As a result, the light scattered by the target can be modulated. The scattered light provides the feedback signal to a control loop. Experimentally, the control loop applies a sinusoidal modulation over a small phase range (dithering) and compares the phase of the sinusoidal driving signal and the scattered feedback signal. Depending on whether the scattered signal is in phase or out of phase with the driving signal, the control loop will shift the phase offset of the dithering one way or the other, which eventually will force the dithering to oscillate near the constructive interference position. To compensate air turbulence, one certainly needs to control more than two modes. A parallel phase-modulation method was developed to dither and control many phase elements in parallel, which is named multidither COAT [22]. The concept is shown in Figure 3.1b. Basically, we can simultaneously modulate all the phase elements. To avoid signal cross talk between these phase elements, each of the phase elements is modulated at a unique frequency. The frequency difference needs to be equal to or greater than the inverse of the overall system response time. Amazingly, multidither COAT works well in practice to ensure all phase elements to have constructive interference at the target location although each of the phase elements is controlled by an independent control loop, unaware of other phase elements.

The scheme employed in multidither COAT is of great interest to wavefront compensation for TPM as it requires no wavefront sensor. In general, the various wavefront compensation methods can be categorized to two groups. One is the sensor based measurement [23, 24], which involves using a wavefront sensor to measure wavefront distortion. This is the same scheme used in astronomical AO. For TPM, the wavefront measurement needs to be performed at NIR wavelength. However, typical fluorescence emission is in the visible range. Injection of NIR emission fluorophores is typically required. As the trend of multiphoton microscopy is to move to even longer wavelengths (e.g., 1.04, 1.3, and 1.7 μm), it would be increasingly difficult to use fluorophore based wavefront measurement. The other group is the sensorless wavefront measurement [25, 26], as in multidither COAT. In comparison, sensorless methods are simpler, i.e., requiring neither wavefront sensor nor NIR emission fluorophores. Moreover, sensorless methods can flexibly work at any wavelength. For practical applications, simplicity and flexibility are greatly desired.

However, modification is needed to convert multidither COAT for practical TPM applications. There are two major differences between air turbulence and biological tissue-induced wavefront correction. One is the system response time. For air turbulence compensation, kHz update rate is needed. For biological systems, 1–10 Hz update rate is typically fast enough. For certain tissues such as the skull of an adult mouse brain, the wavefront correction can work for hours without any update. The other difference is the wavefront complexity. Biological tissue can introduce much more complicated wavefront distortion, which may require ~1000 or more spatial modes.

3.3   Results

3.3.1   IMPACT

Taking these differences and the practical biological imaging requirement into consideration, we developed a modified multidither COAT [27]. First, instead of building 1000 independent phase modulators and control loops, we employed a MEMS-based segmented deformable mirror (Boston Micromachine Corp) for phase modulation. We used a computer to control the phase modulation. Contrary to the sinusoidal small range dithering, we applied a linear phase ramp from 0 to 2π. As such, we can obtain the highest interference contrast regardless of the initial phase distortion value. Secondly, instead of modulating all phase elements simultaneously, we modulated half of the phase pixels simultaneously while keep the other half stationary. This adjustment is particularly important for TPM. If we modulate all pixels simultaneously, the initial nonlinear signal will be too weak. Keeping half of the pixels stationary, we can start the measurement at a descent SNR. We show a typical pattern for dividing the phase pixels into two groups in Figure 3.2. Such a splitting method allows us to uniformly spread the phase pixels across the pupil plane and take advantage of the available excitation numerical aperture and also avoid forming any side lobes as the distribution of the phase pixels is random. Third, the feedback signal has nonlinear dependence on light intensity. As we will

Figure 3.2 A typical pattern for splitting the phase pixels to two groups.

discuss later, this point is very crucial to focusing inside random scattering tissue, as the nonlinearity can force light to converge onto a diffraction-limited spot regardless of the distribution of fluorophores. To determine the phase values of N pixels, we typically use $4N$ modulations (minimum value needed is $2N + 1$). If the MEMS update rate is F Hz, we typically assign the modulation frequency over the range of $F/4$-$F/2$ Hz. The overall measurement time was typically $4N/F$ (a shorter time is possible). For a 500 pixel MEMS running at 8 kHz update rate, three iterations of measurement will take 0.75 second. A major advantage of the modified parallel modulation scheme over the pixel by pixel measurement is the greatly improved SNR. Given the same number of signal photons, it is straightforward to show the large SNR advantage of parallel modulation [27], in a sense similar to the SNR advantage of spectral domain OCT (either with a spectrometer and a broadband source, or a swept source and a single detector) over time domain OCT.

The first experiment was to focus light through random scattering media (a glass diffuser) onto a point target. Without wavefront compensation, the light transmission through the glass diffuser appeared as random speckles (Figure 3.3a). We selected an arbitrary location as the guide star and used a photodiode to detect its signal. At the end of the parallel modulation, we used a Fourier transform to map the signal onto the frequency domain, in which we could directly read out the amplitude and phase of every modulated phase pixel (each pixel corresponds to one frequency). Although the wavefront (phase values of all pixels) seemed random as shown in Figure 3.3b, it helped to form a high quality focus through the diffuser (Figure 3.3c). This experiment showed that the multidither COAT method worked for not only low-order wavefront distortion as in air aberration but also high-order wavefront distortion as in the case of random light scattering.

Focusing onto a diffraction-limited guide star is straightforward. However, such a capability may be inadequate for practical imaging applications. In TPM, a main challenge is that the fluorescence features in most cases are not diffraction-limited spots and sometimes the fluorophores can occupy a large continuous volume. Therefore,

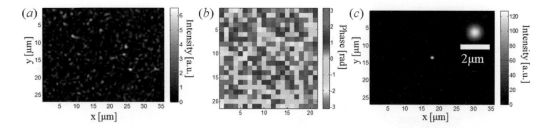

Figure 3.3 (**a**) Light transmitted through a glass diffuser (random speckles). (**b**) The compensation phase pattern acquired by the modified multidither COAT system. (**c**) Transmitted light after the phase pattern was applied. The inset shows the magnified view of the focus. Images are from Optics letters, 36(6) (2011): 870–872. A black and white version of this figure appears in some formats. For the color version, please refer to the plate section.

the wavefront measurement needs to work well even if there is no ideal guide star. Through numerical simulation and experiments, we figured out a recipe for handling this problem. The recipe has two ingredients, iterative feedback and nonlinearity. Iterative feedback has been widely used in science and engineering. Take the laser cavity as an example. Before the start of lasing, the fluorescence emission of the laser gain medium occupies a large number of spatial modes. However, only one (or few) mode can be directed back by the cavity for stimulated emission, which further strengths the emission along this mode. Eventually, all emission is converges to the modes defined by the cavity. Even in the implementation of COAT, iterative feedback is also important when there are multiple-scattering targets, which eventually forces the light to focus onto the brightest target. The iterative feedback has also been explored in time reversal experiments, which has been successfully applied to ultrasound imaging [28, 29].

Through numerical simulations, we found that the focus can converge to a diffraction-limited spot a lot faster if the signal (i.e., fluorescence emission) has a nonlinear dependence on the excitation light intensity. Here is a simplified picture to explain this effect. Suppose we have two guide stars of equal brightness. If the signal has linear dependence on light intensity, there will be no signal difference if the laser is focused onto a single guide star or both of them. Consequently, the COAT method cannot ensure the laser is focused onto a diffraction-limited spot, which is needed for high quality imaging. However, the situation is changed if the signal has quadratic or higher order dependence on light intensity (e.g., $2^2 > 1^2 + 1^2$). As the COAT process is to increase the overall signal level, the existence of nonlinearity will favor the formation of a single diffraction-limited focus. As the signal in multiphoton microscopy inherently has nonlinear dependence on light intensity (e.g., second-order nonlinearity in two-photon excitation, third-order nonlinearity in three-photon excitation), the combination of nonlinearity and iterative feedback is expected to work well in practice. We named this method Iterative Multi-Photon Adaptive Compensation Technique (IMPACT) [30].

To confirm the numerical simulation, we carried out the following experiments to see if IMPACT could force light to converge to a single focus without good guide stars. We set up a TPM with a segmented MEMS deformable mirror placed at the pupil plane

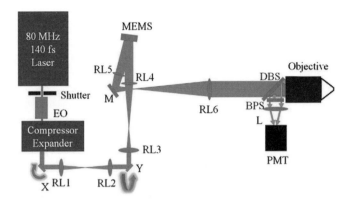

Figure 3.4 Setup that combined IMPACT and TPM. The light source was an 80 MHz Ti:Sapphire oscillator. The laser pulse duration was 140 fs, whose power was regulated by an electro-optic modulator (EO). A pulse compressor compensates the material dispersion in the optical path. The x and y galvo mirrors and the MEMS were all imaged to the pupil plane of the objective lens by relay lenses (RL). A dichroic beam splitter (DBS) directed the fluorescence emission to a band pass filter (BPS) and a GaAsP PMT. A lens (L) formed a reduced image of the objective lens pupil onto the active area of the PMT. The image is from Proceedings of the National Academy of Sciences 109.22 (2012): 8434–8439.

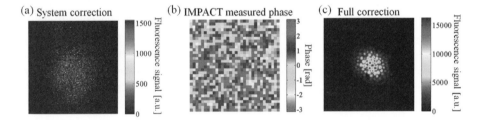

Figure 3.5 (**a**) TPM imaging of fluorescent beads through a mouse skull. Due to random wavefront distortion, the signal was weak and the structural information was lost. (**b**) The compensation phase pattern acquired by three iterations of IMPACT. (**c**) Wavefront compensation allows high resolution imaging through the mouse skull. The same area was clearly imaged with greatly improved signal level. Images are from Proceedings of the National Academy of Sciences 109.22 (2012): 8434–8439. A black and white version of this figure appears in some formats. For the color version, please refer to the plate section.

of the objective lens, as shown in Figure 3.4. In the first experiment, we compared the wavefront acquired with fluorescent beads and a dye cell as the guide star. With the dye cell, the fluorescence signal can be generated at any location in the volume. We showed that IMPACT could converge in three iterations to form a diffraction-limited focus, similar to the results obtained with fluorescence beads as the guide star. In the next test, we put fluorescent beads under a piece of mouse skull (Figure 3.5). To image a mouse brain *in vivo*, a craniotomy is required to expose the brain tissue, which however can trigger a substantial immune response and cause artifacts to the subsequent studies.

Thinning the skull to tens of micron can alleviate the problem [31]. However, the skull tissue can grow and form a scar in the thinning site, causing problems for long-term studies. It would be ideal to noninvasively image through the intact skull. However, the skull can cause severe wavefront distortions. Without proper wavefront compensation, the signal was weak and the structural information was lost, as shown in Figure 3.5a. Using IMPACT, we could measure the wavefront distortion (Figure 3.5b) and restore the diffraction-limited resolution and improve the signal by one to two orders of magnitude (Figure 3.5c). These experiments suggested that IMPACT could indeed work well for focusing light through highly scattering tissue, which led us to explore *in vivo* imaging. Using IMPACT, we directly imaged through the skull of a transgenic mouse whose neurons express YFP. The results are summarized in Figure 3.6. With IMPACT, we could accurately measure the wavefront and performed high-resolution imaging of the dendrite. The very fine dendritic spine neck and spine head were clearly resolved (Figures 3.6a and 3.6b). Without compensating the skull induced wavefront distortion, the entire dendrite was almost invisible (Figure 3.6c). The near random wavefront (Figure 3.6d) made a major difference to the imaging results.

A powerful method in neuroscience is the calcium imaging of neurons. To perform calcium imaging through the intact skull (Figure 3.7), we labeled the neurons with dsRed and GCaMP6. Using the dsRed signal, we performed IMPACT measurements. Afterward, we performed calcium imaging with the compensation wavefront. An interesting feature of skull is that the wavefront distortion could remain unchanged for hours. Therefore, we only need a single IMPACT measurement before the subsequent optical imaging.

Besides transcranial optical imaging, we also employed IMPACT for imaging the hippocampus of mouse brain. The imaging results (Figure 3.8) [32] show that IMPACT can help improve both the signal strength and restore resolution.

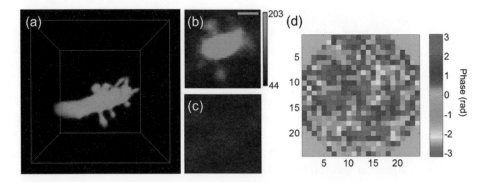

Figure 3.6 (**a**) 3D rendering of dendrite imaged through intact skull of adult mice *in vivo* with IMPACT. (**b**) 2D cross section of a dendrite. The scale bar is 2 μm. (**c**) The same 2D cross section imaged without compensating the skull induced wavefront distortion. (**d**) Wavefront measured by IMPACT. Images are from Optics express 22.20 (2014): 23786–23794. A black and white version of this figure appears in some formats. For the color version, please refer to the plate section.

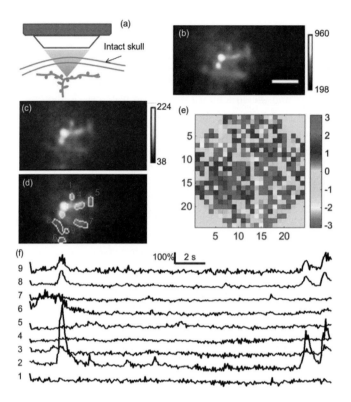

Figure 3.7 *In vivo* calcium imaging of neurons through the intact skull. (**a**) Experiment scheme. (**b, c**) dendrites and spines labeled by dsRed and GCaMP6 at 176 μm below the skull surface. The scale bar is 5 μm. (**d**) Regions of interest used for extracting calcium signals. (**e**) IMPACT measured wavefront. (**f**) Calcium dynamics at the regions of interest. Images are from Optics express 23.5 (2015): 6145–6150. A black and white version of this figure appears in some formats. For the color version, please refer to the plate section.

A general issue of wavefront correction based imaging is that a single correction can only work for a limited imaging field of view. After all, the source of wavefront distortion is the 3D distributed refractive index variation. The 2D information encoded in a wavefront is insufficient to fully address the spatially varying distortion. We use one example (Figure 3.9) to illustrate this problem. We put 1 micron fluorescent beads under a 400 μm thick fixed brain slice. Without proper wavefront correction, we could not image through the fixed tissue. Using wavefront correction, we could form a high quality image (diffraction-limited resolution, low background). However, the good imaging condition was valid for a limited field of view. To form a large imaging field of view, we repeated the IMPACT measurement and imaging for a number of locations and then reconstructed a large field of view image by stitching these small field of view images together. Such a procedure is sufficient for imaging slowly changing systems such as the plasticity of neurons. But for imaging fast dynamics, such as the calcium or even voltage signals, we will need methods that can provide simultaneous wavefront correction over a large imaging volume.

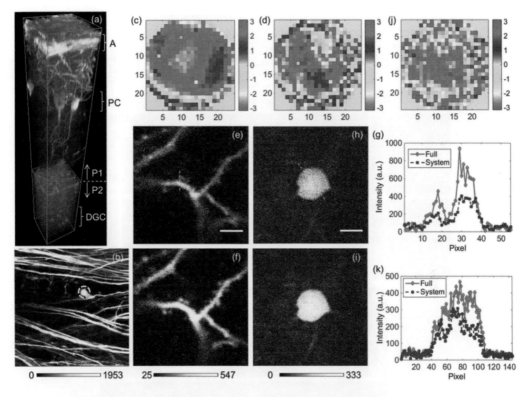

Figure 3.8 Hippocampus imaging. (**a**) The 3D rendered imaging volume from CA1 to dentate gyrus. (**b**) Myelinated axons at 18 μm depth. The cell body of a microglia is highlighted by a dashed circle. (**c**) System compensation at 398 μm depth. (**d**) Full compensation at 398 μm depth. (**e, f**) The corresponding images acquired with system and full compensation, respectively. (**g**) The signal comparison along the dashed line in (**e**). (**h, i**) The images of dentate granule cells at 579 μm depth acquired with system (**c**) and full compensation (**j**). (**k**) Signal comparison along the dashed line in (**h**). The scale bar is 10 μm. Images are from Optics express 23.5 (2015): 6145–6150. A black and white version of this figure appears in some formats. For the color version, please refer to the plate section.

3.3.2   Conjugation AO

Multiconjugate adaptive optics (MCAO) has been developed in astronomical telescope applications to deal with very similar issues [33]. Most AO systems put the deformable mirror at the pupil plane by default. Such a configuration essentially assumes a spatially invariant wavefront distortion, which is not true in most cases. In contrast, MCAO treats the wavefront distorting media as layers of aberration and images multiple deformable mirrors onto these aberration layers, which achieved huge successes in both large diameter telescopes [33] and ophthalmology [34] applications. Using numerical simulations [35], we investigated the advantage of MCAO over pupil plane correction for deep tissue imaging. The simulations suggest that MCAO outperforms pupil plane AO both for

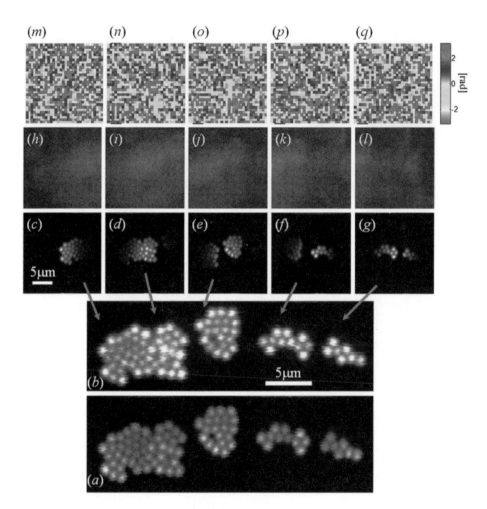

Figure 3.9 (**a**) One micron beads imaged under 400 μm thick fixed mouse cortex slice (fixed cortex is more scattering than *in vivo* conditions) with wavefront compensation. The image is constructed from several small field of view images. (**b**) The same sample imaged with the cortex tissue removed. (**c–g**) The small field of view images used for constructing the image in (**a**). (**h–l**) The corresponding images without wavefront compensation. (**m–q**) The corresponding compensation wavefront. Images are from Proceedings of the National Academy of Sciences 109.22 (2012): 8434–8439. A black and white version of this figure appears in some formats. For the color version, please refer to the plate section.

layered aberrations and for thick tissue, which motivated us to explore conjugation AO for practical applications.

Experimentally, we modified the IMPACT based TPM system by placing the deformable mirror onto a translation stage (Figure 3.10). The system was designed such that the translation affected neither beam direction nor overall optical path length, which allowed us to freely image the deformable mirror onto the proper tissue depth. As discussed previously, noninvasive transcranial optical imaging could be possible with

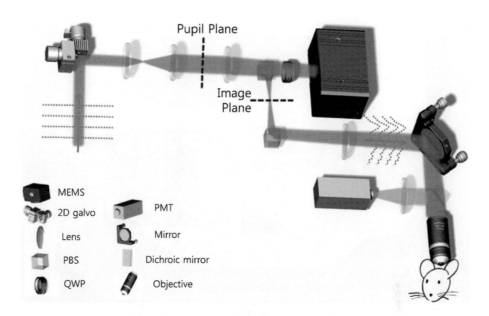

Figure 3.10 Experiment design for single conjugation AO. The image is from Proceedings of the National Academy of Sciences 112.30 (2015): 9236–9241.

high-order wavefront correction. However, the achieved imaging field of view is very constrained. Using single conjugation AO, we aimed to improve the imaging field of view. Experimentally, we observed a ~15 fold field of view improvement during transcranial TPM (Figure 3.11). With the improved field of view, we performed *in vivo* imaging of microglia (Figure 3.12), a type of resident immune cell in the brain [36]. It is worth noting that the same wavefront compensation could be used for a long time, as shown by the time-lapse images (Figure 3.12e).

3.3.3 PRISM

Besides multiphoton microscopy, the recipe of IMPACT can also be used for ultrafast laser pulse measurement and compensation. Essentially, ultrafast pulse is the in-phase coherent superposition of continuous optical waves over a broad spectrum. Propagation in dielectric material can suffer from material dispersion, which leads to spectral phase distortion. If the spectral phase for all colors is a linear function of frequency, we have the so-called transform-limited laser pulse, the shortest possible laser pulses. This is very similar to microscopy. If the phase for all the wave vectors is flat, we have the so-called diffraction-limited focus. With wavefront distortion, the focus is larger and the peak intensity is reduced. Recognizing this similarity, we decided to apply IMPACT, the method we developed in AO, for ultrafast pulse compression.

An issue with ultrafast pulse measurement is that no detector is fast enough to keep up with the femtosecond scale pulse duration. Therefore, no matter what detector is used, the measured signal is a representation of the optical signal integrated over one

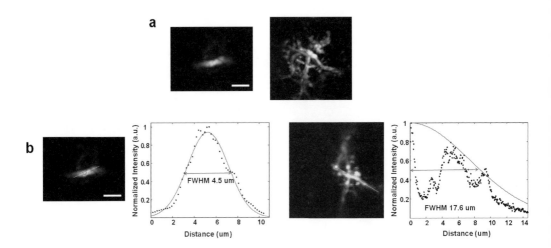

Figure 3.11 (a) Compare pupil plane AO (left) and single conjugation AO (right) for transcranial TPM. (b) Curve fitting shows the field of view advantage of single conjugation AO. The imaging area is improved by ∼15 fold. Scale bar 5 μm. Images are from Proceedings of the National Academy of Sciences 112.30 (2015): 9236–9241.

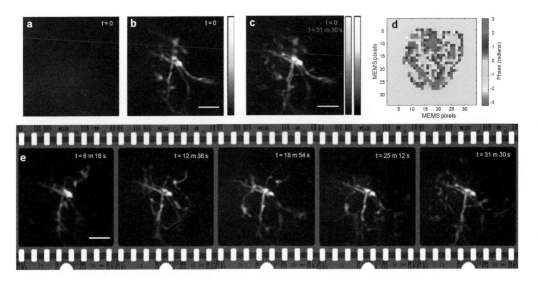

Figure 3.12 (a, b) Images without and with wavefront compensation, respectively. (c) Images recorded at 0 and 30.5 minutes shown in different colors. (d) The compensation wavefront used during imaging. (e) Time lapse images. Images are from Proceedings of the National Academy of Sciences 112.30 (2015): 9236–9241. A black and white version of this figure appears in some formats. For the color version, please refer to the plate section.

or multiple pulse durations. This is very similar to the case of using IMPACT to form a focus onto an extended guide star (not a point source), in which the detected signal is integrated over a certain volume. Both the numerical simulation and the experimental data showed that the combination of iterative feedback and nonlinearity can force the

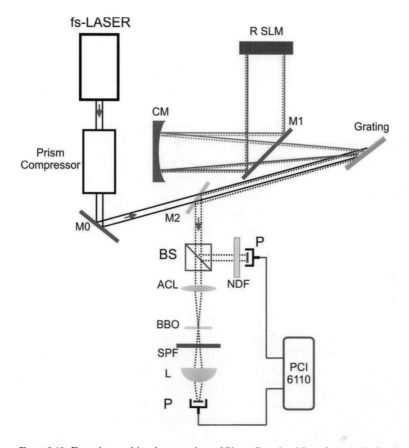

Figure 3.13 Experimental implementation of Phase Resolved Interferometric Spectral Modulation (PRISM). M, mirror, CM, concave mirror, BS, beam splitter, ACL, achromatic lens, SPF, short pass filter, L, lens, P, photodiode detector. The image is from Optics express 19.14 (2011): 12961–12968.

system to converge to a diffraction-limited focus, which led us to apply the same scheme to measure and compensate ultrafast pulse dispersion.

The experiment setup is shown in Figure 3.13. Using a 4f configuration (grating, concave mirror, linear 1D SLM, spaced at one focal length), we mapped the femtosecond pulse onto the spectral domain (or Fourier domain), where its spectral phase (the phase of each color) can be precisely controlled by a liquid crystal–based phase-only modulator. We focused the laser pulse onto a BBO crystal to generate second harmonic signals as the feedback signal. We also used a photodiode to directly monitor the laser power. Using the photodiode signal, we could effectively eliminate the laser power noise from the measurement and obtain only the nonlinear signal variation as a function of phase modulation.

The phase-modulation procedure is very similar to that of IMPACT. Essentially, we kept half of the phase elements stable while simultaneously modulating the other half of the phase pixels. At the end of the parallel phase modulation, we performed a Fourier transform of the measured signal and obtained the phase and amplitude of each of the modulated pixels in parallel. Experimentally, we used three iterations to measure and compensate the pulse dispersion. Due to the nature of spectral domain phase modulation, we named the method Phase-Resolved Interferometric Spectral Modulation (PRISM) [37].

Compared to other ultrafast pulse characterization methods, PRISM has several advantages. First, PRISM measures and compensates the pulse dispersion at the same time. Well-established methods such as FROG [38] and SPIDER [39] only measure but do not compensate pulse dispersion. Second, with the MEMS linear SLM array, PRISM can operate at high speed and the entire process of measurement and compensation takes less than one second. Third, any nonlinear signal (coherent or incoherent, second order or third order) can be used as the feedback signal for the PRISM measurement. With FROG, SPIDER, or MIIPS [40], a coherent nonlinear signal is required to extract the pulse information. PRISM is more flexible in the signal source. For example, we have directly used the two-photon fluorescence signal from the objective focus for PRISM, which is very helpful for working with high numerical aperture short working distance objective lenses.

To evaluate the performance of PRISM, we performed both a numerical simulation and an experiment. The simulation results are summarized in Figure 3.14, in which we consider two different scenarios. One is the low-order phase distortion, as in the chirped pulse caused by material dispersions. We added sufficient quadratic spectral phase distortion such that the pulse duration was stretched by a factor of 12.5 and the peak intensity was reduced to 8%. After one iteration of PRISM, the peak intensity was increased to 50% of the transform-limited value. After the second iteration, the pulse was essentially transform-limited. The third iteration made negligible difference. We further examined the residual spectral phase, which was flat across the spectrum. The other case we considered was the very high order (random) spectral phase distortion, which is often caused by artificially engineered structures. One example is the commonly used dielectric mirror whose spectral phase distortion deviates from the low-order quadratic or cubic shape. In the simulation, we applied a random spectral phase distortion (Figure 3.14g) and the peak intensity of the pulse was reduced to 4% of the transform-limited value (Figure 3.14h). After one iteration of PRISM, the peak intensity was increased to 50% of the transform-limited value, very similar to the case of quadratic spectral phase distortion. After the second iteration, the pulse was essentially transform-limited and again the third iteration made negligible difference and the final spectral phase was flat across the spectrum. These simulations suggested that PRISM can work well regardless of the complexity of the spectral phase distortion.

To confirm with the simulation, we used the setup in Figure 3.13 to carry out PRISM based pulse compression and compared the results with a commercially available MIIPS measurement system [40]. The results are summarized in Figure 3.15. In the first measurement, we used PRISM and MIIPS to measure the existing material dispersion (Figure 3.15a). Within the power spectrum range of the laser pulse (Figure 3.15b), the two measurements agreed very well. In the next measurement, we artificially used

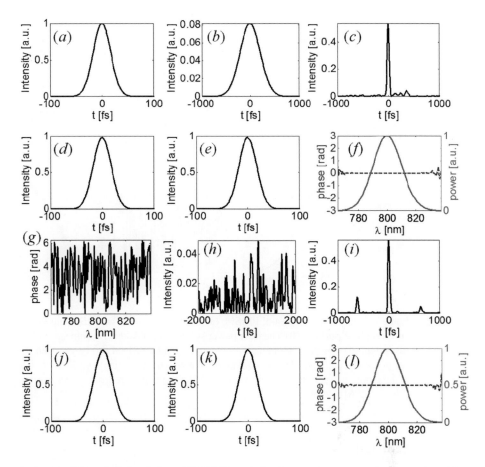

Figure 3.14 Numerical simulation of PRISM based ultrafast measurement and compression. (**a**) Transform limited laser pulse. (**b**) Pulse chirped by quadratic spectral phase distortion. (**c-e**) Compressed pulse after one, two, and three iterations of PRISM. (**f**) Residual spectral phase distortion. (**g**) Apply random spectral phase distortion. (**h**) Distorted pulse in time domain due to random phase distortion. (**i–k**) Compressed pulse after one, two, and three iterations of PRISM. (**l**) Residual spectra phase distortion. Images are from Optics express 19.14 (2011): 12961–12968.

the SLM in the setup to add random spectral phase distortion (essentially we added phase background to the SLM, Figure 3.15c). Using PRISM, we determined the compensation phase profile (Figure 3.15d). Combining the compensation profile and the added phase distortion, we got a very simple curve (Figure 3.15e). After removing the offset (zeroth order) and the slope (first order), the spectral phase was actually the same as the existing system dispersion (Figure 3.15f). These measurements confirm that PRISM can measure both low-order spectral phase distortions caused by material dispersion and very high order distortions, such as that caused by artificial dielectric structures.

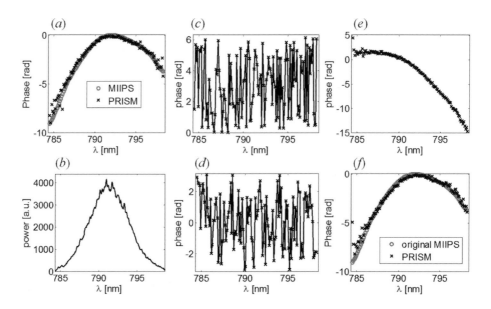

Figure 3.15 Experimental test of PRISM. (**a**) Comparison with the MIIPS and PRISM measured spectral phase profiles. (**b**) The laser power spectrum. (**c**) Added random spectral phase distortion. (**d**) Compensation phase profile measured by PRISM. (**e**) Unwrapped summation of the phase in **c** and **d**. (**f**) The spectral phase in **e** with 0th (baseline) and 1st order (slope) removed and the original spectral phase measured by MIIPS. Images are from Optics express 19.14 (2011): 12961–12968.

3.3.4   OCM Guide Star

So far, we have mainly talked about nonlinear emission (TPM, SHG) based phase profile measurement, for either spatial phase (IMPACT) or spectral phase (PRISM). However, the nonlinear emission may not always be the ideal signal source. In neuroscience, a widely used tool is the calcium imaging of neurons. The fluorescence of the calcium indicators can fluctuate as a result of neuronal activity. Therefore using the signal from the calcium indicator may cause artifacts in wavefront measurements. Additional labeling such as the dsRed used in our transcranial calcium imaging could certainly be used to provide the feedback signal. But to be widely usable in the neuroscience community, the ideal wavefront measurement method should require no additional labeling other than the needed calcium or voltage indicator.

A widely available and strong intrinsic signal would be a better choice for feedback. One example of intrinsic signal is the Second Harmonic Generation (SHG). However, SHG signal is typically very weak in the mouse neocortex. We therefore explored another widely available and strong intrinsic signal for wavefront measurement, the backscattered light from tissue. The state of the art method for backscattering measurement is the optical coherence tomography (OCT) [41]. The high NA and single axial plane configuration of OCT is the optical coherence Microscopy (OCM) [42]. Due to the combined effects of confocal detection and low-coherence interferometry,

OCM can work at large depth in the brain and imaging beyond 1 mm depth has been demonstrated in experiment [42]. One major advantage of OCM is that the signal is available all over the cortex. Therefore OCM signal-based wavefront measurement can be performed at arbitrary regions and is even more flexible than the fluorescence based method.

Prior to experiments, we used numerical simulation to investigate the performance of OCM based wavefront correction in scattering tissue (Figure 3.16). We investigated scattering media of different scattering path lengths (4.5 and 6). Without wavefront correction, the focus was severely distorted. After three iterations, the focus quality was greatly improved: higher peak intensity (Strehl ratio) and less background. For the six scattering path lengths case, although there were still remaining speckles outside the laser focus, the peak intensity was improved by threefold, which could provide a nine fold signal improvement for TPM.

We used the system shown in Figure 3.17 to carry out the measurement [43]. The key of OCM is low coherent interferometry, which can be done either in the time or the spectral domain. Although for OCT measurement in the spectral domain methods have advantages in SNR, there is no difference for OCM measurement as there is only a single axial point. We used a broadband Ti:Sapphire oscillator as the broadband light source and used a piezo actuator for time domain interferometry. A Wollaston prism combined the coherent back scattering signal and the reference signal for balanced detection. For deep imaging, there is a large amount of scattering background. It is therefore important to employ a detector of larger full well charge capacity. We used a balanced detector to

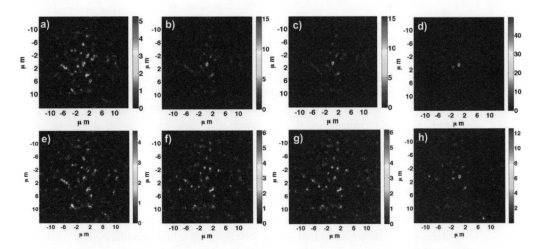

Figure 3.16 Numerical simulation of using iterative feedback and OCM signal for wavefront compensation. (**a–d**) focusing through scattering medium of 4.5 scattering path lengths before and after 1, 2, 3 iterations of measurement, respectively. (**e–h**) The corresponding results for scattering medium of 6 scattering path lengths. Images are from Optics Express 20.15 (2012): 16532–16543. A black and white version of this figure appears in some formats. For the color version, please refer to the plate section.

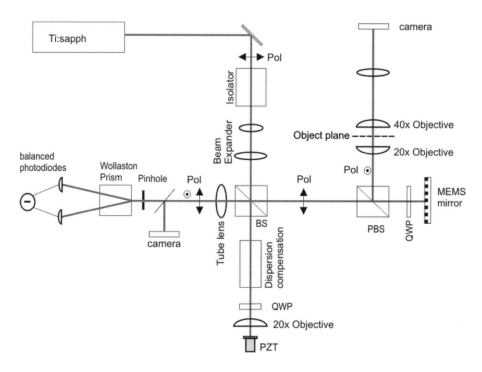

Figure 3.17 Experimental design of OCM based wavefront compensation. Pol, polarization, PBS, polarizing beam splitter, QWP, quarter wave plate, PZT, piezo actuator with a mirror attached to the front. The image is from Optics Express 20.15 (2012): 16532–16543.

collect the two output signals from the Wollaston prism, which helped to suppress the power noise of the light source.

Experimentally, we used a fixed brain slice as the sample (more difficult to image than *in vivo* conditions). The coherent backscattering within the brain tissue provided adequate feedback signal for the wavefront measurement. Without wavefront correction, the focus at large depth was degraded and the iterative wavefront measurement could effectively compensate the wavefront distortion and improve the focus quality (Figure 3.18).

3.3.5   k-Space Measurement

So far, all the wavefront measurements were based on our modified Multidither COAT system, in which we keep half of the pixels stable while simultaneously modulating (with a linear phase ramp) the other half. This method is sufficiently fast and allows millisecond scale measurement time per spatial mode. Toward even higher speed operation, we also developed a drastically different wavefront modulation scheme which currently can achieve a microsecond scale measurement time per spatial mode. The experimental design of the first demonstration [44] is shown in Figure 3.19. Using two

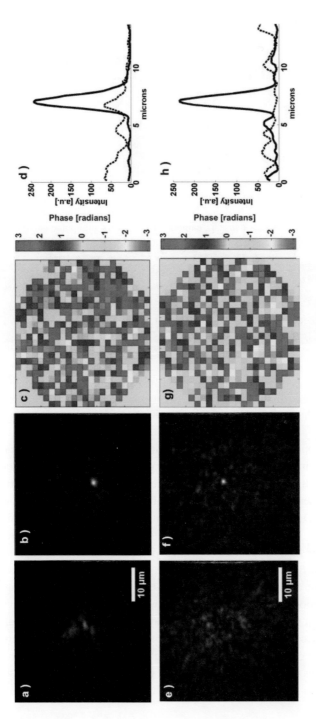

Figure 3.18 Focusing onto scattering sources in fixed brain slice. (**a, b**) Focus at 300 μm depth without and with wavefront compensation. (**c**) Wavefront compensation profile obtained by three iterations of measurement. (**d**) Cross-section of the focus with (solid line) and without (dashed line) wavefront compensation. (**e–f**) The corresponding results for focusing at 500 μm depth. Images are from Optics Express 20.15 (2012): 16532–16543. A black and white version of this figure appears in some formats. For the color version, please refer to the plate section.

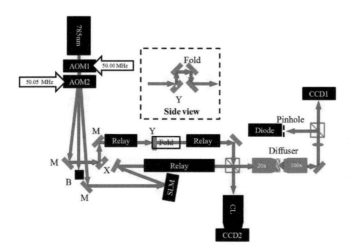

Figure 3.19 Experimental demonstration of spatial frequency domain modulation (k-space modulation) based wavefront measurement. AOM, acousto-optic modulator, M, mirror, B, beam block, Relay, relay lens pair, CL, camera lens, X, horizontal plane galvo mirror, Y, vertical plane galvo mirror. The image is from Optics express 19.4 (2011): 2989–2995.

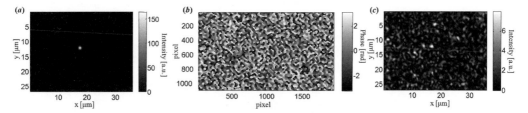

Figure 3.20 (**a**) Focus formed through glass diffuser by *k*-space modulation based wavefront compensation. (**b**) Compensation wavefront. (**c**) Focus without wavefront compensation. Images are from Optics express 19.4 (2011): 2989–2995.

acousto-optic modulators, we generated two beams with 50 kHz beating frequency. We used a pair of galvo mirrors (X and Y galvos) to control the orientation (coordinate in the spatial-frequency domain or *k*-space) and used the SLM to control the spatial phase of the other beam. A beam splitter combined the two beams, which were focused onto a diffuser. During the measurement, we used the galvo to quickly visit a number of coordinates in *k*-space. Fourier transformation of the target signal provided the amplitude and phase for each coordinate, which yielded the optical field in *k*-space. A 2D Fourier transform finally yielded the amplitude and phase (wavefront) in space. We displayed the wavefront on the SLM to compensate the wavefront distortion (Figure 3.20). Iterative optimization can also be implemented with this *k*-space modulation scheme. At the end of the first measurement, we can display the compensation wavefront on the SLM during the second round of measurement and then update the wavefront on the SLM for the next measurement. In our latest implementation, we set the beating frequency of the two beams to 2 MHz and used a pair of galvo mirrors for 2D raster scanning. We fed the signal to a high-speed lock-in amplifier (HF2LI, Zurich Instrument) for demodulation which yielded the phase and amplitude. Using such a configuration, we reduced the measurement time per spatial mode to only 1 µs. Using a higher beating frequency and a resonant galvo for raster scanning, we can potentially reduce the measurement time to ~100 ns per spatial mode. Of course, ultimately the SNR (whether there are sufficient photons) determines the measurement speed in practice.

3.3.6 Break Sound Resolution Limit

Besides microscopy applications, the iterative feedback scheme also finds other applications. Here we discuss two experimental projects in our lab. One is to image in the diffusive regime using an ultrasound guide star and achieve superresolution (breaking the diffraction limit of ultrasound waves). The other is to transmit energy efficiently through random scattering medium.

Being able to focus inside random scattering medium does not guarantee imaging. To perform imaging, one needs the capability of focusing at arbitrary locations, not just at the guide star position. For diffusive regime imaging, ultrasound has been explored as the guide star (45–48). The sound modulation can cause a frequency shift of light, which can be captured by interferometry (holography) means. Using a phase-conjugation device, we can precisely generate light focusing at the ultrasound focus location. Given sufficient optical gain (the power ratio of the phase-conjugation beam and the initial ultrasound-modulated beam), one can use the scheme (Figure 3.21) to perform fluorescence imaging. To have a 3D confined modulation zone, we need to employ pulsed ultrasound and pulsed light. For example, we can use 10 kHz repetition rate for both the ultrasound pulse and the laser pulse. An important factor to consider is the signal aliasing as the signal (ultrasound modulation) and the sampling (optical pulse) are at the same rate. We came up with a simple solution to handle this problem. Experimentally, we set the ultrasound carrier frequency to (50 MHz + 5 kHz). For the 10 kHz pulse train, every pair of consecutive pulses are out of phase by π. Therefore, as long as an even number of pulses are captured, the aliasing between the unmodulated light and the modulated light will be canceled. We can therefore clearly measure the interference of the modulated light and the reference light (e.g., set to 50 MHz + 4990 Hz).

Although this method shows good promise for increasing the imaging depth, it has several major limitations. First, to achieve good spatial resolution (e.g., neuron cell body is ~10 µm), high-frequency ultrasound is required. However, the ultrasound attenuation and the delicate high-frequency transducer both lead to weak ultrasound modulation. As a result, the achievable ultrasound modulation efficiency in a realistic configuration (focusing in scattering tissue, as opposed to focusing in clear water sandwiched by or behind a tissue phantom) is very low. Consequently, a long measurement time is required. Second, biological tissue is inherently dynamic. The intracellular and intercellular motion can easily change the optical wavefront. Therefore, high-speed measurement is required, which is in conflict with the weak ultrasound-modulated signal. Third, although we can focus light onto the ultrasound beam, the imaging inherently

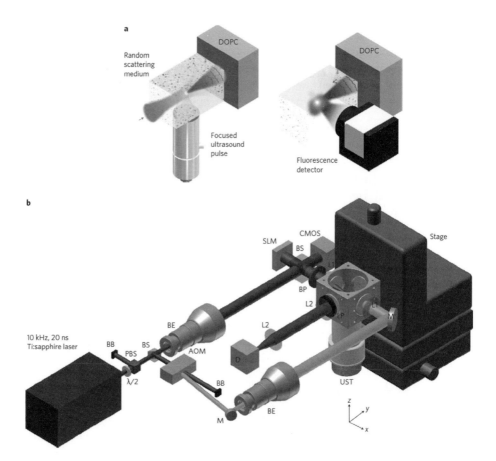

Figure 3.21 (**a**) Scheme of ultrasound pulse guided digital optical phase conjugation. (**b**) Experiment design. The image is from Nature photonics 6.10 (2012): 657–661.

lacks 3D sectioning capability (the linear excitation is not 3D confined and nonlinear excitation is impractical). Moreover, the spatial modes provided by commercially available SLMs are still far from enough to provide a high quality focus (very low Strehl ratio). Therefore, the obtained images will contain a huge background signal in realistic imaging conditions. For SNR demanding measurements such as the calcium imaging in neuroscience [8], the shot noise generated by the background excitation can easily overwhelm the in-focus signal variation due to neuronal activity.

Facing these major problems, we developed an iterative modulation and time reversal scheme [47]. Essentially, the product of the sound modulation profile and the laser focus profile determines the profile of ultrasound-modulated light. Therefore, if we repeatedly apply ultrasound modulation and optical time reversal (phase conjugation), the profile of ultrasound-modulated light will become the ultrasound profile to the Nth power, where N is the number of iterations. If the ultrasound modulation profile is Gaussian, this means that the spatial resolution can be improved by the square root of N (Figure 3.22). We confirmed this intuitive prediction through both numerical simulation

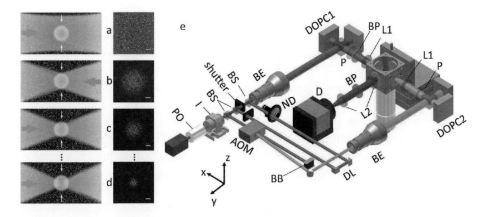

Figure 3.22 (**a–d**) Iterative ultrasound modulations and time reversed focusing lead to super-resolution and greater SNR. (**e**) Experimental implementation. PO, Pockels cell, I, isolator, BS, beam splitter, BE, beam expander, ND, neutral density filter, DL, delay line, P, polarizer, BP, bandpass filter, L, lens, D detector. Images are from Scientific reports 2 (2012): 748. A black and white version of this figure appears in some formats. For the color version, please refer to the plate section.

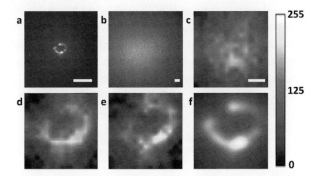

Figure 3.23 (**a**) A fluorescence sample imaged without scattering medium. (**b**) The same sample imaged after being embedded in a scattering medium. The scale bar in **a** and **b** is 100 μm. (**c–e**) The sample in the scattering medium imaged with 1, 5, 9 iterations of ultrasound modulation and time reversal, respectively. The scale bar in **c** is 20 μm. (**f**) The image in **a** 2D convolved with a 12 μm Gaussian profile for comparison with **e**. Images are from Scientific reports 2 (2012) 748.

and experiments. We show the initial superresolution imaging results in Figure 3.23 (ultrasound wavelength = 30 μm).

3.3.7 Power Transmission through Scattering Media

Besides imaging, the iterative feedback regime can also help find ways to transmit an optical signal more efficiently through random scattering medium [49]. Essentially, linear wave propagation in random scattering medium can be treated as a matrix problem [50, 51]. The scattering medium, a matrix, converts the array of input optical modes to the array of output optical modes. An interesting feature of the random matrix is that its eigenvalue follows the Marchenko–Pastur distribution. A small percentage of the eigenmodes have high transmission coefficients. If we can determine these eigenmodes, we will be able to transmit optical wave more efficiently. The key is how to determine these eigenmodes quickly for practical applications. We can certainly fully characterize the matrix and compute the eigenmodes (a combination of many spatial modes with proper amplitude and phase). However, even for an area with 100×100 modes, it will take 10,000 measurements, which is far from practical. Here, we also apply the iterative feedback scheme to quickly find the high transmission eigenmodes and improve the signal transmission. Intuitively, if we send a wave through the scattering medium, the transmission will contain a higher percentage of high transmission eigenmodes than the initial input because the scattering medium works as a filter: the low transmission eigenmodes suffer higher loss. Therefore, if we can perform time reversal and let the wave propagate back and forth a few times through the scattering medium, we can converge the wave toward the combination of high transmission eigenmodes.

Experimentally, we use the setup shown in Figure 3.24 to implement iterative time reversal. Same as in previous experiments, we used a CMOS camera with large full well charge capacity and the same pixel size as that of the SLM. In five iterations, we can increase the power transmission by 2.7. Besides random scattering media, we also explored using iterative time reversal to let the light beam avoid absorbers. Similar to low eigenmodes, light that travels through the absorber experiences higher loss. Therefore, iterative time reversal can help force the light to take the path without absorbers. Overall, we can rely on iterative time reversal to quickly find the path with high transmission in only a few iterations [49].

3.3.8   Fast Volumetric Imaging

Besides imaging through highly scattering tissue, wavefront control can also find applications in high-speed volumetric imaging. Essentially, laser beam scanning is equivalent to changing the wavefront at the objective pupil plane. For example, a linear phase slope can move the laser focus in the transverse plane, and a defocusing wavefront can shift the focus in the axial direction. Recently, our lab has taken advantage of a high-speed ultrasound lens to implement fast axial scanning and perform continuous volumetric imaging (Figure 3.25). A key benefit of fast volumetric imaging is that the system is robust for imaging awake and behaving animals. For example, the motion along the axial direction can cause signal artifacts in 2D calcium imaging. With fast volumetric imaging, as long as the feature of interest is captured within the volume, we can always remove the animal induced movement via postprocessing of the image. Besides calcium imaging of awake and behaving animals, an important application is to image the blood flow (*in vivo* multicolor imaging flow cytometry), taking advantage of the kHz cross-sectional frame rate [4, 52].

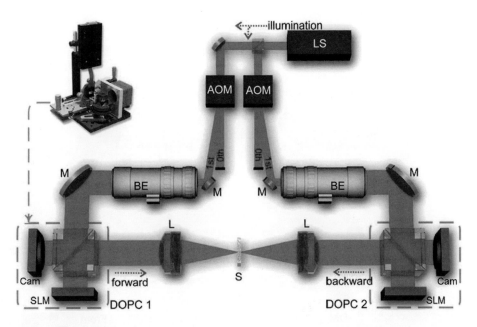

Figure 3.24 Experimental implementation of iterative time reversal through random scattering medium. LS, laser source, M, mirror, BE, beam expander, DOPC, digital optical phase conjugation system. The mechanical model of the DOPC system is also shown. The image is from Scientific reports 4 (2014) 5874.

3.4   Discussion

In summary, we have discussed the application of wavefront control for deep tissue imaging. The applications include imaging the neocortex and the hippocampus at large depth, transcranial TPM and fast volumetric imaging. Besides imaging, the operation principle (the combination of nonlinearity and iterative feedback) can also work for ultrafast laser pulse measurement and compensation. The iterative feedback can also help to achieve superresolution in the ultrasound guided time reversal based imaging and deliver optical power more efficiently through random scattering media.

To have the broadest impact in biomedical sciences, we need to make these methods robust and simple to use. In fact, all microscopy systems should have AO incorporated. Even for imaging cultured cells, the refractive index variation within cells can cause problems for high NA objective lenses. However, AO based imaging systems are still rarely used in biology research. Take AO for TPM based calcium imaging as an example, we list the performance requirement for AO TPM to have a broad impact in neuroscience. First, the AO method must be flexible in working wavelengths (e.g., 0.93, 1.04, 1.3, 1.7 μm). This allows the imaging of various fluorophores with TPM and with three-photon excitation as well. Second, the AO systems need to provide automated operation. Researchers without training in physical science or optical engineering should be able

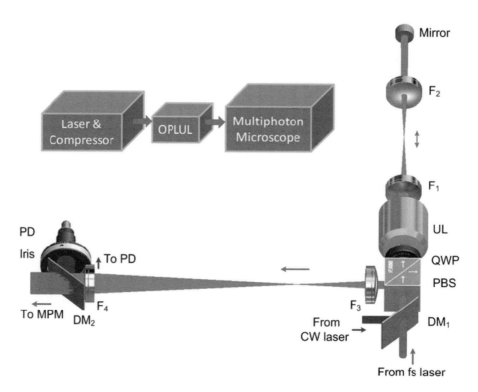

Figure 3.25 System design for fast volumetric imaging. F, lens, UL, ultrasound lens, QWP, quarter wave plate, PBS, polarizing beam splitter, DM, dichroic mirror, PD, photodiode. These element form the optical phase locked ultrasound lens (OPLUL), which is inserted between the laser source and a conventional TPM system to implement fast volumetric imaging. The image is from Nature methods 12.8 (2015): 759–762.

to use the system on a daily basis. Third, the AO system needs to perform simultaneous large volume wavefront correction, which is important for large volume calcium imaging. Fourth, the AO method should require no additional labeling other than the needed calcium indicator. Fifth, the AO system must be able to tolerate motion for awake and behaving animal imaging. These properties can greatly lower the barrier to broad application of AO TPM. Certainly, how useful a method is can be evaluated by whether the method solves an important problem. Customization of the optical measurement for a specific study is always required.

3.5   References

[1] Wilt BA, Burns LD, Ho ETW, Ghosh KK, Mukamel EA, Schnitzer MJ. Advances in light microscopy for neuroscience. Annual Review of Neuroscience. 2009;32:435–506.
[2] Svoboda K, Yasuda R. Principles of two-photon excitation microscopy and its applications to neuroscience. Neuron. 2006 Jun 15;50(6):823–39.

[3] Heintzmann R, Gustafsson MGL. Subdiffraction resolution in continuous samples. Nature Photonics. 2009 Jul;3(7):362–4.

[4] Kong L, Tang J, Little JP, Yu Y, Laemmermann T, Lin CP, et al. Continuous volumetric imaging via an optical phase-locked ultrasound lens. Nature Methods. 2015 Aug;12(8):759-U166.

[5] St-Pierre F, Marshall JD, Yang Y, Gong Y, Schnitzer MJ, Lin MZ. High-fidelity optical reporting of neuronal electrical activity with an ultrafast fluorescent voltage sensor. Nature Neuroscience. 2014;17(6):884–9.

[6] Scanziani M, Häusser M. Electrophysiology in the age of light. Nature. 2009;461(7266):930–9.

[7] Hiraoka Y, Sedat JW, Agard DA. The use of a charge-coupled device for quantitative optical microscopy of biological structures. Science. 1987;238(4823):36–41.

[8] Yasuda R, Nimchinsky EA, Scheuss V, Pologruto TA, Oertner TG, Sabatini BL, et al. Imaging calcium concentration dynamics in small neuronal compartments. Science's STKE. 2004 2004 Feb;2004(219):pl5-pl.

[9] Pawley J. Handbook of Biological Confocal Microscopy. 3rd ed. Springer US, 2006.

[10] Huisken J, Swoger J, Del Bene F, Wittbrodt J, Stelzer EHK. Optical sectioning deep inside live embryos by selective plane illumination microscopy. Science. 2004 Aug 13;305(5686):1007–9.

[11] Keller PJ, Schmidt AD, Wittbrodt J, Stelzer EH. Reconstruction of zebrafish early embryonic development by scanned light sheet microscopy. Science. 2008;322(5904):1065–9.

[12] Denk W, Svoboda K. Photon upmanship: Why multiphoton imaging is more than a gimmick. Neuron. 1997 Mar;18(3):351–7.

[13] Denk W, Strickler JH, Webb WW. 2-Photon laser scanning fluorescence microscopy. Science. 1990 Apr 6;248(4951):73–6.

[14] Helmchen F, Denk W. Deep tissue two-photon microscopy. Nature Methods. 2005 Dec;2(12):932–40.

[15] Stoll S, Delon J, Brotz TM, Germain RN. Dynamic imaging of T cell-dendritic cell interactions in lymph nodes. Science Signaling. 2002;296(5574):1873.

[16] Chung K, Deisseroth K. CLARITY for mapping the nervous system. Nature methods. 2013;10(6):508–13.

[17] Horton NG, Wang K, Kobat D, Clark CG, Wise FW, Schaffer CB, et al. *In vivo* three-photon microscopy of subcortical structures within an intact mouse brain. Nature Photonics. 2013 Mar;7(3):205–9.

[18] Babcock HW. The possibility of compensating astronomical seeing. Publications of the Astronomical Society of the Pacific. 1953:229–36.

[19] Leith EN, Upatniek J. Holographic imagery through diffusing media. Journal of the Optical Society of America. 1966;56(4):523-&.

[20] Feinberg J, Hellwarth RW. Phase-conjugating mirror with continuous-wave gain. Optics Letters. 1980;5(12):519–21.

[21] Yariv A, Yeh P. Phase conjugate optics and real-time holography. IEEE Journal of Quantum Electronics. 1978;14(9):650–60.

[22] Bridges WB, Brunner PT, Lazzara SP, Nussmeie Ta, Omeara TR, Sanguine Ja, et al. Coherent optical adaptive techniques. Applied Optics. 1974 1974;13(2):291–300.

[23] Tao X, Crest J, Kotadia S, Azucena O, Chen DC, Sullivan W, et al. Live imaging using adaptive optics with fluorescent protein guide-stars. Optics Express. 2012 Jul 2;20(14):15969–82.

[24] Bifano T. Adaptive imaging: MEMS deformable mirrors. Nature Photonics. 2011;5(1): 21–3.

[25] Debarre D, Botcherby EJ, Booth MJ, Wilson T. Adaptive optics for structured illumination microscopy. Optics Express. 2008 Jun 23;16(13):9290–305.

[26] Debarre D, Botcherby EJ, Watanabe T, Srinivas S, Booth MJ, Wilson T. Image-based adaptive optics for two-photon microscopy. Optics Letters. 2009 Aug 15;34(16):2495–7.

[27] Cui M. Parallel wavefront optimization method for focusing light through random scattering media. Optics Letters. 2011 Mar 15;36(6):870–2.

[28] Prada C, Thomas JL, Fink M. The iterative time-reversal process – analysis of the convergence. Journal of the Acoustical Society of America. 1995 Jan;97(1):62–71.

[29] Fink M, Cassereau D, Derode A, Prada C, Roux P, Tanter M, et al. Time-reversed acoustics. Reports on Progress in Physics. 2000 Dec;63(12):1933–95.

[30] Tang J, Germain RN, Cui M. Superpenetration optical microscopy by iterative multiphoton adaptive compensation technique. Proceedings of the National Academy of Sciences of the United States of America. 2012 May 29;109(22):8434–9.

[31] Grutzendler J, Kasthuri N, Gan W-B. Long-term dendritic spine stability in the adult cortex. Nature. 2002;420(6917):812–6.

[32] Kong LJ, Cui M. *In vivo* neuroimaging through the highly scattering tissue via iterative multi-photon adaptive compensation technique. Optics Express. 2015 Mar;23(5):6145–50.

[33] Hardy JW. Adaptive Optics for Astronomical Telescopes: Oxford University Press; 1998.

[34] Thaung J, Knutsson P, Popovic Z, Owner-Petersen M. Dual-conjugate adaptive optics for wide-field high-resolution retinal imaging. Optics Express. 2009;17(6):4454–67.

[35] Wu TW, Cui M. Numerical study of multi-conjugate large area wavefront correction for deep tissue microscopy. Optics Express. 2015 Mar;23(6):7463–70.

[36] Park JH, Sun W, Cui M. High-resolution *in vivo* imaging of mouse brain through the intact skull. Proceedings of the National Academy of Sciences of the United States of America. 2015 Jul;112(30):9236–41.

[37] Wu T-w, Tang J, Hajj B, Cui M. Phase resolved interferometric spectral modulation (PRISM) for ultrafast pulse measurement and compression. Optics Express. 2011 Jul 4;19(14):12961–8.

[38] O'shea P, Kimmel M, Gu X, Trebino R. Highly simplified device for ultrashort-pulse measurement. Optics Letters. 2001;26(12):932–4.

[39] Iaconis C, Walmsley IA. Spectral phase interferometry for direct electric-field reconstruction of ultrashort optical pulses. Optics Letters. 1998;23(10):792–4.

[40] Lozovoy VV, Pastirk I, Dantus M. Multiphoton intrapulse interference. IV. Ultrashort laser pulse spectral phase characterization and compensation. Optics Letters. 2004;29(7):775–7.

[41] Fujimoto JG, Brezinski ME, Tearney GJ, Boppart SA, Bouma B, Hee MR, et al. Optical biopsy and imaging using optical coherence tomography. Nature Medicine. 1995 Sep;1(9):970–2.

[42] Srinivasan VJ, Radhakrishnan H, Jiang JY, Barry S, Cable AE. Optical coherence microscopy for deep tissue imaging of the cerebral cortex with intrinsic contrast. Optics Express. 2012;20(3):2220–39.

[43] Fiolka R, Si K, Cui M. Complex wavefront corrections for deep tissue focusing using low coherence backscattered light. Optics Express. 2012 Jul 16;20(15):16532–43.

[44] Cui M. A high speed wavefront determination method based on spatial frequency modulations for focusing light through random scattering media. Optics Express. 2011 Feb 14;19(4):2989–95.

[45] Xu X, Liu H, Wang LV. Time-reversed ultrasonically encoded optical focusing into scattering media. Nature Photonics. 2011 Mar;5(3):154–7.

[46] Fiolka R, Si K, Cui M. Parallel wavefront measurements in ultrasound pulse guided digital phase conjugation. Optics Express. 2012 Oct 22;20(22):24827–34.

[47] Si K, Fiolka R, Cui M. Breaking the spatial resolution barrier via iterative sound-light interaction in deep tissue microscopy. Scientific Reports. 2012 Oct 19;2.

[48] Si K, Fiolka R, Cui M. Fluorescence imaging beyond the ballistic regime by ultrasound-pulse-guided digital phase conjugation. Nature Photonics. 2012 Oct;6(10):657–61.

[49] Hao X, Martin-Rouault L, Cui M. A self-adaptive method for creating high efficiency communication channels through random scattering media. Scientific Reports. 2014;4.

[50] Popoff S, Lerosey G, Carminati R, Fink M, Boccara A, Gigan S. Measuring the transmission matrix in optics: an approach to the study and control of light propagation in disordered media. Physical Review Letters. 2010;104(10):100601.

[51] Fink M. Time reversed acoustics. Physics Today. 1997 Mar;50(3):34–40.

[52] Kong L, Tang J, Cui M. Multicolor multiphoton *in vivo* imaging flow cytometry. Optics Express. 2016;24(6):6126–35.

4 Zonal Adaptive Optical Microscopy for Deep Tissue Imaging

Cristina Rodriguez and Na Ji

4.1 Introduction

The invention of optical microscopy made it possible to discover and study structures and phenomena invisible to or unresolvable by the naked eye. Because of the wave nature of the light, the resolution of a conventional microscope is limited by diffraction. The finite size of the focus makes a point object appear in the image as a smeared-out blob, as described by the point spread function (PSF) of the imaging system. In practice, microscopes operate at the diffraction limit only under strict conditions, such as specific immersion media and coverglass thickness. Furthermore, the optical properties of the specimen and the immersion media for which the microscope objective is designed have to be matched. This is rarely the case for most biological samples, whose very own constituent structures (e.g., proteins, nuclear acids, and lipids) give rise to an inhomogeneous refractive index distribution, which induces wavefront aberrations and leads to a degradation of image quality that further deteriorates with imaging depth. These aberrations depend on the specimen, and can be region-specific within the same specimen. Due to their variability and the difficulty to predict their form *a priori*, sample-induced aberrations cannot be corrected by a fixed optical design.

By dynamically measuring the accumulated wavefront distortions and correcting for them using an active optical component, adaptive optics (AO) can recover diffraction-limited performance at depth. In this chapter we describe several zonal AO methods and demonstrate their applicability to laser scanning fluorescence microscopy. As opposed to other schemes discussed in this book that target the scattered light, our approaches act on the focus-forming ballistic light and correct for aberrations. This allows us to compensate for complex aberrations, such as those encountered during *in vivo* imaging of the mouse brain, and recover diffraction-limited imaging performance over large imaging volumes.

4.2 Adaptive Optics in Microscopy

To recover diffraction-limited performance, AO methods measure the distorted wavefront(s) involved in image formation and employ a wavefront modulator to introduce a wavefront distortion that cancels the aberrations from the sample [1, 2]. The implementation of AO in microscopy depends on the specific microscopy modality and how image formation is attained. For example, in laser scanning microscopy (e.g., confocal and multiphoton laser scanning microscopy), images are obtained by scanning the focus of the excitation laser through the sample and recording the signal at each position. Any sample-induced aberrations distort the wavefront of the excitation light and prevent a diffraction-limited focal spot from being formed, resulting in degradations of signal and resolution. For multiphoton microscopes, where the signal is detected by a nonimaging detector, aberration correction needs to take place in the illumination path only. In a confocal microscope, aberration correction needs to be implemented in both the illumination path (providing a tighter excitation confinement) and the fluorescence emission path (ensuring the in-focus fluorescence passes through the confocal pinhole), which can be achieved by using the same wavefront correction device placed in a common path. In contrast, in a widefield microscope, aberration correction is usually only applied to the emitted fluorescence, which has to propagate through the aberrating specimen before reaching the camera where image formation takes place.

A number of AO methods have been developed for optical microscopy [3–6]. These methods primarily differ in how the aberration is measured and may be classified into indirect and direct wavefront sensing methods (Chapters 1 and 2, respectively). In direct wavefront sensing, a dedicated wavefront sensor, such as the Shack-Hartmann wavefront sensor, is used to directly measure the wavefront aberrations using light from isolated point-like sources [7, 8]. This information is then used by a wavefront-shaping device to compensate for the specimen-induced aberration. Because wavefront aberration is obtained via a single measurement, direct sensing and correction can operate at high speed (e.g., milliseconds). Such an approach has been employed extensively in ground-based telescopes [9] to compensate for aberrations introduced by the Earth's atmosphere to incoming wavefronts, allowing the formation of diffraction-limited images of astronomical objects (see Chapter 2). It has also been applied to retinal imaging to correct for the aberrations intrinsic to the eyes themselves [10], with the resulting high-resolution retinal images providing information on retinal structures, functions, and diseases [11]. Accurate aberration measurements with direct wavefront sensing, however, are only possible when the sample under investigation does not strongly scatter the light whose wavefront is being measured. In samples such as the mammalian brain, direct wavefront sensing at depth requires near-infrared (NIR) light (see Section 4.3).

For scattering samples, indirect wavefront sensing approaches can be applied. Typically only requiring a wavefront modulator but not a wavefront sensor, indirect wavefront sensing methods can be simpler to incorporate into existing microscopes. Various approaches to finding the optimal corrective wavefront have been developed [4, 5] (also see, for example, Chapter 1). Below in Sections 4.4 and 4.5, we describe in detail a class of pupil-segmentation-based or zonal indirect wavefront sensing method that relies on similar physical principles to those employed in SH wavefront sensors.

Finally, we briefly discuss the most commonly used wavefront correction devices in AO microscopy: deformable mirrors (DM) and liquid-crystal spatial light modulators (SLM). A DM consists of a flexible reflective membrane, either continuous or segmented, whose surface can be actively controlled. Light reflecting off the DM acquires a phase aberration that is determined by the surface shape of the DM. DMs have high bandwidths (typically > 1 kHz) and their operation is independent of wavelength and light polarization. Typical DMs for AO microscopy have $\sim$100s of actuators. Liquid-crystal SLMs are devices made of 100,000s or even millions of liquid crystal cells that each imparts a phase offset to the wavefront impinging on it. These phase shifts are introduced via changes in the refractive indices of the liquid crystal cells by means of applied electric fields. The large number of pixels allows for the correction of complex wavefront aberrations. Liquid-crystal SLMs, however, operate only within a narrow wavelength range ($\sim$100s nm) and for a specific polarization. Furthermore, their refresh rates are considerably slower ($\sim$60–300 Hz). Generally, wavefront correction devices should be chosen based on the complexity as well as the spatial and temporal variabilities of the aberrations that need to be corrected.

4.3   Zonal Adaptive Optical Two-Photon Microscopy with Direct Wavefront Sensing

A common direct wavefront sensing approach in AO microscopy involves measuring the wavefront reflected or emitted from inside the sample (e.g., emitted fluorescence) [7, 8, 12–21]. After propagating through the sample and emerging from the back aperture of the microscope objective, the light reaches a Shack-Hartmann (SH) wavefront sensor where its wavefront is measured (see Chapter 2). A SH sensor belongs to zonal sensors, in that the wavefront is divided into segments, with each described by a local phase gradient and a phase offset. In a SH sensor, the local gradient of each segment is determined from the displacement of the focus of the corresponding light ray from the aberration-free position on a camera located at the focal plane of a lenslet array (Figures 4.1a and 4.1b). The phase offset of each segment is then calculated by assuming a continuous wavefront [22]. With both the local gradients and phase offsets known, the shape of the wavefront is obtained. This information is then used, sometimes in a closed loop, to control the wavefront modulation device in order to minimize wavefront aberration before image formation.

For two-photon fluorescence microscopy [23] in thick tissues, optimal imaging performance requires the excitation light to form a diffraction-limited focus (Figure 4.1c). Specimen refractive index mismatches distort the wavefront of the incoming excitation light and lead to a dim, enlarged focus (Figure 4.1d). In order to recover optimal imaging performance, the wavefront of the excitation light needs to be modified before entering the focusing objective, so that all rays entering the back pupil, even after deflection and phase retardation by the inhomogeneous refractive index distribution within the sample, still reach the focal point, constructively interfere, and form a diffraction-limited focus (Figure 4.1e).

In two-photon fluorescence microscopy, the wavefront distortion of the excitation light on its way to the focus can be approximated by the aberration accumulated by the emitted fluorescence on its way out of the sample. The 3D localized two-photon fluorescence excitation provides an isolated "nonlinear guide star" [16] for direct wavefront sensing with a SH sensor. Placing the sensor in a descanned geometry and

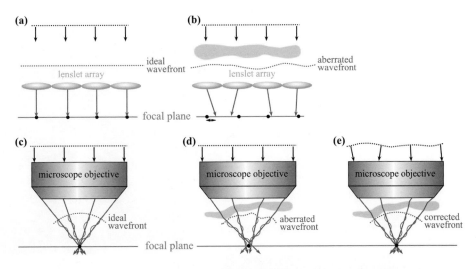

Figure 4.1 An aberration-free wavefront leads to centered focal spots in (**a**) a Shack-Hartmann wavefront sensor and (**c**) a point-scanning microscope (e.g., two-photon fluorescence). (**b, d**) Wavefront distortions, caused by the presence of sample-induced aberrations, lead to displacements of the focal spots. (**e**) A point-scanning microscope with precompensation for sample-induced aberrations. The black dots indicate the focal position of the ideal, nonaberrated rays. The sinusoidal curves denote the phase relationship among the rays.

averaging the aberration over a small area further improves the accuracy of wavefront measurement [24, 17]. Even though working well in transparent samples such as zebrafish larvae (Figure 2.13 in Chapter 2) [17], this approach fails if the light from the nonlinear guide star is strongly scattered on its way out of the sample, as accurate direct wavefront sensing depends on ballistic light reaching the wavefront sensor and forming clear images of the guide star for each wavefront segment. In highly scattering samples, such as the mouse brain, not only is this ballistic component reduced, but a diffuse background is also generated, overwhelming the ballistic component and preventing the wavefront from being measured accurately.

By making use of the reduced tissue scattering of NIR guide stars, the applicability of direct wavefront sensing can be extended to tissues that strongly scatter visible light. In particular, we demonstrated *in vivo* two-photon microscopy imaging down to 760 μm inside the mouse brain [18] using both synthetic and genetically encoded NIR fluorophores.

A schematic of our microscope is shown in Figure 4.2**a**. NIR excitation from a pulsed Ti:Sapphire laser (Coherent, Chameleon Ultra II) is expanded and collimated by a pair of lenses (F1 = 50 mm and F2 = 500 mm). The collimated beam slightly overfills the aperture of a deformable mirror (DM, Alpao, DM 97-15), which is conjugated to two scanning galvo mirrors (X and Y; model 6215H; Cambridge Technology Inc.) and the back pupil of the objective (Nikon, CFI Apo LWD 25XW, 1.1 NA and 2-mm working distance) using three pairs of achromatic relay lenses operating in the 2F1 + 2F2 configuration (focal lengths: F1 = 300 mm and F2 = 100 mm between DM and galvo Y; F1 = 85 mm and F2 = 85 mm between galvo Y and X; and F1 = 100 mm and F2 = 400 mm

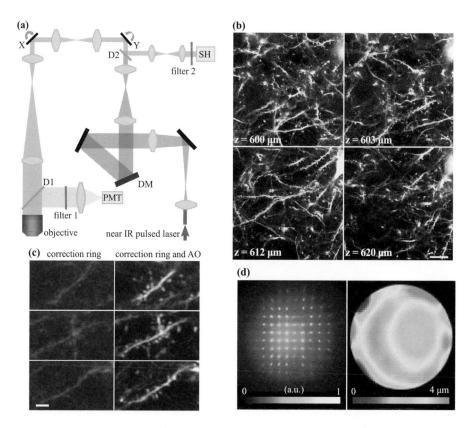

Figure 4.2 AO correction via direct wavefront sensing improves morphological imaging deep inside the cortex of a living mouse [18]. (**a**) Schematics of two-photon microscope: near-infrared pulsed excitation is reflected off a deformable mirror (DM) and scanned by a pair of galvanometer mirrors (X and Y) before entering the rear pupil of a 1.1 NA water-dipping objective. The DM, galvos X and Y, and the objective rear pupil are all mutually conjugate, so that the phase pattern from the deformable mirror is stationary at the rear pupil even while scanning. The emitted fluorescence is collected by the objective, reflected from a dichroic mirror (D1), and focused onto a photomultiplier tube (PMT). The fluorescence from the guide star is descanned by the galvos, separated from the excitation light by a dichroic beamsplitter (D2), and sent to a Shack-Hartmann (SH) sensor. (**b**) 120×120 μm field-of-view single-plane two-photon excited fluorescence images of neurons in a Thy1-YFPH mouse at 600–620 μm below pia after AO correction. Scale bar, 20 μm. (**c**) Two-photon excited fluorescence images of dendrites at 606, 606.5, and 608.5 μm depth taken with objective correction ring adjustment only (left) and correction ring adjustment in addition to AO (right). Scale bar, 5 μm. (**d**) SH sensor image (left) for a near-infrared guide star produced by two-photon excitation of injected indocyanine green (ICG), and the corresponding corrective wavefront (right). Adapted by permission from Macmillan Publishers Ltd: Nature Communications [18], copyright 2015. A black and white version of this figure appears in some formats. For the color version, please refer to the plate section.

between galvo X and the objective rear pupil). The objective focuses the excitation beam into the sample and collects the excited fluorescence. The fluorescence is reflected by a dichroic mirror D1 (Semrock, FF665-Di02-25×36) placed immediately after the objective and focused by a lens (F = 75 mm) onto a photomultiplier tube (PMT, Hamamatsu, H7422-40). For AO correction, the fluorescence from the guide star is descanned by galvos X and Y and separated from the excitation light by dichroic mirror D2 (Semrock, FF875-Di01-25×36), before being relayed by a pair of lenses (F1 = 100 mm and F1 = 150 mm) to a lenslet array (Edmund Optics, 64-483) conjugate with the objective rear pupil. The SH camera (Andor iXon3 897 EMCCD) is placed at the focal plane of the lenslet array to record an array of images of the guide star. The AO system is operated in an open-loop manner, with the DM only controlling the wavefront of the excitation light.

With a single wavefront measurement 600 μm below pia, dendritic spines can be clearly resolved over a large field of view of 120 × 120 μm (Figure 4.2b). Most of these spines were invisible without the signal gain and resolution enhancement provided by AO (Figure 4.2c). The corrective wavefront (Figure 4.2d, right) contained large elements of coma, astigmatism, and spherical aberrations caused by the refractive index mismatch between water (objective's immersion media) and the brain tissue, as well as higher-order aberration modes from the inhomogeneous refractive index distribution within the brain itself. These inhomogeneities are evident in the SH image (Figure 4.2d, left), from the intensity and shape variations of the individual guide star images.

4.4   Pupil-Segmentation AO Microscopy with Single-Segment Illumination

To use the visible fluorescent signal for wavefront sensing in scattering tissue at depth, we need to employ indirect wavefront sensing methods. In this section, we describe a pupil-segmentation AO method based on physical principles similar to those utilized by SH sensors. A zonal approach by nature, it differs from the wavefront measurement scheme in a SH sensor in that, rather than measuring the wavefront segments in parallel and thus being susceptible to tissue scattering as in a SH sensor, it measures the wavefront segments serially, making it applicable to strongly scattering samples such as the mouse brain.

The physical picture behind this scheme is simple and intuitive (Figure 4.1c): a diffraction-limited focus is formed when light rays converge on a common point with a common phase, resulting in optimal constructive interference. When light rays traveling toward the focus encounter refractive index inhomogeneities, both their direction and phase are modified in such a way that they no longer intersect at a common spot and their interference is not optimal (Figure 4.1d). By measuring image shifts when different pupil subregions are illuminated, we obtain the displacement of each ray from the desired focal spot (Figure 4.1d), from which we can determine the phase gradient of each wavefront segment [25], similar to the way that phase gradients are determined from the displacement of guide star images in a SH sensor (Figure 4.1b). The phase

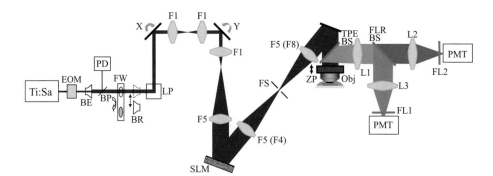

Figure 4.3 Schematics of our AO two-photon fluorescence microscope [25]. From left to right: Ti:Sapphire laser (Ti:Sa); electro-optical modulator (EOM); 2× beam expander (BE); beam pickoff (BP) that reflects 3% of the light into a photodiode (PD); neutral-density filter wheel (FW); motorized 2× beam reducer (BR); two-axis laser positioner (LP); X galvanometer (X); pair of telecentric f-θ relay lenses (F1); Y galvanometer (Y); pair of telecentric f-θ relay lenses (F1 and F5); spatial light modulator (SLM); pair of telecentric f-θ relay lenses (F5 for Zeiss objective configuration; F4 and F8 for Nikon objective configuration); field stop (FS); dichroic beamsplitter (TPE BS); water dipping objective (Obj) mounted on a Z-piezo stage (ZP); sample (S); fluorescence collimating lens (L1); dichroic beamsplitter (FLR BS); two lenses (L2, L3) to focus fluorescence through two filter sets (FL1, FL2) onto two photomultiplier tubes (PMT). Adapted by permission from Macmillan Publishers Ltd: Nature Methods [25], copyright 2010

of each segment can then be obtained either by direct interference measurements or through reconstruction algorithms similar to those developed for astronomical AO [26].

We applied this pupil-segmentation-based approach to two-photon fluorescence microscopy and demonstrated diffraction-limited imaging performance in a variety of biological and nonbiological samples, exhibiting different types of aberrations [25, 27]. A schematic of our microscope setup is shown in Figure 4.3. Near-infrared femtosecond pulses generated by a titanium-sapphire laser (Chameleon Ultra II; Coherent Inc.) are raster-scanned in 2D by a pair of galvanometers (X and Y; model 6215H; Cambridge Technology Inc.), which are optically conjugate to each other with two custom-made 30-mm-focal-length telecentric f-θ lenses (F1) (Special Optics). A third lens (F1) and a custom-made 150-mm-focal-length telecentric f-θ (F5) lens (Special Optics) conjugate the Y galvanometer to a liquid-crystal phase-only SLM (1,920 × 1,080 pixels; PLUTO-NIR; Holoeye Photonics AG). The SLM is itself conjugated either by a pair of lenses (F5) to the rear pupil of a 20×, NA 1.0 water-dipping objective (W Plan-Apochromat; Carl Zeiss Inc.) or by a custom-made 120-mm-focal-length telecentric f-θ (F4) lens and a custom-made 240-mm-focal-length telecentric f-θ (F8) lens (Special Optics) to the rear pupil of a 16×, NA 0.8 water-dipping objective (LWD 16 W; Nikon Corp.). Conjugation of the SLM to the objective rear pupil insures that the corrective phase pattern applied to the SLM is transferred to the rear pupil and remains stationary during beam scanning. The objective is mounted on a piezo-flexure stage (P-733.ZCL; Physik Instrumente, GmbH) for 2D and 3D imaging in the axial direction. The fluorescence signal is split into red and green components and detected by two photomultiplier tubes (PMT). Other important components of our setup include a motorized beam reducer (BR) (56C-30-1X-4X, Special Optics) and a 2D laser positioning system (LP) made of two beam-steering mirrors mounted on two fast translation stages (M-663; Physik Instrumente, GmbH). Together, they allow the excitation energy to be concentrated on individual wavefront segments during the deflection measurement of light rays. The entire optical path was designed and optimized with ray-tracing software, OSLO (Sinclair Optics, Inc.) and Zemax (Zemax Development Corp.).

The phase-only SLM is used for both measuring and correcting aberrations. With 1,920 × 1,080 pixels, it can be readily divided into hundreds of mechanically uncoupled, independent subregions, each with a smoothly varying linear phase ramp. With phase wrapping, it can produce >100 waves of phase change and >60 wavelengths.mm^{-1} phase gradients. A global phase ramp is applied to separate the specularly reflected and undiffracted light, which cannot be controlled, from the large fraction of light modulated by the SLM. During focal displacement/phase gradient measurements, to illuminate only one pupil segment, we turn "off" all other segments by applying a binary phase grating consisting of alternate rows of 0 and π phase shift, which diffracts most of the light to a field stop at an intermediate image plane (FS in Figure 4.3), where it is blocked. For wavefront correction, compensatory phase gradients are applied to subregions of the SLM (thus wavefront segments) to tilt the rays so that they converge on the same focal spot after passing the aberrating sample. The phases of all rays are controlled by superimposing constant phase offsets to wavefront segments so that the light rays interfere constructively at the focus.

4.4.1   Aberration Measurement Algorithm

In the first step, where local phase gradients introduced by sample aberration are measured, only one segment of the back pupil is illuminated at a time and the resulting two-photon fluorescence image of the specimen is recorded. The light ray corresponding to the illuminated segment is deflected by the refractive index inhomogeneities along its path, and as a result the fluorescence image is shifted laterally relative to a reference image taken with the full pupil illuminated. The wavefront gradient can be calculated from the image displacement and the focal length of the objective, and a compensatory phase pattern can be applied to the corresponding subregion of the SLM to cancel out the direction shift introduced by the sample. This process is repeated for all segments until the full pupil is sampled and compensatory phase gradients on the SLM cause all rays to intersect at a common focal spot (Figures 4.4a–4.4e). To measure the image shift in this approach, one can use either isolated fluorescent beads embedded in the sample or the intrinsic fluorescence from the sample itself (preferable for most biological applications).

The next step involves determining the relative phases of all rays that maximize constructive interference at the focal region. This can be done either by phase reconstruction algorithms or by direct measurement. Phase reconstruction algorithms usually assume that the wavefront is continuous. The method we adopted [26] calculates the phases through an iterative algorithm from the phase gradients determined in the first step. This

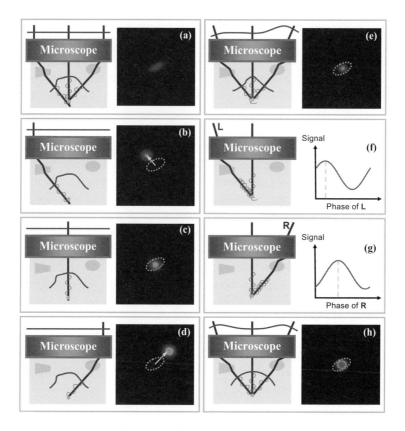

Figure 4.4 Schematics illustrating the pupil-segmentation-based AO approach using three subregions, represented by three light rays [25]. (**a**) Aberrated wavefront, due to refractive index inhomogeneities in the sample (illustrated by the blobs), leads to an aberrated image. (**b–d**) Images acquired with the left, center, and right subregions illuminated, respectively, allow the tilt of each ray to be measured from the displacement of the image. (**e**) Light rays intersect at a common point, after proper phase gradients are applied at the SLM to steer each ray. Optimal phase offset (dashed vertical line) is determined from the interference of the (**f**) left and (**g**) right light ray with the central reference ray. (**h**) Final corrected wavefront and recovered diffraction-limited focus provide an aberration-free image. If phase reconstruction is used, the phase measurement steps (**f** and **g**) are skipped. Reprinted with permission from Macmillan Publishers Ltd: Nature Methods [25], copyright 2010

approach, however, has several shortcomings. First, errors in the wavefront gradient measurement of a single segment can propagate into neighbouring segments. Secondly, not all aberration wavefronts are spatially continuous. Moreover, even when the wavefront is continuous, if the size of the pupil segments is too large to sample the wavefront sufficiently, the best-fit wavefront may not be continuous across segment boundaries, causing the reconstruction methods to perform suboptimally.

The phase offsets that would allow all rays to constructively interfere can be measured directly. Assigning one light ray as the reference (with unknown phase θ_r), we add an additional phase $\Delta\theta$ to another ray (with unknown phase θ_1) while monitoring the signal at the focal spot, which is determined by the focal intensity $I = 2 + 2\cos(\Delta\theta + \theta_1 - \theta_r)$. Fitting the signal variation curve obtained by stepping the value of $\Delta\theta$ from 0 to 2π (usually 5 evenly spaced phase values are chosen), we obtain the $\Delta\theta$ that gives the maximal signal (i.e. $\Delta\theta = \theta_r - \theta_1$). Repeating this process for all other rays relative to the same reference (Figures 4.4f–4.4g), we obtain the phase additions that would allow all rays to be in phase and thus interfere constructively at the laser focus (Figure 4.4h). Measuring the phases directly yields a smaller residual wavefront error, but at the cost of taking more images and thus slowing down the process.

We can speed up the phase measurement process by employing the concept of multi-dither coherent optical adaptive technique (COAT) [28–30]. Consider again the two-ray interference picture described above: if, instead of stopping after one cycle of signal variation, we continue to vary $\Delta\theta$ at step size ω, the laser intensity then varies as $I = 2 + 2\cos(\omega t + \theta_1 - \theta_r)$. To determine the optimal phase $\Delta\theta = \theta_r - \theta_1$ is to find the phase of the function $\cos(\omega t + \theta_1 - \theta_r)$. One way to do this is by Fourier-transforming the time-dependent signal and reading out the phase at the frequency $\omega/2\pi$. To measure the phases of multiple light rays in parallel, we tag each ray with a distinct modulation frequency, and obtain their respective $\Delta\theta$s by reading out the phases of the Fourier-transformed signal at their distinct modulation frequencies. Experimentally, we modulate the phases of half of the pupil segments at a distinct frequency ω_i, keeping the phases of the other half, which serves as a reference, unchanged (Figure 4.5). After determining the phase offsets for the modulated half of the segments, the reference and modulated segments are swapped and the procedure repeated, so as to determine the phases of the rest of the segments, and obtain the final corrective wavefront [31]. Speeding up the phase measurement process to tens of milliseconds, this method also

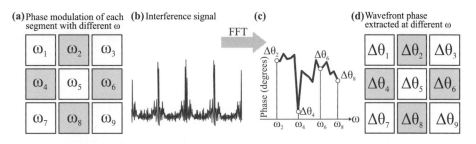

Figure 4.5 Schematics of multidither COAT used to directly measure the phases of pupil segments [31]. (**a**) We first modulate the phases of half of the segments (e.g., gray segments) at distinct frequencies ω_is while keeping the phases of the other half (e.g., white segments), which serves as the reference, constant. (**b**) The modulated rays and the reference ray interfere and modulate the focal intensity, resulting in two-photon fluorescence signal variation. (**c–d**) The phase for each modulated segment $\Delta\theta_i$ is read out from the Fourier transform of the signal trace at its modulation frequency. (**a–c**) is repeated on the other half of the segments to obtain their phases. Reprinted with permission from [31], *Optics Express*.

benefits from the fact that the measurement noise power density often decreases with frequency (e.g., 1/f noise), thus better signal-to-noise ratio can be obtained by moving the phase detection away from DC into nonzero modulation frequency. With this approach the phase for each wavefront segment is determined independently, as opposed to the phase reconstruction approaches. It is worth noting, however, that depending on the initial aberration in the system, the reference phases in the two steps could be different, in which case several iterations of the phase measurement can be repeated until the reference phases in both steps are the same.

4.4.2   System Aberration Correction

In order to isolate the aberrations originating from the sample, it is necessary to characterize the aberration intrinsic to the optical system. In our setup (Figure 4.3), most of the system aberration comes from the SLM, whose silicon substrate has a potato-chip-shaped surface profile that causes a peak-to-valley variation of $\sim 1.7\lambda$ at 850 nm. This aberration causes a fourfold signal reduction on a 500-nm diameter fluorescent bead immersed in water (Figure 4.6), the immersion media for which the Zeiss objective is designed. After applying our pupil-segmentation-based AO algorithm, by segmenting the pupil into 36 segments and using direct phase measurement, the full width at half maximum of the bead images in both lateral and axial directions approached their diffraction-limited values.

4.4.3   Aberration Correction in Nonbiological Samples

To illustrate the effectiveness of our pupil-segmentation-based AO approach, the water-dipping 1.0 NA Zeiss objective was used to image 500-nm-diameter fluorescent beads without the immersion water. Highly distorted images are obtained as a result, as shown in Figure 4.7a. AO correction results in an eightfold increase in the two-photon signal, and both the lateral and axial resolution approaching the diffraction limit (Figures 4.7b and 4.7c). Spherical aberration dominates the aberration, with a peak-to-valley wavefront error of 20 wavelengths (1 wavelength equals 2π phase shift, Figure 4.7d). The corrective wavefront remained valid over a 98×98 μm field of view. This example illustrates an important fact about AO in the context of microscopy: active wavefront shaping can be applied to extend the regime of optimal operation of a microscope, with AO transforming the water-dipping objective into an air objective while maintaining diffraction-limited performance.

As a second example, we placed a 500-nm-diameter bead on the inside surface of an air-filled glass capillary tube; a geometry that introduces a highly asymmetric wavefront aberration composed mostly of coma and astigmatism (Figure 4.7e). AO correction leads to a signal increase of about 3.5-fold, and near diffraction-limited resolution (Figures 4.7f and 4.7g).

In the examples to follow, "No AO" images were taken after system aberration correction.

4.4.4 Aberration Correction for *In Vivo* Brain Imaging

Visualizing biological structures and processes *in vivo* is one of the most important applications of optical microscopy, where studying physiological processes in their natural state at depth requires tissue-induced optical aberrations to be corrected. The AO approach described above succeeded in recovering diffraction-limited performance at a depth of 450 μm in the mouse cortex *in vivo* [27] (Figure 4.8). The corrective wavefront obtained at one location can improve image quality over a surrounding volume of hundreds of microns in each dimension. Additionally, the brain-induced aberrations were found to remain unchanged for several hours.

The signal and resolution improvements were found to be particularly strong for fine neurites, with many neurites detectable and resolvable only after AO correction (Figure 4.8d). The same corrective wavefront leads to higher signal gain for fine dendritic processes than larger cell bodies (Figures 4.8g–4.8i), where an enlarged focal volume caused by aberrations results in more fluorophores being excited inside the cell bodies, partially compensating for the decreased focal intensity. (Similarly, the two-photon fluorescence signal in a fluorescent sea, in the absence of aberrations, is independent of the

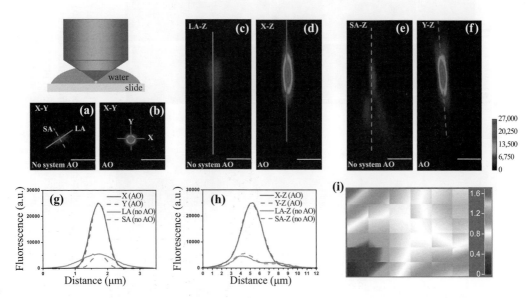

Figure 4.6 Correcting for system aberration improves images of 500-nm diameter beads immersed in water, the designed immersion media of the microscope objective [25]. Lateral and axial images before (**a, c, e**) and after (**b, d, f**) AO correction. Z denotes the axial direction and LA and SA denote the long and short axes, respectively, of the bead image before correction. (**g**) Intensity profiles along the lines drawn in the lateral plane (as shown in (**a**) and (**b**)) and (**h**) along lines drawn in the axial planes (as shown in (**c–f**)). (**i**) The final corrective wavefront for system aberration when using a Zeiss 20×, 1.0 NA objective, in units of wavelength ($\lambda = 850$ nm). Scale bar: 2 μm. Adapted by permission from Macmillan Publishers Ltd: Nature Methods [25], copyright 2010. A black and white version of this figure appears in some formats. For the color version, please refer to the plate section.

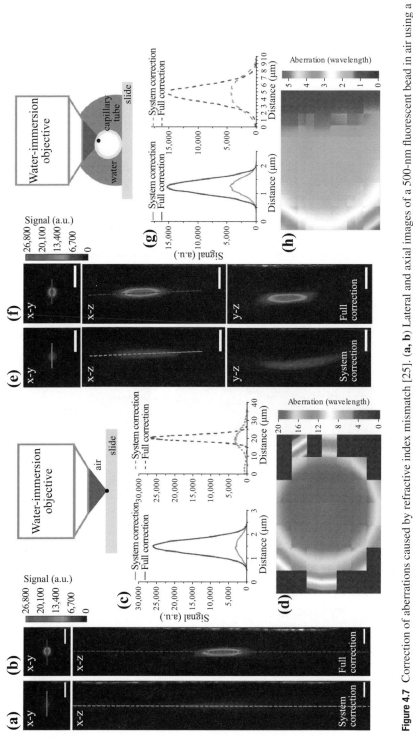

Figure 4.7 Correction of aberrations caused by refractive index mismatch [25]. (**a**, **b**) Lateral and axial images of a 500-nm fluorescent bead in air using a water-dipping objective, with correction for only system aberrations (**a**) or all aberrations (**b**). (**c**) Signal profiles in the lateral (x–y) and axial (x–z) planes along the solid and dashed lines in (**a**) and (**b**). (**d**) Final corrective wavefront, after subtraction of system aberrations, obtained with 49 independent subregions and direct phase measurement. Approximately 20 wavelengths of predominantly spherical aberration were corrected. (**e**, **f**) Images of a bead in an air-filled capillary tube in water, with system aberration correction only (**e**) and full AO correction (**f**). (**g**) Signal profiles in the lateral (x–y) and axial (x–z) planes along the solid and dashed lines in (**e**) and (**f**). The final corrective wavefront in units of excitation light wavelength (850 nm), after subtraction of system aberrations, obtained with 36 independent subregions and direct phase measurement. Scale bar: 2 μm. Reprinted with permission from Macmillan Publishers Ltd: Nature Methods [25], copyright 2010. A black and white version of this figure appears in some formats. For the color version of this figure, please refer to

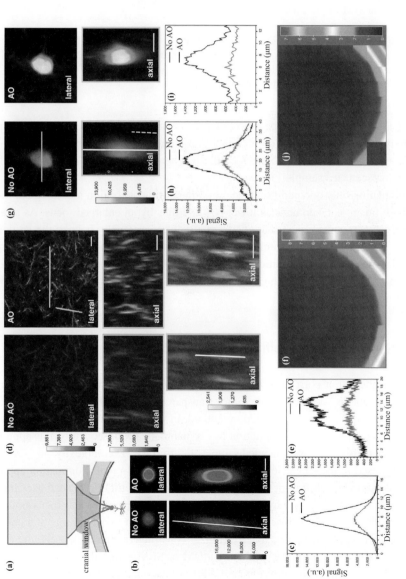

Figure 4.8 AO improves imaging quality in the mouse brain *in vivo* [27]. (**a**) Schematic of the geometry for *in vivo* imaging in the mouse brain, showing the cranial window embedded in the skull. (**b**) Lateral and axial images of a 2-μm-diameter bead 170 μm below the brain surface before and after AO correction. (**c**) Axial signal profiles along the white line in **b** before and after AO correction. (**d**) Lateral and axial images of GFP-expressing dendritic processes over a field centered on the bead in **a**. (**e**) Axial signal profiles along the white line in **d**. (**f**) Measured aberrated wavefront in units of excitation wavelength. (**g**) Lateral and axial images of GFP-expressing neurons 110 μm below the surface of the brain with and without AO correction. (**h**) Axial signal profiles along the white line in **g**. (**i**) Axial signal profiles along the dashed line in **g**. (**j**) Aberrated wavefront measured in units of excitation wavelength. Scale bars: 2 μm in **b** and 10 μm elsewhere. Reprinted with permission from [27] PNAS, copyright (2012) National Academy of Sciences. A black and white version of this figure appears in some formats. For the color version, please refer to the plate section.

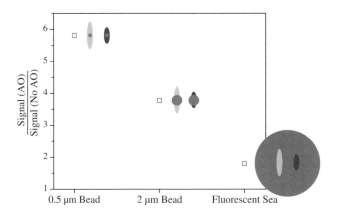

Figure 4.9 Effect of AO correction on fluorescent features of different sizes [27]. The picture next to each (square) data point shows the size of the fluorescent feature (circles) relative to the uncorrected and corrected excitation focal volumes (ovals). Reprinted with permission from [27] PNAS, copyright (2012) National Academy of Sciences

NA of the focusing microscope objective [32].) This principle is demonstrated experimentally in Figure 4.9, where correcting for aberrations introduced by a cranial window results in a signal increase of $5.8\times$ for a 0.5-μm bead, $3.8\times$ for a 2-μm bead, and $1.8\times$ for a fluorescent sea sample.

Illuminating only one pupil segment at a time during aberration measurement means that each image is taken under a lower NA and thus by an enlarged focus. As a result, the images can contain contributions from structures originally beyond the volume of excitation under full-pupil illumination, making image shift measurements in densely labeled samples difficult. Additionally, the reduced NA demands an increase in the excitation power, ultimately limiting the application of this method to superficial depths in scattering tissues.

4.5   Pupil Segmentation AO with Full-Pupil Illumination

An ideal aberration correction method should use structures intrinsic to the sample, under any labeling density, inside either transparent or strongly scattering media. We developed a new aberration correction method, in which the entire pupil is illuminated at all times [33] thus the full excitation NA is maintained. Consequently, it is applicable to both sparsely and densely labeled samples of arbitrary 3D complexity.

As opposed to the single segment illumination, the full-pupil illumination approach does not use image shift to determine the local wavefront tilt, but instead uses fluorescence signal strength. The physical picture is still the same: an ideal focus is formed at the focal plane of a microscope objective when all light rays converge at a common point and with a relative phase that maximizes constructive interference (Figure 4.10a). By raster scanning one of the rays using a DM or SLM across the aberrated reference focus formed by the other rays (Figure 4.10b), we can measure the varying interference

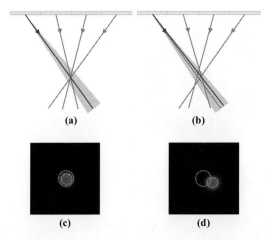

Figure 4.10 Schematics depicting the pupil-segmentation method with full-pupil illumination [33]. (**a**) An ideal focus has all rays (arrowed lines) intersect at the same point. (**b**) In the presence of aberrations, rays do not intersect at the same point. Scanning one of the rays (e.g. leftmost ray) through a range of angles (shaded cone) causes a modulation of the focal intensity, and consequently on the signal strength. (**c**) Plotted as an image, these data exhibit an intensity extremum centered (dashed circle) over the scan region, if the ray already intersects the ideal focus. (**d**) A shift of the extremum from the center indicates a local wavefront tilt. The dashed line in **b** depicts the original ray direction, while the solid line depicts the direction at which signal extremum is reached. Reprinted with permission from [33], *Optics Letters*.

between them and thus the modulation of the focal intensity by monitoring the fluorescence signal strength. This modulation can be visualized by plotting the signal as a function of the 2D displacement of the light ray in the focal plane. The local wavefront gradient of this particular ray can be determined by measuring the additional tilt needed for maximal or minimal constructive interference, which happens when the ray intersects the reference focus. If there are no aberrating inhomogeneities along the path of the scanned ray, this 2D signal-versus-displacement map will show a maximum intensity at its center (Figure 4.10c), corresponding to zero applied wavefront gradient. If, on the other hand, the scanned ray travels through inhomogeneities that distort its wavefront, this map will take one of three forms: (1) a shifted maximum, if the scanned ray interferes constructively with the fixed rays (Figure 4.10d); (2) a shifted minimum, if the interference is destructive; and (3) a somewhat flat or biphasic 2D profile if the relative phase between scanned and fixed rays is near $\pm\pi/2$. In the latter case, an additional $\pi/2$ phase offset is applied to the active pupil segment, and the measurement is repeated, yielding an image with an extremum. For any of the three scenarios, the displacement of the extremum from the center of the map indicates the additional tilt that needs to be applied for this ray to intersect the reference focus.

After sequentially computing the local wavefront gradient of all segments across the entire back pupil, the phases of each segment are determined, either by direct measurement or through algorithmic reconstruction, as explained in Section 4.4.

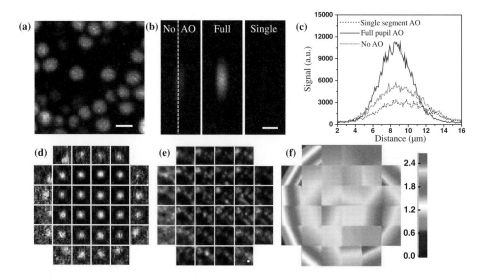

Figure 4.11 Comparison of single-segment illumination and full-pupil illumination AO in a densely labeled sample [33]. (**a**) Maximal intensity projection across a depth of 30 μm in a dense sample of fluorescent beads. (**b**) Axial images of a 2-μm-diameter bead without AO, with full-pupil illumination, and with single-segment illumination AO. (**c**) Signal profile along the dotted line in **b**. (**d**) Signal modulation during full-pupil illumination AO for different pupil segments. (**e**) Images measured with single-pupil illumination AO. (**f**) Aberration in units of wavelength. Scale bar: 2 μm. Reprinted with permission from [33], *Optics Letters*. A black and white version of this figure appears in some formats. For the color version, please refer to the plate section.

4.5.1   Comparison of Single-Segment and Full-Pupil Illumination Schemes

We demonstrated the ability of the full-pupil illumination approach to attain accurate AO correction in densely labeled samples [33]. Figure 4.3 shows a schematic of the two-photon fluorescence microscope. Scanning of each individual ray was accomplished by applying a series of phase ramps to the corresponding segment on the SLM. Figure 4.11 shows the AO correction of an applied aberration using the signal from a dense aggregate of 2-μm-diameter fluorescent beads in agarose (Figure 4.11a). Under full-pupil illumination, the shift of extrema in the 2D signal-versus-displacement map can be measured unambiguously (Figure 4.11d), resulting in improved signal and resolution (Figures 4.11b and 4.11c); whereas with single-segment illumination, out-of-focus features lead to inaccurate displacement measurements (Figure 4.11e), resulting in an erroneous correction that led to reduced signal (Figures 4.11b and 4.11c) when compared to the uncorrected case.

Due to the sequential nature of this approach, one major drawback is its slow execution speed. Furthermore, large aberrations may cause highly distorted reference foci, which would require this procedure to be repeated multiple times, with increasingly diffraction-limited reference foci. Furthermore, detecting the small signal variations (sometimes only ∼1%–2%) caused by the interference of each scanned ray and the remaining fixed rays requires long integration times, slowing down the process further and rendering this AO scheme impractical for *in vivo* applications, mainly due to sample photobleaching and motion.

4.5.2 Multiplexed Aberration Measurement

To speed up the aberration measurement, gradients of multiple wavefront segments can be determined in parallel through frequency multiplexing [28]. Here, the contribution of each ray in the final interference signal is derived by modulating either its intensity or phase with a characteristic frequency ω, Fourier-transforming the signal trace, and plotting the Fourier magnitude at the characteristic frequency of each ray in a 2D signal-versus-displacement map.

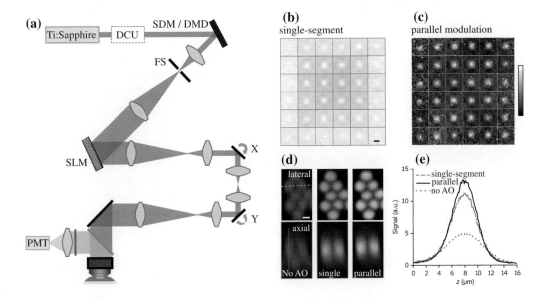

Figure 4.12 (**a**) Schematics of our AO two-photon fluorescence microscope [34]. From top to bottom: Ti:Sapphire laser; optional dispersion compensation unit (DCU) to be used with DMD; micromirror device (DMD) or segmented deformable mirror (SDM); field stop (FS) at an intermediate image plane between SDM/DMD and liquid crystal spatial light modulator (SLM); X galvanometer (X); Y galvanometer (Y); photomultiplier tube (PMT). (**b, c**) Signal-versus-displacement maps for 36 pupil segments obtained from the fluorescence signal directly by modulating N = 1 pupil segment at a time (**b**), and from the Fourier magnitudes of the signal trace obtained by modulating N = 18 pupil segments in parallel with a micromirror device (**c**). Each map is normalized between 0 and its maximum value. (**d**) Lateral and axial (along dashed line) images of 2-μm-diameter beads obtained without AO (left), with the corrective wavefront from the single-segment method (center), and with the corrective wavefront from the multiplexed method (right). (**e**) Signal profiles along the vertical line in **d**. The parallel modulation method speeds up the aberration measurement by 18× when compared to the single-segment modulation method [34]. Scale bar: 2 μm. Signal intensity scales range from high (light) to low (dark). Reprinted with permission from Macmillan Publishers Ltd: Nature Methods [34], copyright 2014

A schematic of the two-photon microscope is shown in Figure 4.12a. Near-infrared femtosecond pulses at 900 nm generated by a titanium-sapphire laser (Ti:Sapphire, Chameleon Ultra II, Coherent) are reflected off either by a segmented deformable mirror (SDM, Iris AO) or a high-speed digital micromirror device (DMD, V4100, ViALUX Messtechnik+Bildverarbeitung) following a dispersion compensation unit (DCU). A pair of achromat doublets (AC300-080-B and AC254-500-B, Thorlabs) are used to conjugate the DMD or SDM to a liquid-crystal, phase-only spatial light modulator (SLM, PLUTO-NIR, Holoeye Photonics), which is further conjugated to a pair of galvanometers (X and Y, 3-mm beam aperture, 6215H, Cambridge Technology) and the back pupil plane of a 40× or a 16×, 0.8 NA water-dipping objective (CFI Apo 40XW NIR, CFI LWD 16XW, Nikon) using three pairs of achromat doublets (AC254-150-B and AC254-060-B, AC508-080-B and AC508-080-B, AC508-75-B and 014-1310, Thorlabs and OptoSigma). A field stop (FS) is located at the intermediate image plane between the DMD/SDM and the SLM to block unwanted diffraction orders. The microscope objective is mounted on a single-axis piezo stage (P-725.4CD, Physik Instrumente) for taking axial images. In the detection path, the fluorescence signal reflected from a dichroic beam splitter (BS, FF665-Di02-25x36, Semrock) is collected by a photomultiplier tube (PMT, H7422-40, Hamamatsu) after passing through an emission filter (FF01-680/SP, Semrock). Ray-tracing software Zemax (Zemax 12 EE, Radiant Zemax) is used to optimize the optical design.

The DMD is a bistable spatial light modulator capable of intensity modulation, consisting of $1,024 \times 768$ movable micromirrors, where each mirror can be individually deflected $\pm 12°$. The DMD is divided into segments through precision mapping to the SLM segments. Because femtosecond pulses are not monochromatic, they are spatially dispersed by the DMD. To compensate for this, we built a dispersion compensation unit (DCU) based on a single-prism approach (for more details, refer to [34]). The SDM is made of piston-tip-tilt mirror segments and is operated in two modes: by modulating the piston value of each mirror segment, the phase of the light ray at the corresponding pupil segment is modulated; by tilting each mirror segment so that the reflected ray gets blocked by the field stop (Figure 4.12a), the light intensity at the corresponding pupil segment is modulated. Both types of modulation affect the interference between the light ray and the reference focus. Without the need for dispersion compensation, the SDM provides better power throughput. It also has the potential to replace the SLM for aberration correction and could thus simplify the setup substantially. For the SDM, the number of pupil segments is fixed at 37 in our experiment, whereas for the DMD, having significantly more independent actuators, the number of pupil segments can be varied.

Besides speeding up the aberration measurement, our multiplexed aberration-measurement approach also improves the corrective performance by moving the measurement away from DC into nonzero modulation frequencies, which results in a better signal-to-noise ratio (e.g., noise related to fluctuations in the excitation laser power is much smaller at higher frequencies). This is illustrated in Figure 4.12b–4.12e, where the previously described approach of using the fluorescence signal directly by

(3) Repeat 1 time for the other half of pupil segments

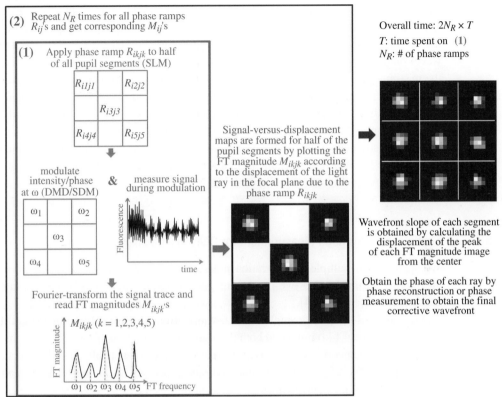

Figure 4.13 Schematics of the multiplexed aberration measurement method [34]. Reprinted with permission from Macmillan Publishers Ltd: Nature Methods [34], copyright 2014

scanning one ray across the aberrated focus at a time, is compared to the multiplexed aberration measurement with parallel modulation of light ray intensity.

4.5.3   Multiplexed Aberration Measurement Algorithm

Figure 4.13 illustrates the multiplexed aberration-measurement algorithm. Experimentally, the laser focus is parked at one sample location, and the fluorescence signal from this location is used for aberration measurement. For half of the pupil segments, raster scanning across the reference focus is accomplished by applying a phase ramp R_{ikjk} to segment k, by means of an SLM. Initially (step 1 in Figure 4.13), a phase ramp R_{ikjk} is randomly chosen from a set of N_R phase ramps, which together form a 2D grid in the focal plane, along which each ray is scanned. Using either a digital micromirror device (DMD) or a segmented deformable mirror (SDM), conjugated to the SLM (Figure 4.12a), the intensity or phase of these rays is modulated at a distinct frequency ω_k for a time duration T while monitoring the resultant interference signal. To separate the contribution from each pupil segment, the recorded trace is Fourier transformed and the Fourier magnitudes M_{ikjk} are read out at each distinct modulation frequency ω_k. After this procedure is repeated N_R times, 2D signal-versus-displacement maps are constructed for the first half of segments by plotting the magnitudes M_{ikjk}, according to the displacement of each ray in the focal plane (step (2) in Figure 4.13). The corrective wavefront tilt for each particular pupil segment is determined from the displacement of the maximum from the center of the map. While for phase modulation there is always a maximum present in the map, for intensity modulation, if the relative phase between the ray and the reference focus is near $\pi/2$, a clear maximum would be absent from the resulting map, in which case a $\pi/2$ phase offset is added to the corresponding segment and the measurement redone. This procedure is repeated for the other half of segments to obtain the local wavefront gradients of all pupil segments, with an overall integration time of $2 \times N_R \times T$ (step (3) in Figure 4.13).

4.5.4   Multiplexed Aberration Measurement for Deep Tissue Imaging *In Vivo*

We applied this multiplexed aberration measurement method to measure and correct aberrations during *in vivo* two-photon imaging of *Caenorhabditis elegans*, zebrafish larva, and mouse brain [34].

As Figure 4.14 shows, a single corrective wavefront (Figure 4.14c) obtained at the cell body (Figure 4.14a) improved both the signal and resolution of layer 2/3 neurons in the primary visual cortex of a mouse transfected with the genetically encoded calcium sensor GCaMP6s [35], with many fine neuronal processes becoming resolvable only after AO correction (Figure 4.14b).

The fluorescence background signal, or "neuropil background," arising from the spatially extended diffusive labeling of the neurites changed little after AO correction. The reason for this is that the loss of focal intensity due to an aberrated focus is compensated by exciting more fluorophores with the enlarged focus in the spatially extended neuropil, as discussed earlier. This phenomenon has important implications for *in vivo* calcium imaging experiments, in which the detected fluorescence change caused by neuronal activity may originate from two sources: the individual neuronal structure of interest (e.g., the brighter soma, axons, and dendrites) and its surrounding, diffusely stained neuropil, with the latter signal often uncorrelated with the stimulus feature and considered a contamination [36]. As Figures 4.14d and 4.14e show, AO improves the strength of the desired signal much more than the neuropil contamination, resulting in the detection of more calcium events and a more accurate determination of orientation selectivity of dendrites (Figure 4.14e). The implications of such an effect were made clear in an experiment that aimed to characterize the orientation tuning properties of thalamic boutons in the primary visual cortex of awake mice [37]. Here, with a 340-μm-thick cranial window, 70% of all imaged boutons appeared to be non-responsive to visual stimuli and only 7% satisfied orientation-selective criteria. With a thinner cranial window of 170 μm , 31% of boutons were found to satisfy orientation-selective criteria, in contrast to 48% orientation-selective boutons as determined when the same boutons were imaged after AO correction.

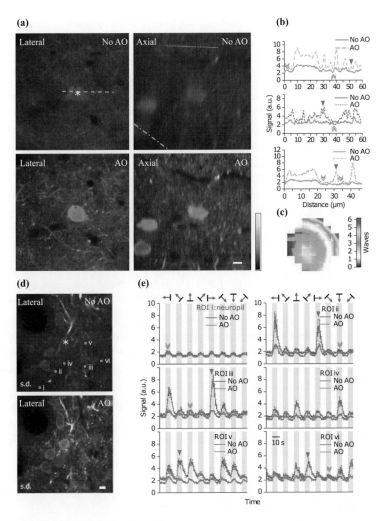

Figure 4.14 AO correction with multiplexed aberration measurement at layer 2/3 of a GCaMP6s-expressing mouse brain *in vivo* [34]. (**a**) Lateral and axial images at 115 μm below dura without and with AO correction (representative images of 33 experiments from 10 mice). (**b**) Signal profiles along the dashed, dotted, and dash-dotted lines in **a**. Arrowheads label the structures rendered invisible by the diffusely labeled background (triple arrowheads). (**c**) Corrective wavefront in units of waves. (**d**) Lateral images at 150-μm depth without and with AO correction (representative images of 18 experiments from 10 mice). Pixel brightness reflects the s.d. of pixel values across 800 calcium imaging frames, with larger s.d. values indicating higher neuronal activity. (**e**) Fluorescence signals reflecting calcium transients measured without and with AO correction at six regions of interest (ROIs; white squares i–vi in **d**) under visual stimulations of gratings moving in eight different directions. Arrowheads mark the example responses at the preferred directions of grating motion; triple arrowheads label example responses dominated by neuropil contamination. Gray bars denote the duration with stimuli on. Signals are averages of 5 trials of calcium imaging with the error bars indicating the s.e.m. Signals at the asterisks were used to measure aberration through multiplexed intensity modulation via DMD. Scale bar: 10 μm. Reprinted with permission from Macmillan Publishers Ltd: Nature Methods [34], copyright 2014. A black and white version of this figure appears in some formats. For the color version, please refer to the plate section.

With increasing imaging depth, the excitation laser power has to be increased to compensate for the loss of ballistic photons at the focus due to scattering. Eventually, however, the laser intensity outside of the focal region can be high enough to excite more superficially located fluorescently labeled structures (typically near the brain surface), resulting in a background signal that could ultimately overwhelm that generated in the focus, thus limiting the imaging depth [38]. This out-of-focus signal is not affected by the presence of aberrations and it is therefore not increased by AO correction. The in-focus signal, on the other hand, strongly depends on the focus quality. This is illustrated in Figures 4.15a and 4.15b, where dendritic spines (385 µm inside a Thy1-YFP-H mouse brain) initially invisible in the uncorrected image due to the out-of-focus background, became clearly visible after AO correction. The improvement in both signal and resolution allowed us to reduce the excitation power (e.g., from 75 to 40 mW; Figure 4.15a), and still obtain higher quality images with a better-suppressed out-of-focus background.

With the synaptic-level resolution achievable with AO, it was possible to resolve submicrometer-sized spines on the basal dendrites of layer 5 pyramidal neurons at depth. As shown in Figures 4.15c and 4.15e, with AO correction the signal from dendritic spines at 458- to 500-µm depth increased above the background, making these structures easily identifiable in the corrected images. For a more densely labeled brain (wild-type mouse with CGaMP6s expression), a single correction improved image quality at 427- to 547-µm depth, with fine neuronal processes and even the much larger somata going from invisible to clearly resolvable after AO correction (Figure 4.15f). Additionally, calcium imaging of fine neuronal processes and somata greatly improved with AO at a depth of 490 µm. As shown in Figures 4.15g and 4.15h, structures whose calcium activity was undetectable before aberration correction, showed clear visually evoked calcium transients after AO correction.

4.6   Discussion

In this chapter we described a few zonal AO approaches, applied to correct sample-induced aberrations in two-photon fluorescence microscopy. Using a direct wavefront sensing approach, with a SH wavefront sensor, along with the reduced scattering of near-infrared guide stars, we demonstrated *in vivo* two-photon imaging down to 700 µm inside the mouse brain.

Based on physical principles similar to those applied in direct wavefront sensing using SH sensors, we described several indirect AO approaches utilizing visible fluorescence in strongly scattering samples. Given that an ideal focus is formed when all light rays entering the rear pupil of a microscope objective intersect at a common point and with the same phase, these methods first measure the deviation of these light rays from the desired focus and, after applying compensatory phase gradients to the corresponding wavefront segments and making the rays intersect at the focus, measure and correct their relative phases to make them constructively interfere at the focus.

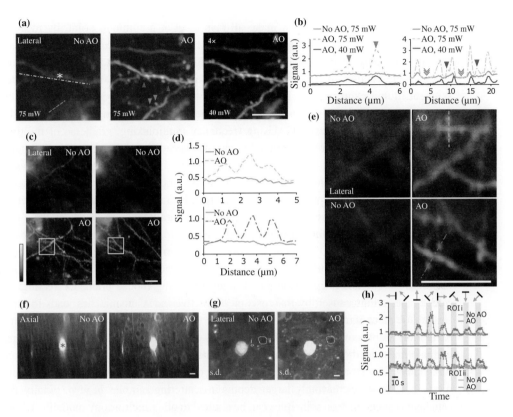

Figure 4.15 AO correction with multiplexed aberration measurement at layers 4 and 5 of the mouse brain *in vivo* [34]. (**a**) Maximal-intensity projections of dendrites at 376–395 μm below dura measured without and with AO at excitation powers 75 mW and 40 mW, respectively. The gain of the image taken at 40 mW is increased 4×. (**b**) Line intensity profiles along the dashed orange and dashed-dotted cyan lines in **a**. Purple and orange arrowheads label dendritic spines invisible before AO correction. Red triple arrowheads indicate the out-of-focus background level at 75-mW excitation. (**c**) Maximal-intensity projections of basal dendrites at 473–481 μm (left) and 481–490 μm (right). (**d**) Line intensity profiles along the dotted yellow and dashed-dotted purple lines in **e**. (**e**) Images taken inside the yellow squares in **c** at 480-μm (top) and 484-μm (bottom) depths. (**f**) Axial images at 427- to 547-μm depth. (**g**) From **f**, lateral images at 490-μm depth. Pixel brightness reflects the s.d. of pixel values across 800 calcium imaging frames. (**h**) Calcium transients as in Figure 4.14**e** measured without and with AO correction at the two regions of interest in **g** (ROI i, dendrite; ROI ii, soma). Signals are averages of 5 trials of calcium imaging with the error bars indicating the s.e.m. In (**a, f**), corrective wavefront was measured at the asterisk; in **c**, correction was at 30 μm above the imaging volume. In **a, c**, phase modulation with SDM was performed; in **f**, intensity modulation with SDM. For (**a–e**), thy1-YFP-H mice were used (representative images of 12 experiments from 10 mice); (**f–h**), wild-type mouse expressing GCaMP6s (representative images of 5 experiments from 5 mice). Scale bar: 10 μm. Reprinted with permission from Macmillan Publishers Ltd: Nature Methods [34], copyright 2014. A black and white version of this figure appears in some formats. For the color version, please refer to the plate section.

In the first approach, we first make all rays meet at a common point, by measuring image shifts when individual segments of the back pupil are illuminated. Next, the optimal phase of each ray is determined either by interference measurements or phase reconstruction. Alternatively, to avoid the reduced NA associated with single-segment illumination AO, we illuminate the full pupil and recover an ideal focus by scanning one ray around the focus formed by the remaining rays while monitoring the variation on the signal strength. Using frequency multiplexing, gradients of multiple wavefront segments can be determined in parallel, speeding up the aberration measurement and improving the signal-to-noise ratio. The method of phasing the light rays to ensure constructive interference at the focus is also improved by applying the concept of multidither coherent optical adaptive technique. By tagging each ray with a different frequency, we measure the phase of multiple rays in parallel, which speeds up the phase measurement process to tens of milliseconds.

Using our indirect AO method, we correct for aberrations and recover diffraction-limited imaging performance from both biological and nonbiological samples, including hundreds of microns deep inside the strongly scattering mouse brain. Correcting the optical aberrations caused by macroscopic refractive index mismatches leads to large isoplanatic patch sizes (e.g., in mouse brain, $168 \times 168 \times 120 \, \mu m^3$ in Figure 4.15f) and temporally stable corrective wavefront patterns, which is essential for the routine application of these methods in biological experiments. The development of faster wavefront modulators as well as brighter and near-infrared fluorophores would also allow these methods to reach increasingly larger depths in scattering samples *in vivo*. Finally, the same pupil-segmentation principles can be applied to other microscopy modalities, such as those based on linear fluorescence excitation, three-photon excitation, and harmonic generation.

4.7   References

[1] Tyson RK. Principles of Adaptive Optics. San Diego, CA: Academic Press; 1991.

[2] Babcock HW. Adaptive optics revisited. Science. 1990;249(4966):253–257.

[3] Booth MJ. Adaptive optics in microscopy. Philos Trans R Soc A. 2007;365(1861):2829–2843.

[4] Kubby JA. Adaptive Optics for Biological Imaging. Boca Raton, FL: CRC Press; 2013.

[5] Booth MJ. Adaptive optical microscopy: the ongoing quest for a perfect image. Light Sci Appl. 2014 04;p. e165.

[6] Booth M, Andrade D, Burke D, Patton B, Zurauskas M. Aberrations and adaptive optics in super-resolution microscopy. Microscopy. 2015 May;64(4):251–261.

[7] Azucena O, Crest J, Cao J, Sullivan W, Kner P, Gavel D, et al. Wavefront aberration measurements and corrections through thick tissue using fluorescent microsphere reference beacons. Opt Express. 2010 Aug;18(16):17521–17532.

[8] Tao X, Azucena O, Fu M, Zuo Y, Chen DC, Kubby J. Adaptive optics microscopy with direct wavefront sensing using fluorescent protein guide stars. Opt Lett. 2011 Sep;36(17):3389–3391.

[9] Hardy JW. Adaptive Optics for Astronomical Telescopes; 1998.

[10] Porter J, Queener H, Lin J, Thorn K, Awwal A. Adaptive Optics for Vision Science: Principles, Practices, Design, and Applications. Hoboken, NJ: Wiley Interscience; 2006.

[11] Godara P, Dubis AM, Roorda A, Duncan JL, Carroll J. Adaptive optics retinal imaging: emerging clinical applications. Opt Vis Sci. 2010;87(12):930–941.

[12] Azucena O, Crest J, Kotadia S, Sullivan W, Tao X, Reinig M, et al. Adaptive optics wide-field microscopy using direct wavefront sensing. Opt Lett. 2011 Mar;36(6):825–827.

[13] Jorand R, Le Corre G, Andilla J, Maandhui A, Frongia C, Lobjois V, et al. Deep and clear optical imaging of thick inhomogeneous samples. PLoS ONE. 2012 Mar;7(4):e35795.

[14] Rueckel M, Mack-Bucher JA, Denk W. Adaptive wavefront correction in two-photon microscopy using coherence-gated wavefront sensing. Proc Nat Acad Sci USA. 2006 Jun;103(46):17137–17142.

[15] Cha JW, Ballesta J, So PTC. Shack-Hartmann wavefront-sensor-based adaptive optics system for multiphoton microscopy. J Biomed Opt. 2010 Jun;15(4):046022.

[16] Aviles-Espinosa R, Andilla J, Porcar-Guezenec R, Olarte OE, Nieto M, Levecq X, et al. Measurement and correction of *in vivo* sample aberrations employing a nonlin-ear guide-star in two-photon excited fluorescence microscopy. Biomed Opt Express. 2011 Oct;2(11):3135–3149.

[17] Wang K, Milkie DE, Saxena A, Engerer P, Misgeld T, Bronner ME, et al. Rapid adaptive optical recovery of optimal resolution over large volumes. Nat Meth. 2014 Jun;11(6): 625–628.

[18] Wang K, Sun W, Richie CT, Harvey BK, Betzig E, Ji N. Direct wavefront sensing for high-resolution *in vivo* imaging in scattering tissue. Nat Commun. 2015 Apr;6:7276.

[19] Tao X, Fernandez B, Azucena O, Fu M, Garcia D, Zuo Y, et al. Adaptive optics confocal microscopy using direct wavefront sensing. Opt Lett. 2011 Apr;36(7):1062–1064.

[20] Rahman SA, Booth MJ. Direct wavefront sensing in adaptive optical microscopy using backscattered light. Appl Opt. 2013 Aug;52(22):5523–5532.

[21] Tao X, Crest J, Kotadia S, Azucena O, Chen DC, Sullivan W, et al. Live imaging using adaptive optics with fluorescent protein guide-stars. Opt Express. 2012 Jun;20(14):15969–15982.

[22] Southwell WH. Wave-front estimation from wave-front slope measurements. J Opt Soc Am. 1980 Aug;70(8):998–1006.

[23] Denk W, Strickler J, Webb W. Two-photon laser scanning fluorescence microscopy. Science. 1990;248(4951):73–76.

[24] Hofer H, Artal P, Singer B, Aragón JL, Williams DR. Dynamics of the eye's wave aberration. J Opt Soc Am A. 2001 Mar;18(3):497–506.

[25] Ji N, Milkie DE, Betzig E. Adaptive optics via pupil segmentation for high-resolution imaging in biological tissues. Nat Meth. 2010 02;(2):141–147.

[26] Panagopoulou S, Neal D. Zonal matrix iterative method for wavefront reconstruction from gradient measurements. J Refract Surg 2005;21(5).

[27] Ji N, Sato TR, Betzig E. Characterization and adaptive optical correction of aberrations during *in vivo* imaging in the mouse cortex. Proceedings of the National Academy of Sciences. 2012;109(1):22–27.

[28] Bridges WB, Brunner PT, Lazzara SP, Nussmeier TA, O'Meara TR, Sanguinet JA, et al. Coherent optical adaptive techniques. Appl Opt. 1974 Feb;13(2):291–300.

[29] O'Meara TR. The multidither principle in adaptive optics. J Opt Soc Am. 1977 Mar;67(3):306–315.

[30] O'Meara TR. Theory of multidither adaptive optical systems operating with zonal control of deformable mirrors. J Opt Soc Am. 1977 Mar;67(3):318–325.

[31] Liu R, Milkie DE, Kerlin A, MacLennan B, Ji N. Direct phase measurement in zonal wavefront reconstruction using multidither coherent optical adaptive technique. Opt Express;(2):1619–1628.

[32] Xu C, Webb WW. Topics in Fluorescence Spectroscopy, vol. 5. Berlin: Springer; 1997.

[33] Milkie DE, Betzig E, Ji N. Pupil-segmentation-based adaptive optical microscopy with full-pupil illumination. Opt Letters;(21):4206–4208.

[34] Wang C, Liu R, Milkie DE, Sun W, Tan Z, Kerlin A, et al. Multiplexed aberration measurement for deep tissue imaging *in vivo*. Nat Meth. 10;(10):1037–1040.

[35] Chen TW, Wardill TJ, Sun Y, Pulver SR, Renninger SL, Baohan A, et al. Ultrasensitive fluorescent proteins for imaging neuronal activity. Nature. 2013 Jul;499(7458):295–300.

[36] Goebel W, Helmchen F. *In vivo* calcium imaging of neural network function. Physiology. 2007 Dec;22(6):358.

[37] Sun W, Tan Z, Mensh BD, Ji N. Thalamus provides layer 4 of primary visual cortex with orientation- and direction-tuned inputs. Nat Neurosci. 2016 Feb;19(2):308–315.

[38] Theer P, Hasan MT, Denk W. Two-photon imaging to a depth of 1000 μm in living brains by use of a Ti:Al2O3 regenerative amplifier. Opt Lett. 2003 Jun;28(12):1022–1024.

Part III

Focusing Light through Turbid Media Using the Scattering Matrix

5 Transmission Matrix Approach to Light Control in Complex Media

Sylvain Gigan

5.1   Introduction

Propagation of light in an inhomogeneous environment, such as biological tissues, gives rise to elastic scattering of the light waves, due to the variations of the refractive index of the material. In the context of imaging, we have seen that the scattered fraction of the light is usually considered as useless for imaging, since the unpredictable perturbations of the light path make it difficult to recover the initial emission pattern coming from the object to be imaged. A large fraction of the effort of the biomedical optical imaging community has relied on trying to get rid of scattered light and form the image using ballistic light, i.e., light that has not been scattered by tissues. However, it is well understood that ballistic light decreases exponentially with depth; the characteristic length of this decay being the so-called scattering mean free path l_s. For typical tissues in the visible or near-infrared range, l_s is usually considered to be on the order of 100 microns. This means that imaging with ballistic light is usually limited to the first few 100 microns at most and becomes exponentially difficult, with only a handful of techniques reaching 1 mm. Beyond a couple of millimeters, all light can be considered to have been multiply scattered and useless for imaging.

However, the recent paradigm of wavefront shaping has, in the last decade, permitted researchers to overcome this limitation by showing that it was possible to recover a focus after propagation through extremely strongly scattering materials, and exploiting multiply scattered light, and coherent sources of monochromatic light, i.e., lasers. In particular, it relies on the fact that coherent light generates, after multiple scattering, a complex yet deterministic interference pattern: the speckle. The initial approaches relied either on an optimization algorithm [1] or on digital phase conjugation [2]. In the first approach, a detector at the target position and an iterative algorithm allowed convergence to an optimal wavefront, while on the other, a point source at the target positions could propagate through a complex medium and be detected and subsequently digitally phase-conjugated by means of a spatial light modulator. This phase-conjugate beam would re-propagate through the material and retrace its path to the initial source position, generating a focus. Both approaches relied on common properties of the propagation of waves and elastic scattering: the linearity and time-invariance of the propagation equations. The new insight brought by these two experiments was that these properties were valid even in the multiple scattering regime, in strong analogy with what was observed for other waves, such as radar and ultrasound [3]. This means that a dielectric structure, however complex, scattering, and thick, can be described as a linear transformation of the field. This problem is also well known in mesoscopic physics, where this linear transformation is described by a matrix S, the scattering matrix, that connects all input field modes to all output field modes impinging and propagating away from the volume of interest. Here, when considering complex structures in the context of imaging, it is usually convenient to consider a slab geometry, where the medium has two sides, and one considers the light that impinges on one side of the medium and is collected on the other side. In this geometry, the quantity connecting a set of input modes to a set of output modes is the transmission matrix T, and is a subpart of the full scattering matrix of the system. Correspondingly, one can consider the reflection matrix R, connecting input and output modes on the same side of a medium. Here we will consider transmission matrices only, but most of the concepts and techniques can be immediately translated to the reflection geometry.

While transmission and scattering matrices have been a convenient way to describe wave propagation in complex media in mesoscopic physics, being able to describe important quantities such as total transmission, intensity statistics, etc., it has mainly been studied in this domain from a statistical point of view, i.e., to extract average quantities. Extracting the exact matrix of a given system was never even considered.

In this chapter, we will describe how, leveraging on spatial light modulator technologies and the new possibilities offered by digital holography, it is possible to experimentally measure this quantity in the optical domain, not in a statistical sense, but for a particular realization of disorder, i.e., a given system, which will be for most of the experiments described in this chapter as an ideal multiply scattering medium, essentially a layer of white paint, but could in principle be biological tissues. Once this information is known, the problem of recovering an image is not bound to be carried using ballistic photons. Indeed, even in the diffusive regime, the result of the propagation of a field can be deterministically predicted for scattered light. In particular, the optimal wavefront that will generate a focus on one or several output modes can be easily extracted from the matrix. We will discuss the various initial implementations of this concept and its applications for focusing and imaging. Recent developments have shown different ways of either simplifying the procedure or expanding its capabilities to new domains, for instance, the spectral domain.

5.2   Techniques

5.2.1   The Transmission Matrix

The basic idea behind the measurement of the transmission matrix of a given linear system is to send a series of light patterns as inputs, to record the response of the medium to it, and to determine from this set of input-output response pairs a best estimate of the transmission matrix of the medium. The range of inputs that can be generated and the amount of information that can be collected depends on the source and modulator and on the detector type, respectively. An important condition for this technique to be

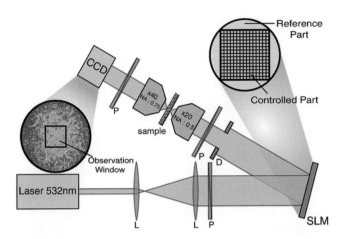

Figure 5.1 Schematic of the experiment of the first measurement of the transmission matrix in optics. The laser beam is expanded and reflected off a SLM. The phase-modulated beam is focused on the multiple-scattering sample and the output intensity speckle pattern is imaged by a CCD camera: lens (L), polarizer (P), diaphragm (D). Reprinted figure with permission from [4]. Copyright (2010) by the American Physical Society.

relevant, as is evident from the method above, is that a large number of sequential steps are required to acquire all this information, and the medium has by definition to be static during this time, since the transmission matrix is extremely sensitive to the exact realization of disorder. We will discuss this condition in more detail.

In a typical experiment, for instance, the one described in Figure 5.1, a laser is used to shine on the medium but is first shaped by means of a spatial light modulator device, and the transmitted light is collected on the far side of the medium on a CCD camera. A set of optics (microscope objectives, relay lenses, polarizers, mirrors, diaphragms) are also present on the optical path.

Let us first describe how a transmission matrix between a set of N pixels of the modulator and a set of M pixels of a detector can be measured, taking as an example the first experimental realization reported in [4] and shown in Figure 5.1. In practice, the modulator is often a spatial light modulator, and in most of the first experiments, these modulators were liquid-crystal-based and phase-only modulators, i.e., the phase of each pixel could be varied between 0 and 2π in discrete steps. Obtaining the transmission matrix means in practice measuring the relation between input fields and output fields, i.e., finding the complex coefficients t_{mn} linking the field on the nth pixel at the input to the mth pixel at the output:

$$E_{out} = \sum_m t_{mn} E_{in}$$

The transmission matrix $T = (t_{mn})$ is in practice not only the transmission matrix of the medium, but more precisely of the full optical system between the modulator and the detector, including all the optical elements (microscope objectives, polarizers, etc.). However, depending on the exact optical configuration, this can correspond to

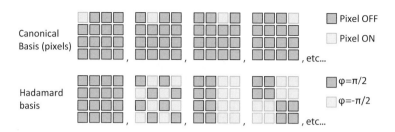

Figure 5.2 different basis of input modes can be used to learn the TM, depending on the type of Spatial light modulator used. Here two example, the canonical basis, and the Hadamard basis.

simple situations; for instance, if the modulator or the detector plane is conjugated to input or output surface of the material, then pixels correspond to spatial positions on the sample, while if the modulator or detector is in a Fourier plane, then pixels correspond approximately to k-vectors of the medium.

In order to measure the transmission matrix, a set of input wavefronts has to be displayed, and in order to accurately describe all possible inputs, this set of modes must constitute a complete orthonormal basis. For instance, the canonical basis would correspond to modulating individual pixels. With phase-only SLMs, the solution chosen in [4] was to use a different basis, the so-called Hadamard basis, that was ideally adapted to phase-only modulation (see Figure 5.2), although it is restricted to input dimensions of specific size, typically $N = 2^p$.

5.2.2　Determining the Output Field

For each of these input spatial modes, the corresponding output field must be determined on all output modes. However, in the optical domain, detectors are intensity detectors, meaning that amplitude and phase have to be indirectly accessed, using digital holography. This consists in interfering the unknown light beam to be measured with another known coherent beam that is used as a reference, the amplitude and phase of the unknown beam being extracted from the interference. A very convenient way to measure in a single image the field is a scheme called off-axis holography [5], where an angularly tilted plane wave from the same laser is used as a reference and produces spatial fringes. In the 2D-spatial Fourier transform of the intensity images, the interference term between reference and signal beam is spatially shifted from the 0th order, therefore the amplitude and phase information can be recovered from a single image (see Figure 5.3). Another possible technique is called phase-shifting holography [6]. It consists in having an on-axis reference plane wave, in this case the interference pattern can be obtained by appropriately combining several images (at least 3) while phase-shifting the relative phase between the reference and the signal beam. It therefore multiplies the measurement time by at least 3, but allows using the full resolution of the detector.

However, both techniques have a drawback, which is that it requires interferometric stability of an external reference beam over the whole measurement time, i.e., typically

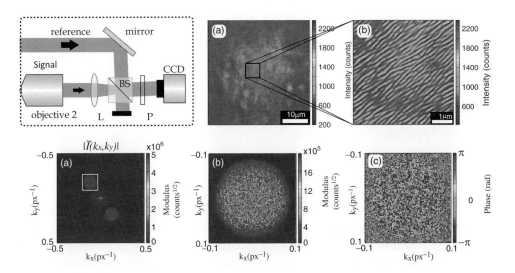

Figure 5.3 Scheme for off-axis holographic detection of a speckle field. (top) experimental setup and CCD image of the hologram (**a–b**). (bottom) (**a**) 2D spatial Fourier transform of the CCD image, the interference terms is spatially separated and the 1st diffraction order enclosed in the white square allow recovering the amplitude (**b**) and phase (**c**) of the speckle Adapted from Akbulut, PhD thesis, University Twente NL. A black and white version of this figure appears in some formats. For the color version, please refer to the plate section.

over minutes, which can be challenging experimentally. In [4], an alternative approach was proposed, that is basically phase-shifting interferometry, but where the reference beam is not an external plane wave that needs to be phase-stable during the measurement time but is instead a fraction of the input light that is left unmodulated by the SLM. It therefore produces on the far side a static reference speckle that can be used as a reference to extract the amplitude and phase of the modulated light. It corresponds basically to having a co-propagating interferometer. It has however the drawback that the reference speckle is not a plane wave with a well-defined amplitude and phase that can be factored out of the interference term, but a speckle. This is however not detrimental in many cases, be it for focusing, imaging, or modal analysis, and can be used advantageously in many practical situations. In principle, it is also possible to use an external reference plane wave to measure this reference in a single shot, and remove the ambiguity.

In terms of measurement time, the transmission matrix requires at least a set of input-output measurements equal to the number of input pixels N to record a basis. The speed of a CCD camera typically ranges in the 10–100 Hz range (but some kHz or above frame-rate cameras are available). But most phase-only liquid crystals modulators are typically in the 10–20 Hz range for proper operation, due to the inherent time it takes for the liquid crystal to reorientate, therefore the limiting factor is usually the modulation part. For instance, measuring a N=1024 TM was taking several minutes in [4]. Using galvanometric scanning mirrors, and exploiting a 500 Hz CCD, the measurement in 6 seconds of $N = 3000$ input modes was reported in [7]. However, such scanners do not allow generating arbitrary wavefronts, and an SLM had to be added to the system for the purpose of displaying well-defined wavefronts (see Figure 5.4).

5.3   Results

5.3.1   Focusing

Once a matrix has been acquired, and as long as the medium is stable, the transmission matrix accurately describes the propagation from the input modes (pixels or k-vectors) to the detector pixels. It is therefore possible to exploit this information to either (1) calculate the input mode that will produce a given output mode or (2) extract information on the input by measuring the output, both being relevant to imaging.

The first situations allows to calculate for instance the input wavefront that will focus on a given position. For this purpose, the optimal wavefront can be computed by phase conjugating the target field:

$$\mathbf{E_{in}} = \mathbf{T}^{\dagger}\mathbf{E_{target}}$$

However, due to the fact that the SLMs used in many experiments are phase-only modulators, it is not possible to exactly send this mode. Fortunately, it was demonstrated in acoustics [9] that the phase was the most important information, and the final result when focusing is equivalent to phase and amplitude control, with just a factor 2 penalty in signal-to-noise ratio. Therefore, for an input plane wave on the spatial light modulator, focusing can be very efficiently done using a phase-mask corresponding to taking the phase of the equation above, discarding the amplitude information. This solution corresponds in practice, when focusing to a single point, to putting the contributions from every pixel in phase at the corresponding position at the output, i.e., it is equivalent to the solution obtained after iterative optimization [1] (see also Chapter 8). Just like for iterative optimization, in the ideal case, for perfect multiply scattering medium, the efficiency of the focusing depends only on the number N of input spatial degrees of freedom (pixels), more precisely, the signal-to-noise of the focus to background intensity can be written as

$$\mathrm{SNR} = \frac{\pi}{4}N$$

where the average background intensity is not modified by the optimization process, a result that can be modified in the case of existence of mesoscopic speckle correlations [10]. It is also worth pointing out that although most experiments to date have exploited the transpose conjugate of the transmission matrix for focusing, other operations are also possible, depending on the goal.

5.3.2   Imaging and Inverse Problem

Beyond focusing, the knowledge of the transmission matrix can be exploited for imaging an object, i.e., to recover, from the output pattern E_{out} the unknown object pattern E_{obj}.

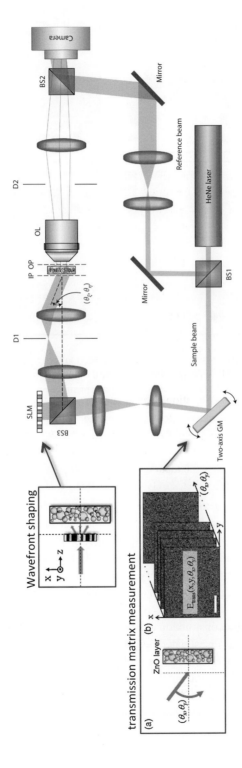

Figure 5.4 hybrid approach to TM measurement : two axis fast galvanometric scanning mirrors (GM) allow rapid scanning of incident k angles on the sample, and recording of the TM using off-axis holography using a fast CCD camera. A slow SLM can then be used to send complex wavefront on the sample, for imaging or mode injection. Adapted by permission from Macmillan Publishers Ltd: Nature Photonics [7], copyright 2012, and with permission from [8], copyright (2011) by the American Physical Society.

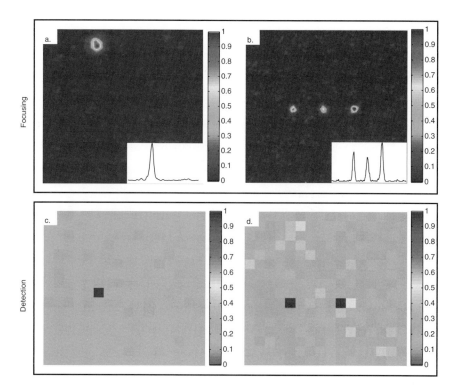

Figure 5.5 Experimental results of focusing and detection using phase conjugation. Results obtained are for N = M = 256. We show (**a**) typical output intensity pattern for one-spot target focusing and (**b**) multiple target focusing. We present similar results for the detection of (**c**) one point and (**d**) two points before the medium. The point objects are synthesized using the SLM. Reprinted figure with permission from [4]. Copyright (2010) by the American Physical Society. A black and white version of this figure appears in some formats. For the color version, please refer to the plate section.

However, the strategy to recover the object can be a bit more complex. The simplest strategy is to use time-reversal, i.e., to reconstruct the image field E_{img} by "virtual" phase conjugation, i.e., by computing $E_{img} = T^\dagger E_{out}$. Since $E_{out} = TE_{obj}$, it means the image and the original object are related via $E_{img} = T^\dagger TE_{obj}$. Experimental results are presented in Figure 5.5 for very simple input object corresponding to one or two input pixels lit. The reconstruction allows detecting the corresponding input pixels, but is relatively noisy.

More advanced reconstruction can be achieved, as was demonstrated in [8] and [11], where inversion was investigated, i.e., using the pseudo-inverse of T, noted T^{-1}, or regularized operators derived from the pseudo inverse, to take into account the noise in the matrix. In essence, even a small experimental noise in the measurement of the transmission matrix can impact dramatically the reconstruction, since the smallest singular values in T will be the dominant terms in T^{-1}. Figure 5.6 shows some imaging results.

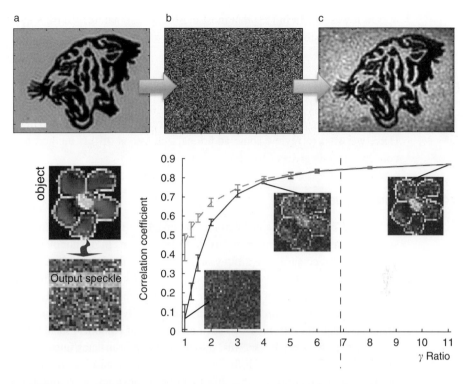

Figure 5.6 Image reconstruction by inverting the transmission matrix. Top (adapted from [8]): (**a**) original object, (**b**) transmitted speckle, and (**c**) reconstructed object using T^{-1}. Scale bar 10 μm. bottom (adapted from [11]) (left) original object and resulting speckle (right) reconstruction fidelity, as a function of the output to input pixel ratio $\gamma = M/N$, for two reconstruction operators : T^{-1} (continuous line) and Mean Square Operator MSO (dashed line) taking into account the experimental noise in the measurement of T. While both approaches are equivalent for high γ, for γ close to unity where noise impacts more dramatically the transmission channels, T^{-1} is unable to retrieve the object but MSO stills performs relatively well.

5.3.3   Mode Study

Finally, beyond imaging and focusing, the transmission matrix can also be exploited to study the transmission modes of the systems. It has been shown in [4] that the transmission matrix of a multiply scattering medium showed clear signatures of random matrices, in particular in the distribution of their singular value distribution. In addition, the transmission matrix also gives access to the input/output vectors corresponding to each of the transmission eigenchannels. This was for instance exploited in [12] to enhance the energy transmission between input and output modes by up to a factor 4. Also, it is known from mesoscopic theory that the transmission modes of a complex medium have some very interesting properties. These properties should be accessible within the matrix, however, one usually has access in practice to only a small fraction of the total input and output modes, due in particular to the finite numerical aperture of the optics. It was shown in [13] that most mesoscopic signatures vanish when measuring a partial transmission matrix.

5.4   Recent Developments

The transmission matrix concepts have been fruitfully applied to imaging and focusing. Here, we want to review some additional extensions of this method to new imaging concepts, new degrees of freedom, and new systems.

5.4.1   Compressive Imaging

In "simple" imaging reconstruction, the image is reconstructed from the measured speckle using a simple operator, as described above. In Tikhonov regularization, for instance, the reconstruction simply aims at finding the initial objects, minimizing the mean square error. However, there is no *a priori* information about the object.

The recent paradigm of compressive sensing has been introduced as a way to reconstruct a *sparse* or *compressible* object, with only few measurements (fewer than the Shannon-Nyquist limit) provided each measurement contains global information about the object. In mathematical terms, this requirement translates into a condition that the measurement basis is very different from the sparsity basis of the object [14]. Knowing the measurement basis, the object can be reconstructed. A complex medium, that strongly mixes the input wavefront, naturally provides an ideal playground for compressive sensing, as each local measurement at the output results from very complex projections at the input, due to the strong mixing due to the multiple scattering process. This was demonstrated in [15], and a result is shown in Figure 5.7.

5.4.2   Polarization-Resolved Transmission Matrix

The first transmission matrix measurements considered only one input polarization and one output polarization. However, propagation through a scattering material mixes polarizations, as always in a deterministic way. As a consequence, it can be interesting to measure and control not only a single polarization but both, therefore accessing a larger amount of control. Optimal control of both polarizations has been performed first in [10] but control and detection of both polarization, allowing access to the vectorial transmission matrix, has been demonstrated in [16], the corresponding matrix being of size $2N \times 2M$.

5.4.3   Spectrally Resolved Transmission Matrix

Another very interesting degree of freedom is the spectrum. Indeed, the whole treatment described above is fully monochromatic. However, it is well known that a complex medium has a spectrally varying response: the transmitted speckle depends on the wavelength. This spectral dependence can be linked to the traversal time of the medium τ_m.

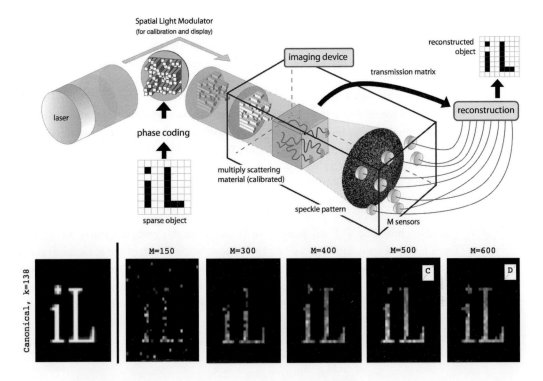

Figure 5.7 Principle of compressive sensing imaging using the transmission matrix of a complex medium (adapted from [15]). (top) Principle of the setup, exploiting only $M < N$ detectors; it is possible to reconstruct a sparse object. (bottom) Experimental reconstruction; the object has $N = 1020$ pixels but has only $k = 138$ nonzero coefficients. Reconstruction with M ranging from 150 to 600 shows clear sub-Nyquist reconstruction.

If spatially coherent but broadband light is incident on such a medium (such as from a SLED or from a femtosecond laser), then the resulting transmitted light can have a complex spectral behavior. When observing the transmitted light on a intensity detector, one will observe an intensity pattern which is the incoherent sum of several speckles for each independent speckle, i.e., a speckle with a lower contrast [17, 18]. It is however possible to measure the speckle as a function of wavelength and then to recover this spectral dependence [19]. In turn, this spectral dependence means that, for short pulses, the pulse will be stretched in time up to an average time τ_m. Optimization wavefront shaping has been shown to allow refocusing not only spatially but also in time [20, 21].

In order to access and control the spectral and temporal degrees of freedom by the matrix method, it is necessary to measure a temporally or spectrally resolved matrix. Time-gated matrices are described in Chapters 7 and 11; therefore we will focus our presentation here on spectrally resolved matrix measurement. These measurements have been presented in [22]. In this experiment, a Ti-Sa laser is used successively as a tunable CW source, then as a broadband source around 800 nm. In the first step, the medium spectral bandwidth is determined by scanning the wavelength without wavefront shaping, and measuring the spectral correlation of the transmitted speckle, which for the thin layer of ZnO powder used in the experiment was typically on the order of a nanometer, which corresponds to a dwell time on the order of a picosecond. This gives the spectral resolution with which the measurement must be performed to capture the full spectral response of the medium. In a second step, several monochromatic transmission matrices are measured, tuning the wavelength of the laser, and choosing the spectral resolution in accordance with the spectral resolution of the medium. In this way, a multispectral transmission matrix (MSTM) is acquired (see Figure 5.8).

Once the MSTM is known, it is possible to exploit it for spectral control of broadband light, over the range of wavelength that has been measured. For instance, is it possible to focus only a specific wavelength of the broadband light, by selecting a single monochromatic TM and using it for focusing. In this case, obviously, the resulting focus has a spectral bandwidth given by the correlation length of the medium. But more advanced functions are also possible. For instance, it is possible to compute the input wavefront that will focus different wavelength at several positions. This wavefront can simply be obtained by summing the complex amplitudes for each specific wavefront

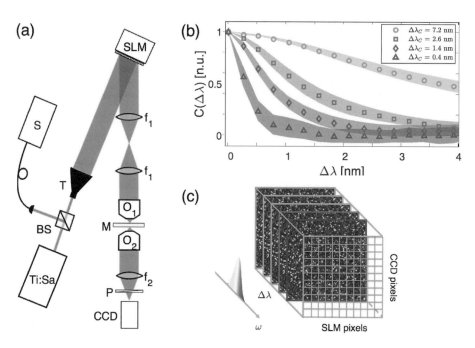

Figure 5.8 Measurement of a multispectral transmission matrix. (**a**) Experimental scheme (BS: beamsplitter, T: telescope, S: spectrometer, M: medium, P: polarizer). The setup is basically similar to [4], except the laser, which is a Ti-Sa laser that can be used either as a tunable laser or as a mode-locked femtosecond broadband source. (**b**) Spectral correlation of speckle transmitted through the medium for different thicknesses. (**c**) Representation of the MSTM, collection of several monochromatic transmission matrices, for several wavelengths spanning over the bandwidth of the Ti-Sa laser when operating in mode-locked regime. Adapted from [22].

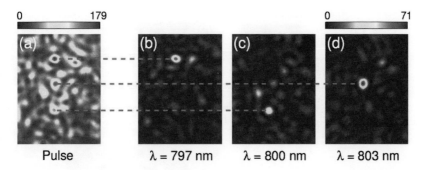

0   179   0   71

(a)   (b)   (c)   (d)

Pulse   λ = 797 nm   λ = 800 nm   λ = 803 nm

Figure 5.9 Spectral focusing using the MSTM. By displaying the phase mask corresponding to the sum of several phase-conjugate solutions focusing different spectral component at different positions, it is possible to disperse a broadband pulse in an arbitrary way. Left: A broadband pulse is focused in 3 spots. Right: By setting the laser in CW mode and scanning the wavelength, leaving the SLM mask untouched, one can observe that each focus corresponds to a different wavelength, effectively using the medium and SLM as a generalized reconfigurable grating. Adapted from [22]. A black and white version of this figure appears in some formats. For the color version, please refer to the plate section.

for each individual focus obtained by phase conjugation, then displaying this solution on the SLM (or simply displaying the phase of this sum, if the SLM is phase-only). The results are summarized on Figure 5.9. With this technique, temporal control is still elusive due to the fact that the relative phase between the spectral component remain unknown, but the measurement of a full MSTM with a reference would provide such a control.

5.4.4   Transmission Matrix of a Multimode Fiber

Multimode fibers (MMF) have been studied since the early days of wavefront shaping. The first studies have shown that it was possible to re-form a focus, be it via phase conjugation [23], optimization [24, 25], or digital phase conjugation [26, 27]. The knowledge of the wavefront that can focus light at any output position of the fiber is very close to being the transmission matrix information, except for the missing relative phase between output positions (with similarities with [4]). The transmission matrix measurement of a MMF was first demonstrated in [28]. The main result is shown in Figure 5.10. The transmission matrix is first measured as in [8]. In a second step, full field imaging is performed from the proximal end only, by exploiting the transmission matrix in both ways: first, by calculating the resulting illumination of the object from an input wavefront, and second to recover the object reflectivity from the retroreflected light. As shown in Figure 5.10, a single step does not allow one to reconstruct the object, because the illumination is speckly, due to the scrambling during propagation in the fiber of the input light. However, excellent reconstruction is achieved when averaging over several illumination patterns.

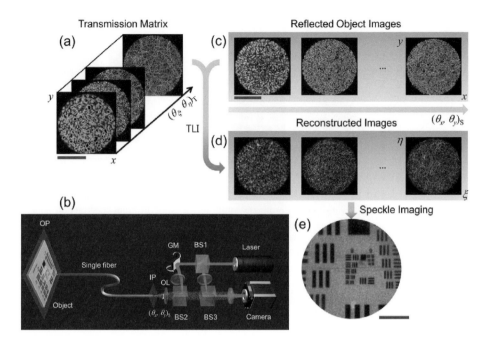

Figure 5.10 Measuring and exploiting the transmission matrix of a multimode fiber (from [28]).
(**a**) Transmitted pattern for different input angles, from which the transmission matrix is
obtained. (**b**) Setup for imaging; the input angles and the reflected patterns are both measured
from the proximal end. (**c**) and (**d**) Reflected patterns and reconstructed speckle images. (**e**)
Averaging all reconstructed images to recover the object.

5.4.5   New Measurement Techniques

In most implementations, the TM measurement has been performed using liquid crystal phase-only SLMs and requires digital holography for field measurement, which is
respectively slow and cumbersome to implement. However, several new exciting results
have shown that some of these constraints could be softened.

First, digital micromirror devices, which are binary amplitude modulators, have
been shown to allow efficient matrix reconstruction by direct binary modulation [29],
although not at the full speed (23 kHz) allowed by the modulator. However, this
approach still used off-axis holography and a reference arm to infer the transmission
matrix. In [30], this detection scheme was hugely simplified by showing that the transmission matrix could be inferred from intensity-only measurements, i.e., without the
need for a reference (neither external nor co-propagating), therefore simplifying the
setup to its simplest form (Figure 5.11).

5.5   Future Directions

We have described in this chapter how a transmission matrix of a complex medium can
be measured and exploited for imaging applications. This concept had also numerous

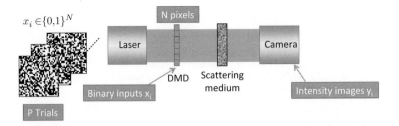

Figure 5.11 Principle of phase-retrieval TM measurement (adapted from [30]). A set of P input binary patterns is projected on the medium, and the resulting intensities y_i are recorded. Reconstructing the transmission matrix T means solving $y_i = |Tx_i|$ for all i, which is a phase-retrieval problem.

other applications, in particular as a fundamental tool to study mesoscopic light propagation in complex media. It has been extended to image "inside" rather than "through" a medium, by using the photoacoustic effect (see [31] and Chapter 7). Due to the inherently large number of measurement steps, the transmission matrix (and the optimization approaches as well) might not be the most straightforward tool for rapid imaging in biological media (which always have a limited stability time), compared to other approaches developed in Chapters 3, 9, and 12. It remains nonetheless a priceless tool to better understand light propagation in complex media. The recent emergence of the memory effect [32, 33, 34], by allowing scanning a focus without recalculating the wavefront, provides an interesting extension to minimize the measurement time. Novel SLM technologies, such as DMD, or fast MEMS, or simplifications of the techniques as described in [30], could also help the penetration of the technique for imaging applications in the biomedical world.

5.6   References

[1] Vellekoop IM, Mosk AP. Focusing coherent light through opaque strongly scattering media. Optics Letters. 2007;32(16):2309.

[2] Cui M, Yang C. Implementation of a digital optical phase conjugation system and its application to study the robustness of turbidity suppression by phase conjugation. Optics Express. 2010 Feb;18(4):3444–3455.

[3] Fink M. Time Reversed acoustics. Physics Today. 1997;50(3):34.

[4] Popoff SM, Lerosey G, Carminati R, Fink M, Boccara AC, Gigan S. Measuring the transmission matrix in optics: an approach to the study and control of light propagation in disordered media. Physical Review Letters. 2010 Mar;104(10):100601.

[5] Cuche E, Marquet P, Depeursinge C. Spatial filtering for zero-order and twin-image elimination in digital off-axis holography. Applied Optics. 2000 Aug;39(23):4070–4075.

[6] Yamaguchi I, Zhang T. Phase-shifting digital holography. Optics Letters. 1997 Aug;22(16):1268–1270.

[7] Kim M, Choi Y, Yoon C, Choi W, Kim J, Park QH, et al. Maximal energy transport through disordered media with the implementation of transmission eigenchannels. Nature Photonics. 2012 Sep;6(9):581–585.

[8] Choi Y, Yang TD, Fang-Yen C, Kang P, Lee KJ, Dasari RR, et al. Overcoming the diffraction limit using multiple light scattering in a highly disordered medium. Physical Review Letters. 2011 Jul;107(2):023902.

[9] Derode A, Tourin A, Fink M. Random multiple scattering of ultrasound.I.Coherent and ballistic waves. Physical Review E. 2001 Aug;64(3):036605.

[10] Vellekoop IM, Mosk AP. Universal optimal transmission of light through disordered materials. Physical Review Letters. 2008 Sep;101(12):120601.

[11] Popoff S, Lerosey G, Fink M, Boccara AC, Gigan S. Image transmission through an opaque material. Nature Communications. 2010 Sep;1:81.

[12] Choi W, Mosk AP, Park QH, Choi W. Transmission eigenchannels in a disordered medium. Physical Review B. 2011 Apr;83(13):134207.

[13] Goetschy A, Stone AD. Filtering random matrices: the effect of incomplete channel control in multiple scattering. Physical Review Letters. 2013 Aug;111(6):063901.

[14] Donoho DL. Compressed sensing. IEEE Transactions on Information Theory. 2006 Apr;52(4):1289–1306.

[15] Liutkus A, Martina D, Popoff S, Chardon G, Katz O, Lerosey G, et al. Imaging with nature: compressive imaging using a multiply scattering medium. Scientific Reports. 2014 Jul;4:5552.

[16] Tripathi S, Paxman R, Bifano T, Toussaint KC Jr. Vector transmission matrix for the polarization behavior of light propagation in highly scattering media. Optics Express. 2012 Jul;20(14):16067–16076.

[17] Goodman JW. Some fundamental properties of speckle. Journal of the Optical Society of America. 1976 Nov;66(11):1145–1150.

[18] Curry N, Bondareff P, Leclercq M, van Hulst NF, Sapienza R, Gigan S, et al. Direct determination of diffusion properties of random media from speckle contrast. Optics Letters. 2011 Sep;36(17):3332–3334.

[19] McCabe DJ, Tajalli A, Austin DR, Bondareff P, Walmsley IA, Gigan S, et al. Spatio-temporal focusing of an ultrafast pulse through a multiply scattering medium. Nature Communications. 2011 Aug;2:447.

[20] Katz O, Small E, Bromberg Y, Silberberg Y. Focusing and compression of ultrashort pulses through scattering media. Nature Photonics. 2011 May;5(6):372–377.

[21] Aulbach J, Gjonaj B, Johnson PM, Mosk AP, Lagendijk A. Control of light transmission through opaque scattering media in space and time. Physical Review Letters. 2011;106(10):103901.

[22] Andreoli D, Volpe G, Popoff S, Katz O, Gresillon S, Gigan S. Deterministic control of broadband light through a multiply scattering medium via the multispectral transmission matrix. Scientific Reports. 2015 May;5.

[23] Yariv A. On transmission and recovery of three-dimensional image information in optical waveguides. Journal of the Optical Society of America. 1976 Apr;66(4):301–306.

[24] Bianchi S, Di Leonardo R. A multi-mode fiber probe for holographic micromanipulation and microscopy. Lab on a Chip. 2012;12(3):635.

[25] imr T, Dholakia K. Shaping the light transmission through a multimode optical fibre: complex transformation analysis and applications in biophotonics. Optics Express. 2011 Sep;19(20):18871–18884.

[26] Papadopoulos IN, Farahi S, Moser C, Psaltis D. Focusing and scanning light through a multimode optical fiber using digital phase conjugation. Optics Express. 2012 Apr;20(10):10583.

[27] Papadopoulos IN, Farahi S, Moser C, Psaltis D. High-resolution, lensless endoscope based on digital scanning through a multimode optical fiber. Biomedical Optics Express. 2013 Jan;4(2):260–270.

[28] Choi Y, Yoon C, Kim M, Yang TD, Fang-Yen C, Dasari RR, et al. Scanner-free and wide-field endoscopic imaging by using a single multimode optical fiber. Physical Review Letters. 2012 Nov;109(20):203901.

[29] Kim D, Choi W, Kim M, Moon J, Seo K, Ju S, et al. Implementing transmission eigen-channels of disordered media by a binary-control digital micromirror device. Optics Communications. 2014 Nov;330:35–39.

[30] Dremeau A, Liutkus A, Martina D, Katz O, Schalke C, Krzakala F, et al. Reference-less measurement of the transmission matrix of a highly scattering material using a DMD and phase retrieval techniques. Optics Express. 2015 May;23(9):11898.

[31] Chaigne T, Katz O, Boccara AC, Fink M, Bossy E, Gigan S. Controlling light in scattering media non-invasively using the photoacoustic transmission matrix. Nature Photonics. 2014 Jan;8(1):58–64.

[32] Vellekoop IM, Aegerter CM. Scattered light fluorescence microscopy: imaging through turbid layers. Optics Letters. 2010 Apr;35(8):1245–1247.

[33] Katz O, Heidmann P, Fink M, Gigan S. Non-invasive single-shot imaging through scattering layers and around corners via speckle correlations. Nature Photonics. 2014 Oct;8(10):784–790.

[34] Schott S, Bertolotti J, Leger JF, Bourdieu L, Gigan S. Characterization of the angular memory effect of scattered light in biological tissues. Optics Express. 2015 May;23(10):13505.

6 Coupling Optical Wavefront Shaping and Photoacoustics

Emmanuel Bossy

6.1 Introduction

In this chapter, we introduce and review recent research works based on coupling optical wavefront shaping and photoacoustics. Coupling these two research fields, which have until recently developed rather independently, offers mutual advantages for both fields: on the one hand, the photocoustic effect provides a powerful sensing mechanism for optical wavefront shaping techniques, while on the other hand, photoacoustic imaging can greatly benefit from optical wavefront shaping techniques. This chapter first introduces the principle of photoacoustics relevant in the context of this chapter, including an introduction to the photoacoustic effect and a brief overview of the principles of biomedical photoacoustic imaging. We then first review the recent works on photoacoustic-guided optical wavefront shaping, either with optimization or transmission-matrix approaches, and finally illustrate how optical wavefront shaping can be applied to develop minimally invasive photoacoustic microendoscopy mith multimode fibers. Some parts of this chapter have been adapted from a recent review article on coupling photoacoustics and coherent light [1].

6.2 Principles of Photacoustics

6.2.1 The Photoacoustic Effect

The photoacoustic (also called optoacoustic) effect consists in the conversion of light absorption into acoustic emission, via thermo-elastic stress-generation. Historically, it was first discovered and investigated in 1880 by Alexander Graham Bell who experimented with long-distance sound transmission [2]. The photoacoustic effect was exploited first in the field of optical absorption spectroscopy, in gas [3] and condensed matter [4, 5], and was later introduced for biomedical applications in the mid 1990s [6, 7, 8]. In the context of photoacoustic imaging of soft biological tissue, one of the simplest and widely used theoretical description of the photoacoustic effect can be summarized by the following equation [9, 10]:

$$\left[\frac{\partial^2}{\partial t^2} - c_s^2 \nabla^2 \right] p(\mathbf{r}, t) = \Gamma \mu_a(\mathbf{r}) \frac{\partial I}{\partial t}(\mathbf{r}, t) \tag{6.1}$$

where $p(\mathbf{r}, t)$ is the photoacoustic pressure field and $I(\mathbf{r}, t)$ is the optical intensity. $\Gamma = \frac{\beta c_s^2}{\rho}$ (with β the thermal expansion coefficient, ρ the mass density and c_s the speed of sound) is the dimensionless Grüneisen coefficient, which typical value in tissue at physiological temperature is 0.2. The Grüneisen coefficient reflects the stress generation produced by the temperature rise following nonradiative absorption by optical absorbers. The message carried by Eq. 6.1 is that temporally varying light absorption induces acoustic waves, therefore called photoacoustic waves. As a consequence, the photoacoustic effect allows sensing optical absorption and/or optical intensity via sound detection. In particular, for acoustically transparent media such as soft biological tissue, it provides as powerful tool to image optical absorption, as now widely exploited in photoacoustic imaging of soft biological tissue.

6.2.2   Photoacoustic Imaging

In the past decade, photoacoustic imaging has been one of the fastest growing biomedical imaging modality [10, 11, 12, 13, 14]. As mentioned above, photoacoustic imaging relies on the photoacoustic effect to couple light absorption to acoustic detection, and therefore provides image of optical absorption. The general principle of photoacoustic imaging is the following: the sample to be imaged is illuminated by pulsed light (for most implementations), and acoustic waves generated from illuminated absorbing regions (generally in the ultrasound frequency range) are detected by acoustic sensors. The contrast provided by photoacoustic imaging is therefore optical in essence.

The resolution of photoacoustic imaging in biological tissue depends mostly on the targeted imaging depth. Photoacoustic imaging of biological tissue was first proposed in the 1990s to overcome the loss of optical resolution caused by multiple scattering of light [6, 7, 8]. Because biological tissue are transparent (nonscattering or very weakly scattering) to ultrasound, as opposed to light, one can use photoacoustic waves to provide images of optical absorption with the resolution of ultrasound. In this regime where the illumination is spatially distributed over the sample to be imaged (usually because of multiple scattering of light), the resolution is dictated by the ultrasound detection and photoacoustic imaging is thus qualified as *acoustic-resolution* photoacoustic imaging. Ultimately, the resolution and imaging depth of acoustic-resolution photoacoustic imaging is limited by the frequency-dependent absorption of ultrasound in tissue (on the order of 0.5 dB/cm/MHz), with a typical depth-to-resolution ratio of about 100 [11, 13]. Figure 6.1(a) shows an acoustic-resolution photoacoustic image of the skin vasculature of the abdomen of a mouse, obtained noninvasively *in vivo* with a typical resolution around 100 μm a few mm deep in tissue.

As opposed to acoustic-resolution photoacoustic imaging, *optical-resolution* photoacoustic imaging (often called OR-PAM for optical-resolution photoacoustic microscopy) usually relies on the ability to focus light, while the detection of ultrasound has no relevant spatial resolution. In OR-PAM, it is the size of the optical focus spot, which is usually raster-scanned across the sample, which dictates the

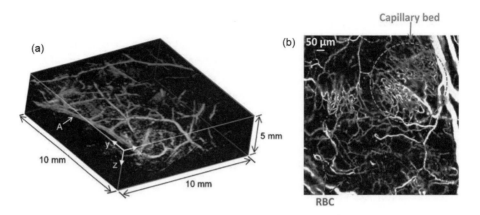

Figure 6.1 (**a**) Example of *acoustic-resolution* photoacoustic imaging of blood vasculature *in vivo* (vasculature in the skin around the abdomen of a mouse). The resolution is of the order of 100 μm and the imaging depth is of the order of a few mm. Reprinted with permission from [15], IOP. (**b**) Example of *optical-resolution* photoacoustic imaging of blood vasculature *in vivo* (mouse ear). The resolution is of the order of a few μm and the imaging depth is of the order of a few hundred μm. Reprinted with permission from [16], *Optics Letters*.

resolution [17, 16, 13]. OR-PAM is therefore limited for noninvasive imaging to shallow depths for which light still propagates ballistically (depths of typically a few hundred microns). OR-PAM is quite similar to other purely optical scanning-based microscopy techniques based on focused illumination (2-photon microscopy and confocal microscopy for instance), except that it is specifically sensitive to optical absorption (as opposed to scattering). Figure 6.1(b) shows an example of optical-resolution photoacoustic image obtained noninvasively *in vivo* on a mouse ear. Optical-resolution photoacoustic imaging may also be performed at large depth at the cost of invasiveness, with endoscopic approaches [18, 19, 20].

6.2.3   Rationale for Coupling Photoacoustics and Optical Wavefront Shaping

As illustrated in many chapters of this book, optical wavefront shaping generally requires a measurement of the optical intensity, either for optimization or to measure an optical transmission matrix. As introduced above, the photoacoustic effect provides a means to sense light intensity at optical absorbers. Because soft biological tissues are mostly transparent to ultrasound, photoacoustic sensing/imaging provides a way to noninvasively sense light intensity via remote detection of ultrasound, and thus a way to perfom photoacoustic-guided wavefront shaping, as illustrated in Section 6.3. While photoacoustics can therefore be exploited to perform optical wavefront shaping, photoacoustic imaging may also benefit from optical wavefront shaping techniques: Section 6.4 illustrates how optical wavefront shaping through multimode optical fibers can be used to develop new minimally invasive photoacoustic microendoscopes, building on the recent developments in optical endoscopy with multimode fibers.

6.3   Photoacoustic-Guided Optical Wavefront Shaping

All implementations of optical wavefront shaping require some feedback signal from the targeted region. A feedback mechanism for optical wavefront shaping should provide some sensing of the optical intensity. Appropriate mechanisms include direct intensity measurement with a camera or optical detector, or the use of some "guide star" following the approach in adaptive optics for astronomy. While the use of a camera or detector limits wavefront shaping toward the region *outside* the scattering media [21, 22, 23, 24], the "guide star" approach may be implemented *inside* a scattering sample. Fluorescent or second-harmonic "guide stars" have been successfully investigated as feedback mechanisms [25, 26] (see also Chapter 8), but these approaches, in addition to being invasive, only allow focusing in the vicinity of a single static target. Ultrasound tagging via the acousto-optic effect is a promising approach that offers dynamic and flexible control, which has been the subject of several recent investigations [27, 28, 29, 30, 31, 32, 33] (see also Chapter 12). In this approach, the selective detection of tagged photons followed by phase-conjugation allows to backpropagate light toward the ultrasound focus where the tagging via acousto-optic modulation occured. This has the advantage to allow single shot digital phase conjugation, i.e., finding the optimal wavefront to refocus on the guide-star without a long learning process (like optimization or transmission matrix), and therefore refocusing in a single refresh frame of the spatial light modulator. This was, for instance, demonstrated by Liu and coworkers who demonstrated focusing in tissues with 5.6 ms decorrelation time [33]. Although this approach is in principle compatible with *in vivo* imaging, the activation of a local guide star by acoustic tagging is limited to a single ultrasound focal zone, and scanning is required to focused light at various directions, requiring in turns long acquisition times.

As introduced in Section 6.2, the photoacoustic effect is sensitive to the absorption of optical energy, and therefore provides a mechanism to sense both the optical absorption and the optical intensity inside multiple scattering media. Based on its sensitivity to optical absorption, photoacoustic-guided wavefront shaping was first investigated for *ultrasound* wavefront shaping, to focus acoustic waves toward optical absorbers with time-reversal approaches [34, 35]. Photoacoustic-guided wavefront shaping was later applied in the context of acousto-optic imaging in the presence of acoustic aberrations [36]. In 2011, Kong and coworkers first demonstrated the use of the photoacoustic effect as a feedback mechanism for *optical* wavefront shaping [37], triggering significant research efforts toward photoacoustic-guided optical wavefront shaping. Analogous to wavefront shaping with the other feedback mechanisms introduced above, two main approaches, optimization or transmission matrix, have been used to implement photoacoustic-guided optical wavefront shaping (PA-WFS).

6.3.1   Optimization-Based Approach

In the first demonstration by I. Vellekoop and A. Mosk of optical focusing through a scattering slab with optical wavefront shaping [21], an optimization-based approach was

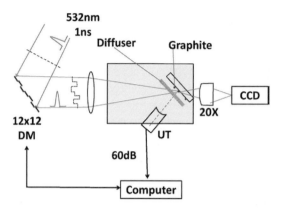

Figure 6.2 Experimental setup used by Kong and coworkers to demonstrate optical wavefront shaping with a deformable mirror (DM) through a scattering media with photoacoustic feedback [37]. A glass slide covered with absorbing graphite particles was placed behind the scattering layer, and a high-frequency ultrasound transducer (UT) was focused on the absorbing slide to measure the photoacoustic signal from its focal region. The photoacoustic signal was used as a feedback signal for the optimization procedure driving the DM. The CCD camera was only used here to verify the light intensity distribution on the absorbing layer after the optimization. Reprinted with permission from [37], *Optics Letters*.

used that relied on the measurement of light intensity with a CCD camera (see Chapter 8 for more details). A few years after this pioneer work, Kong and colleagues reproduced a similar experiment based on a photoacoustic-guided optimization approach [37], as illustrated in Figure 6.2. The target plane consisted of a glass layer covered with graphite particles, placed behind the scattering layer. Different concentrations and types of absorbers were used to demonstrate photoacoustic-guided wavefront shaping: the authors first demonstrated optical tracking and focusing toward the 41 μm-diameter focal zone of a 75 MHz ultrasound transducer with a homogeneously absorbing layer, in clear water. Experiments with single microparticles (10 μm or 50 μm in diameter) isolated within the 90-μm-diameter focal zone of a 40 MHz ultrasound transducer confirmed that the enhancement of the optimized photoacoustic signal decreased with the number of optical speckle grains (with grain size about 2 μm) within the absorbing target, in qualitative agreement with what is predicted for the optical enhancement factor. Typical enhancement for the photoacoustic signal ranged from 5 to 10, with the larger enhancements observed for the smallest particles.

This first work was rapidly followed by several investigations of photoacoustic-guided optical wavefront shaping [38, 39, 40, 41, 42, 43, 44]. In the works discussed below, the experimental setups are similar to that introduced by Kong and coworkers: in particular, photoacoustic feedback signals are measured from speckle patterns produced in a free-space geometry after propagation *through* a scattering sample. Importantly, the size of the speckle grains is systematically adjusted to match the typical dimension of the ultrasound focus by setting the distance between the scattering sample and the measurement plane on the properties of optical speckle patterns. The spatial light modulators or

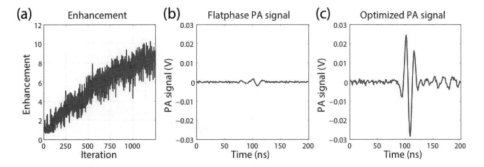

Figure 6.3 Illustration of the photoacoustic signal enhancement obtained with optimization-based photoacoustic-guided optical wavefront shaping. (**a**) Evolution of the photoacoustic enhancement with the optimization process, based on a genetic algorithm. (**b**) Photoacoustic signal prior to wavefront shaping. (**c**) Enhanced photoacoustic signal obtained for the optimal input wavefront. Reprinted with permission from [38], *Optics Express*.

deformable mirrors used to perform wavefront shaping were used in a reflection configuration, as in Figure 6.2. Following the approach proposed in [37], Caravaca-Aguirre and coworkers used a genetic algorithm to perform PA-WFS and enhance the light intensity behind a scattering layer by one order of magnitude [38], as illustrated in Figure 6.3. Chaigne and coworkers further demonstrated that the large bandwidth of photoacoustic signals could be exploited in the frequency domain to adjust the dimensions of the photoacoustic focal zone [39]. By iterative optimization of the highest frequency components (55–70 MHz band) of the broadband photoacoustic signals measured with a transducer with central frequency 27 MHz, the authors obtained a photoacoustic enhancement factor of about ×12, higher than the enhancement obtained with optimization in lower frequency bands (ranging from ×4 to ×8) or from peak-to-peak amplitude measurements (×8), as illustrated in Fig. 6.4. To maximize the sensitivity of photoacoustic measurement to phase modulation of the light beam, the optimization algorithm used a Hadamard basis vectors as the basis for the input modes (instead of the canonical pixel basis) of 140-element deformable mirror. Moreover, by simultaneously monitoring the evolution of the speckle pattern during the optimization process, it was confirmed experimentally that the optimization with the highest photoacoustic frequencies lead to a tighter optical focus than what was obtained by optimization with the lower frequency components [39].

A key advantage of the photoacoustic effect as a feedback mechanism is that the sensing may be performed simultaneously over the whole measurement volume, by use of imaging ultrasound arrays. With a spherical matrix array of 256 piezoelectric transducers, Deán-Ben and coworkers demonstrated photoacoustic-guided optical wavefront shaping by optimizing photoacoustic signals from selected targets of a 3D photoacoustic image, by means of a genetic algorithm [44], as illustrated on Figure 6.5. PA-WFS is usually limited in speed by either the laser pulse repetition frequency or the refresh rate of the adaptive optics device. In the context of photoacoustic flowmetry, Tay and coworkers investigated the potential of digital micromirror devices (DMD), which are

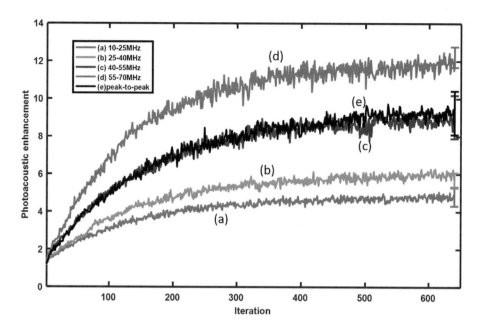

Figure 6.4 Evolution of the photoacoustic enhancement factor by optical wavefront shaping
during the optimization process, with different feedback values. In addition to the usual
peak-to-peak photoacoustic amplitude as feedback signal, RMS values over several frequency
bands computed from a Fourier analysis were used as alternative photoacoustic feedback values.
For each feedback quantity, the photoacoustic enhancement factor was computed by normalizing
the optimized quantity by its value under homogeneous illumination. Reprinted with permission
from [39], *Optics Letters*.

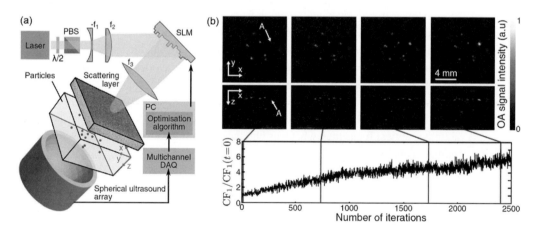

Figure 6.5 3D photoacoustic-guided optical wavefront shaping. (**a**) Experimental setup, based on
a spherical matrix array of 256 piezoelectric transducers. (**b**) Illustration of photoacoustic signal
enhancement toward a selected absorber (indicated by arrow A), as a function of the number of
iterations in the genetic algorithm used to perform optical wavefront shaping. Reprinted with
permission from [44], *Optics Letters*.

binary amplitude modulators, toward rapid PA-WFS [43]: a combination of Hadamard multiplexing with multiple binary-amplitude illumination patterns was implemented to perform wavefront shaping based on the photoacoustic signal measured with a 10 MHz spherically focused transducer, and an intensity enhancement of a factor 14 was obtained. Although the DMD refresh rate was as high as 22 kHz, the optimization approach remained very long (typically two hours) because of a SNR issue. This study, however, demonstrated the potential of using DMD for PA-WFS.

One specific feature of photoacoustic sensing for optical wavefront shaping arises from the possibility to create an optical focus smaller than the ultrasound resolution [45, 46] (see Figure 6.6), thus opening the possibility for superresolution photoacoustic imaging. When several optical speckle grains are present within the ultrasound resolution spot, the feedback signal mixes the information coming from individual speckles. However, based on the nonuniform spatial sensitivity across the ultrasound focal region, it has been shown that the spatially non-uniform photoacoustic feedback tends to localize the optimized optical intensity to a single speckle smaller than the acoustic focus, by preferentially weighting the single optical speckle closest to the center of the ultrasound focus during the optimization (see Figure 6.6a). As a consequence, an optical enhancement factor of 24 was reported for the optimized optical grain, about three times higher than the photoacoustic enhancement factor which averages the optical enhancement over all the optical speckles present in the focal spot [45]. While this effect was first reported in the context of linear photoacoustics, where the photoacoustic amplitude is proportional to the absorbed optical intensity as described by Eq. 6.1, Lai and coworkers introduced a nonlinear PA-WFS scheme with a dual-pulse illumination approach [46]. In short, this approach exploits the change in photoacoustic conversion efficiency between two consecutive intense illuminations to produce a feedback signal that is nonlinearly related the optical intensity: the first illumination pulse creates a photoacoustic signal that is linearly related to the optical intensity, but also changes the value of the Grüneisen coefficient Γ involved in the second illumination pulse. The change in the Grüneisen coefficient is caused by the temperature increase that follows the first illumination pulse [47, 46]. As a consequence, the feedback signal defined as the difference of the photoacoustic amplitudes of the two consecutive pulses varies nonlinearly with the optical intensity. As a result, optimization based on such a nonlinear feedback signal strongly favors focusing toward a single optical speckle grain rather than distributing the optical intensity evenly over all the speckle grains inside the acoustic focus spot. This effect had first been demonstrated with optical wavefront shaping based on nonlinear feedback from two-photon fluorescence [48, 49]. With nonlinear PA-WFS, Lai and coworkers achieved focusing to a single optical speckle grain 10 times smaller than the acoustic focus, with an optical intensity enhancement factor of $\sim$6000 and a photoacoustic enhancement factor of $\sim$60 (see Figure 6.6b).

6.3.2   The Photoacoustic-Transmission Matrix

As opposed to the approach based on optimization, which requires optimizing for each targeted point, the transmission-matrix approach is based on a prior calibration step

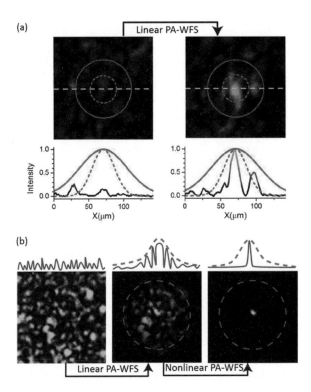

Figure 6.6 Illustration of subacoustic optical focusing with photoacoustic-guided wavefront shaping with homogeneously absorbing samples, adapted from [45] and [46]. (**a**) The circles show the approximate filtered transducer focal region (80 MHz, –6 dB, dashed line) and focal spot size at the frequency peak of the detected photoacoustic response (50 MHz, –6 dB, solid line). Left: optical speckle field (intensity) without optimized wavefront. Right: optical focus (intensity) generated by the optimized wavefront. The authors proposed that the subacoustic optical focusing is achieved thanks to the nonuniform spatial response of the ultrasound transducer that would favor optical modes at the center [45]. Adapted by permission from Macmillan Publishers Ltd: Nature Communications [45], copyright 2015. (**b**) By using nonlinear photoacoustic-guided wavefront shaping, Lai and coworkers performed subacoustic optical focusing with a final optical enhancement factor of ~ 6000 [46]. Linear PA-WFS first provided focusing with an enhancement of ~ 60, and subsequent nonlinear PA-WFS provided an additional factor of ~ 100. Adapted by permission from Macmillan Publishers Ltd: Nature Photonics [45], copyright 2015. Figure (**a**) adapted with permission from [45], 2015 NPG.

to measure the transmission matrix, before the matrix can be used to compute inputs (typically phase masks) corresponding to any desired output patterns. A detailed presentation of the transmission-matrix is given in Chapter 5. In this section, we present the photoacoustic-transmission matrix [41, 40], which was introduced for photoacoustic-guided wavefront shaping soon after the first demonstrations based on optimization approaches. In short, the photoacoustic-transmission matrix approach uses a photoacoustic imaging system to measure optical intensity instead of an optical detector (usually a CMOS or CCD camera), thanks to the fact that the amplitude of photoacoustic waves is proportional to the optical intensity at the absorbers (cf. Eq.6.1).

As for optical measurement of the transmission matrix, a phase-shifting inter-ferometric approach can be used to retrieve the optical field amplitudes from the photoacoustic measurement of the optical intensity (see Chapter 5, Section 5.2.2), by modulating the light intensity. The modulation of optically absorbed light turns into a modulation of the photoacoustic signal, which corresponds to the sum of acoustic emissions of all absorbers contained in the probed resolution cell volume. As in the all-optical transmission-matrix measurement, thanks to the linearity of the photoacous-tic generation process, the photoacoustic signal from each ultrasound resolution cell is cosine-modulated when the input modes are phase-shifted, and the photoacoustic trans-mission matrix can be retrieved from the measurement of the amplitude and phase of such modulation [41].

The photoacoustic transmission matrix [41] approach was first implemented with the time-resolved photoacoustic signal from a single-element transducer (i.e., a 1D pho-toacoustic image), with a time-to-space conversion simply derived by scaling time to distance by the known speed-of-sound in the medium. In this case, each output mode of the photoacoustic transmission matrix is defined by a temporal slot on the photoa-coustic signal (equivalently a pixel of the 1D photoacoustic image). By measuring the amplitude modulation of the photoacoustic signal in each of this time slot, when phase shifting each input mode (SLM pixel, or any vector in the input basis chosen to describe the SLM array), one can retrieve the photoacoustic transmission matrix, as illustrated on Figure 6.7. Once the photoacoustic transmission matrix is measured, it is then pos-sible as for the all-optical approach to appropriately combine input modes on the SLM to selectively enhanced the photoacoustic signal at any desired location on the photoa-coustic trace, as illustrated by Figure 6.8. This first proof-of-principle demonstration with 1D photoacoustic signals was quickly followed by the measurement of a 2D pho-toacoustic transmission matrix [40]. The use of a linear ultrasound array provided 2D photoacoustic images (reconstructed from individual photoacoustic time traces by use of a conventional beam forming approach), whose reconstructed pixels were used to define the output modes of the photoacoustic transmission matrix. Figure 6.9 illustrates how light can be focused toward a target zone (red box) of the conventional photoacoustic image.

There are several differences between the photoacoustic transmission matrix and its all-optical counterpart. One main difference is then that, unlike for an optical camera, the resolution of the photoacoustic camera is usually not able to resolve a single optical mode (i.e., speckle grain). In such a case, the photoacoustic signal amplitude from a given position results from several optical modes within the ultrasound resolution cell (determined by the numerical aperture and the frequency response of the acoustic trans-ducer). This situation is equivalent to that with an optical camera for which each pixel averages several speckles grains. When the targeted area contains several speckle grains, the global effect of a phase modulation of a single input mode is decreased compared to that obtained for a single speckle grain, as the phases on each speckle grain are uncor-related. As a consequence, the possibility to detect intensity modulation in the target region depends on the signal-to-noise ratio and decreases with the number of indepen-dent speckle grains in the detection area. Because the frequency-dependent attenuation

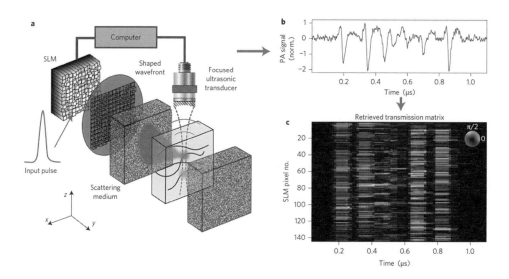

Figure 6.7 Illustration of the 1D photoacoustic-transmission matrix approach. (**a**) Schematic of the experimental setup: a spatial light modulator (SLM) is used to shape the optical wavefront from a pulsed laser. The absorbing sample is located behind a scattering medium. A spherically focused single-element ultrasound transducer is used to collect photoacoustic signals for each input mode on the SLM. (**b**) A typical photoacoustic signal. Each absorber appears on the signal as a bipolar pulse, whose amplitude is proportional to the amount of light at the absorber. In this case, where time can be directly converted to space via the known speed of sound, one temporal slot on the photoacoustic signal defines one output mode of the photoacoustic transmission matrix, corresponding to one position along the ultrasound beam. (**c**) Phase-shifting holography approaches, as described in Chapter 5, Section 5.2.2, and measurement of the corresponding amplitude modulation on the photoacoustic signal, allows retrieving the photoacoustic-transmission matrix. Adapted by permission from Macmillan Publishers Ltd: Nature Photonics [41], copyright 2014. A black and white version of this figure appears in some formats. For the color version, please refer to the plate section.

in tissue limits the frequency range of ultrasound imaging system to a few tens of MHz, the corresponding resolution is usually in the range of a few tens of μm. *Inside* multiple scattering media, the typical dimension of the optical speckle grain size is half the optical wavelength. As a consequence, there will always be a large number of optical speckle grains in the acoustic resolution cell, which limits the relative amplitude modulation measured to acquire the transmission matrix. In the end, it is thus the signal-to-noise that will limit the ability to apply the photoacoustic transmission matrix *inside* scattering media. So far, all the reported proof-of-concept demonstration have used SNR-favorable situations, by enlarging the optical speckle grains via free space propagation, which is only possible after light propagation *through* scattering sample and not inside scattering sample. Another important difference with the all-optical transmission matrix is that photoacoustic sensing may only be performed where optical absorption is present: the photoacoustic transmission matrix may therefore considered as a partial version of the full all-optical transmission matrix, limited to absorbing regions. It nonetheless remains highly valuable as absorbing regions are indeed the ones targeted by laser therapy or

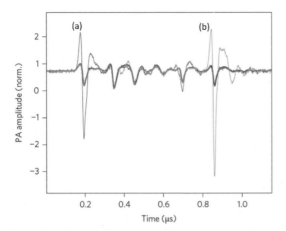

Figure 6.8 Illustration of selective photoacoustic enhancement with the photoacoustic transmission matrix. The trace (**a**) corresponds to the SLM input pattern chosen to maximize the signal on the closest absorber (signal at time 0.2 μs), the trace (**b**) corresponds to the SLM input pattern chosen to maximize the farthest absorber (signal at time 0.9 μs). The signal enhancement, typically a factor 6, reflects light focusing toward the targeted absorber. Adapted by permission from Macmillan Publishers Ltd: Nature Photonics [41], copyright 2014.

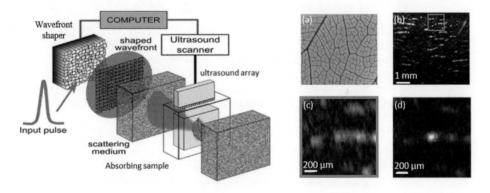

Figure 6.9 Illustration of the 2D photoacoustic-transmission matrix approach. Left: Schematic of the experimental setup. The photoacoustic transmission matrix is defined between the SLM basis (input modes) and the pixels of the 2D photoacoustic image (output modes). Right: (**a**) White-light photograph of the leak skeleton sample. (**b**) Conventional photoacoustic image of the sample. (**c**) Zoom onto the blue inset in (**b**). The red zone defines a target region for focusing with optical wavefront shaping. (**d**) Photoacoustic image of the same area as (**c**), after the SLM masks was set to maximize the photoacoustic signal in the red target zone, based on the knowledge of the photoacoustic transmission matrix. Reprinted with permission from [40], *Optics Letters*. A black and white version of this figure appears in some formats. For the color version, please refer to the plate section.

fluorescent imaging techniques. Apart from this limitation to absorbing region, the photoacoustic transmission matrix may be exploited as its all-optical counterpart, by taking advantage of the power of linear algebra. It was for instance shown by Chaigne and colleagues that a singular value decomposition of the photoacoustic transmission matrix can provide a way to automatically identify individuals absorber on a photoacoustic image [41].

6.4 Wavefront Shaping for Minimally Invasive Photoacoustic Microendoscopy

The development of photoacoustic techniques was made possible by the advent of laser technologies, which provide powerful and varied (choice of wavelength, pulsed/CW operation) sources of light. In the biomedical photoacoustic imaging community, the most common implementations are based on laser sources, mostly for their high energy per pulse and tunability (although less expensive alternative approaches are being investigated, such as those based on light emitting diodes [50, 51]). Beyond energy, the coherence properties of laser sources also make it possible to exploit optical wavefront shaping in the context of photoacoustic imaging. Optical wavefront shaping has recently been investigated to enhance photoacoustic imaging, toward both noninvasive photoacoustic imaging and endoscopic imaging [1]. In this section, we review recent research works toward the development of minimally invasive photoacoustic endomicroscopy by use of optical wavefront shaping through multimode waveguides [52, 53, 54]. We first introduce basics on light propagation and focusing through multimode fibers and then illustrate how the corresponding concepts can be applied to minimally invasive photoacoustic microendoscopy with digital phase conjugation.

6.4.1 Introduction to Light Focusing through Multimode Optical Fibers

Multimode optical fibers are optical waveguide that can transport light beam via several different propagation modes, i.e., modes that have different transverse spatial structures across the waveguide and different propagation constants [56]. As opposed to bundles of single mode fibers, multimode optical fibers can carry a much greater number of optical mode for a given total cross-sectional area. For a given optical wavelength λ_0, the number N of transverse modes in multi-mode optical fibers is given approximately by [56]

$$N \sim \frac{V^2}{2} = \frac{1}{2}\left(2\pi \frac{a}{\lambda_0}\mathrm{NA}\right)^2 \tag{6.2}$$

The V parameter $2\pi \frac{a}{\lambda_0}\mathrm{NA}$ is therefore the relevant parameter that governs the number of mode, with NA the numerical aperture of the fiber. Because the various propagation modes propagate with different propagation constants, or equivalently different phase velocities, their relative phases are rapidly shuffled upon propagation, and the output pattern after propagation through a multimode fiber is generally a random-like speckle pattern [55]. Multimode fibers have thus generally been considered as inappropriate systems for optical imaging, and endoscopic imaging system uses bundles of single-mode fibers rather than multimode fibers, although resulting in larger devices for a given number of modes. Although output patterns from multimode fibers look seemingly random, the propagation remains deterministic as long as the physical parameters of the fiber (shape, dimensions, temperature, etc.) remain stable, and it was recently demonstrated in such situation that optical wavefront shaping could be used to propagate controlled patterns through multimode fibers [55, 57, 58, 59, 60]. In particular, by performing optimization-based optical wavefront shaping with a spatial light modulator, Di Leonardo and Bianchi demonstrated optical focusing at the output of a multimode fiber [55], as illustrated in Figure 6.10. Čižmár and Dholakia further demonstrated that a wavefront shaping technique in conjunction with the knowledge of a transmission matrix can be used to shape the intensity pattern at will at the far side of the fiber [57]. Digital phase conjugation was also shown to be a powerful tool to focus and scan light in order to image at the end of a multimode fibers [59, 60]. The principle of digital phase conjugation is detailed in the following section. Turtaev and colleagues recently reviewed the significant research efforts toward minimally invasive optical imaging with multimode fibers by use of optical wavefront shaping [61].

6.4.2   Minimally Invasive Photoacoustic Microendoscopy with Digital Phase Conjugation

In this section, we illustrate how focusing through multimode fibers with digital phase conjugation (DPC) can be exploited toward minimally invasive photoacoustic

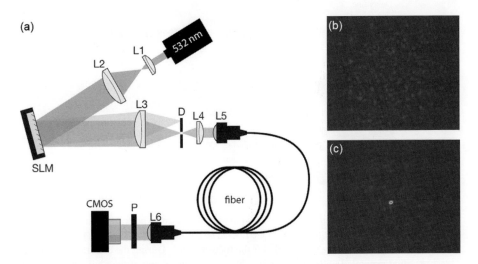

Figure 6.10 Focusing through a multimode fiber with optimization-based optical wavefront shaping. (**a**) Experimental setup: a phase-only spatial light modulator (SLM) is used to optimize the intensity measured with a CMOS camera at the output of a multimode fiber. (**b**) Speckle pattern measured at the fiber output without phase modulation. (**c**) Focal spot measured at the fiber output after applying the optimized phase mask on the SLM. Reprinted with permission from [55], *Optics Express*.

microendoscopy. A typical DPC setup is shown in Figure 6.11. Focusing light with digital phase conjugation consists of two steps, a calibration step and a focusing step.

During the calibration step, a focused spot is produced at the distal end of the fiber (at the location of the future sample to be imaged) by the calibration block shown in Figure 6.11. The light from this focused spot propagates in the multimode waveguide, and the interference pattern between the corresponding speckle pattern at the proximal side and the light beam from the imaging arm (used during this calibration step as a reference arm) is recorded on the CMOS camera, in order to measure holographically the field amplitude of the speckle pattern. One interference pattern is recorded for each position of the calibration focus (controlled here by galvanometric mirrors), resulting in a stack of holographic measurements of each speckle field as the final result of the calibration step.

During the focusing step, the calibration block is removed from the setup and replaced by the sample to be imaged. Thanks to the time-reversal invariance of optical wave propagation (provided the multimode fiber remains stable during the whole process), the phase conjugation of the previously measured field amplitude automatically leads to refocus light at the distal end of the fiber. To do so, the spatial light modulator (which

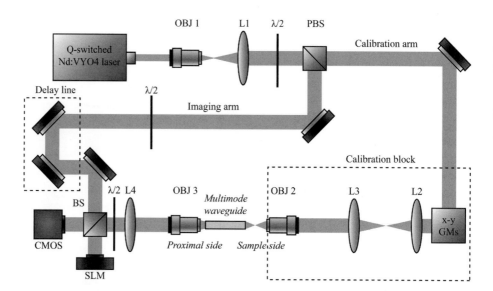

Figure 6.11 Schematic of the experimental setup used to perform focusing with digital phase conjugation through a multimode fiber. The calibration block is only used for the calibration step and is replaced by the sample for the focusing step. During the calibration step, a focused spot is first generated with the calibration block at the distal end of the fiber, and the corresponding speckle field at the proximal side is recorded holographically on the CMOS camera by interference with the reference beam (imaging arm). For the focusing step, the phase of the speckle field amplitude previously derived from the camera measurement is conjugated and loaded on the SLM. Thanks to time-reversal invariance of the optical propagation, the SLM provides the phase mask required to refocus the light from the imaging arm to the distal end of the fiber. Reprinted with permission from [54], *Biomedical Optics Express*.

pixel positions are mirrored with beamsplitter BS to those of the CMOS camera where the field was measured) is used to imprint the imaging beam with the phase conjugate of the field previously measured during the calibration step.

To work with the nanosecond pulses used for photoacoustic imaging, a delay line is required to temporally match the interference between the beam propagating through the calibration block and the imaging/reference beam. By successively programming the SLM with each phase-conjugated wavefront corresponding to each calibration position, a focused spot can be raster-scanned at the sample plane to obtain an image, provided that some interaction between the sample and the focused spot can be measured. High-resolution imaging with DPC through multimode fibers was first demonstrated by measuring fluorescence from stained neural cells [60]. In this case, the image was built by measuring the amount of fluorescence for each position of the focal spot. Here, we illustrate how the photoacoustic effect can be used analogously to sense and image optical absorption. A schematic of the experimental setup used to demonstrate the principle of optical-resolution photoacoustic imaging through a multi-mode fiber is shown in Figure 6.12. The distal tip of the multimode fiber was immersed in water, in order to couple the photoacoustic waves emitted by the absorbing sample to a single-element 20 MHz ultrasound transducer. To enhance the detection sensitivity, a focused transducer with a typical beamwidth around 150 µm was used to detect the photoacoustic waves. DPC was used to raster-scan a diffraction-limited focal spot at the distal end of the fiber, and optical-resolution photoacoustic images were obtained by plotting the peak-to-peak amplitude of the signal for each scan position. The photoacoustic image of a knot made with two absorbing wire is shown in Figure 6.13(b). The measured resolution was 1.5 µm, dictated by the diffraction limit given by the numerical aperture of the fiber.

While promising, the results shown in Figure 6.13 were obtained in highly idealized configuration, where the photoacoustic waves propagate in water to reach the transducer. In biological tissue, the high-frequency photoacoustic waves generated by nanosecond focused pulses would be very strongly attenuated because of the large absorption of ultrasound at high frequencies (on the order of 0.5 dB/cm/MHz). This would prevent

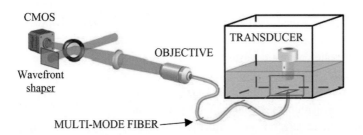

Figure 6.12 Schematic of the experimental setup used to perform optical-resolution photoacoustic imaging through a multimode fiber. Digital phase conjugation (DPC) was used at the fiber proximal end to generate and raster-scan a diffraction-limited focused spot at the distal end after propagation trough the fiber. A 20 MHz ultrasound transducer was used to detect photoacoustic waves emitted by absorbing samples placed at the distal end. Figure adapted with permission from [52], AIP.

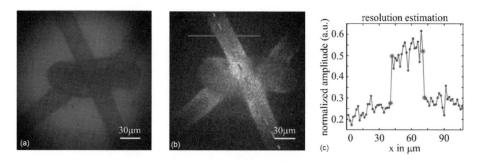

Figure 6.13 Optical-resolution photoacoustic imaging through a multimode fiber. (**a**) White light photograph of a knot made with two absorbing wires, located at the distal end of the fiber. (**b**) Optical-resolution photoacoustic image of the same knot. (**c**) Cross-section along the horizontal line in (**b**), illustrating the resolution performance. Figure adapted with permission from [52], AIP.

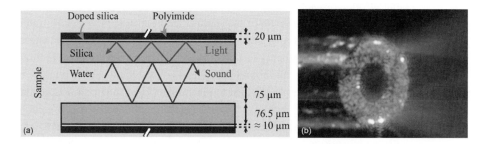

Figure 6.14 A water-filled silica capillary as a dual optical-in acoustic-out waveguide (left). Schematic illustrating light and sound waveguiding based on the principles of a step-index waveguide (right). Illustration of the optical speckle observed at the output of the capillary waveguide after scrambled multimodal light propagation in the silica capillary. Figure adapted with permission from [53], AIP and from [54], *Biomedical Optics Express*.

from the possibility to remotely detect photoacoustic waves generated deep inside tissue. To circumvent this limitation, Simandoux and coworkers proposed to use a fluid-filled silica capillary as a dual-waveguide, used as a multimode optical waveguide to transport and focus light toward the sample and used as an acoustic waveguide to collect and transport photoacoustic waves toward the proximal end [53]. Figure 6.14 illustrates how both light and sound are guided through the capillary, based on the principles of a step-index waveguide. Multimodal light propagation through the capillary is strictly similar to propagation in a conventional multimode step-index fiber. To couple light excitation to sound detection at the fiber proximal side, an optical/acoustic beam splitter was made with a glass slide, as illustrated in Figure 6.15 [54]. An optical-resolution photoacoustic image of an absorbing wire placed 50 μm away from the capillary distal end (so that the sample could be properly illuminated given the numerical aperture of the capillary), obtained with digital phase conjugation through a silica capillary 330 μm in diameter, is shown in Figure 6.15(b). The resolution in this case was approximately 10 μm, as illustrated in Figure 6.15(c). This results was obtained with the distal tip of the capillary free of any optical or acoustical components, only required at the proximal tip, therefore resulting in a minimally invasive device for photoacoustic microendoscopy.

6.5   Conclusions and Future Directions

In this chapter, we have reviewed the coupling of optical wavefront shaping and photoacoustics.

The photoacoustic effect offers a powerful mechanism to sense optical intensity noninvasively via optical absorption. The optical wavefront shaping methods initially developed with purely optical feedback mechanism (direct intensity measurement, fluorescence), such as optimization-based light focusing or transmission matrix measurements, were therefore quickly investigated by use of photoacoustics as feedback mechanism. The main advantage of indirect light intensity measurements via the photoacoustic effect is based on the possibility to sense absorbed light inside scattering media such as biological tissue, which are mostly transparent to ultrasound. However,

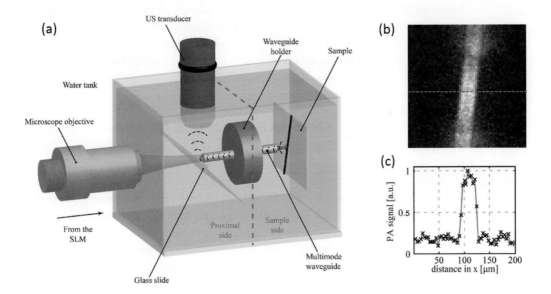

Figure 6.15 Optical-resolution photoacoustic microscopy at the tip of a silica capillary. (**a**) Principle of light and sound coupling at the proximal tip: the shaped optical wavefront is injected into the silica tubing after propagation through a transparent glass slide. The glass slide is used as an ultrasound reflector to direct the photoacoustic waves collected through the fluid core of the capillary to the ultrasound transducer. (**b**) Photoacoustic image of an absorbing wire placed 50 μm away from the capillary tip. The image was obtained by raster-scanning optical spots, by using the appropriate optical wavefront computed from digital phase conjugation. (**c**) Cross-section along the dotted line in (**b**), illustrating the resolution around 110 μm, dictated by the numerical aperture of the capillary tubing. Reprinted with permission from [54], *Biomedical Optics Express*.

the preliminary results obtained so far with photoacoustic-guided wavefront shaping are limited to situations where photoacoustic sensing took place outside the scattering media, in order to control the optical speckle grain size and to insure a sufficient signal-to-noise ratio. The demonstration of photoacoustic-guided optical wavefront shaping inside turbid media remains a challenge, due to the large mismatch between the size of the acoustic resolution and the optical speckle grain, which limits the amplitude modulation of photoacoustic signals. Because photoacoustic imaging probes light intensity only at optical absorbers, samples with sparse distribution of absorption (such as microcapillary networks, or exogenous contrast agents) can provide situations where the numbers of *absorbed* optical speckle grains within the acoustic resolution cell could be small enough to sense intensity modulation with a sufficient signal-to-noise ratio. Increasing the ultrasound frequency range to improve the acoustic resolution is also one potential option, but this will require the development of very sensitive high-frequency transducers designed specifically for photoacoustic sensing, in order to overcome the ultrasonic absorption at depth in tissue.

Photoacoustic-guided optical wavefront shaping shares with other optical wavefront shaping techniques the technological limitations of the currently available devices in terms of speed or number of degrees of freedom. Apart from digital micromirrors (but which are limited to binary amplitude modulation), fast systems (in the kHz range or more) for phase modulation have a number of degrees of freedom (typically about a thousand pixels), and the refreshment rate of million-pixel devices are limited to a few tens of Hz. The important recent research efforts in the field of complex optical wavefront shaping are expected to promote further technological development toward devices that could be used for realistic biomedical applications, including photoacoustic-guided wavefront shaping.

While coupling photoacoustic and optical wavefront shaping was first investigated by exploiting the photoacoustic effect as a feedback mechanism, it was quickly realized that optical wavefront shaping was also a powerful tool to enhance the capabilities of photoacoustic imaging. Very preliminary results have shown how optical focusing with optical wavefront shaping could enhanced the resolution of photoacoustic imaging through scattering sample [45]. However, the possibility to *focus* light inside scattering media has not proven yet an efficient way to *image* such media, beyond the very specific case where the memory effect can be exploited [62]. Going from focusing through tissue to imaging remains a general challenge for optical wavefront shaping, including photoacoustic-related technique. In this chapter, we have illustrated how optical wavefront shaping through very thin multimode optical waveguides could be applied to develop minimally invasive photoacoustic microendoscopy, building on recent ideas initially developed for purely optical endoscopy. In this case, the approach is based on the fact that a multimode waveguide behaves analogously to a complex multiple scattering medium, but as one that can be entirely precalibrated before being inserted into tissue for imaging. Optical wavefront shaping-assisted photoacoustic microendoscopy will undoubtedly both motivate and benefit from the new technological developments mentioned above, and could hopefully lead to practical biomedical instruments in a close future.

Acknowledgments

This book chapter was written by the author during his sabbatical leave from the Ecole Supérieure de Physique et Chimie Industrielles (ESPCI Paris, France) to the Ecole Polytechnique Fédérale de Lausanne (EPFL, Switzerland), before his move to University Grenoble Alpes. The author gratefully acknowledges funding from EPFL through a Visiting Professor Fellowship, in Pr. Psaltis's Optics Laboratory and Pr. Moser's Laboratory of Applied Photonics Devices. The author also thanks Nicolino Stasio from the Optics Laboratory for his help with adapting several of the figures in this chapter.

6.6   References

[1]   Bossy E, Gigan S. Photoacoustics with coherent light. Photoacoustics. 2016;4(1):22–35.

[2]   Bell AG. On the production and reproduction of sound by light. American Journal of Science. 1880;(118):305–324.

[3]   Kreuzer L, Patel C. Nitric oxide air pollution: detection by optoacoustic spectroscopy. Science. 1971;173(3991):45–47.

[4]   Rosencwaig A. Photoacoustic spectroscopy of solids. Optics Communications. 1973;7(4):305–308.

[5]   Patel C, Tam A. Pulsed optoacoustic spectroscopy of condensed matter. Reviews of Modern Physics. 1981;53(3):517.

[6]   Kruger RA. Photoacoustic ultrasound. Medical physics. 1994;21(1):127–131.

[7]   Karabutov A, Podymova N, Letokhov V. Time-resolved laser optoacoustic tomography of inhomogeneous media. Applied Physics B. 1996;63(6):545–563.

[8]   Oraevsky AA, Jacques SL, Tittel FK. Measurement of tissue optical properties by time-resolved detection of laser-induced transient stress. Applied Optics. 1997;36(1):402–415.

[9]   Diebold G, Sun T, Khan M. Photoacoustic monopole radiation in one, two, and three dimensions. Physical Review Letters. 1991;67(24):3384.

[10] Wang LV, Wu Hi. Chapter 12: Photoacoustic tomography. In: Biomedical optics: principles and imaging. John Wiley & Sons; 2012. .

[11] Ntziachristos V, Razansky D. Molecular imaging by means of multispectral optoacoustic tomography (MSOT). Chemical reviews. 2010 May;110(5):2783.

[12] Beard P. Biomedical photoacoustic imaging. Interface focus. 2011;1(4):602–631.

[13] Wang LV, Hu S. Photoacoustic tomography: *in vivo* imaging from organelles to organs. Science. 2012;335(6075):1458–1462.

[14] Wang LV, Gao L. Photoacoustic microscopy and computed tomography: from bench to bedside. Annual Review of Biomedical Engineering. 2014;16:155.

[15] Zhang E, Laufer J, Pedley R, Beard P. *In vivo* high-resolution 3D photoacoustic imaging of superficial vascular anatomy. Physics in Medicine and Biology. 2009;54(4):1035.

[16] Hu S, Maslov K, Wang LV. Second-generation optical-resolution photoacoustic microscopy with improved sensitivity and speed. Optics Letters. 2011;36(7):1134–1136.

[17] Maslov K, Zhang HF, Hu S, Wang LV. Optical-resolution photoacoustic microscopy for *in vivo* imaging of single capillaries. Optics Letters. 2008;33(9):929–931.

[18] Hajireza P, Shi W, Zemp R. Label-free *in vivo* fiber-based optical-resolution photoacoustic microscopy. Optics Letters. 2011;36(20):4107–4109.

[19] Hajireza P, Harrison T, Forbrich A, Zemp R. Optical resolution photoacoustic microendoscopy with ultrasound-guided insertion and array system detection. Journal of Biomedical Optics. 2013;18(9):090502.

[20] Yang JM, Li C, Chen R, Rao B, Yao J, Yeh CH, et al. Optical-resolution photoacoustic endomicroscopy *in vivo*. Biomedical Optics Express. 2015;6(3):918–932.

[21] Vellekoop IM, Mosk A. Focusing coherent light through opaque strongly scattering media. Optics Letters. 2007;32(16):2309–2311.

[22] Popoff S, Lerosey G, Carminati R, Fink M, Boccara A, Gigan S. Measuring the transmission matrix in optics: an approach to the study and control of light propagation in disordered media. Physical Review Letters. 2010;104(10):100601.

[23] Katz O, Small E, Bromberg Y, Silberberg Y. Focusing and compression of ultrashort pulses through scattering media. Nature Photonics. 2011;5(6):372–377.

[24] Katz O, Small E, Silberberg Y. Looking around corners and through thin turbid layers in real time with scattered incoherent light. Nature Photonics. 2012;6(8):549–553.

[25] Hsieh CL, Pu Y, Grange R, Laporte G, Psaltis D. Imaging through turbid layers by scanning the phase conjugated second harmonic radiation from a nanoparticle. Optics Express. 2010;18(20):20723–20731.

[26] Vellekoop I, Van Putten E, Lagendijk A, Mosk A. Demixing light paths inside disordered metamaterials. Optics Express. 2008;16(1):67–80.

[27] Xu X, Liu H, Wang LV. Time-reversed ultrasonically encoded optical focusing into scattering media. Nature Photonics. 2011;5(3):154–157.

[28] Wang YM, Judkewitz B, DiMarzio CA, Yang C. Deep-tissue focal fluorescence imaging with digitally time-reversed ultrasound-encoded light. Nature Communications. 2012;3:928.

[29] Si K, Fiolka R, Cui M. Breaking the spatial resolution barrier via iterative sound-light interaction in deep tissue microscopy. Scientific Reports. 2012;2.

[30] Si K, Fiolka R, Cui M. Fluorescence imaging beyond the ballistic regime by ultrasound-pulse-guided digital phase conjugation. Nature Photonics. 2012;6(10):657–661.

[31] Judkewitz B, Wang YM, Horstmeyer R, Mathy A, Yang C. Speckle-scale focusing in the diffusive regime with time reversal of variance-encoded light (TROVE). Nature Photonics. 2013;7(4):300–305.

[32] Tay JW, Lai P, Suzuki Y, Wang LV. Ultrasonically encoded wavefront shaping for focusing into random media. Scientific Reports. 2014;4.

[33] Liu Y, Lai P, Ma C, Xu X, Grabar AA, Wang LV. Optical focusing deep inside dynamic scattering media with near-infrared time-reversed ultrasonically encoded (TRUE) light. Nature Communications. 2015;6.

[34] Bossy E, Daoudi K, Boccara AC, Tanter M, Aubry JF, Montaldo G, et al. Time reversal of photoacoustic waves. Applied Physics Letters. 2006;89(18):184108.

[35] Funke AR, Aubry JF, Fink M, Boccara AC, Bossy E. Photoacoustic guidance of high intensity focused ultrasound with selective optical contrasts and time-reversal. Applied Physics Letters. 2009;94(5):054102.

[36] Staley J, Hondebrink E, Peterson W, Steenbergen W. Photoacoustic guided ultrasound wavefront shaping for targeted acousto-optic imaging. Optics Express. 2013 dec;21(25):30553.

[37] Kong F, Silverman RH, Liu L, Chitnis PV, Lee KK, Chen YC. Photoacoustic-guided convergence of light through optically diffusive media. Optics Letters. 2011;36(11):2053–2055.

[38] Caravaca-Aguirre AM, Conkey DB, Dove JD, Ju H, Murray TW, Piestun R. High contrast three-dimensional photoacoustic imaging through scattering media by localized optical fluence enhancement. Optics Express. 2013;21(22):26671–26676.

[39] Chaigne T, Gateau J, Katz O, Boccara C, Gigan S, Bossy E. Improving photoacoustic-guided optical focusing in scattering media by spectrally filtered detection. Optics Letters. 2014;39(20):6054–6057.

[40] Chaigne T, Gateau J, Katz O, Bossy E, Gigan S. Light focusing and two-dimensional imaging through scattering media using the photoacoustic transmission matrix with an ultrasound array. Optics Letters. 2014;39(9):2664–2667.

[41] Chaigne T, Katz O, Boccara AC, Fink M, Bossy E, Gigan S. Controlling light in scattering media non-invasively using the photoacoustic transmission matrix. Nature Photonics. 2014;8(1):58–64.

[42] Aguirre AMC, Conkey DB, Dove JD, Ju H, Murray TW, Piestun R. Three-dimensional photoacoustic imaging through scattering media. In: Biomedical Optics 2014. Optical Society of America; 2014. p. BS3A.53. Available from: www.opticsinfobase.org/abstract.cfm?URI=BIOMED-2014-BS3A.53.

[43] Tay JW, Liang J, Wang LV. Amplitude-masked photoacoustic wavefront shaping and application in flowmetry. Optics Letters. 2014;39(19):5499–5502.

[44] Deán-Ben XL, Estrada H, Razansky D. Shaping volumetric light distribution through turbid media using real-time three-dimensional opto-acoustic feedback. Optics Letters. 2015;40(4):443–446.

[45] Conkey DB, Caravaca-Aguirre AM, Dove JD, Ju H, Murray TW, Piestun R. Super-resolution photoacoustic imaging through a scattering wall. Nature Communications. 2015;6:7902.

[46] Lai P, Wang L, Tay JW, Wang LV. Photoacoustically guided wavefront shaping for enhanced optical focusing in scattering media. Nature Photonics. 2015.

[47] Wang L, Zhang C, Wang LV. Grueneisen relaxation photoacoustic microscopy. Physical Review Letters. 2014;113(17):174301.

[48] Tang J, Germain RN, Cui M. Superpenetration optical microscopy by iterative multiphoton adaptive compensation technique. Proceedings of the National Academy of Sciences. 2012;109(22):8434–8439.

[49] Katz O, Small E, Guan Y, Silberberg Y. Noninvasive nonlinear focusing and imaging through strongly scattering turbid layers. Optica. 2014;1(3):170–174.

[50] Allen TJ, Beard PC. Pulsed near-infrared laser diode excitation system for biomedical photoacoustic imaging. Optics Letters. 2006;31(23):3462–3464.

[51] Allen TJ, Beard PC. Light emitting diodes as an excitation source for biomedical photoacoustics. In: Photons Plus Ultrasound: Imaging and Sensing 2013, vol. 8581, p. 85811F. International Society for Optics and Photonics; 2013.

[52] Papadopoulos IN, Simandoux O, Farahi S, Huignard JP, Bossy E, Psaltis D, et al. Optical-resolution photoacoustic microscopy by use of a multimode fiber. Applied Physics Letters. 2013;102(21):211106.

[53] Simandoux O, Stasio N, Gateau J, Huignard JP, Moser C, Psaltis D, et al. Optical-resolution photoacoustic imaging through thick tissue with a thin capillary as a dual optical-in acoustic-out waveguide. Applied Physics Letters. 2015;106(9):094102.

[54] Stasio N, Shibukawa A, Papadopoulos IN, Farahi S, Simandoux O, Huignard JP, et al. Towards new applications using capillary waveguides. Biomedical Optics Express. 2015;6(12):4619–4631.

[55] Di Leonardo R, Bianchi S. Hologram transmission through multi-mode optical fibers. Optics Express. 2011;19(1):247–254.

[56] Saleh BE, Teich MC, Saleh BE. Fundamentals of photonics, vol. 22. Wiley New York; 1991.

[57] Čižmár T, Dholakia K. Shaping the light transmission through a multimode optical fibre: complex transformation analysis and applications in biophotonics. Optics Express. 2011;19(20):18871–18884.

[58] Choi Y, Yoon C, Kim M, Yang TD, Fang-Yen C, Dasari RR, et al. Scanner-free and wide-field endoscopic imaging by using a single multimode optical fiber. Physical Review Letters. 2012;109(20):203901.

[59] Papadopoulos IN, Farahi S, Moser C, Psaltis D. Focusing and scanning light through a multimode optical fiber using digital phase conjugation. Optics Express. 2012;20(10):10583–10590.

[60] Papadopoulos IN, Farahi S, Moser C, Psaltis D. High-resolution, lensless endoscope based on digital scanning through a multimode optical fiber. Biomedical Optics Express. 2013;4(2):260–270.

[61] Turtaev S, Leite IT, Čižmár T. Multimode fibres for micro-endoscopy. Optofluidics, Microfluidics and Nanofluidics. 2015;2(1):31–35.

[62] Bertolotti J, van Putten EG, Blum C, Lagendijk A, Vos WL, Mosk AP. Non-invasive imaging through opaque scattering layers. Nature. 2012;491(7423):232–234.

7 Imaging and Controlling Light Propagation Deep within Scattering Media Using a Time-Resolved Reflection Matrix

Youngwoon Choi, Sungsam Kang, and Wonshik Choi

7.1 Introduction

Over the past decades, optical techniques have been widely applied to life science and biomedicine, and found their good use in various areas such as diagnostics, surgery and therapy [1]. They have been particularly successful in interrogating either thin specimens or superficial layers of thick biological tissues. When it comes to dealing with targets embedded deep within tissues, however, their performance has been significantly undermined [2]. The major reason is that light wave experiences multiple-scattering events while it travels through biological tissues. A major fraction of incident wave loses both original incident momentum and energy when it reaches target depth. The loss in momentum results in the reduction of spatial resolving power, and that in energy leads to the degradation in the efficiency of biosensing, photo-treatment, and deep tissue imaging techniques relying on the so-called snakelike photons such as photoacoustic imaging and diffuse optical tomography [3–6].

The length scale of loss for momentum is different from that for energy. The momentum loss takes place at a length scale of scattering mean free path (MFP; l_s). The intensity of the unperturbed wave preserving original momentum is exponentially attenuated each time it travels by MFP of a scattering medium. Considering that typical MFP of biological tissues is around 100 μm [7], the intensity of unperturbed wave that can reach 1 mm deep is only a factor of $10^{(-10)}$ of the original intensity, and the rest serves as a background noise. As a result, the working depth of high-resolution imaging is largely limited to a few MFPs, which is evidenced by the relation between spatial resolution and imaging depth of various optical imaging techniques shown in Figure 7.1.

On the other hand, the length scale of energy decay is governed by the transport mean free path (TMFP; l_t), which is the propagation distance required for the wave to be fully randomized in its direction. For biological tissues, TMFP is typically 10 times larger than MFP. And the decay in light energy delivered to the target depth is linear rather than exponential with respect to TMFP [18, 19], suggesting that the energy delivery may be much less demanding than the imaging. But still, poor light penetration to the depth of tens of TMFPs causes difficulty in *in vivo* sensing and light therapy.

For the past decades, many studies have been conducted to address the limitations of optical techniques associated with the detrimental effects of multiple light scattering

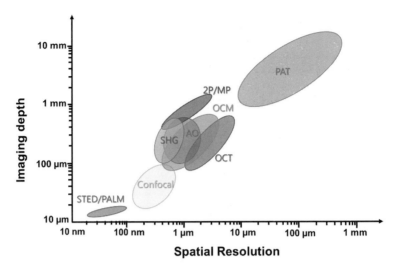

Figure 7.1 Relation between spatial resolution and imaging depth for various microscopic techniques. PAT: photoacoustic tomography [3, 4]; 2P/MP: two-photon/multiphoton microscopy [8, 9]; OCT: optical coherence tomography [10]; OCM: optical coherence microscopy [11]; AO: adaptive optics [12, 13]; SHG: second harmonic generation microscopy [14]; Confocal: confocal microscopy [15]; STED: stimulated emission depletion microscopy [16]; PALM: Photoactivated localization microscopy [17].

explained above. In particular, those studies that took the advantages of the deterministic treatment of scattering events have drawn great interest [20–35]. For instance, feedback control of the incident wavefront was used to generate a focus [23, 24, 32] or perform imaging through a scattering layer [21, 35]. The transmission/reflection matrix of a scattering medium was measured for the image delivery through scattering media, and endoscopic imaging through unconventional image guiding media such as multimode fibers [21, 35]. In addition, the eigenchannels of these matrices were exploited for the enhanced light energy delivery and imaging [27, 32–34]. Similar approaches have been taken in the ultrasound-modulated optical imaging in which wavefront sensing and control were administrated for the acoustically modulated optical waves [26, 30]. Most of these studies, while they promise great potential, dealt with the wave propagation in the steady state, thereby ruling out the temporal degree of freedom. In fact, a new trend has been emerging which attempted to take the advantage of the temporal response of the interaction between waves and scattering media. For example, the knowledge of the temporal response makes it possible to sort out multiple-scattered waves in time, which are otherwise treated in total. With this knowledge at hand, various studies demonstrated temporal focusing of waves through scattering media [24, 36], enhanced light energy delivery to the target depth [20], and improvement in imaging depth [37].

In this chapter, we introduce two noteworthy methods for exploring the use of the so-called time-resolved reflection matrix (TRRM) of the scattering medium [20, 37]. TRRM is made of the amplitude and phase maps of reflected waves taken at specific arrival time and for various angles or positions of illumination. It provides us with unprecedented amount of information covering both spatial input-output correlation and temporal response. With the vast amount of information at hand, studies have been conducted to relieve the limitations of imaging depth and energy delivery that the multiple light scattering imposes. In section 2, we describe a time domain approach for measuring TRRM and its application for dramatically improving imaging depth that maintains diffraction-limited spatial resolution [37]. Spectral domain approach of measuring TRRM is introduced in section 3 along with its application for enhancing energy delivery to the target depth by the implementation of time-dependent eigenchannels [20]. Future perspectives discussed in section 4 will conclude this chapter.

7.2   Time Domain Measurements of the Time-Resolved Reflection Matrix Measurements

In this section, a time domain approach of measuring TRRM is introduced along with its important applications for high-resolution and deep tissue imaging [37]. The approach measures wide-field phase and amplitude maps of reflected waves for a selected arrival time while scanning the angle of illumination, and constructs TRRM from a set of these measurements. To obtain a high-resolution image for a target object embedded within the scattering medium, it is necessary to extract single-scattered waves, which are the waves scattered only a single time by the target object and therefore retain object information, hidden in the background of multiple-scattered waves. As will be explained in the following, single-scattered waves possess two unique properties – depth-specific flight time and the transverse momentum conservation. TRRM provides enough information to make use of these properties of single-scattered waves and makes it possible to enhance single-scattered waves embedded within a background of random multiple-scattered waves. Using what is called the collective accumulation of single-scattering (CASS) microscopy, the method can identify targets embedded at a depth of about 11.5 l_s with almost no loss of spatial resolution.

7.2.1   The Necessity of TRRM for High-Resolution Deep Tissue Imaging

The reflection signal from scattering medium can be categorized into single-scattered waves $\vec{E}^S$ and multiple-scattered waves $\vec{E}^M$ as depicted in Figure 7.2a. The single-scattered waves are the components of reflected waves that are scattered only once by the target object, but not at all by the scattering medium. Therefore, these single-scattered waves carry object information. The rest are multiple-scattered waves, which serve as background noise and cause degradation in resolving power. With the increase of the target depth, the intensity of single-scattered waves decays exponentially at the length scale of MFP. For example, an object embedded in a scattering medium with a depth of z, the intensity of single-scattered waves is attenuated by a factor of e^{-2z/l_s}. Here the factor 2 at the exponent accounts for the round trip to the target depth. As a result, it is increasingly more difficult to resolve the fine details of target as it is embedded deeper inside a scattering medium.

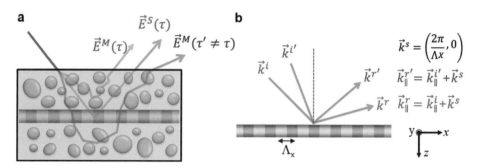

Figure 7.2 Experimental schematic for the collective enhancement of single-scattered waves. (**a**) Classification of reflected waves. $\vec{E}^S(\tau)$: Electric field of a wave scattered a single time by a target object located at a depth of $c\tau/2$, where τ and c are the time-of-flight from the surface and the speed of light in the medium, respectively. $\vec{E}^M(\tau)$: Multiple-scattered waves with the same time-of-flight as the single-scattered waves. $\vec{E}^M(\tau' \neq \tau)$: Multiple-scattered waves with time-of-flight different from τ. (**b**) In-plane momentum conservation property of the single-scattered waves. The $\vec{k}^i$ and $\vec{k}^r$ are wave vectors of incident and reflected waves, respectively, for one illumination angle, and the $\vec{k}^{i'}$ and $\vec{k}^{r'}$ are those for another illumination angle. The $\vec{k}^i_\parallel, \vec{k}^r_\parallel, \vec{k}^{i'}_\parallel$ and $\vec{k}^{r'}_\parallel$ are the components of the associated wave vectors parallel to the x-y plane. The reflected waves gain the lateral momentum of $\vec{k}^S = \left(\frac{2\pi}{\Lambda_x}, 0\right)$ of the target object, where Λ_x is the period of the structure along the x-axis, irrespective of incident momentum. Adapted by permission from Macmillan Publishers Ltd: Nature Photonics [37], copyright 2015.

To overcome the detrimental effect of multiple light scattering, it is necessary to selectively extract single-scattered waves among the strong background of multiple-scattered waves. As illustrated in Figure 7.2a, the arrival time of single-scattered waves is determined by the target depth such that $\tau = 2z/c$. On the contrary, the arrival time of multiple-scattered waves spreads over wide temporal range. Therefore, time-gated detection selecting the arrival time of single-scattered waves can filter out a large fraction of multiple-scattered waves having different flight time from the single-scattered waves. However, there still remain significant multiple-scattered waves which arrive at the same flight time as single-scattered waves. For target depths deeper than a few MFPs, these multiple-scattered waves that have the same arrival time as single-scattered waves can still dominate over the single-scattered waves. Therefore, the time-gated detection alone may not guarantee enough signal-to-noise ratio (SNR).

To further enhance the imaging depth, the spatial input-output correlation of single-scattered waves was introduced. The input and output of single-scattered waves should preserve the in-plane momentum along x-y plane as shown in Figure 7.2b. The output angle of single-scattered waves have a specific correlation with incidence angle, while that of the multiple-scattered waves is random. More specifically, the in-plane momentum difference between input ($\vec{k}^i_\parallel$), and output ($\vec{k}^r_\parallel$) wave is determined by the target object. For example, if the target is a periodic structure with period of Λ_x, then the in-plane momentum difference induced by the target is determined by $\vec{k}^S = \left(\frac{2\pi}{\Lambda_x}, 0\right)$. Therefore, the reflection wavevector is given by $\vec{k}^r_\parallel = \vec{k}^i_\parallel + \vec{k}^S_\parallel$. If the reflected waves are projected onto the in-plane momentum difference space (2D momentum space defined by $\vec{k}_\parallel^r - \vec{k}_\parallel^i$), then the single-scattered waves should be mapped on a fixed position at $\vec{k}^s$ regardless of the incidence angle. On the other hand, the momentum differences of the multiple-scattered waves are distributed over broad range of spatial frequency. Therefore electric field maps of the single-scattered waves taken at various incidence angles can be coherently accumulated on the momentum difference space. If one uses N_{tot} different incidence angles, then the amplitude of single-scattered waves increase by N_{tot}, and consequently its intensity grows with N_{tot}^2. On the contrary, the intensity of multiple-scattered waves scales with N_{tot} as they are incoherently added. Therefore, the signal-to-background ratio, which is the relative intensity of single-scattered waves to multiple-scattered waves, is increased by a factor of N_{tot}, and so is the SNR. For this reason, this approach is called collective accumulation of single-scattered waves (CASS) microscopy. The term "collective" was used since the signal increase for single-scattered waves is similar to that of the superradiance induced by the collective radiation of electric dipoles [38].

7.2.2   Experimental Setup

To make use of the properties of single-scattered waves, TRRM of a scattering medium needs to be measured. For this purpose, a reflection-mode interferometric microscope was constructed (Figure 7.3). A superluminescent diode laser (center wavelength 800 nm, spectral bandwidth 25 nm) was used as a low-coherence light source to provide temporal gating. The output from a laser was divided into sample and reference waves at a beam splitter (BS1). In the incidence beam path, a spatial light modulator (SLM) was inserted at the conjugate plane of sample plane to control the incidence angle. Typically,

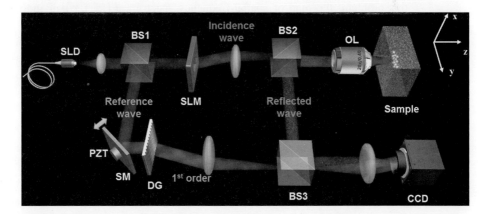

Figure 7.3 Experimental schematic diagram of CASS microscope. SLD: superluminescent diode laser, OL: objective lens, BS1, BS2 and BS3: beam splitters, SLM: spatial light modulator (working in the reflection mode, but indicated here as a transmission mode for simplicity), DG: diffraction grating. An aperture was used to select the first-order diffracted wave, SM: path length scanning mirror, and CCD: camera. Adapted by permission from Macmillan Publishers Ltd: Nature Photonics [37], copyright 2015.

2500 phase ramps were written on the SLM to cover the maximum number of orthogonal input modes for the view field of $70 \times 70 \ \mu m^2$. The incidence waves were delivered to a sample with an objective lens ($20\times$, 0.4NA). The reflected waves from the sample was collected by the same objective lens, and then delivered to the CCD camera placed on the conjugate image plane. The reference wave was combined with reflected wave at the BS3 placed in front of the CCD camera. With a scanning mirror in the reference wave beam path, the path length of the reference beam was tuned to match that of the reflected wave within the coherence length ($11.4 \ \mu m$), which corresponds to 78 fs time-gating window. A diffraction grating (DG) was used for implementing off-axis digital holography from which the complex field map of the reflected wave can be obtained [39].

7.2.3   Construction of the Time-Resolved Reflection Matrix

In this section, we explain the way to construct TRRM. The time-gated complex field maps $u(x, y; i, \tau)$ of reflected wave can be measured using the experimental setup in Figure 7.3, where i stands for the ith phase ramp written on the SLM. Figures 7.4a and 7.4b show representative phase ramps written on the SLM and its corresponding time-gated complex field maps (amplitude), respectively. By taking the 2D Fourier transform of these measured complex field maps with respect to spatial coordinates, the spatial-frequency spectrum $U\left(\vec{k}_{\parallel}^r; i, \tau\right)$ of the reflected waves were computed. For ideal imaging system, we can construct the reflection matrix $U\left(\vec{k}_{\parallel}^r; \vec{k}_{\parallel}^i, \tau\right)$ directly from the calculated spectrum by assuming that the incidence angle is determined by phase ramps on the SLM. However, in practice, the incident waves at the sample plane were not pure plane waves as the realization of phase ramps in the SLM is not ideal. To address this problem, the map of spatial-frequency spectrum $J\left(\vec{k}_{\parallel}^i; i, \tau\right)$ of individual phase ramps were measured separately. Then, the product $U\left(\vec{k}_{\parallel}^r; i, \tau\right) \cdot J^{-1}\left(\vec{k}_{\parallel}^i; i, \tau\right)$ results the time-resolved reflection matrix $U\left(\vec{k}_{\parallel}^r; \vec{k}_{\parallel}^i, \tau\right)$ connecting the momenta of incident and reflected waves (Figure 7.4c).

7.2.4   Image Reconstruction from TRRM

With the time-gated reflection matrix $U\left(\vec{k}_{\parallel}^r; \vec{k}_{\parallel}^i, \tau\right)$, we can now make use of the two properties of single-scattered waves. The first property, which is the depth-dependent flight time, can be used by selecting the specific arrival time τ associated with the target depth. In this way, we can collect the single-scattered waves located at the target depth and rule out a large fraction of multiple-scattered waves with different flight time from that of single-scattered waves at the same time. Figure 7.5 shows typical intensity reflectance of a target embedded in a $11.5l_s$-thick scattering medium measured with respect to the arrival time. A peak at 11 ps corresponds to the reflection signal which have the arrival time $\tau = 2z/c$. If we apply 76 fs time-gating window on the peak, then we can rule out a major part of multiple-scattered waves. In the measurement, the

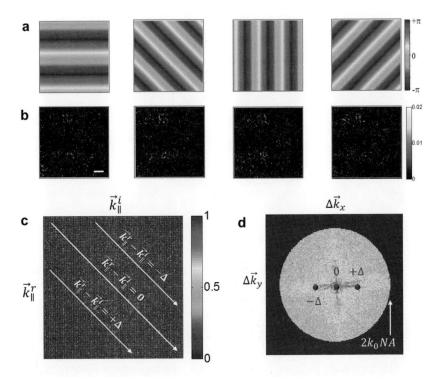

Figure 7.4 Measurement and construction of the time-gated reflection matrix. (**a** and **b**) Representative phase ramps written on the SLM and the time-gated amplitude maps of the reflected waves, respectively. The phase maps were simultaneously acquired, but not shown here. The color bars in (**a**) and (**b**) indicate phase in radians and amplitude in arbitrary unit, respectively. Scale bar, 10 μm. (**c**) Time-gated reflection matrix, U, constructed from 2500 complex field maps. Column and row indices are $\vec{k}_\parallel^i$ and $\vec{k}_\parallel^r$, respectively. The color bar measures amplitude in arbitrary unit. (**d**) The result of mapping U on 2D in-plane momentum difference space. Colors indicates amplitude in log-scale. Adapted by permission from Macmillan Publishers Ltd: Nature Photonics [37], copyright 2015. A black and white version of this figure appears in some formats. For the color version, please refer to the plate section.

intensity of multiple-scattered waves is reduced to about 1/300 of the original intensity by applying this time-gating window. Note that peaks around 8 ps in Figure 7.5 comes from the multiple reflections in the optical elements of the microscope system.

We can then make use of the second property of the single-scattered waves, which is the in-plane momentum conservation, to selectively extract the single-scattered waves from TRRM. Since the reflected waves include both single and multiple-scattered waves, the obtained matrix can be decomposed as, $U = U_S + U_M$, where U_S and U_M are single- and multiple-scattering matrix, respectively. From the in-plane momentum conservation described in Figure 7.2b, the single-scattering matrix should be connected to the target object. If we let the 2D object function of the target as $f(x, y)$, and its associated object transfer function as $F\left(\vec{k}^S\right)$ which is the 2D Fourier transform of $f(x, y)$, then single-scattering matrix $U_S\left(\vec{k}_\parallel^r; \vec{k}_\parallel^i\right)$ should be given by the following equation.

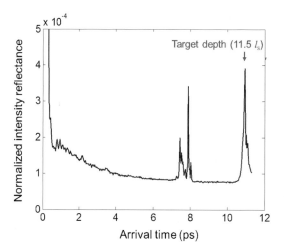

Figure 7.5 Flight time–dependent reflection signal from a target embedded in a thick scattering medium. Data were obtained by a USAF target under $11.5l_s$-thick scattering medium. Data were displayed after normalizing the area under the curve.

$$U_S\left(\vec{k}_\parallel^r;\vec{k}_\parallel^i,\tau\right) \approx F\left(\vec{k}_\parallel^r - \vec{k}_\parallel^i\right) = F\left(\vec{k}^S\right). \tag{7.1}$$

It implies that the matrix elements of U_S with the same in-plane momentum difference $\vec{k}^S$ originate from the same spatial-frequency component of F. On the other hand, the matrix elements of U_M at the same matrix index is statistically random. Therefore, if we add all matrix elements of U with the same in-plane momentum difference, $\vec{k}^S$ then the single-scattered waves originated from $F\left(\vec{k}^S\right)$ is accumulated coherently while that of multiple-scattered waves is randomly added. If we let the number of matrix elements involved in the summation as $N\left(\vec{k}^S\right)$, then the summation becomes,

$$\sum_{\vec{k}_\parallel^i} U\left(\vec{k}_\parallel^r;\vec{k}_\parallel^i\right) \approx N\left(\vec{k}^S\right) F\left(\vec{k}^S\right) + \sum_{\vec{k}_\parallel^i} U_M. \tag{7.2}$$

In the equation, $\sum_{\vec{k}_\parallel^i}$ stands for the summation of matrix elements which have the same in-plane momentum difference $\vec{k}^S$.

Experimental implementation of Eq. (7.2) is straightforward. From the measured TRRM shown in Figure 7.4c, one can map the reflected wave on the in-plane momentum difference space (Δk-space). For instance, the diagonal line in Figure 7.4c corresponds to zero in-plane momentum difference ($\vec{k}^S = (0,0)$). Therefore, all the complex numbers at the diagonal were added and then the resulting value was assigned to the $(0, 0)$ of Δk-space (Figure 7.4d). Similarly, the addition of the shifted diagonal components in Figure 7.4c can be assigned to the associated point in Δk-space. The resulting map of reflected waves in the Δk-space corresponds to the operation of Eq. (7.2).

From Eq. (7.2), we can recover the object transfer function as,

$$F\left(\vec{k}^S\right) = \left(\sum_{\vec{k}_\parallel^i} U - \sum_{\vec{k}_\parallel^i} U_M\right) / N\left(\vec{k}^S\right) \approx \frac{1}{N\left(\vec{k}^S\right)}\sum_{\vec{k}_\parallel^i} U, \tag{7.3}$$

with the assumption that the summation of multiple-scattered waves is negligible compared to the coherently added single-scattered waves. This assumption works fine when the average intensity of multiple-scattered waves is much smaller than $N\left(\vec{k}^S\right)$ times that of the single-scattered waves. Therefore, $N\left(\vec{k}^S\right)$ works as a weighting function. The number of $N\left(\vec{k}^S\right)$ for a given $\vec{k}^S$ should be determined by the autocorrelation of aperture function of the system. The maximum value of $N\left(\vec{k}^S\right)$ is equal to the total number of incident angles, N_{tot} at $\vec{k}^S = (0,0)$. And $N\left(\vec{k}^S\right)$ is distributed up to $\left|\vec{k}^S\right| = 2NA \times k_0$, where k_0 is the wavenumber in vacuum. Figure 7.6 displays the map of $N\left(\vec{k}^S\right)$ for $N_{tot} = 2500$ case. As shown in the figure, $N\left(\vec{k}^S\right)$ is reduced with increasing $\left|\vec{k}^S\right|$. Therefore, high spatial frequency is more prone to the multiple scattering noise, which results in the reduction in spatial resolution.

7.2.5 Experimental Results for Demonstrating Deep Tissue and High-Resolution Imaging

7.2.5.1 Image Performance Depending on the Thicknesses of Scattering Layers

Figure 7.7 shows the experimental results for various thicknesses of scattering layers placed on top of the USAF target. The scattering layer was made by dispersing polystyrene beads of 1 μm diameter in the Polydimethylsiloxane (PDMS). The MFP of this scattering layer was about 102 μm, which is similar to that of typical biological

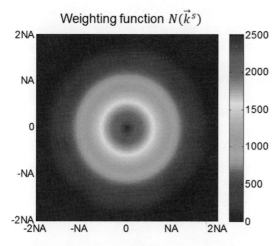

Figure 7.6 The map of $N\left(\vec{k}^S\right)$ in k-space for the case of $N_{tot} = 2500$. Because the object vector $\vec{k}^S$ is determined by the difference between the wave vectors of the incident and reflected waves, $\vec{k}^S$ is distributed within a circle of radius 2NA with weighting factor displayed in the figure. In labeling the horizontal and vertical axes, we dropped the factor k_0 for simplicity. A black and white version of this figure appears in some formats. For the color version, please refer to the plate section.

tissues. For a comparison, we present a single time-gated images for normal illumination, and angular compounding images [40] obtained by the summation of the intensity of time-gated images for each thickness (Figures 7.7a and 7.7b). Individual time-gated images cannot identify the fine details of the target even for $1.60l_s$ case. With an increase of thickness, the image contrast of angular compounding image decreased, and at $9.7l_s$, it failed to identify the ninth group of USAF target whose smallest line pair separation was $1.56\ \mu m$. On the contrary, CASS microscopy showed a good image contrast for the ninth group of the target for up to $11.5l_s$-thick scattering layer.

The difference between the angular compounding and CASS image can be easily figured out by comparing the ratio between the signal intensity from the Chrome coating of USAF target (square box I in Figure 7.7a) and that of background region (square box II in Figure 7.7a). Both single- and multiple-scattered waves contributed to the signal in box I, while only multiple-scattered waves contribute to the signal in box II. The images in Figure 7.7 were normalized by the mean intensity of square box II such that the color scale visualizes the ratio. With the increase of thickness, this ratio quickly decreases for the case of angular compounding method. On the other hand, in CASS image, this ratio remained about 100 for $11.5l_s$-thick samples due to the collective accumulation of the single-scattered waves.

To quantify the performance of CASS microscopy in the collective enhancement of single-scattered waves, we investigated the intensity growth of reconstructed image at the region of square boxes I and II (Figure 7.7a) depending on the number of incident angles. As shown in the dashed curves in Figure 7.8, the intensity at the region II grows with N_{tot} for both cases of angular compounding and CASS microscopy. On the other hand, in square box I, single-scattered waves were dominant for the reconstructed image. Therefore, the intensity of the CASS image grew much faster than that of the angular compounding image (solid curves in Figure 7.8). We observe that the single-scattering intensity grows with $N_{tot}^{1.82}$, which is close to the expectation of N_{tot}^2. Meanwhile, the same plot for angular compounding image shows the growth rate of $N_{tot}^{0.92}$. The SNR is determined by the ratio between the intensity of the single-scattered waves and the intensity fluctuation of the multiple-scattered waves. For the case of the angular compounding, the fluctuation of multiple-scattering grows with $\sqrt{N_{tot}}$ as a result of the addition of independent random speckles. Therefore, SNR scales with $\sqrt{N_{tot}}$. For the case of CASS image, multiple-scattered waves forms single speckle pattern which results in the growth of intensity fluctuation with N_{tot} [41]. In the meanwhile, the intensity of single-scattered waves grows with N_{tot}^2. Therefore, SNR of CASS image grows in proportional to N_{tot}, which is $\sqrt{N_{tot}}$ times better than the angular compounding method.

7.2.5.2   Imaging of Gold Particles Embedded within Scattering Medium

In the previous sections, we discussed the performance of CASS microscopy with USAF target under various thicknesses of scattering layers. To demonstrate with more general targets, we prepared gold particles of $2\ \mu m$ diameter embedded in between two $8.3l_s$-thick scattering layers. As a point of reference, we took clean target images without covering upper scattering layer as displayed in Figures 7.9a–7.9c. As shown

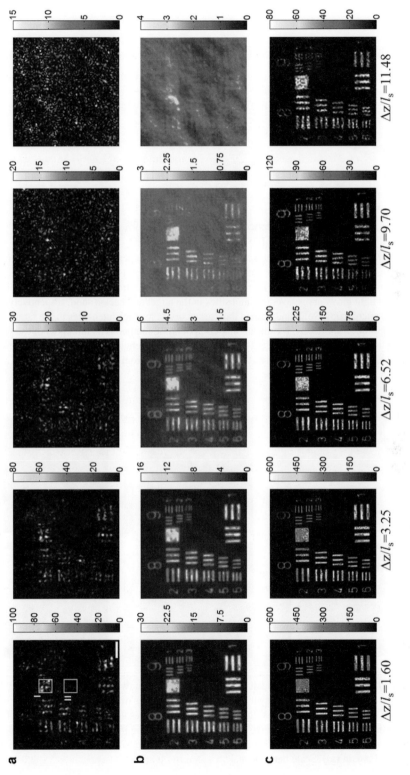

Figure 7.7 Demonstration of near diffraction limit imaging in a thick scattering medium. (**a**) Single-shot images under normal illumination. From the left, the thickness of the scattering layer was $1.60l_s$, $3.25l_s$, $6.52l_s$, $9.70l_s$, and $11.48l_s$, respectively. (**b**) Angular compounding images for the same thicknesses in (**a**). (**c**) Reconstructed images using CASS for the same thicknesses in (**a**). All images were normalized by the mean values in the region indicated by square box II where there was no chrome coating. Because mostly multiple-scattered waves were present in this region, the color bars correspond to the intensity ratio of single- to multiple-scattered waves. Scale bar, 10 μm. Adapted by permission from Macmillan Publishers Ltd: Nature Photonics [37], copyright 2015.

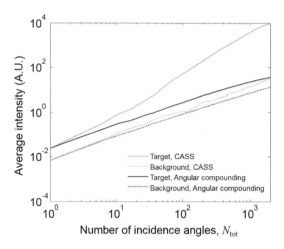

Figure 7.8 Performance of single-scattering enhancement depending on the number of incidence angles, N_{tot}. The average intensities are plotted for the target (square box I in Figure 7.7a) and background (square box II in Figure 7.7a) before the application of normalization. Solid and dashed light gray curves: average intensities in the target and background, respectively, for CASS microscopy. Solid and dashed dark gray curves: average intensities in the target and background, respectively, for the angular compounding method. The thickness of the scattering layer used in this plot was 9.7 l_s. The slopes of the solid light and dark gray lines were measured to be 1.82 and 0.92, respectively. Adapted by permission from Macmillan Publishers Ltd: Nature Photonics [37], copyright 2015.

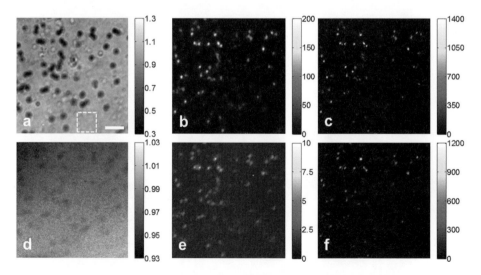

Figure 7.9 Imaging gold-coated silica beads placed between two scattering layers. (**a–c**) Images without covering a scattering layer on the top. (**d–f**) Images of target beads embedded between two 8.3l_s-thick scattering layers. (**a, d**) Transmission images taken by illuminating LED lights from the bottom of the sample. (**b, e**) Angular compounding images. (**c, f**) CASS images. Color scales indicate intensity normalized by the background intensity where there were no gold beads (dashed box in (**a**)). Scale bar: 10μm.

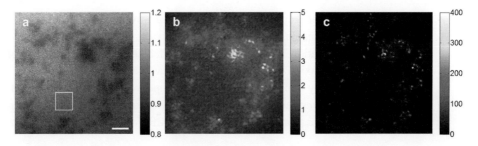

Figure 7.10 Imaging of 2-μm-diameter beads embedded in a thick rat brain tissue. (**a**) Transmission image taken with the light source (light-emitting diode, λ = 780 nm) illuminating the specimen from the bottom. The target sample was 2 μm-diameter gold-coated beads sandwiched between two 560 μm-thick rat brain slices. (**b**) Angular compounding image for the same sample in (**a**). (**c**) CASS image for the same sample in (**a**). Scale bar, 10 μm. Color scales were normalized by the background region containing no beads indicated by a white box in (**a**). Adapted by permission from Macmillan Publishers Ltd: Nature Photonics [37], copyright 2015.

in the figure, all three images clearly identified the individual particles with high contrast.

Figures 7.9d–7.9f shows the images taken at the same position after covering a thick scattering layer on the top. Unlike the previous case, the image quality of both transmission image and angular compounding image was significantly degraded by the multiple-scattered waves. However, CASS image still showed a comparable image quality to the case without upper scattering layer. The correlation between Figures 7.9c and 7.9f was measured to be 0.91, which proved that we could obtain the correct image with accuracy better than 90% even in the presence of $8.3l_s$-thick scattering layer.

7.2.5.3   Imaging Targets Embedded within Biological Tissues

To demonstrate that CASS microscopy is applicable to real biological tissues, we imaged gold particles embedded in between two slices of 560 μm-thick rat brain tissues. These tissues were extracted from a three days-old Sprague Dawley rat. Similar to previous experimental results, CASS image (Figure 7.10c) shows much better image quality compared with transmission image by LED light (Figure 7.10a) and angular compounding image (Figure 7.10b). Only the CASS image could resolve individual particles. This confirmed that CASS technique can be useful for biological studies.

7.3   Spectral Domain Measurements of the Time-Resolved Reflection Matrix

In this section, we introduce a method of measuring time-resolved reflection matrix (TRRM) of a turbid medium using spectral domain detection method and its application for enhancing light energy delivery in the reflection geometry [20]. Specifically, the point of illumination of low-coherence light source was scanned by a spatial light modulator, and time-resolved detection was performed by the interference between reflected waves from a scattering medium and reference wave at the spectrometer as was done in the spectral domain OCT [42]. From a set of these measurements taken at various points of illumination, TRRM can be constructed and a particular input channel can be identified that maximizes the backscattered waves arriving after a specific flight time. By experimentally implementing this unique input wave using a spatial light modulator, light energy delivery enhancement of up to 20% was demonstrated to a target depth of 376.6 μm within a scattering medium whose MFP was a 35.9 μm.

7.3.1   Spectral Domain Recording of Time-Resolved Reflection Matrix for a Disordered Medium

In order to record a sample's reflection matrix, we need to send light waves through various independent input channels and detect the corresponding reflected complex wave fields. The separate incident channels can be implemented by generating linear phase ramps of different gradients and orientations at the back aperture of the objective lens. Various arrival times in the time-gated measurements serve as independent detection channels. An experimental setup to realize this scheme is depicted in Figure 7.11. Light from a broadband laser source (mode-locked Ti:Sapphire laser) with center wavelength

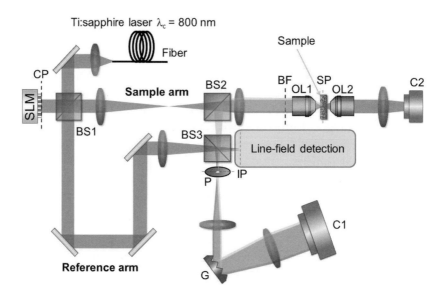

Figure 7.11 Schematic of experimental setup. The setup employs a coherent time-gated detection system in a reflection geometry. The interference of reference beam and backscattered light from the sample is measured using a spectrometer composed of a diffraction grating (G) and a line scan camera (C1). A line-field detection setup is attached at the second detection port of BS3. SLM: spatial light modulator, BSi:ith beam splitter, P: pinhole, OL2: objective lens (20×, 0.4NA), C2: 2D camera, BF: back focal plane of the objective lens (OL1, 20×, 0.4NA), CP: SLM plane conjugated to BF, SP: sample plane, IP: image plane conjugated to SP. (Figure and caption adopted from [20].)

$\lambda_c = 800$ nm and bandwidth $\Delta\lambda = 16.7$ nm (FWHM) is sampled by a beam splitter (BS1) and subsequently reflected by a spatial light modulator (SLM, Hamamatsu,
LCOS-SLM X10468–02) operating in a phase-only control mode. The light reflected
by the SLM is then transmitted through the same beam splitter and illuminates a turbid
medium via a 4-f imaging system and an objective lens (OL1, 20×, 0.4NA). A turbid medium made of polydimethylsiloxane (PDMS) and 4.5-μm-diameter polystyrene
beads were used as a test sample. The mean scattering free path of the turbid sample l_s
was measured to be 35.9 ± 0.2 μm and the transport mean free path $l_t = 394.3 \pm 2.0$ μm
through ballistic wave propagation detection [31]. The measured sample thickness L was
376.6 μm, which is the thickest sample used in these experiments, corresponding to an
average $2L/l_s = 21.0$ scattering events ($2L/l_t = 1.41$ reduced scattering events) for
light reflected from the farthest surface in the sample. The other portion of the laser
output directly transmitted through BS1 is combined with the backscattered light from
the turbid medium at BS3 to serve as a reference beam. The interference signal is measured using a custom-built spectrometer comprising a diffraction grating (G) and a line
scan camera (C1: Basler, Sprint L2048–39km). The measured spectral fringe pattern is
Fourier transformed to acquire a time-resolved complex field, which carries both the
amplitude and phase information of the backscattered light returning from the sample.
The temporal resolution of the spectrometer was characterized as $\Delta t = 20.3$ fs with a
total time span of 12.4 ps.

A spatial degree of freedom for the detection is constrained by a pinhole (P, $\phi =$
50 μm) positioned in the detection arm at a plane conjugate to the sample plane. Thus,
this detection system measures the space-constrained and time-resolved response of a
turbid medium. We have also implemented a time-resolved line-field detection setup
at the second detection port of BS3 to observe 1-dimensional field distribution of the
backscattered light from the sample (see more details in Figure 7.14). At the opposite
side of the sample, an imaging system composed of an objective lens OL2 (20×, 0.4NA),
a tube lens, and a camera C2 are positioned to observe the transmitted light through the
sample.

Each incident channel is generated by writing an appropriate linear phase ramp to a
square active area (350 × 350 pixels) of the SLM. This subregion of the SLM fills the
entire back aperture of the objective lens. Each phase ramp generates a field $E_{in}\left(\vec{k}_i\right)$
on the sample plane where $\vec{k}_i$ is the wave vector for incident light and i is the index of
the phase ramp. To construct TRRM, we choose a subset of the orthonormal basis of
phase ramps with an integer number of cycles in the vertical and horizontal directions.
As shown in Figure 7.12a, we apply one of the phase ramps on the SLM and measure
complex amplitude of the backscattered signal (Figure 7.12b). Next, the complex field is
detected for another phase ramp in the set. The measured complex fields have different
speckle formation due to sample turbidity. Only a single measurement is required for
each channel of the illumination because of the complex field recording via the interferometric detection. This efficiently increases the measurement speed compared to a
phase-stepping approach [29]. The maximum number of phase ramps used in the experiment is 1600; the total scanning time is about 160 seconds, a quantity that is determined

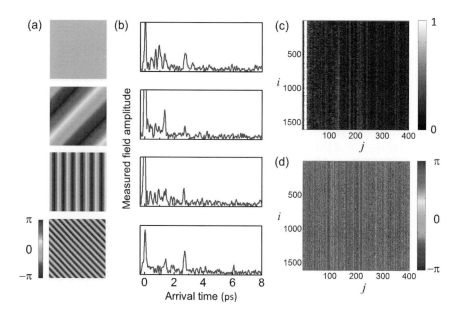

Figure 7.12 Measurement of TRRM. (**a**) Representative phase ramps written on the SLM to generate illumination channels. Each region represents a 350 × 350-pixel square array. These ramps constitute an orthogonal set of incident channels. Color bar: phase in radians. (**b**) Time-resolved reflection signal from a turbid medium for each incident channel illuminated by the phase ramps shown in (**a**). The arrival time is represented as a round-trip delay of light. Reflection from the first surface of a turbid sample is set as 0. Only amplitude components are shown here. (**c**) Amplitude part of the constructed TRRM. Indices i and j represent incident channel (phase ramp) and detection channel (time bin) respectively. Color bar: arbitrary units. (**d**) Phase portion of the measured TRRM. Color bar: phase in radians. (Figure and caption adopted from [20].) A black and white version of this figure appears in some formats. For the color version, please refer to the plate section.

primarily by the access time of the SLM ($\sim$50 msec). After measuring reflected fields for all the phase ramps in the basis set, we construct TRRM, $r\left(\vec{k}_i, j\Delta t\right)$, where index j represents jth time bin. Figures 7.12c and 7.12d show the amplitude and phase of the measured TRRM.

7.3.2 Optimization of the Energy of the Reflected Waves at the Target Arrival Time and Position

The incident field that maximizes the backscattered wave at a specific arrival time, $t_0 = j_0 \Delta t$, can be calculated by a superposition of the input channels as follows:

$$E_{opt}\left(t_0\right) = \sum_{i=1}^{N} a_i \left(j_0 \Delta t\right) E_{in}\left(\vec{k}_i\right) \tag{7.4}$$

where N is the total number of input channels. From the measured reflection matrix, we can determine the following complex amplitude for ith phase ramp,

$$a_i\,(j_0\Delta t) = r^* \left(\vec{k}_i, j_0\Delta t\right) \Bigg/ \sqrt{\sum_{i'=1}^{N} \left|r\left(\vec{k}_{i'}, j_0\Delta t\right)\right|^2} \qquad (7.5)$$

In Eq. (7.5), the denominator on the right-hand side is a normalization factor that accounts for the conservation of input power. By applying the complex amplitudes of all the phase ramps with their amplitudes assigned in Eq. (7.5), we can obtain the amplitude of the reflected signal optimized at the arrival time $j_0\Delta t$:

$$\left|S_{opt}\,(j_0\Delta t)\right| = \sqrt{\sum_{i=1}^{N} \left|r\left(\vec{k}_i, j_0\Delta t\right)\right|^2} \qquad (7.6)$$

Note that the optimization process aligns the complex phasors of backscattered waves for various input channels each multiplied by a complex weighting factor.

In order to experimentally realize this optimization process, we implement a superposed wave of illumination channels that satisfies Eq. (7.4). The insets in Figures 7.13a–7.13c show the phase patterns applied to the incident light that maximize the backscattered signal at the arrival times indicated as dashed lines in respective figures. Only the phase distribution of the calculated complex superposition is generated experimentally with a flat amplitude profile since the SLM is operated in a phase-only control mode. The blue lines in Figures 7.13a–7.13c represent the experimentally measured reflection signal. Compared with a single-channel illumination with no wavefront control (black lines), the amplitude of the reflected wave is dramatically increased at the target arrival times. When $N = 1000$, the maximum achievable enhancement of the reflected energy is 335 at the target arrival time and position.

In order to assess the accuracy of our experimental optimization, we estimate the reflected wave reconstructed from the measured TRRM. For the input wave given by Eq. (7.4) and Eq. (7.5), which optimizes the intensity at arrival time of $t_0 = j_0\Delta t$, the complex field associated with backscattered wave at an arbitrary arrival time $t = j\Delta t$ is calculated to be

$$\left|S_{opt}\,(j\Delta t)\right| = \sum_{i=1}^{N} r^* \left(\vec{k}_i, j_0\Delta t\right) r\left(\vec{k}_i, j\Delta t\right) \Bigg/ \sqrt{\sum_{i'=1}^{N} \left|r\left(\vec{k}_{i'}, j_0\Delta t\right)\right|^2} \qquad (7.7)$$

The green lines in Figures 7.13a–7.13c show the expected reflected waves reconstructed using Eq. (7.7). These are the signals we can obtain if we implement both amplitude and phase patterns on the SLM. Therefore, the degree of potential enhancement is higher than that achieved in our experiment. In order to account for the effect of the phase-only control used in our experiment, we eliminated the amplitude part of input wave and used only the phase part for calculating the output wave. In doing so, we determined the expected reflected waves for the phase-only control cases (red lines). The expected signal enhancement in this case is now very similar to that observed experimentally.

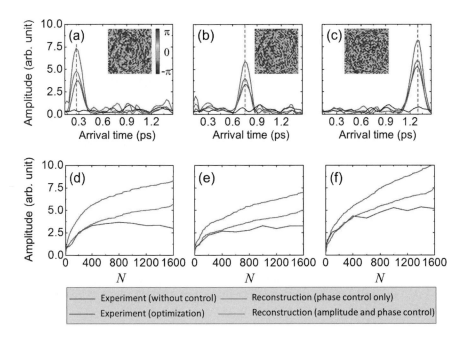

Figure 7.13 Optimization of reflected wave at a target arrival time. (**a–c**) Reflected light signals after optimization with $N = 1000$ for arrival times of (**a**) 0.27 ps, (**b**) 0.75 ps, and (**c**) 1.28 ps, respectively. Insets are phase patterns of incident waves used for optimization. The black lines: single-channel, time-gated signal without wavefront control. Blue lines: reflected wave for the experimentally generated incident wave by phase-only mode of SLM. The green lines: theoretically expected reflection signal estimated by Eq. (7.7). Red lines: same as green lines but with the phase-only input. Color bar: phase in radians. (**d–f**) Signal enhancement at the same arrival times as (**a**), (**b**), and (**c**), respectively, as a function of number of channels N. (Figure and caption adopted from [20].) A black and white version of this figure appears in some formats. For the color version, please refer to the plate section.

Next, we systematically investigate the enhancement of the signal amplitude at the target arrival time as a function of N, the number of input channels we used. Figures 7.13d–7.13f show the peak signal in Figures 7.13a–7.13c with increasing N. The two theoretical expectations (green and red curves) of TRRM reconstruction steadily increase; they achieve good agreement with a curve taking the form N^{α}, with α about 0.5 [23]. This finding also conforms to the theoretical predictions. In our experiment the backscattered signal is enhanced fairly well up to $N \sim 400$ and is in excellent agreement with the theoretical TRRM reconstruction, especially that using phase-only control. Beyond $N \sim 400$, however, the experimental curve exhibits saturation, unlike the theoretical reconstruction. The main reason for this discrepancy is the inability of the SLM to generate finer patterns. As N increases, the optimization pattern becomes finer. As a result, the effect of diffraction at the boundaries of the finite-sized pixels of the SLM deteriorates the realization of optimized waves [27]. Uneven spectral response of the SLM [24] and mechanical perturbation of the system may also account for some of the discrepancies.

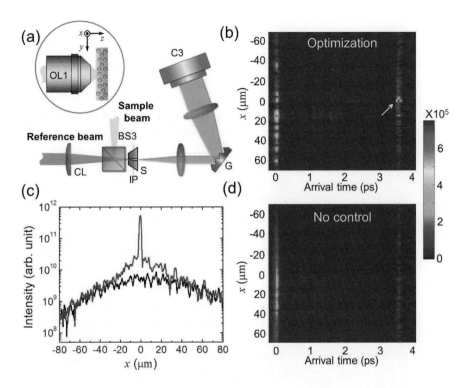

Figure 7.14 Spatiotemporal focusing of reflected wave. (**a**) Experimental schematic for the 1-D line-field detection setup. CL: cylindrical lens, S: slit, C3: 2-D camera. The inset represents spatial coordinate system at the sample plane. (**b**) Representative measured amplitude of the reflected light after optimization with $N = 1600$ phase ramps. A strong focused spot is generated at the target arrival time and the pinhole position (indicated with an arrow: $t_0 = 3.6$ ps, $x = 0$). (**c**) Ensemble averaged intensity profiles along the target arrival time $\tau = 3.6$ ps in (**b**) and (**d**). Black and blue lines represent the profiles before and after optimization, respectively. (**d**) The same as (**b**), but before the optimization. Color bar: arbitrary units. (Figure and caption adopted from [20].) A black and white version of this figure appears in some formats. For the color version, please refer to the plate section.

7.3.3   Observation of Spatiotemporal Focusing of Reflected Light

Next, we explore the field distribution of backscattered light from the turbid medium before and after the optimization. The experimental schematic is depicted in Figure 7.14a. After recording the TRRM using 1600 input channels, we chose a specific time bin corresponding to the direct flight to the back surface of the turbid medium, and constructed the input pattern that optimizes the intensity of the detected signal for the selected time bin. We then measured time-resolved complex field at the image plane (IP, dashed line in Figure 7.11 and Figure 7.14a) along the line including the target point using the separate line-field detection setup shown in Figure 7.14a. The time-resolved reflected waves measured at 15 different positions of the sample are averaged so as to achieve an ensemble average of various realization of the turbid medium (with the same thickness and scattering properties). Figure 7.14b shows one of the representative amplitude profiles of the backscattered wave along the line with respect to the arrival time. The optimized spot, which is indicated by an arrow, is clearly visible at the target arrival time ($\tau = 3.6$ ps) and corresponding location of the pinhole ($x = 0\ \mu$m). This result illustrates the spatiotemporal focusing of the returning wave at the pinhole location. For comparison, we applied a flat pattern on the SLM and measured the backscattered wave along the line (Figure 7.14d) and confirmed that the optimized spot at the pinhole location had disappeared. The spatial line profile of the ensemble averaged intensity distribution at the target arrival time is plotted in Figure 7.14c.

The peak intensity after the optimization is about two orders of magnitude higher than that before the optimization. Moreover, the observed intensity in the region surrounding the target optimization position also increases to a certain extent as well. This background intensity augmentation is known to be an effect of single-channel optimization [22]. Furthermore, when we calculate the energy of the reflected light at the target arrival time by assuming the azimuthal symmetry of scattered waves, we observe 41% increment in reflected energy after the optimization. This indicates that the optimization process delivers more light energy to the target arrival time. One important implication of this observation is that the optimization process can enhance light energy delivery to the *sample depth* associated with the target arrival time. This is most feasible for a medium with moderate turbidity because the arrival time is directly linked to the depth at which light is reflected to a large extent.

7.3.4   Verifying Enhanced Light Energy Delivery to the Target Depth

In order to unambiguously support our claim that the optimization of reflected wave corresponding to a specific arrival time enhances light energy delivered to the associated depth, we measure transmitted energy at the far side of the turbid medium after the optimization. The experimental schematic is shown in Figure 7.15a where a camera is

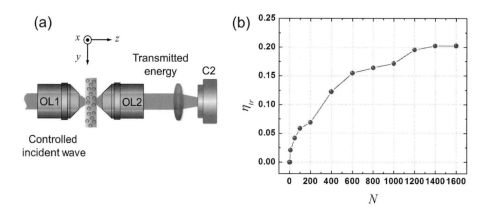

Figure 7.15 Transmission enhancement. (**a**) Experimental schematic. Transmitted energy is measured before and after optimization. (**b**) The transmission energy enhancement is measured after the optimization as a function of N. It is equal to about half of the reflection enhancement. (Figure and caption adopted from [20].)

installed at the opposite side of the illumination (C2 in Figure 7.11). After optimizing the backscattered wave at the arrival time corresponding to the back surface of the sample, the energy enhancement (η_{tr}), defined as the ratio of the net increased transmission energy after the optimization to the transmission energy under no control, is obtained as a function of the number of phase ramps N used in the optimization process (Figure 7.15b). The enhancement increases up to about 20% as N is increased. Since the back surface of the turbid medium serves as a partial reflector, the transmitted energy is proportional to the amount of light delivered to the back surface. Therefore, the increase in measured transmission confirms that the energy delivery to the target depth is indeed enhanced.

For experimental verification, we repeated the same experiment with several samples of different thickness but same scattering properties. The number of reduced scattering events, $2L/l_t$, ranges between 0.44 and 1.91, covering weak to moderate turbidity. As shown in Figure 7.16, the measured enhancement depends on the sample thickness. For example, with the thinnest sample ($2L/l_t = 0.44$) the reflection enhancement at the target arrival time and the corresponding transmission enhancement were measured to be 100% and 41% respectively, the best experimental demonstration reported so far. These values can be further improved if more input channels are used with improved accuracy for wavefront shaping. In an ideal condition where full solid angle of input channels is covered, our theoretical model predicts more than a fourfold increase in the energy delivery. As the degree of sample disorder (or turbidity) increases, the light

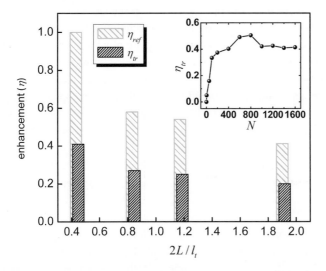

Figure 7.16 Reflection enhancement (η_{ref}) and transmission enhancement (η_{tr}) via single-channel optimization process for various sample thicknesses. The target arrival time for the reflection signal is set in a way that the corresponding depth coincides with the back surface of each sample. For the reflection enhancement, only the energy contained in the target time bin is considered. However, total energy is taken into account for the transmission case. The inset represents the transmission energy enhancement for the thinnest sample as a function of N. The final enhancement is 41% for $N = 1600$. (Figure and caption adopted from [20].)

energy delivery becomes less efficient. This is due to the increased discrepancy between arrival time and corresponding depth.

The enhancement of reflected light energy at the target arrival time by means of the optimization process can be understood by an analogy with the single-channel optimization performed in the transmission geometry [22, 25]. In the transmission case, the optimization of intensity at a single point in the output plane enhances the total energy at the entire output plane. In the reflection configuration, as in our experiment, the optimization for a single point represented by a specific time bin enhances light energy delivered to the entire plane associated with the same time bin, which was evidenced in Figure 7.14c. Since the arrival time and the depth are related to a certain extent in a moderately turbid medium, the enhancement of light waves with a particular time-of-flight will lead to the enhanced light energy delivery to the corresponding depth.

7.4   Summary and Future Outlook

In this chapter, we have presented the use of time-resolved reflection matrix in relieving the limitations of imaging depth and light energy delivery imposed by multiple light scattering. On the one hand, TRRM made it possible to collectively enhance single-scattered waves from a target object embedded in thick scattering media, which has resulted in the improvement of the depth of diffraction-limited optical imaging up to 11.5 times the scattering mean free path. Because the method works for coherent light scattering process, it itself is not applicable to fluorescence imaging, but may be combined with two-photon or three-photon imaging techniques to exploit various contrast mechanisms for specimens. High-resolution imaging at an unprecedented target depth will lead to advances in the life sciences and biomedicine, in which the shallow imaging depth of conventional optical microscopy is a major obstacle. In addition, the ability to distinguish single- and multiple-scattered waves will open new approaches to studying the physics of the interaction of light with complex media.

On the other hand, the use of the eigenchannels of TRRM led to the experimental optimization of the backscattered wave in time and space. The optimization dramatically enhanced the magnitude of the reflected signal from the target arrival time and consequently increased delivered energy to the corresponding depth inside a scattering medium. Since the method employs a reflection configuration rather than a transmission geometry, it is highly relevant to medical applications such as deep tissue *in vivo* imaging [43–45] and efficient light delivery for photo-therapy.

For future perspectives, follow-up studies would be necessary for improving the data acquisition speed of TRRM in order to cope with the perturbation of scattered waves caused by the dynamics of living specimen. At the moment, the acquisition of 2500 angle-dependent wide-field images takes ∼250 s when a liquid crystal–based spatial light modulator is used. This technical issue may be resolved by the use of a high-speed wavefront-shaping device as well as a fast camera. With the proper dealing with this speed issue, we may foresee a broad use of TRRM in various areas of life science and biomedicine.

7.5   References

[1] D. B. Murphy, *Fundamentals of Light Microscopy and Electronic Imaging.* Wiley-Liss, 2001.

[2] V. Ntziachristos, "Going deeper than microscopy: the optical imaging frontier in biology," *Nature Methods,* vol. 7, pp. 603–614, 08//print 2010.

[3] L. V. Wang, "Prospects of photoacoustic tomography," *Medical Physics,* vol. 35, pp. 5758–5767, 2008.

[4] L. V. Wang and S. Hu, "Photoacoustic tomography: *in vivo* imaging from organelles to organs," *Science,* vol. 335, pp. 1458–1462, 2012.

[5] A. Gibson, J. Hebden, and S. R. Arridge, "Recent advances in diffuse optical imaging," *Physics in Medicine and Biology,* vol. 50, p. R1, 2005.

[6] V. Ntziachristos, A. G. Yodh, M. Schnall, and B. Chance, "Concurrent MRI and diffuse optical tomography of breast after indocyanine green enhancement," *Proceedings of the National Academy of Sciences,* vol. 97, pp. 2767–2772, 2000.

[7] F. A. Duck, *Physical Properties of Tissues: A Comprehensive Reference Book.* Elsevier Science, 1990.

[8] W. Denk, J. Strickler, and W. Webb, "Two-photon laser scanning fluorescence microscopy," *Science,* vol. 248, pp. 73–76, 1990.

[9] P. Theer, M. T. Hasan, and W. Denk, "Two-photon imaging to a depth of 1000 μm in living brains by use of a Ti:Al2O3 regenerative amplifier," *Optics Letters,* vol. 28, pp. 1022–1024, 2003.

[10] D. Huang, E. A. Swanson, C. P. Lin, J. S. Schuman, W. G. Stinson, W. Chang, *et al.*, "Optical coherence tomography," *Science,* vol. 254, pp. 1178–1181, 1991.

[11] V. J. Srinivasan, H. Radhakrishnan, J. Y. Jiang, S. Barry, and A. E. Cable, "Optical coherence microscopy for deep tissue imaging of the cerebral cortex with intrinsic contrast," *Optics Express,* vol. 20, pp. 2220–2239, 2012.

[12] N. Ji, D. E. Milkie, and E. Betzig, "Adaptive optics via pupil segmentation for high-resolution imaging in biological tissues," *Nature Methods,* vol. 7, pp. 141–147, 2009.

[13] M. Rueckel, J. A. Mack-Bucher, and W. Denk, "Adaptive wavefront correction in two-photon microscopy using coherence-gated wavefront sensing," *Proceedings of the National Academy of Sciences,* vol. 103, pp. 17137–17142, 2006.

[14] C. L. Hsieh, Y. Pu, R. Grange, G. Laporte, and D. Psaltis, "Imaging through turbid layers by scanning the phase conjugated second harmonic radiation from a nanoparticle," *Optics Express,* vol. 18, pp. 20723–20731, 2010.

[15] J. B. Pawley, *Handbook of Biological Confocal Microscopy,* 2nd ed. Plenum Press, 1995.

[16] S. W. Hell and J. Wichmann, "Breaking the diffraction resolution limit by stimulated emission: stimulated-emission-depletion fluorescence microscopy," *Optics Letters,* vol. 19, pp. 780–782, 1994.

[17] E. Betzig, G. H. Patterson, R. Sougrat, O. W. Lindwasser, S. Olenych, J. S. Bonifacino, *et al.*, "Imaging intracellular fluorescent proteins at nanometer resolution," *Science,* vol. 313, pp. 1642–1645, 2006.

[18] A. B. Davis and A. Marshak, "Photon propagation in heterogeneous optical media with spatial correlations: enhanced mean-free-paths and wider-than-exponential free-path distributions," *Journal of Quantitative Spectroscopy and Radiative Transfer,* vol. 84, pp. 3–34, 2004.

[19] K. Busch, C. M. Soukoulis, and E. N. Economou, "Transport and scattering mean free paths of classical waves," *Physical Review B,* vol. 50, pp. 93–98, 1994.

[20] Y. Choi, T. R. Hillman, W. Choi, N. Lue, R. R. Dasari, P. T. C. So, *et al.*, "Measurement of the time-resolved reflection matrix for enhancing light energy delivery into a scattering medium," *Physical Review Letters,* vol. 111, 243901, 2013.

[21] Y. Choi, T. D. Yang, C. Fang-Yen, P. Kang, K. J. Lee, R. R. Dasari, *et al.*, "Overcoming the diffraction limit using multiple light scattering in a highly disordered medium," *Physical Review Letters,* vol. 107, 023902, 2011.

[22] I. M. Vellekoop and A. P. Mosk, "Universal optimal transmission of light through disordered materials," *Physical Review Letters,* vol. 101, Sep 19 2008.

[23] I. M. Vellekoop and A. P. Mosk, "Focusing coherent light through opaque strongly scattering media," *Optics Letters,* vol. 32, pp. 2309–2311, 2007.

[24] J. Aulbach, B. Gjonaj, P. M. Johnson, A. P. Mosk, and A. Lagendijk, "Control of light transmission through opaque scattering media in space and time," *Physical Review Letters,* vol. 106, 2011.

[25] W. Choi, A. P. Mosk, Q. H. Park, and W. Choi, "Transmission eigenchannels in a disordered medium," *Physical Review B,* vol. 83, 2011.

[26] B. Judkewitz, Y. M. Wang, R. Horstmeyer, A. Mathy, and C. Yang, "Speckle-scale focusing in the diffusive regime with time reversal of variance-encoded light (TROVE)," *Nature Photonics,* vol. 7, pp. 300–305, 2013.

[27] M. Kim, Y. Choi, C. Yoon, W. Choi, J. Kim, Q.-H. Park, *et al.*, "Maximal energy transport through disordered media with the implementation of transmission eigenchannels," *Nature Photonics,* vol. 6, p. 581, 2012.

[28] S. Popoff, G. Lerosey, M. Fink, A. C. Boccara, and S. Gigan, "Image transmission through an opaque material," *Nature Communications,* vol. 1, p. 81, 2010.

[29] S. M. Popoff, G. Lerosey, R. Carminati, M. Fink, A. C. Boccara, and S. Gigan, "Measuring the transmission matrix in optics: an approach to the study and control of light propagation in disordered media," *Physical Review Letters,* vol. 104, 2010.

[30] X. Xu, H. Liu, and L. V. Wang, "Time-reversed ultrasonically encoded optical focusing into scattering media," *Nature Photonics,* vol. 5, pp. 154–157, 2011.

[31] Z. Yaqoob, D. Psaltis, M. S. Feld, and C. Yang, "Optical phase conjugation for turbidity suppression in biological samples," *Nature Photonics,* vol. 2, pp. 110–115, 2008.

[32] W. Choi, M. Kim, D. Kim, C. Yoon, C. Fang-Yen, Q.-H. Park, *et al.*, "Preferential coupling of an incident wave to reflection eigenchannels of disordered media," *Scientific Reports,* vol. 5, 2015.

[33] M. Kim, W. Choi, Y. Choi, C. Yoon, and W. Choi, "Transmission matrix of a scattering medium and its applications in biophotonics," *Optics Express,* vol. 23, pp. 12648–12668, 2015.

[34] M. Kim, W. Choi, C. Yoon, G. H. Kim, S.-H. Kim, G.-R. Yi, *et al.*, "Exploring anti-reflection modes in disordered media," *Optics Express,* vol. 23, pp. 12740–12749, 2015.

[35] Y. Choi, C. Yoon, M. Kim, T. D. Yang, C. Fang-Yen, R. R. Dasari, *et al.*, "Scanner-free and wide-field endoscopic imaging by using a single multimode optical fiber," *Physical Review Letters,* vol. 109, 203901, 2012.

[36] O. Katz, E. Small, Y. Bromberg, and Y. Silberberg, "Focusing and compression of ultrashort pulses through scattering media," *Nature Photonics,* vol. 5, pp. 372–377, 2011.

[37] S. Kang, S. Jeong, W. Choi, H. Ko, T. D. Yang, J. H. Joo, *et al.*, "Imaging deep within a scattering medium using collective accumulation of single-scattered waves," *Nature Photonics,* vol. 9, pp. 253–258, 2015.

[38] R. Dicke, "Coherence in spontaneous radiation processes," *Physical Review,* vol. 93, p. 99, 1954.

[39] Y. Choi, T. D. Yang, K. J. Lee, and W. Choi, "Full-field and single-shot quantitative phase microscopy using dynamic speckle illumination," *Optics Letters,* vol. 36, pp. 2465–2467, 2011.

[40] A. E. Desjardins, B. J. Vakoc, W. Y. Oh, S. M. Motaghiannezam, G. J. Tearney, and B. E. Bouma, "Angle-resolved optical coherence tomography with sequential angular selectivity for speckle reduction," *Optics Express,* vol. 15, pp. 6200–6209, 2007.

[41] J. W. Goodman, *Statistical Optics.* Wiley, 2000.

[42] S. Yun, G. Tearney, J. de Boer, N. Iftimia, and B. Bouma, "High-speed optical frequency-domain imaging," *Optics Express,* vol. 11, pp. 2953–2963, 2003.

[43] M. J. Booth, M. A. Neil, R. Juskaitis, and T. Wilson, "Adaptive aberration correction in a confocal microscope," *Proceedings of the National Academy of Sciences of the United States of America,* vol. 99, pp. 5788–5792, 2002.

[44] P. N. Marsh, D. Burns, and J. M. Girkin, "Practical implementation of adaptive optics in multiphoton microscopy," *Optics Express,* vol. 11, pp. 1123–1130, 2003.

[45] M. Rueckel, J. A. Mack-Bucher, and W. Denk, "Adaptive wavefront correction in two-photon microscopy using coherence-gated wavefront sensing," *Proceedings of the National Academy of Sciences of the United States of America,* vol. 103, pp. 17137–17142, 2006.

Part IV

Focusing Light through Turbid Media Using Feedback Optimization

8 Feedback-Based Wavefront Shaping

Ivo M. Vellekoop

8.1 Introduction

Telescopes, cameras, and microscopes all rely on the notion that light propagates along a straight path. However, when light propagates through a turbid medium, its direction gets scrambled by scattering on microscopic particles and other inhomogeneities. As a result, no sharp focus is formed. For long, it was believed that light scattering posed a fundamental limitation on the penetration depth and resolution of all optical methods [1]. The paradigm to use ballistic (nonscattered) light for imaging has been pushed to extreme limits with the ability to separate ballistic and scattered light through spatial selection [2–4] or coherence gating [5], and through the advances of multiphoton microscopy [6]. Even so, these methods still rely on ballistic light to form an image. Since the amount of ballistic light decreases exponentially with depth, increasing the penetration depth of a microscope without compromising its resolution is exponentially hard. Even with the most advanced multiphoton microscope, the typical penetration depth in biological tissue is less than a millimeter [7, 8]. For long, the only options to image at larger depths were diffuse optical tomography and related methods; approaches that unfortunately suffer from a severely reduced resolution [9, 10].

In the last few years, however, it is becoming increasingly clear that scattering needs not be a fundamental limitation for imaging. In 2007 Allard Mosk and I developed a technique, now called wavefront shaping, that can be used to focus light through [11] or even inside scattering objects [12] (see Figure 8.1). Even though light scatters in a complex and initially unknown way, the scattering process is linear and entirely deterministic as long as the medium is static. By shaping the incident wavefront in just the right way, light can be focused anywhere, even deep inside the most strongly scattering materials.

The ability to focus light at any desired position has a huge potential for applications. The field of wavefront shaping is under rapid development, and proof-of-concept applications in deep tissue microscopy [13–17], endoscopy [18–21], optical trapping [22], superresolution imaging [23, 24], nano-positioning [25], and cryptography [26, 27] have been demonstrated.

Even though light can be focused anywhere once the correct incident wavefront is known, finding this matching wavefront is far from trivial. When a detector can be placed at the desired focus, a simple feedback scheme (Figure 8.1) can be used. For most applications, however, a different approach is required. In the last couple of years, a large

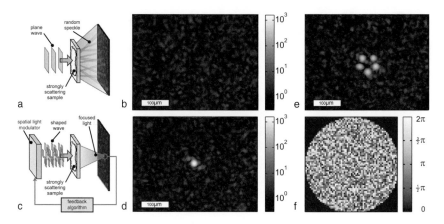

Figure 8.1 Principle of wavefront shaping. (**a**) When a scattering sample is illuminated with a plane wave, the scattered light forms a disordered interference pattern known as laser speckle. (**b**) Measured intensity of speckle pattern (logarithmic color scale). (**c**) A spatial light modulator is used to shape the wavefront of the incident beam; the light modulator is programmed using feedback from a detector placed behind the object. After completion, the feedback algorithm has found the optimum solution for focusing light through the sample. The intensity in the target has increased by over three orders of magnitude (**d**). (**e**) By combining feedback from multiple points, simple images can be projected through the object. (**f**) Phase of the incident wavefront that was used for measuring (**e**), no correlations can be observed in the incident field, indicating that the sample fully scrambles the incident wave. Adapted with permission from [11], *Optics Letters*. A black and white version of this figure appears in some formats. For the color version, please refer to the plate section.

variety of approaches and algorithms for focusing light in the presence of scattering have been demonstrated. In this chapter, I focus only on the category of feedback-based optical wavefront shaping. Therefore, I will not attempt to cover the rapidly expanding field of optical phase conjugation, nor the foundational work utilizing other wave modalities, such as ultrasound and radio waves. Recent reviews of these fields can be found in Horstmeyer et al. [28] and Mosk et al. [29], respectively. Also, it is worth noting that the wavefront shaping research field has strong roots in mesoscopic scattering physics. An excellent review of the tight connection between mesoscopic scattering theory and wavefront shaping can be found in Ref. [30].

In this chapter, I will first introduce the concept of feedback-based wavefront shaping and some of the fundamental properties of this technique. Then, I will review the different algorithms that can be used for finding the wavefront, and discuss different options for obtaining a feedback signal in the first place. After that, I touch on some of the wave correlations that play a role in wavefront shaping, and briefly link to related research fields. This chapter is concluded with an outlook of future applications.

8.2   Matrix Formalism

When a turbid medium is illuminated with coherent light, the transmitted light forms a disordered interference pattern that is known as laser speckle (see Figure 8.1b). For a given turbid sample it is not possible to predict the speckle pattern in advance. Therefore, most work on multiple scattering is aimed at finding a statistical description for an ensemble of similar samples (e.g., [31, 32]). Often, the term "random scattering" is used. This term, however, is somewhat misleading since for a given sample light scattering is completely deterministic.

Wavefront shaping exploits the fact that light scattering is linear and fully deterministic. Light transport through any linear medium – including absorbing media and magneto-optic materials where time-reversal symmetry is broken – can be described using a general matrix approach:

$$E_b = \sum_a^N t_{ba} E_a \qquad (8.1)$$

where t_{ba} are the elements of the transmission matrix (TM), a are the indices of the components, or modes, of the incident field E_a in some arbitrary basis, and, similarly, b are the indices of the components of the transmitted field E_b, potentially in a different basis; N is the number of incident modes. For simplicity, we assume that both bases are orthonormal.

The matrix approach is a convenient method that is used extensively in multiple-scattering theory [33], where the basis vectors of the incident and outgoing field are called "channels." Any orthogonal basis can be used and, when desired, the basis can be chosen to include both polarization states of the light. The "transmission matrix" Eq. (8.1) can be replaced by a different matrix without loss of generality to model, e.g., reflection, or transport to a point inside the sample, or – in fact – any other linear process.

Describing a scattering medium with a discrete matrix is a natural choice in most experimental situations, where indices a may correspond to pixels on a spatial light modulator (SLM), and indices b to pixels on a camera. Also, this formalism is the most natural way to describe scattering in a multimode waveguide, where the indices correspond to waveguide modes. In some cases, however, it is more natural to use a continuous version of Eq. (8.1), where $t(\mathbf{r}_b, \mathbf{r}_a)$ is a function of the transversal coordinates at the back and front surface of the sample respectively (for an example of this approach, see the supplement of Ref. [34]).

8.3   Wavefront Shaping

In its simplest form, wavefront shaping is concerned with solving the following optimization problem: ***what is the incident field E_a that maximizes transmission into a desired output mode β.*** Here, mode β can represent, e.g., a focus, a plane wave, or any other desired field pattern. Fortunately, this optimization problem is very easy to solve for a linear system when the TM is known. Defining the "intensity"[1] in mode β as $I_\beta \equiv |E_\beta|^2$, we have from Eq. (8.1)

---

[1] It is common practice to use the term "intensity" liberally. In a matrix representation there are no spatial coordinates and "intensity" is a quantity that is proportional to the power (Watts) in the mode: "intensity" and "power" are equivalent. In a continuous field description, however, the distinction must be made, and I

$$I_\beta = \left| \sum_a^N t_{\beta a} E_a \right|^2 \leq \sum_a^N |t_{\beta a}|^2 \sum_{a'}^N |E_{a'}|^2 \tag{8.2}$$

where the Cauchy-Schwartz inequality was used in the second step. The equality only holds when $E_a \propto t_{\beta a}^*$. Assuming that we are not allowed to change the total incident power, $P_{in} \equiv \sum_a^N |E_a|^2$ remains constant. Therefore, the incident field that maximizes transmission into mode E_β is simply given by

$$\hat{E}_a = E_0 t_{\beta a}^* \tag{8.3}$$

with $E_0 = \left(P_{in} / \sum_a^N |t_{\beta a}|^2 \right)^{1/2}$ a normalization constant to ensure that the total incident power remains fixed. This solution represents a global maximum that is unique up to an arbitrary phase factor. Interestingly, this solution is exactly the phase conjugate of a wave propagating from β back to a (see Section 1.7.2 for a brief discussion on the relation with phase conjugation).

Note that Eq. (8.3) is valid for any linear medium, not only for scattering media.

8.3.1   Total Transmission into the Focus

Even though wavefront shaping techniques can increase the intensity at a given point, it is theoretically impossible to focus all incident light to that point. The fraction of the incident power that reaches the focus after optimization is usually far less than 1%. From Eq. (8.2) we can see directly that

$$\frac{I_\beta}{P_{in}} \leq T_\beta \text{ with } T_\beta \equiv \sum_a^N |t_{\beta a}|^2 \tag{8.4}$$

For phase-conjugation experiments this relation represents a simple symmetry: in this context T_β can be interpreted as the fraction of the source power that reaches the phase-conjugation system. If, for example, a phase-conjugation system collects 0.1% of the light from the source, no more than 0.1% of the phase-conjugated light can be sent back to the reconstructed focus. The rest of the light will partially be reflected and absorbed, and partially forms a background speckle around the focus. Usually $T_\beta \ll 1\%$ due to a combination of factors: reflected light is not collected, usually only a single polarization is used, the system has a finite numerical aperture and a finite extent, the sample may absorb some of the light, etc. For feedback-based wavefront shaping the situation is even worse than for phase conjugation, because the number of control element is usually limited to a few thousand for performance reasons. Fortunately, for most applications it is not essential to put a large fraction of the total energy in the focus: achieving a good contrast is sufficient.

will here use the terms "intensity" and "power" consistently with a continuous field representation, where power (W) is intensity (W/m^2) integrated over area.

8.3.2   Enhancement

One of the most common figures of merit is the enhancement η. It is defined as

$$\eta \equiv \frac{\hat{I}_\beta}{I_{\text{ref}}} \tag{8.5}$$

where $\hat{I}_\beta$ is the intensity in the focus after optimizing the incident wavefront, and I_{ref} is the reference intensity. There are several subtly different methods to determine the reference intensity:

8.3.2.1   Initial Intensity in β

A common pitfall is to use the initial intensity in β (before optimization) as a reference. However, this approach is incorrect because the initial intensity is a speckle pattern, so the initial intensity in β is a random number drawn from an exponential distribution, making it unsuitable as a reference value.

8.3.2.2   Average over Incident Field

One of the simplest approaches to define I_{ref} is to measure the target intensity I_β averaged over a large number of random incident wavefronts, all with the same total incident power P_{in}.

$$I_{\text{ref}}^{(E)} = \left\langle \left| \sum_a t_{\beta a} E_a \right|^2 \right\rangle_E = \sum_a^N |t_{\beta a}|^2 \langle |E_a|^2 \rangle_E + \sum_a^N \sum_{a' \neq a}^N t_{\beta a}^* t_{\beta a'} \langle E_a^* E_{a'} \rangle_E \tag{8.6}$$

where $\langle \ \rangle_E$ denotes averaging over all random incident fields. Since these random fields have, by definition, no correlation between components E_a and $E_{a' \neq a}$ the second term vanishes. Assuming that the spatial light modulator is illuminated homogeneously, $\langle |E_a|^2 \rangle_E = P_{\text{in}}/N$, and we find $I_{\text{ref}} = T_\beta P_{\text{in}}/N$. Combining Eqs. (8.4) and (8.5), we find $\eta \leq N$, giving $\eta = N$ for the optimal incident wavefront. In case that the SLM is not illuminated homogeneously, a correction factor can be applied that effectively reduces the number of modes [35].

This method of determining the enhancement is both simple to implement and to describe analytically. However, there are two subtle caveats that make this way of determining the enhancement a less suitable figure of merit in some experimental situations.

Firstly, in experiments the total power that is incident on the sample surface often depends strongly on the pattern that is displayed on the light modulator, mainly because of diffraction and vignetting in the light path. Since the TM includes the transmission from the SLM through the optical system to the sample, the ideal wavefront $E_a \propto t_{\beta a}$ jointly optimizes transmission from the SLM to the sample surface and transmission from the sample surface to the target β. Hence, part of the enhancement that is observed in an experiment may come from the "trivial" fact that the transmission from the SLM to the sample surface is optimized. To avoid this problem, one should implement a reference path to measures how much light reaches the sample (e.g., see Ref. [11]).

Depending on the geometry of the experiment and characteristics of the sample, additional problems may arise with this way of determining the reference intensity. For example, consider the situation of focusing light through a glass slide. Even though it would be easy to achieve an enhancement of a factor 1000 *over a random incident wavefront*, a more relevant number would be the enhancement over a focused wave, which will be minimal. In practice, this method of determining the reference intensity only makes sense if the elements of the optimal wavefront $\hat{E}_a$ are uncorrelated, which may be checked by determining the spatial autocorrelation function of the generated field [11].

8.3.2.3 Averaging over Samples

A different approach was introduced in Ref. [36]. Instead of varying the incident wave to determine I_{ref}, this approach uses the *same incident wavefront* that was used for measuring $\hat{I}_\beta$, but *averages over the ensemble of possible samples*. Ensemble averaging can usually be replaced by translation of the sample, or by time averaging (for dynamic media). Using this method, we find

$$I_{\text{ref}}^{(t)} = \left\langle \left| \sum_a^N t_{\beta a} \hat{E}_a \right|^2 \right\rangle = \sum_a^N \langle |t_{\beta a}|^2 \rangle \left| \hat{E}_a \right|^2 + \sum_a^N \sum_{a' \neq a}^N \langle t_{\beta a}^* t_{\beta a'} \rangle \hat{E}_a^* \hat{E}_{a'} \qquad (8.7)$$

where $\langle\ \rangle$ denotes ensemble averaging. For now, let's assume that $t_{\beta a}$ and $t_{\beta a'}$ are uncorrelated. Then the second term vanishes and by substituting of Eq. (8.3) we have.

$$I_{\text{ref}}^{(t)} = E_0^2 \sum_a^N \langle |t_{\beta a}|^2 \rangle |t_{\beta a}|^2 = P_{in} \frac{\sum_a^N \langle |t_{\beta a}|^2 \rangle |t_{\beta a}|^2}{\sum_a^N |t_{\beta a}|^2} \qquad (8.8)$$

If $\langle |t_{\beta a}|^2 \rangle$ is homogenous over a, we have $\langle |t_{\beta a}|^2 \rangle = \langle T_\beta \rangle / N$, which gives $I_{\text{ref}}^{(t)} = \langle T_\beta \rangle P_{in}/N$, and $\langle \eta \rangle = N$. Otherwise, we need to replace N by

$$N_{\text{eff}} = \frac{\left(\sum_a^N |t_{\beta a}|^2 \right)^2}{\sum_a^N \langle |t_{\beta a}|^2 \rangle |t_{\beta a}|^2} \leq N \qquad (8.9)$$

In other words, modes that contribute little to the intensity in the focus are counted with a lower weight than the other modes (also see [35]). Even though the enhancement that is determined this way will be lower than using the first method, is automatically excludes the "trivial" effects of maximizing transmission toward, and intensity envelope across, the sample surface.

In many scenarios, $t_{\beta a}$ and $t_{\beta a'}$ are not uncorrelated (also see Chapter 13), which means that the second term in Eq. (8.7) does not vanish. The effect of such correlations is similar to the effect of having an intensity envelope on $\langle |t_{\beta a}|^2 \rangle$, and they lead to a further reduction of N_{eff}. Conceptually, this reduction makes sense because in the presence of correlations there are fewer *independent* incident modes.

Considering the trivial example of focusing through a perfectly flat and transparent glass slide, we find $I_{\text{ref}}^{(t)} = \hat{I}_\beta$ and the sensible value of $\eta = 1$.

8.3.2.4 Signal-to-Background Ratio (Contrast)

For many applications, the most relevant figure of merit is the signal-to-background ratio (SBR), also called contrast. The SBR is particularly applicable in the case that the signal is a sharp focus on an otherwise homogeneous speckle background. To determine the SBR, the intensity in the optimized focus is compared to the average background intensity directly around it.

$$
I_{\text{ref}}^{(\beta)} = \left\langle \left| \sum_a^N t_{ba} \hat{E}_a \right|^2 \right\rangle_{b \neq \beta} = \sum_a^N \langle |t_{ba}|^2 \rangle_{b \neq \beta} |\hat{E}_a|^2 + \sum_a^N \sum_{a' \neq a}^N \langle t_{ba}^* t_{ba'} \rangle_{b \neq \beta} \hat{E}_a^* \hat{E}_{a'} \quad (8.10)
$$

where $\langle \, \rangle_{b \neq \beta}$ denotes averaging over the background modes directly around the focus. In the case that that ensemble averaging $\langle |t_{\beta a}|^2 \rangle$ and averaging over the background $\langle |t_{ba}|^2 \rangle_{b \neq \beta}$ are equivalent, Eq. (8.10) reduces to Eq. (8.7), and both metrics are identical. Note, however, that the two averages are not equivalent in general. For instance, when a large fraction of the incident modes is controlled with a high accuracy, intensity correlations between $|t_{ba}|^2$ and $|t_{\beta a}|^2$ start to play an important role. In this case, wavefront shaping increases the background intensity as well as the intensity in the focus [36, 37], and the enhancement and the SBR will differ. Also, in cases that the ballistic component is not negligible, the SBR will differ from η. In the example of a glass slide, for instance, we would have an SBR of infinity.

8.3.2.5 Discussion

The three different ways to define a figure of merit are largely equivalent in most cases. The first method, where random incident fields are used (Eq. (8.6)), is the simplest to treat analytically. The relation $\eta = N$ holds for all media, even for transparent ones.

Despite the simplicity of this definition, however, the enhancement determined this way is not always the most relevant figure of merit. Still, this way of determining the enhancement proves a very useful tool to evaluate how well an experimental system behaves. Typically, it is possible to reach over 50% of the theoretical best enhancement; lower values may indicate a problem in the experimental setup.

To determine only the "relevant" part of the enhancement, one can either average over disorder (Eq. (8.7)) or determine the signal-to-background ratio (Eq. (8.10)). Effectively, these methods replace N by N_{eff}, where modes that have a lower intensity contribution and modes that are correlated with each other are counted with a lower weight. As a result, the "trivial" part of the enhancement is canceled, resulting in a lower, but more relevant, value for η.

Finally, when using the SBR as a figure of merit, one should be aware that wavefront shaping also increases the background intensity, which is an effect of intensity correlations in the TM [36, 37], and one should be careful when there is still a strong ballistic component present.

8.3.3   Imperfect Modulation

So far, we assumed that the SLM perfectly generates $\hat{E}_a$. In practice, however, there are many reasons for the actual field to deviate from this perfect scenario, such as measurement noise or the use of phase-only modulation instead of full phase and amplitude modulation.

To quantify how well the wavefront shaping system performs, we introduce the fidelity $|\gamma|^2$, where γ is the normalized inner product of the generated field E_a with the perfect field $\hat{E}_a$

$$\gamma \equiv \frac{\sum_a^N \hat{E}_a^* E_a}{\sqrt{\sum_a^N \left|\hat{E}_a\right|^2}\sqrt{\sum_a^N |E_a|^2}} \tag{8.11}$$

The fidelity is equal to the fraction of the incident power that is shaped correctly. Values of $|\gamma|^2$ range from 1 (perfect wavefront shaping) to 0 (generated wavefront is perfectly orthogonal to the required wavefront). The intensity in the focus is related to the fidelity by $I_\beta = |\gamma|^2 T_\beta$, as can be seen directly from Eq. (8.2). Consequently, we have

$$\eta = |\gamma|^2 N_{\text{eff}} \tag{8.12}$$

In practice, it is common to modulate either the phase or the intensity of the light. For such cases, the expected enhancement can be calculated by assuming that the sample is strongly scattering. In that case, the elements in a row of the TM are statistically independent and follow a circular Gaussian distribution [32]. An average value for γ can be obtained easily by evaluating Eq. (8.11) for a circular Gaussian distribution of $\hat{E}_a$ and averaging over disorder. Introducing $\alpha \equiv |\langle\gamma\rangle|^2 \in [0, 1]$, we find [38]

$$\langle\eta\rangle = \alpha \left(N_{\text{eff}} - 1\right) + 1 \tag{8.13}$$

where the term of order unity is the result of the slight difference between $|\langle\gamma\rangle|^2$ and $\langle|\gamma|^2\rangle$. A few common scenarios are listed in Table 8.1. Typical enhancements reached in feedback-based experiments currently range from 10 to $> 10^3$, depending mainly on the number of control elements, signal-to-noise ratio of the feedback signal, and the stability of the sample (see Section 1.6.1).

Table 8.1 Relative maximum enhancement α for different wavefront shaping methods.

Light modulator type	α	Reference
Amplitude and phase	1	[36]
Phase only	$\pi/4$	[11]
Intensity only	$1/(2\pi)$	[35]
Binary phase	$1/\pi$	[39, 40]

8.3.4   Overall Transmission

An intriguing aspect of wave scattering is the existence of open transport channels. In 1984, Dorokhov predicted that for each nonabsorbing scattering sample it is possible to construct a wave that is transmitted completely (i.e., experiences zero reflection) [41]. The existence of such open channels is responsible for a large variety of mesoscopic phenomena, including universal conductance fluctuations [33]. Originally studied mostly in the context of electron scattering, the phenomenon universally applies to wave scattering. Important pioneering work was done with microwaves and ultrasound [29], for which the field of the wave can be detected over a large bandwidth relatively easily. For light, an interferometric technique is needed to measure the phase of the wave, and experiments are usually conducted with monochromatic waves. However, the huge advantage of using light is the availability of spatial light modulators (SLMs) with millions of degrees of freedom and detectors with millions of pixels.

The wavefronts corresponding to open channels are singular vectors of the TM of a closed system. In this context, "closed" means that all degrees of freedom of the incident field are controlled, and all degrees of freedom of the transmitted field are resolved, without any light being absorbed or leaking from the system. In an optical experiment, only a subset of the TM of the sample can be determined; usually limited by the numerical aperture of the microscope objectives that are used. Still, the total energy transmission through a sample can be increased if a large fraction of the degrees of freedom in the incident field are controlled [36, 42, 43]. Also, if only a small subset of the TM is known, it is also possible to increase the total transmitted power locally at the detector surface [44].

8.4   Algorithms

The TM for a given object is not known a-priori: it depends on the microscopic refractive index distribution in the sample, which will be different for each specimen. In order to focus light to a point, there are two types of approaches: 1) first measure the TM elements and then calculate the ideal field using Eq. (8.3), or 2) find the optimum field using an iterative optimization algorithm. This section contains a description of the most commonly used approaches for forward wavefront shaping. All algorithms described here are aimed at maximizing η, and they are designed for the situation where the light is completely diffuse.

8.4.1   Transmission Matrix Measurements

This class of algorithms measures the elements of the TM first, and then calculates the optimal wavefront for focusing light onto a point or a set of points. Experimentally, the simplest approach to measure the matrix elements is to use the sample itself as an interferometer. Part of the incident field is kept static and serves as a co-propagating reference beam; resulting in a static speckle pattern (the reference field) at the back of

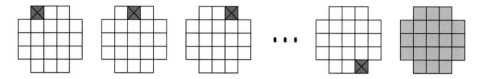

Figure 8.2 Stepwise sequential algorithm. This algorithm measures a single row of the TM of a scattering sample. The light modulator is divided into a series of segments, and the phase of each consecutive segment is swept from 0 to 2π (depicted by the crossed out segment) while monitoring the feedback signal. This way, the phase and amplitude of the elements of the TM row are measured. After all measurements are completed, the wavefront is updated (far right) to display the optimum wavefront for focusing. Reprinted with permission from [38], *Optics Communications.*

the sample. The phase of the matrix elements is determined relative to the phase of this reference field.

This approach is used in the "stepwise sequential" algorithm (see Figure 8.2). For this algorithm a phase-only spatial light modulator is subdivided into N segments, and the phase ϕ_a of each segment a is varied between 0 and 2π consecutively. At the detector, the light from the modulated segment $E_{\beta a}$ interferes with the light originating from all other segments $E_{\beta 0}$.

$$I_\beta(\phi_a) = |E_{\beta a} + E_{\beta 0}|^2 = |E_{\beta a}|^2 + |E_{\beta 0}|^2 + 2|E_{\beta a}E_{\beta 0}|\cos(\phi_{\beta a} + \phi_a) \quad (8.14)$$

with $\phi_{\beta a}$ the phase of the transmission matrix element $t_{\beta a}$ relative to the unknown phase of the reference field. The relative values of $t_{\beta a}$ can be extracted from this measurement by fitting a cosine, typically through a single-frequency Fourier transform. By repeating the measurement for each segment, all elements of a row of the TM can be obtained up to a single common prefactor. After determining the complete row of matrix elements, the light modulator is programmed such that $\phi_a = -\phi_{\beta a}$, which causes constructive interference (focusing) at point β.

Using the field from the nonmodulated segments as a reference beam is only allowed when a small fraction of the segments are modulated at a time, so that background field $E_{\beta 0}$ will be nearly equal for each segment. Unfortunately, in this case the signal-to-noise ratio (SNR) of the measurement may not be optimal. To optimize the performance in the presence of shot noise and technical noise (i.e., fluctuations in the laser intensity, phase jitter in the SLM, etc.), a preoptimization with a low N can be used first to increase the reference intensity [11, 45]. A different approach to improve the SNR of the measurements is to modulate multiple segments at once, for example by choosing a Hadamard basis. In this case, a different approach should be used, such as reserving a part of the light modulator for generating the co-propagating reference beam [46, 47]. Also, instead of measuring the elements of the TM sequentially, it is possible to modulate each pixel with a different frequency and Fourier transform the detector signal to separate the different signals [48, 49].

When the point detector in Figure 8.1 is replaced by a camera, it is possible to measure multiple rows of the TM at once [46]. This technique is extremely powerful as it allows for transmitting images [50, 51], direct imaging [24, 52], raster scanning fluorescence microscopy [18], and maximizing the transmission to a given large-area target [44]. Also, it allows a determination of the eigenvalues of the TM [42, 44, 46], the statistical distribution of which is of fundamental interest in mesoscopic scattering theory (e.g., [33]). When it is sufficient to know the transmission matrix up to a common phase factor for each row, a co-propagating reference beam like described above can be used. If the relative phase between different rows is important, an external reference beam can be used instead (see, e.g., [44]).

8.4.2 Iterative Algorithms

Instead of measuring the matrix elements directly, a variety of iterative approaches can be used to find the optimal wavefront for creating a focus. Where TM approaches only update the wavefront once all measurements are complete, iterative methods update the wavefront after each single step. Iterative approaches have two advantages: 1) the intensity in the target starts increasing right away. 2) in many cases the SNR is better than for TM methods [53]. Interestingly, the type of noise (additive, multiplicative, or shot noise) determines which algorithm is optimal [45].

The fact that the target intensity starts increasing right away is especially important when the medium is dynamic. TM algorithms don't perform well, if at all, if the medium changes before all measurements are completed. Therefore, the number of segments N needs to be adjusted to the speckle decorrelation time of the medium. If the decorrelation time is not known, or not constant, it may be favorable to use an iterative approach to reach the highest possible enhancement without adjusting any parameters for the tissue dynamics [38].

A simple iterative approach is illustrated in Figure 8.3. In this random partitioning algorithm, the pixels of the light modulator are randomly subdivided into two partitions of equal size, and the relative phase of the partitions is adjusted to maximize the feedback signal. Then the procedure is repeated for a different random subset, ad infinitum. When this algorithm is started, the intensity in the focus increases rapidly and the performance is good even with a low initial signal-to-noise ratio. However, after several steps the algorithm starts slowing down since the partitions are not truly orthogonal. Moreover, the required phase corrections shrink over time, so the phase should be determined with a very high accuracy [38]. Any error in determining the phase can be devastating:

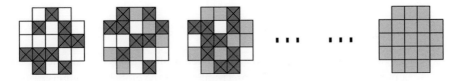

Figure 8.3 Random partitioning algorithm. In this iterative approach, half of the segments of the light modulator are chosen at random (crossed squares). The phase of these selected segments is varied until the maximum intensity is reached. After that, the iteration is repeated with a different subset of selected segments. Reprinted with permission from [38], *Optics Communications.*

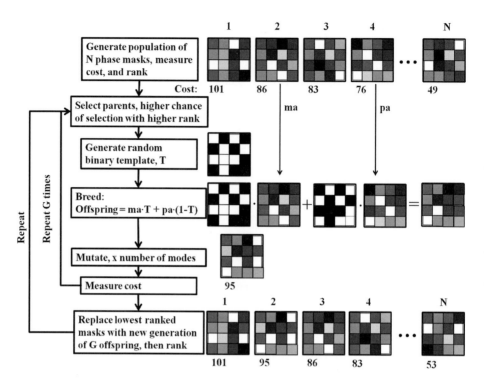

Figure 8.4 Block diagram for the genetic algorithm. For each generation several wavefronts are displayed, and the ones giving the highest enhancement are mixed and mutated to generate the next generation of wavefronts. Reprinted with permission from [53], *Optics Express.*

if a single noisy measurements gives an error of π, the focus disappears abruptly. Therefore, the algorithm should be able to "backtrack" to a previously stored state if the focus is lost.

An interesting category is formed by genetic algorithms, first employed in this context by Conkey et al. [53]. Genetic algorithms start with a pool of randomly generated wavefronts and select the ones with the highest enhancement (see Figure 8.4). These "winning" wavefronts are mixed randomly and modified to generate the next generation of wavefronts. In simulations, genetic algorithms are very robust to additive noise (see Figure 8.5). Like the random partitioning algorithm, convergence does slow down after a certain number of iterations.

8.4.3   Discussion

Regardless of what algorithm is used, ideally all methods converge to the unique global optimum given by Eq. (8.4). For TM methods, the global optimum is reached after performing a single measurement for each row of the TM. The iterative methods generally do not converge to the maximum in an entirely deterministic manner, and their convergence slows down significantly after several iterations. For a strongly scattering sample, all N incident modes are statistically independent. Therefore, any given algorithms will

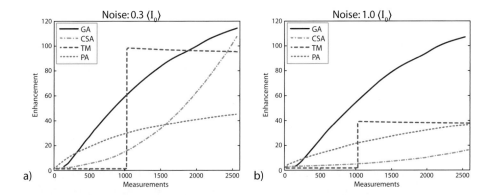

Figure 8.5 Simulated performance of different algorithms as a function of time step. (**a**) With additive Gaussian noise at 30% of the initial average intensity. (**b**) 100% additive noise. GA: genetic algorithm; CSA: continuous sequential algorithm (see [38]); TM: transmission matrix / stepwise sequential algorithm; PA: random partitioning algorithm. Reprinted with permission from [53], *Optics Express*.

require at least $O(N)$ measurements to achieve an enhancement of N. The actual number typically ranges from $N + 1$ [40, 54] to $10N$. As such, iterative methods are not faster than TM methods; their main advantages are the lower sensitivity to noise [53] and their quick recovery after sample movement [38].

8.5   Extended Targets

The algorithms described above are able to find the optimal wavefront for focusing to a spot that is the size of a single speckle. However, when the target area is larger than a single speckle, these methods are not guaranteed to find the optimum solution. The behavior of the stepwise sequential algorithm in this situation can be analyzed easily [38]: the algorithm is expected to find the wavefront that maximizes transmission into the effective target "mode" $E_\beta^{\text{eff}}(r) \equiv D(r)E_0(r)$, with $D(r)$ the envelope of the detection sensitivity and $E_0(r)$ the original speckle field at the start of the algorithm. Therefore, the end result depends on the starting conditions of the algorithm. This behavior, however, is likely to depend on the algorithm that is used. For instance, it was observed that genetic algorithms caused the light to focus onto the brightest speckle [55].

Even though it is still possible to increase the amplitude in E_β^{eff} by the same factor as for a point target, the SBR contrast will be lower. For instance, consider the situation where the envelope $D(r)$ spans an area that supports M optical modes. In this case, increasing the intensity of only one of these modes by a factor η would cause an overall increase of the target signal of a factor of η/M [38]. However, when the M optical modes of the target are spatially very close to each other (as is the case for a simple extended target), the intensity of the modes is strongly correlated, which helps focusing light to the target with a higher contrast than η/M [37].

The difficulty in focusing to multiple spots at once is a fundamental problem that plays a significant role in wavefront shaping with photoacoustic feedback or fluorescence feedback when the target signal comes from an area that spans many optical modes. The same problem is seen in time reversal of ultrasound-encoded light, where typically $N \approx 10^5$ and $M \approx 10^4$, resulting in a typical contrast of around 20 [13, 14, 16].

8.6   Feedback

All algorithms discussed above rely on the availability of a feedback signal to optimize. In the simplest case we can use a camera or point detector placed behind the sample. However, if we want to use wavefront shaping for imaging inside turbid media, a feedback signal from inside the medium is needed. I will now review different approaches for obtaining such a signal.

8.6.1   Fluorescence

A small fluorescent particle can be used as a probe to locally measure the amount of excitation light. The total amount of fluorescent emission is a direct measure for the excitation intensity at the location of the particle. By using the total fluorescence emission as a feedback signal, light was focused onto a 300-nm diameter fluorescent sphere embedded at a depth of $\sim$20 μm inside strongly scattering ZnO pigment, reaching an enhancement of > 20 [12]. Care must be taken that the labeled volume is small (see Section 1.4.4) and that the autofluorescence of the medium is sufficiently low to detect the signal from the probe. Since it is often possible to fluorescently label structures of interest, this method may be used for focusing light on labeled cells or organelles.

8.6.2   Nonlinear Feedback

When a two-photon fluorescence signal is used, the feedback signal is proportional to the square of the intensity in the medium. This property adds two interesting and useful characteristics to the wavefront shaping process. First of all, the feedback signal responds to an improved spatial confinement of the signal, as well as to an improved temporal confinement. In 2011, Katz et al. [56] demonstrated that, as a result, forward wavefront shaping focuses the light both in space and in time. The authors applied the stepwise sequential algorithm for increasing the intensity of a two-photon signal. Even though they did not control the temporal degrees of freedom in the incident light, they observed that wavefront shaping resulted in a near–transform limited transmitted pulse. These findings nicely complement observations in acoustics, where it was shown that light can be focused spatially by controlling temporal degrees of freedom only [57, 58]. See Section 1.6.2 for a more detailed discussion of this phenomenon.

A second useful property of nonlinear feedback is that optimization algorithms make the light converge to a single diffraction-limited spot, even when the area where the feedback is coming from is larger than that. This "autofocusing" property was demonstrated using the second harmonic signal from nanoparticles embedded in the scattering medium [59], two-photon fluorescence from a dye layer hidden behind a scattering medium [56], and nonlinear photoacoustics [60] as nonlinear feedback signals.

A nonlinear feedback signal is proportional to I^n, where I is the local intensity, and n depends on the process that is used. If the feedback signal is coming from a 2-D source (e.g., a thin sheet of 2-photon dye), autofocusing will occur when $n > 1$. When the feedback comes from a homogeneous medium, a stricter condition of $n > 2$ is necessary [61]. The effect of varying n was demonstrated in a beautiful experiment by Sinefeld et al. [62].

Because of autofocusing, the use of nonlinear feedback is a very promising approach to generate a sharp focus with a "fuzzy" feedback signal, thereby breaking through the limitations observed with linear feedback (see Section 1.4.4).

Tang et al. [63] used a two-photon fluorescence signal as a feedback for focusing light inside turbid tissue. In adaptive optics for two-photon microscopy, the two-photon fluorescence signal is often used as a feedback signal to optimize the image quality. However, Tang et al. demonstrated that this approach is even feasible when scattering is so bad that there is no image to start with. In a follow-up experiment, this technique was used for imaging the brain of a living mouse through the intact skull [17].

8.6.3   Photoacoustic Feedback

When an absorbing structure is illuminated with a short laser pulse (usually several nanoseconds), the structure heats up rapidly. The resulting thermal expansion induces an acoustic wave that can be detected with an ultrasound detector, and subsequently be projected back to reconstruct the original structure. Imaging based on this photoacoustic (PA) effect is a rapidly finding biomedical applications [64]. PA imaging combines the specificity of optical absorption spectroscopy with the resolution of ultrasound imaging. Also, it provides a means of obtaining a feedback signal from deep inside a scattering medium without preparing the medium in any special way. The only requirement is that the medium contains absorbing structures.

The use of PA for wavefront shaping was pioneered by Kong et al. [65]. Chaigne et al. [66] extended the method to use the full 3-D photoacoustic reconstruction in order to measure a complete TM for all points inside an absorbing structure simultaneously. Imaging using PA-based feedback was demonstrated by scanning a sample through the focus [67].

Currently, the major limitation of PA-based techniques is the large feedback area, which causes the contrast of the focus to be low (see Section 1.4.4). Demonstrations so far all used a geometry where the speckles in the medium were much larger than the optical wavelength, thereby artificially lowering the number of optical modes in the ultrasound focus. The use of nonlinear photoacoustics [60] offers a promising approach to break through this limitation.

8.6.4   Other Feedback

Any signal that is representative of the light intensity in a spatial or temporal target can be used as feedback. This flexibility allows for a large variety of target functions to be optimized. Apart from the feedback signals discussed above, other examples include the acousto-optic signal [68], and the polarization state of the output mode [69, 70]. A special case of iterative approaches was demonstrated by Nixon et al., who used an optical gain medium as an analog feedback mechanism to find lasing modes through scattering media [71]. Also, a promising direction may be to combine wavefront shaping with coherence gating, in order to achieve depth-selective focusing [72], or to reduce the amount of scattered light [73]. Feedback-based wavefront shaping can basically generate any spatiotemporal mode that one is able to detect; a feature that was convincingly demonstrated by Aulbach et al. [74] who used spatial wavefront shaping in combination with a scattering medium to guide light into any desired spatiotemporal mode.

8.7   Correlations and Dynamics

So far, we have assumed that the medium is perfectly static, that perfectly monochromatic light is used, and that the incident wave has a perfect pointing stability. When one of these conditions is not met, the enhancement will decrease. Fortunately, this decrease is usually quite gradual. In this section, I will discuss what happens when one of these conditions is not met.

8.7.1   Dynamic Media

If the medium is dynamic, its transmission matrix will change over time. Therefore, any feedback algorithm that aims to construct a matching wavefront needs to be fast enough to keep up with the dynamics of the medium, and continuously needs to keep updating the wavefront. If the feedback algorithm is slower than the medium dynamics, the constructed wavefront that is constructed has no relation with the current TM of the medium, and no enhancement will occur.

If a TM method is used, the medium should be stable over the duration of all N measurements. Since all measurements are performed sequentially, one should reduce N to match the decorrelation time of the medium [38]. As a rule of thumb, the maximum enhancement that can be achieved is $\eta_{max} \approx \alpha T_p/T_m$, with T_m the time required for a single measurement, T_p the persistence time (speckle decorrelation time) of the medium, and α was introduced in Section 1.3.2. Most iterative approaches automatically approach this maximum enhancement without adjustment to the sample correlation time [38, 53].

The advance of micromechanical light modulators has enabled wavefront shaping at a rate of over 20 kHz, which is 2000 times faster than the first wavefront shaping experiments (that used twisted nematic liquid crystal light modulators). The use

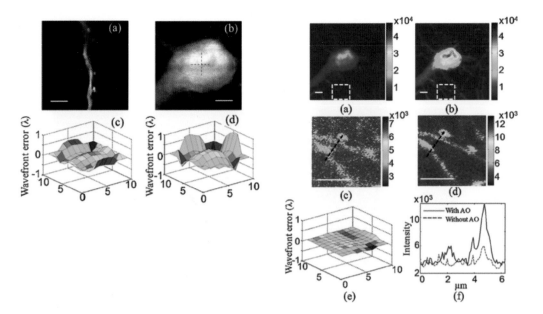

Figure 2.11

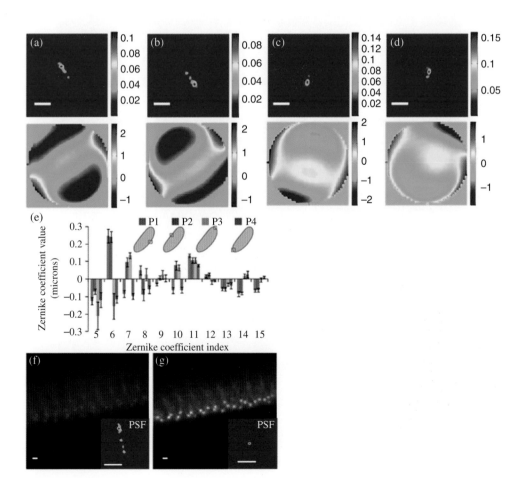

Figure 2.12

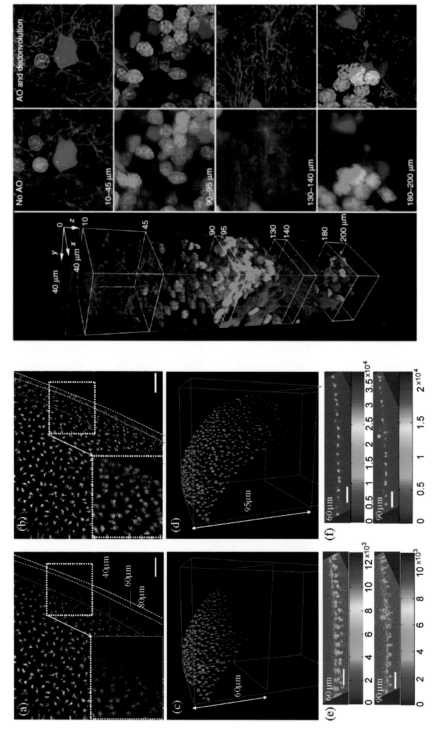

Figure 2.13

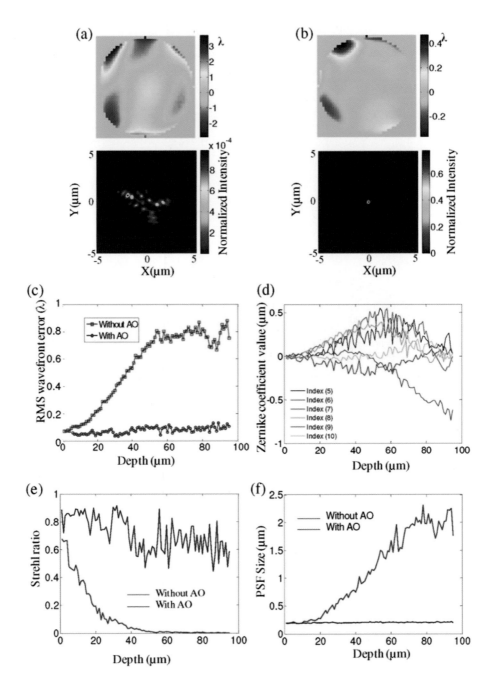

Figure 2.14

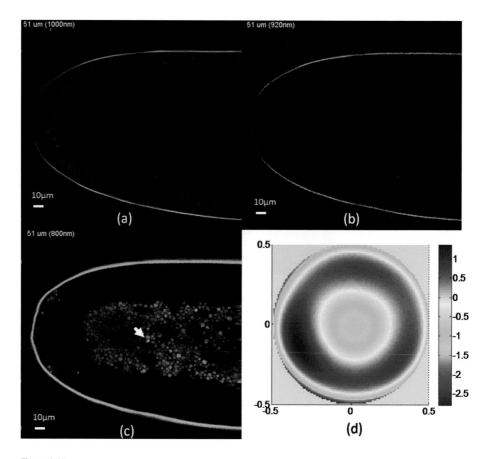

Figure 2.17

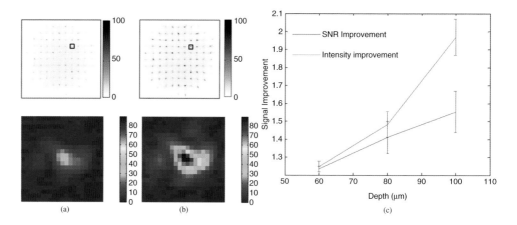

Figure 2.19

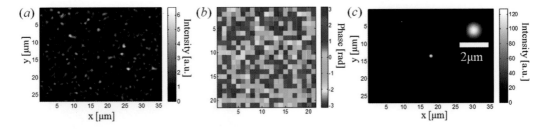

Figure 3.3

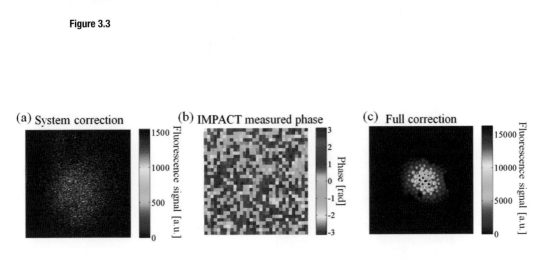

(a) System correction   (b) IMPACT measured phase   (c) Full correction

Figure 3.5

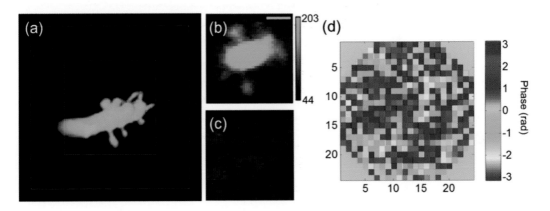

Figure 3.6

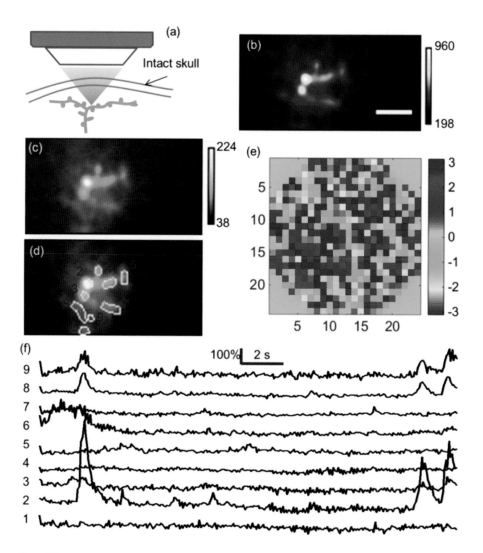

Figure 3.7

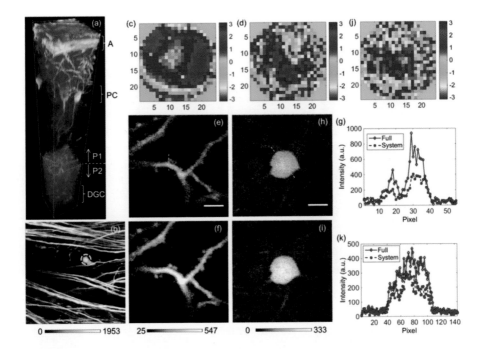

Figure 3.8

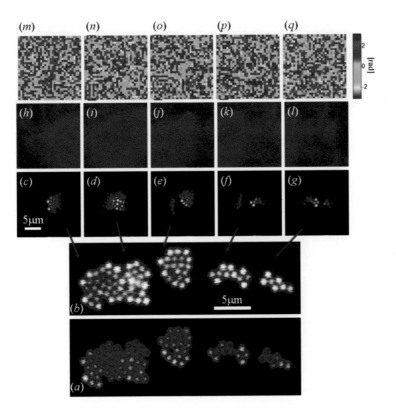

Figure 3.9

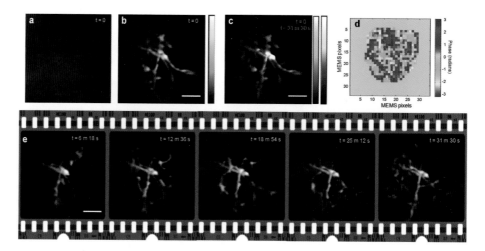

Figure 3.12

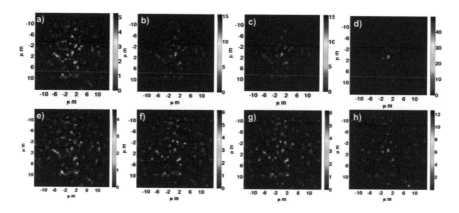

Figure 3.16

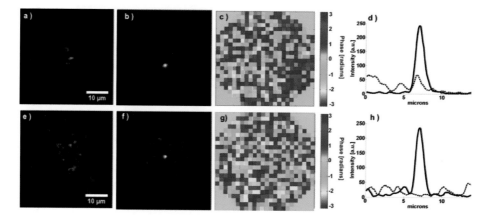

Figure 3.18

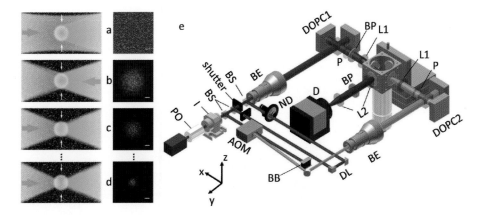

Figure 3.22

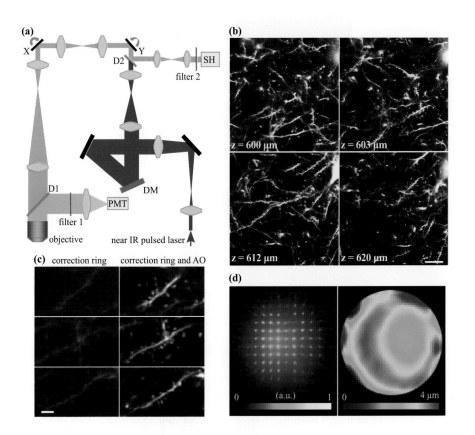

Figure 4.2

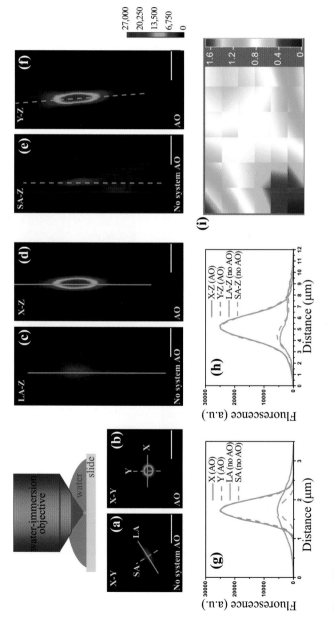

Figure 4.6

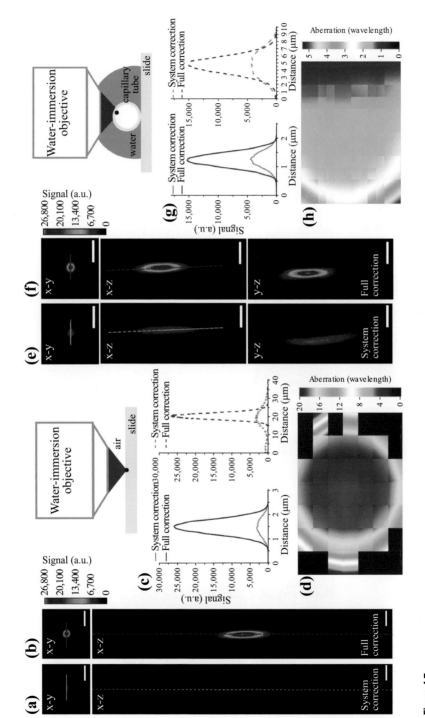

Figure 4.7

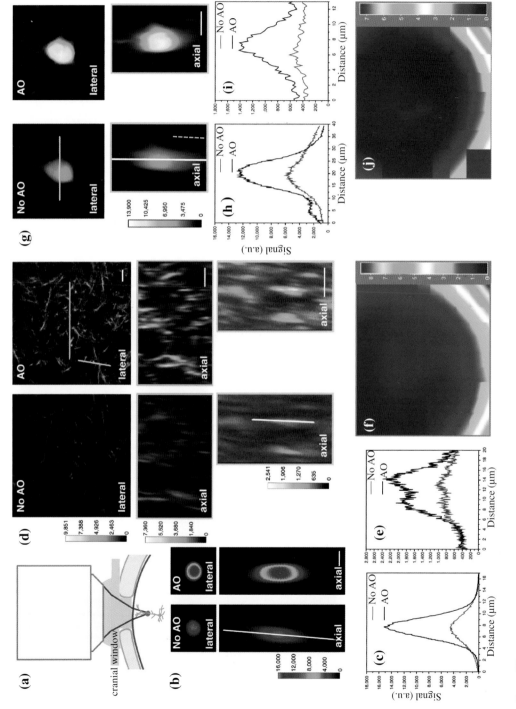

Figure 4.8

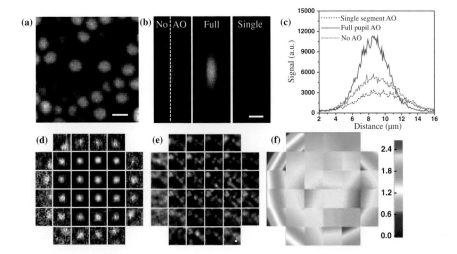

Figure 4.11

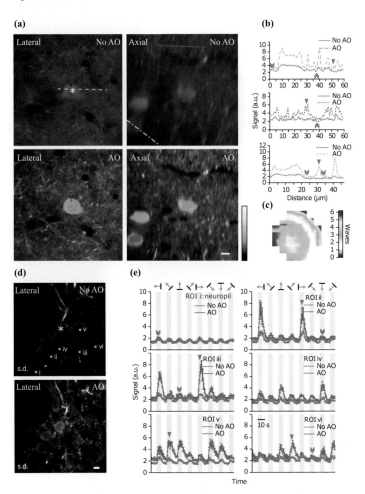

Figure 4.14

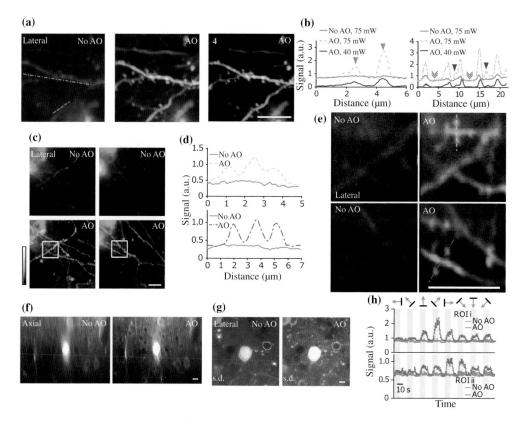

Figure 4.15

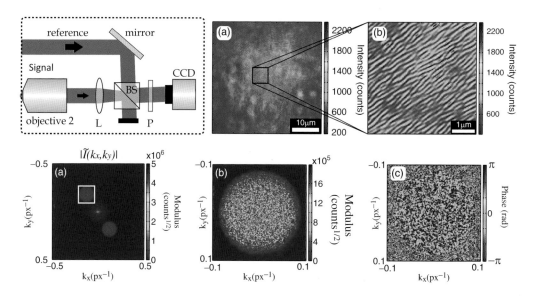

Figure 5.3

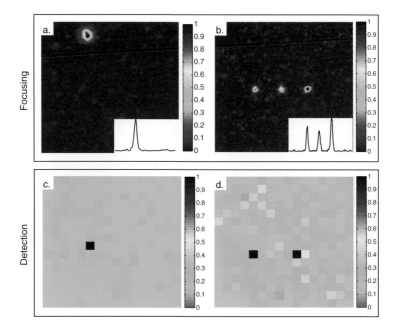

Figure 5.5

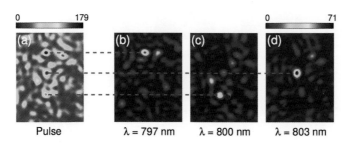

Figure 5.9

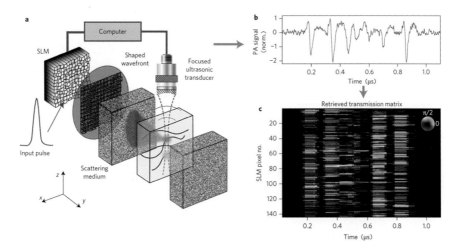

Figure 6.7

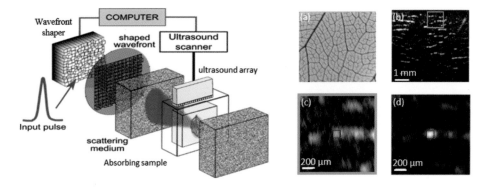

Figure 6.9

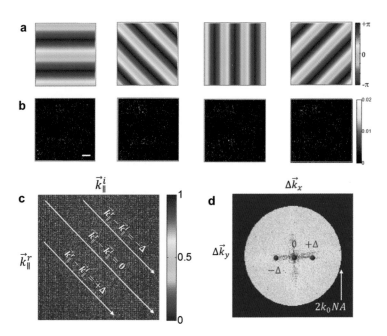

Figure 7.4

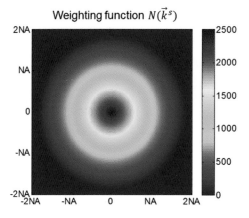

Figure 7.6

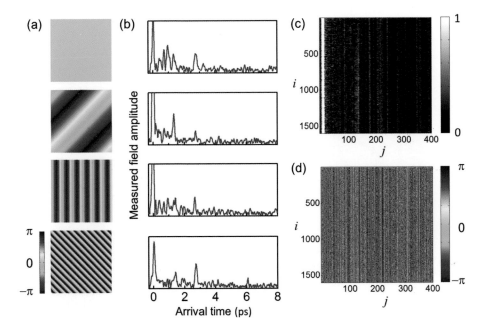

Figure 7.12

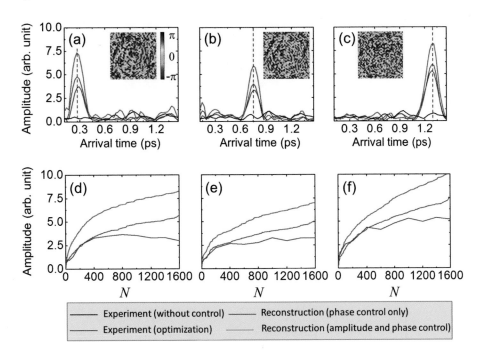

Figure 7.13

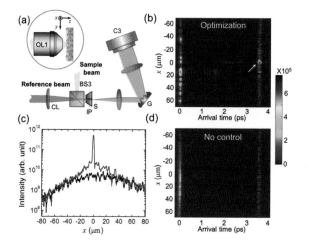

Figure 7.14

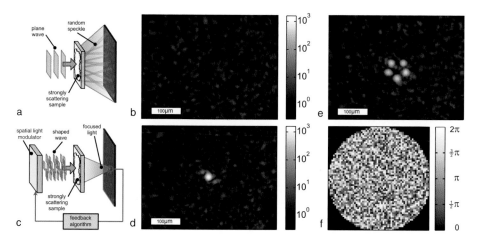

Figure 8.1

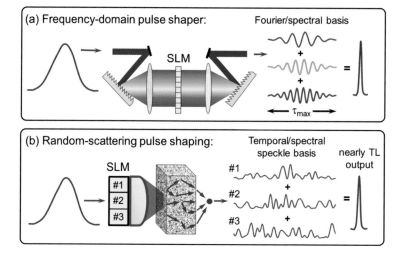

Figure 8.6

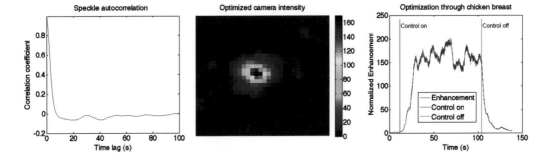

Figure 9.3

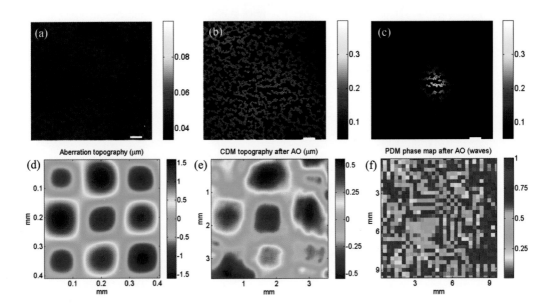

Figure 9.13

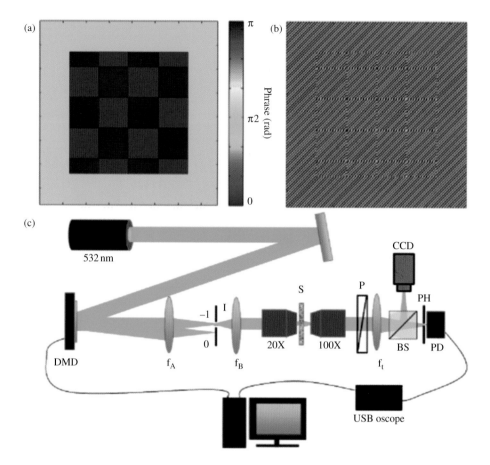

Figure 10.4

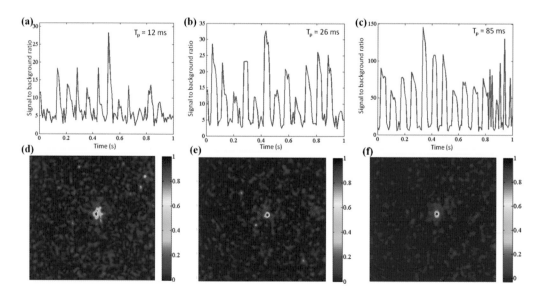

Figure 10.6

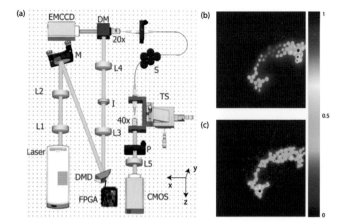

Figure 10.10

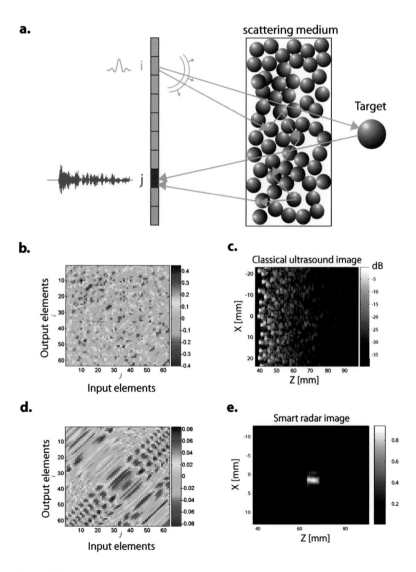

Figure 11.4

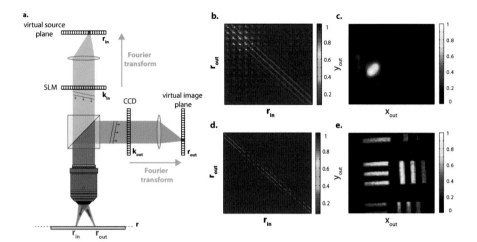

Figure 11.11

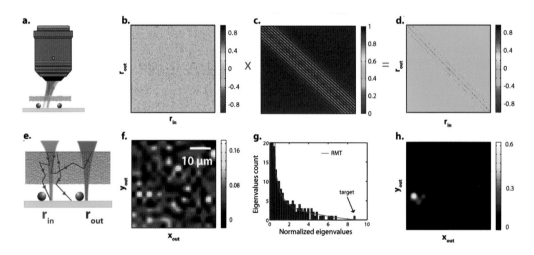

Figure 11.12

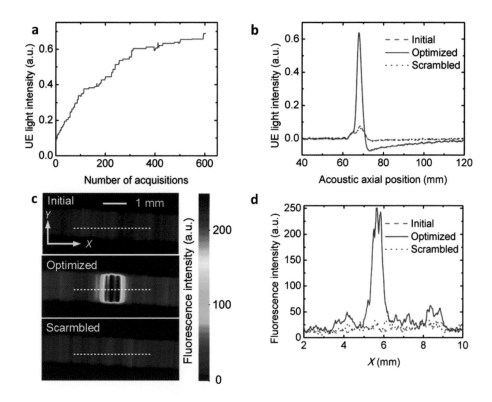

Figure 12.15

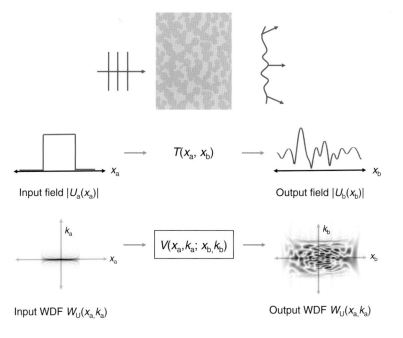

Figure 13.5

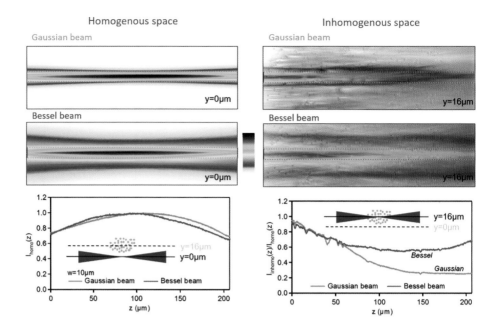

Figure 14.13

Zernike Mode	Targeted Focus wavefront	Adjacent Wavefront	Amplitude Damping Ratio
Defocus	 $s_t = 0.3$	 $s_a = 4.07 \times 10^{-4}$	27.1
Astigmatism	 $s_t = 0.3$	 $s_a = 4.07 \times 10^{-4}$	27.1
Coma	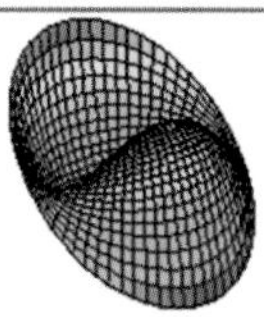 $s_t = 0.3$	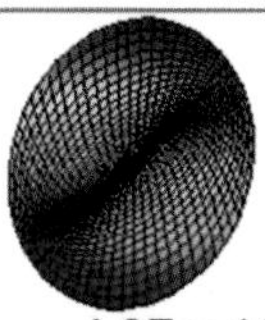 $s_a = 4.07 \times 10^{-4}$	13.4 - 53.0
Spherical Aberration	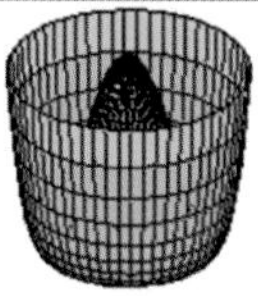 $s_t = 0.3$	 $s_a = 4.07 \times 10^{-4}$	27.1

Figure 16.7

of micromechanical light modulators for light focusing was pioneered by Conkey et al. using a digital micromirror device (DMD) [47], followed by Stockbridge et al. [75] who used a segmented mirror with a comparable speed. DMDs can be used to mimic phase modulation (e.g., [47]), or used in a binary-intensity modulation mode (e.g., [35] or [76]). At such speeds, the PC that is traditionally used for executing the feedback algorithm becomes the main bottleneck. Conkey et al. resolved this bottleneck by using programmable hardware for executing the algorithm and controlling the DMD; thereby eliminating the need for a PC in the feedback loop [77].

Once a matching wavefront is found, the enhancement will decay in time; exactly following the trend of the temporal speckle autocorrelation function [78, 79]. For long, it was believed that the relevant time scale for focusing light through the living skin was less than a millisecond (a typical time scale observed in dynamic speckle measurements). Surprisingly, in *in vivo* experiments it was found that the reconstructed focus remains stable for a much longer time than that, up to several seconds in skin [79, 80], especially when the blood flow is temporarily clamped, and even over half an hour in bone [17]. A possible explanation may be that there are two components to the speckle decorrelation: blood flow (fast), and dynamics of nonperfused tissue (slow).

The dynamics of living tissue is often considered a major hurdle for *in vivo* applications of wavefront shaping. Fortunately, the future is bright: wavefront shaping can be done in several milliseconds, whereas a time scale of seconds is already fast enough for some *in vivo* experiments. Also, techniques based on phase conjugation (see Section 1.7.2) can even be orders of magnitude faster than iterative feedback-based methods.

8.7.2 Broadband Light

When the light is perfectly monochromatic, the optimal wavefront for focusing to a spot follows trivially from Eq. (8.1). For broadband light, however, the situation is slightly more complex since the temporal/spectral dimension needs to be taken into account too. Here, two regimes can be identified. If the bandwidth of the light source is smaller than the Thouless frequency $d\omega$, the system can effectively be considered monochromatic [81, 82]. The Thouless frequency is inversely proportional to the Thouless time: the average time it takes light to diffuse through the sample. An equivalent criterion is to say that the correlation length of the light source should be longer than the typical path length difference in the medium [83].

In the second regime, the bandwidth of the source exceeds the Thouless frequency. As a result, each spatial degree of freedom is associated with several independent spectral degrees of freedom [62]; increasing the dimensionality of the problem by one. In this regime, the scattering medium mixes spatial and temporal degrees of freedom, an effect well known from pioneering work in acoustics [57, 58]. As a result, spatiotemporal focusing can be achieved by only controlling spatial degrees of freedom [56, 59, 74] (see Figure 8.6).

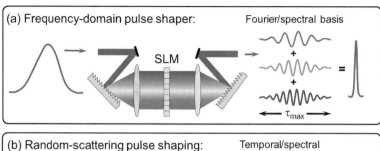

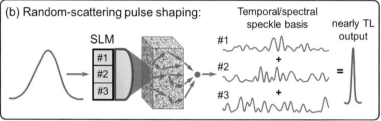

Figure 8.6 Temporal compression using only spatial wavefront shaping. (**a**) In a conventional pulse shaper the pixels of the SLM are coupled to the spectral degrees of freedom by scattering from a grating. (**b**) Similarly, coherent scattering in a random medium couples each SLM pixel to a different linear combination of the spectral degrees of freedom, forming a new random spectral basis that is phase-controlled by the SLM. In both cases, the spectral resolution maximum is determined by the optical path length differences in the medium/shaper. Adapted by permission from Macmillan Publishers Ltd: Nature Photonics [56], copyright 2011. A black and white version of this figure appears in some formats. For the color version, please refer to the plate section.

8.7.3   The Memory Effect

Under some conditions, the reconstructed focus is remarkably robust to a change in the angle of the incident beam. This feature is the direct result of the optical memory effect, a type of wave correlation that has been known since 1988 [84, 85]. The optical memory effect is where a tilt in the incident wavefront results in an equal tilt in the transmitted wavefront. This way the transmitted field can be tilted without significant decorrelation; up to an angle known as the memory effect angle. The memory effect angle is proportional to λ/L, with λ the wavelength and L the thickness of the sample [84–86]. When observed at a distance d away from the back surface of the sample, the transmitted pattern shifts over a distance proportional to $d\lambda/L$.

The memory effect has been exploited for raster scanning fluorescence microscopy of objects hidden behind a scattering layer [23, 87]. Recently, it was discovered that the memory effect can also be used for computational imaging even without creating a focus in the object plane [88, 89].

The field of view of methods based on the memory effect is limited; the method can only be used for small objects placed at a large distance behind a thin sample (object size $w < d\lambda/L$). However, recently it was discovered that a different type of memory effect is present *inside* turbid materials that scatter light in the forward direction predominantly (such as biological tissue) [34]. Also, the memory effect angle for tissue is

Table 8.2 Key differences between adaptive optics and wavefront shaping. Both techniques represent different ends of the scattering spectrum.

Adaptive optics	Wavefront shaping
Ray optics	Multiple scattering, diffraction, interference
Start with distorted image	Start with nothing (speckle)
Paraxial waves	Fully diffuse light
Concept: aberration correction	Concept: constructive interference
1 segment → 1 ray	1 segment → 1 multipath interference pattern
Large field of view	Usually single point correction
Large bandwidth	Narrowband solution

much larger than predicted [90], an observation that can be explained by generalizing both memory effects into a single phenomenon [91]. These findings indicate that it may be possible to extend the concepts of memory effect imaging and scanning microscopy to work inside biological tissue. See Chapter 13 for a further discussion of correlations in scattering media.

8.8   Related Topics

8.8.1   Adaptive Optics

The term adaptive optics (AO) refers to a series of techniques to improve the image quality of an imaging system by correcting for aberrations in the optical path. Conventionally, AO works in the regime where the aberrations are smooth functions of the spatial coordinates, and where light can be described as paraxially propagating beams (e.g., [92]). As a result, most techniques for AO implicitly or explicitly use the smoothness of the aberration correction by expanding the wavefront in Zernike modes, by using a deformable mirror with connected segments, or by using a wavefront sensing approach that requires the wavefront to be relatively flat and free of optical vortices.

In contrast, wavefront shaping was conceived for the regime of strong multiple scattering. In this regime, the wavefront of the light is disordered on a length scale of $\lambda/2$ [93]. Even more importantly, in AO each point at the deformable mirror can usually be associated with a single eikonal ray that propagates to the target. In wavefront shaping, each ray coming from the SLM results in a fully developed speckle pattern. Because this speckle pattern is the result of an infinite series of interfering, multiple-scattered waves [31], it is not even possible to define a single optical path length for propagating from an SLM pixel to the target focus.

The differences between the regimes of AO and wavefront shaping are summarized in Table 8.2. One should note, however, that these regimes just represent different ends of a continuum. Therefore, a strict separation cannot always be made. For example, the multidither coherent optical adaptive technique (COAT) that was developed for in the 1970s [48], is essentially a feedback-based wavefront shaping method. The main difference being that COAT was used in turbulent atmosphere, where light propagation is highly paraxial and only a few elements are needed for an effective correction.

The region where the depth lies between one scattering mean free path, ℓ_{sc}, and one transport mean free path, ℓ_{tr}, can be considered a cross-over regime between these two fields. In biological tissue, this regime spans from $\ell_{sc} \approx 100 \ \mu m$ down to $\ell_{tr} \approx 1$ mm. In the last couple of years, ideas from AO and wavefront shaping are being combined to address specifically this biologically relevant regime (e.g., [94, 95]).

8.8.2   Phase Conjugation

Optical phase conjugation is a well-established research field that started in the early 1970s [96]. Like adaptive optics, most early applications were concerned with correcting smooth aberrations on a paraxially propagating beam [97, 98]. Interest in phase conjugation has renewed after an experiment by Yaqoob et al. [80], where it was shown that phase conjugation can also be used to focus light through biological tissue, for which light propagation is completely diffuse. A further breakthrough was the development of digital phase conjugation that allows digital processing of wavefronts, and permits an almost unlimited intensity gain [99, 100].

In the special case of paraxial propagation through an absorption-free medium, a large fraction of the source light can be collected by the phase-conjugation system ($T_\beta \approx 1$ in Eq. (8.4)). Therefore, images of arbitrary complexity can be reconstructed with a high fidelity. For the case of biological tissue, however $T_\beta \ll 1$, which means that most of the light does not propagate back to the source. Still, the small fraction of light that propagates back to the original source is sufficient to create a focus with a high ($>10^3$) contrast [80].

Phase conjugation uses the time-reversal symmetry of light propagation to propagate scattered waves back to their original source. Since phase conjugation is essentially a single-shot process, it can be much faster than feedback-based methods, allowing the formation of a focus in milliseconds [101–103]. This speed comes at a price: light can only be focused back onto its original source. Although focusing light onto a light source may seem of limited use, the original source can be a nanoparticle that is easily embedded into tissue [39, 100], or a virtual light source created by acousto-optic tagging [13–16], offering very exciting possibilities for deep tissue imaging (see [28] for a review). The distinction between phase conjugation and iterative wavefront shaping is not always sharp: phase conjugation, too, can be applied iteratively [104, 105], in order to focus onto bright reflectors or to further shrink the generated focus [63].

8.8.3   Computational Imaging

Since a ground breaking experiment by Bertolotti et al. [88], the fields of wavefront shaping and computational imaging have experienced a strong cross-fertilization. Bertolotti demonstrated that the optical memory effect can be used in combination with a phase retrieval algorithm to look through thin layers of scattering material. The exciting aspect of these methods is that they do not require any a-priori information, and as such are truly able to see through opaque "walls" [88, 89, 106].

A different class of computational methods requires the TM of the scattering sample to be measured in advance. Once the TM is known, it can be inverted to descramble any transmitted image [21, 24, 46, 50, 52]. Other promising computational approaches are under development, such as the use of spectral encoding [107]. Computational imaging methods "descramble" a speckle pattern in order to look through a scattering medium. Even though no wavefront shaping is required for these methods to work, many of the concepts of wavefront shaping find a place in computational imaging, and vice versa.

8.9   Applications and Outlook

In the last decade years, the field of wavefront shaping has expanded tremendously, and first applications are starting to be demonstrated. Clearly, one of the "holy grails" of the field is to achieve microscopic-resolution imaging inside strongly scattering media. One immediately thinks of biomedical imaging, and how biology, medicine, and neuroscience would change if humans could effectively be made transparent to visible light. Not only biomedical imaging would profit from such a technology: by definition all visible objects scatter light, so the range of possible applications is virtually unlimited.

In order to form an image, focusing light to a single point alone is not sufficient; typically the point needs to be scanned in a controlled manner to allow for, e.g., raster scanning fluorescence imaging. When the scattering medium can be characterized in advance, TM methods can be used for scanning and imaging [18–21, 24, 46, 50, 52]. Even though the need for such a calibration represents a severe restriction, several groups have successfully implemented imaging through multimode fibers [18–21, 52, 108] based on these techniques, with applications in endoscopy. Also, scattering media may be used as a microscope objective for imaging at an extremely high resolution [23, 24]. In addition to TM methods, the optical memory effect can provide the ability to scan a focus in certain geometries [23, 87]. Especially the new anisotropic memory effect [34] opens new possibilities for deep tissue imaging. Finally, new feedback mechanisms are being explored that allow focusing light at arbitrary locations inside scattering tissue [60, 65, 66, 68]. Even though these methods are currently too slow for *in vivo* imaging, there is still plenty of room for improvement both through engineering and through a better understanding of the physics of scattered light and sound/light interaction. These feedback-based methods may someday be combined with phase-conjugation-based techniques for deep tissue microscopy [13, 14, 16].

Imaging is certainly not the only application of wavefront shaping. The ability to manipulate light and to measure the TM of a sample has proven instrumental for fundamental scattering research [36, 42, 46]. In addition, wavefront shaping can be used for generating arbitrary spatiotemporal modes [74], high-resolution focusing [23, 24, 109], or for concentrating light inside nanoscale objects or plasmonic structures [110–112]. In addition, by shaping the incident light, disordered scattering materials can be "programmed" to perform a large variety of optical functions, including beam splitters [113], spectrometers [114], polarization optics [69, 70], and even single-photon wavefront generators [115].

Striking examples of nonimaging applications of wavefront shaping are found in quantum-secure authentication [27], cryptographic key storage [116], and secure communication [26]. These applications use disordered scattering materials as physical unclonable functions (PUF), since no existing method can duplicate a disordered scattering material in such a way that the TMs of the copy and original are identical [117].

In conclusion, the emerging field of wavefront shaping research is full of exciting new ideas with a huge range of potential applications. With creative ideas – combined with advances in engineering, electronics, and theoretical scattering physics – imaging through turbidity may just be one of many applications that the imaging community will be able to realize in the years to come.

Acknowledgments

This chapter is based largely on a review I wrote for *Optics Express* (Ref. [118]). I am supported by the European Research Council under the European Union's Horizon 2020 Programme / ERC Grant Agreement No. [678919].

8.10   References

[1]   A. Ishimaru, "Limitation on image resolution imposed by a random medium," *Appl. Optics,* vol. 17, no. 3, pp. 348–352, 1978.

[2]   M. Minsky, ed., *Microscopy Apparatus.* US patent, 1961.

[3]   P. J. Keller, A. D. Schmidt, J. Wittbrodt, and E. H. K. Stelzer, "Reconstruction of zebrafish early embryonic development by scanned light sheet microscopy," *Science,* vol. 322, pp. 1065–1069, 2008.

[4]   J. Mertz, "Optical sectioning microscopy with planar or structured illumination," *Nature Methods,* vol. 8, no. 10, pp. 811–819, 2011.

[5]   D. Huang *et al.*, "Optical coherence tomography," *Science,* vol. 254, pp. 1178–1181, 1991.

[6]   F. Helmchen and W. Denk, "Deep tissue two-photon microscopy," *Nature Methods,* vol. 2, no. 12, pp. 932–940, 2005.

[7]   P. T. C. So, C. Y. Dong, B. R. Masters, and K. M. Berland, "Two-photon excitation fluorescence microscopy," *Annual Review of Biomedical Engineering,* vol. 2, pp. 399–429, 2000.

[8]   M. Oheim, E. Beaurepaire, E. Chaigneau, J. Mertz, and S. Charpak, "Two-photon microscopy in brain tissue: parameters influencing the imaging depth," *Journal of Neuroscience Methods,* vol. 111, no. 1, pp. 29–37, 2001.

[9]   P. Sebbah, *Waves and Imaging through Complex Media.* Kluwer Academic, 2001.

[10] V. Ntziachristos, "Going deeper than microscopy: the optical imaging frontier in biology," *Nature Methods,* vol. 7, no. 8, pp. 603–614, 2010.

[11] I. M. Vellekoop and A. P. Mosk, "Focusing coherent light through opaque strongly scattering media," *Optics Letters,* vol. 32, no. 16, pp. 2309–2311, 2007.

[12] I. M. Vellekoop, E. G. van Putten, A. Lagendijk, and A. P. Mosk, "Demixing light paths inside disordered metamaterials," *Optics Express,* vol. 16, no. 1, pp. 67–80, 2008.

[13] X. Xu, H. Liu, and L. V. Wang, "Time-reversed ultrasonically encoded optical focusing into scattering media," *Nature Photonics,* vol. 5, pp. 154–157, 2011.

[14] Y. M. Wang, B. Judkewitz, C. A. DiMarzio, and C. Yang, "Deep-tissue focal fluorescence imaging with digitally time-reversed ultrasound-encoded light," *Nature Communications,* vol. 3, p. 928, 2012.

[15] B. Judkewitz, Y. M. Wang, R. Horstmeyer, A. Mathy, and C. Yang, "Speckle-scale focusing in the diffusive regime with time reversal of variance-encoded light (TROVE)," *Nature Photonics,* vol. 7, 2013.

[16] K. Si, R. Fiolka, and M. Cui, "Fluorescence imaging beyond the ballistic regime by ultrasound pulse guided digital phase conjugation," *Nature Photonics,* vol. 6, no. 10, pp. 657–661, 2012.

[17] J.-H. Park, W. Sun, and M. Cui, "High-resolution *in vivo* imaging of mouse brain through the intact skull," *Proceedings of the National Academy of Sciences of the USA,* vol. 112, no. 30, pp. 9236–9241, 2015.

[18] T. Čižmár and K. Dholakia, "Exploiting multimode waveguides for pure fibre-based imaging," *Nature Communications,* vol. 3, p. 1027, 2012.

[19] I. N. Papadopoulos, S. Farahi, C. Moser, and D. Psaltis, "Focusing and scanning light through a multimode optical fiber using digital phase conjugation," *Optics Express,* vol. 20, no. 10, pp. 10583–10590, 2012.

[20] I. N. Papadopoulos, S. Farahi, C. Moser, and D. Psaltis, "High-resolution, lensless endoscope based on digital scanning through a multimode optical fiber," *Biomedical Optics Express,* vol. 4, no. 2, pp. 260–270, 2013.

[21] Y. Choi *et al.*, "Scanner-free and wide-field endoscopic imaging by using a single multimode optical fiber," *Physical Review Letters,* vol. 109, no. 20, 203901, 2012.

[22] T. Čižmár, M. Mazilu, and K. Dholakia, "In situ wavefront correction and its application to micromanipulation," *Nature Photonics,* vol. 4, pp. 388–394, 2010.

[23] E. G. van Putten, D. Akbulut, J. Bertolotti, W. L. Vos, A. Lagendijk, and A. P. Mosk, "Scattering lens resolves sub-100 nm structures with visible light," *Physical Review Letters,* vol. 106, 193905, 2011.

[24] C. Park *et al.*, "Full-field subwavelength imaging using a scattering superlens," *Physical Review Letters,* vol. 113, no. 11, 113901, 2014.

[25] E. G. van Putten, A. Lagendijk, and A. P. Mosk, "Nonimaging speckle interferometry for high-speed nanometer-scale position detection," *Optics Letters,* vol. 37, no. 6, pp. 1070–1072, 2012.

[26] R. Horstmeyer, B. Judkewitz, I. M. Vellekoop, S. Assawaworrarit, and C. Yang, "Physical key-protected one-time pad," *Scientific Reports,* vol. 3, p. 3543, 2013.

[27] S. A. Goorden, M. Horstmann, A. P. Mosk, B. Škorić, and P. W. Pinkse, "Quantum-secure authentication of a physical unclonable key," *Optica,* vol. 1, no. 6, pp. 421–424, 2014.

[28] R. Horstmeyer, H. Ruan, and C. Yang, "Guidestar-assisted wavefront-shaping methods for focusing light into biological tissue," *Nature Photonics,* vol. 9, no. 9, pp. 563–571, 2015.

[29] A. P. Mosk, A. Lagendijk, G. Lerosey, and M. Fink, "Controlling waves in space and time for imaging and focusing in complex media," *Nature Photonics,* vol. 6, pp. 283–292, 2012.

[30] S. Rotter and S. Gigan, "Light fields in complex media: mesoscopic scattering meets wave control," *Reviews of Modern Physics,* vol. 89, 015005, 2017.

[31] M. C. W. van Rossum and T. M. Nieuwenhuizen, "Multiple scattering of classical waves," *Reviews of Modern Physics,* vol. 71, no. 1, pp. 313–371, 1999.

[32] J. W. Goodman, *Statistical Optics.* Wiley, 2000.

[33] C. W. J. Beenakker, "Random-matrix theory of quantum transport," *Reviews of Modern Physics,* vol. 69, pp. 731–808, 1997.

[34] B. Judkewitz, R. Horstmeyer, I. M. Vellekoop, I. N. Papadopoulos, and C. Yang, "Translation correlations in anisotropically scattering media," *Nature Physics,* vol. 11, no. 8, pp. 684–689, 2015.

[35] D. Akbulut, T. J. Huisman, E. G. van Putten, W. L. Vos, and A. P. Mosk, "Focusing light through random photonic media by binary amplitude modulation," *Optics Express,* vol. 19, no. 5, pp. 4017–4029, 2011.

[36] I. M. Vellekoop and A. P. Mosk, "Universal optimal transmission of light through disordered materials," *Physical Review Letters,* vol. 101, 120601, 2008.

[37] C. W. Hsu, A. Goetschy, Y. Bromberg, A. D. Stone, and H. Cao, "Broadband coherent enhancement of transmission and absorption in disordered media," *Physical Review Letters,* vol. 115, no. 22, 223901, 2015.

[38] I. M. Vellekoop and A. P. Mosk, "Phase control algorithms for focusing light through turbid media," *Optics Communications,* vol. 281, 2008.

[39] I. M. Vellekoop, M. Cui, and C. Yang, "Digital optical phase conjugation of fluorescence in turbid tissue," *Applied Physics Letters,* vol. 101, 081108, 2012.

[40] S. N. Chandrasekaran, H. Ligtenberg, W. Steenbergen, and I. M. Vellekoop, "Using digital micromirror devices for focusing light through turbid media," in *SPIE MOEMS-MEMS,* 897905–10. International Society for Optics and Photonics, 2014.

[41] O. Dorokhov, "On the coexistence of localized and extended electronic states in the metallic phase," *Solid State Communications,* vol. 51, no. 6, pp. 381–384, 1984.

[42] S. F. Liew, S. M. Popoff, A. P. Mosk, W. L. Vos, and H. Cao, "Transmission channels for light in absorbing random media: from diffusive to ballistic-like transport," *Physical Review B,* vol. 89, no. 22, 224202, 2014.

[43] S. M. Popoff, A. Goetschy, S. F. Liew, A. D. Stone, and H. Cao, "Coherent control of total transmission of light through disordered media," *Physical Review Letters,* vol. 112, no. 13, 133903, 2014.

[44] M. Kim *et al.,* "Maximal energy transport through disordered media with the implementation of transmission eigenchannels," *Nature Photonics,* vol. 6, no. 9, pp. 581–585, 2012.

[45] H. Yılmaz, W. L. Vos, and A. P. Mosk, "Optimal control of light propagation through multiple-scattering media in the presence of noise," *Biomedical Optics Express,* vol. 4, no. 9, pp. 1759–1768, 2013.

[46] S. M. Popoff, G. Lerosey, R. Carminati, M. Fink, A. C. Boccara, and S. Gigan, "Measuring the transmission matrix in optics: an approach to the study and control of light propagation in disordered media," *Physical Review Letters,* vol. 104, no. 10, 100601, 2010.

[47] D. B. Conkey, A. M. Caravaca-Aguirre, and R. Piestun, "High-speed scattering medium characterization with application to focusing light through turbid media," *Optics Express,* vol. 20, no. 2, pp. 1733–1740, 2012.

[48] W. B. Bridges *et al.,* "Coherent optical adaptive techniques," *Applied Optics,* vol. 13, no. 2, pp. 291–300, 1974.

[49] M. Cui, "Parallel wavefront optimization method for focusing light through random scattering media," *Optics Letters,* vol. 36, no. 6, pp. 870–872, 2011.

[50] S. Popoff, G. Lerosey, M. Fink, A. C. Boccara, and S. Gigan, "Image transmission through an opaque material," *Nature Communications,* vol. 1, p. 81, 2010.

[51] D. B. Conkey and R. Piestun, "Color image projection through a strongly scattering wall," *Optics Express,* vol. 20, no. 25, pp. 27312–27318, 2012.

[52] Y. Choi *et al.*, "Overcoming the diffraction limit using multiple light scattering in a highly disordered medium," *Physical Review Letters,* vol. 107, 023902, 2011.

[53] D. B. Conkey, A. N. Brown, A. M. Caravaca-Aguirre, and R. Piestun, "Genetic algorithm optimization for focusing through turbid media in noisy environments," *Optics Express,* vol. 20, no. 5, pp. 4840–4849, 2012.

[54] A. Drémeau *et al.*, "Reference-less measurement of the transmission matrix of a highly scattering material using a DMD and phase retrieval techniques," *Optics Express,* vol. 23, no. 9, pp. 11898–11911, 2015.

[55] D. B. Conkey, A. M. Caravaca-Aguirre, J. D. Dove, H. Ju, T. W. Murray, and R. Piestun, "Super-resolution photoacoustic imaging through a scattering wall," *Nature Communications,* vol. 6, p. 7902, 2015.

[56] O. Katz, E. Small, Y. Bromberg, and Y. Silberberg, "Focusing and compression of ultra-short pulses through scattering media," *Nature Photonics,* vol. 5, no. 6, pp. 372–377, 2011.

[57] C. Draeger, J.-C. Aime, and M. Fink, "One-channel time-reversal in chaotic cavities: Experimental results," *Journal of the Acoustical Society of America,* vol. 105, no. 2, pp. 618–625, 1999.

[58] C. Draeger and M. Fink, "One-channel time-reversal in chaotic cavities: theoretical limits," *Journal of the Acoustical Society of America,* vol. 105, no. 2, pp. 611–617, 1999.

[59] J. Aulbach, B. Gjonaj, P. Johnson, and A. Lagendijk, "Spatiotemporal focusing in opaque scattering media by wave front shaping with nonlinear feedback," *Optics Express,* vol. 20, no. 28, pp. 29237–29251, 2012.

[60] P. Lai, L. Wang, J. W. Tay, and L. V. Wang, "Photoacoustically guided wavefront shaping for enhanced optical focusing in scattering media," *Nature Photonics,* vol. 9, no. 2, pp. 136–142, 2015.

[61] O. Katz, E. Small, Y. Guan, and Y. Silberberg, "Noninvasive nonlinear focusing and imaging through strongly scattering turbid layers," *Optica,* vol. 1, no. 3, pp. 170–174, 2014.

[62] D. Sinefeld, H. P. Paudel, D. G. Ouzounov, T. G. Bifano, and C. Xu, "Adaptive optics in multiphoton microscopy: comparison of two, three and four photon fluorescence," *Optics Express*, vol. 23, no. 24, pp. 31472–31483, 2015.

[63] J. Tang, R. N. Germain, and M. Cui, "Superpenetration optical microscopy by iterative multiphoton adaptive compensation technique," *Proceedings of the National Academy of Sciences of the USA,* vol. 109, pp. 8434–8439, 2012.

[64] V. Ntziachristos, J. Ripoll, L. V. Wang, and R. Weissleder, "Looking and listening to light: the evolution of whole-body photonic imaging," *Nature Biotechnology,* vol. 23, no. 3, pp. 313–320, 2005.

[65] F. Kong, R. H. Silverman, L. Liu, P. V. Chitnis, K. K. Lee, and Y.-C. Chen, "Photoacoustic-guided convergence of light through optically diffusive media," *Optics Letters,* vol. 36, no. 11, pp. 2053–2055, 2011.

[66] T. Chaigne, O. Katz, A. C. Boccara, M. Fink, E. Bossy, and S. Gigan, "Controlling light in scattering media non-invasively using the photoacoustic transmission matrix," *Nature Photonics,* vol. 8, no. 1, pp. 58–64, 2014.

[67] A. M. Caravaca-Aguirre, D. B. Conkey, J. D. Dove, H. Ju, T. W. Murray, and R. Piestun, "High contrast three-dimensional photoacoustic imaging through scattering media by

localized optical fluence enhancement," *Optics Express,* vol. 21, no. 22, pp. 26671–26676, 2013.

[68] J. W. Tay, P. Lai, Y. Suzuki, and L. V. Wang, "Ultrasonically encoded wavefront shaping for focusing into random media," *Scientific Reports,* vol. 4, 3918, 2014.

[69] Y. Guan, O. Katz, E. Small, J. Zhou, and Y. Silberberg, "Polarization control of multiply scattered light through random media by wavefront shaping," *Optics Letters,* vol. 37, no. 22, pp. 4663–4665, 2012.

[70] J.-H. Park, C. Park, H. Yu, Y.-H. Cho, and Y. Park, "Dynamic active wave plate using random nanoparticles," *Optics Express,* vol. 20, no. 15, pp. 17010–17016, 2012.

[71] M. Nixon *et al.,* "Real-time wavefront shaping through scattering media by all-optical feedback," *Nature Photonics,* vol. 7, no. 11, pp. 919–924, 2013.

[72] J. Jang *et al.,* "Complex wavefront shaping for optimal depth-selective focusing in optical coherence tomography," *Optics Express,* vol. 21, no. 3, pp. 2890–2902, 2013.

[73] H. Yu *et al.,* "Depth-enhanced 2-D optical coherence tomography using complex wavefront shaping," *Optics Express,* vol. 22, no. 7, pp. 7514–7523, 2014.

[74] J. Aulbach, B. Gjonaj, P. M. Johnson, A. P. Mosk, and A. Lagendijk, "Control of light transmission through opaque scattering media in space and time," *Physical Review Letters,* vol. 106, no. 10, 103901, 2011.

[75] C. Stockbridge *et al.,* "Focusing through dynamic scattering media," *Optics Express,* vol. 20, no. 14, pp. 15086–15092, 2012.

[76] X. Tao, D. Bodington, M. Reinig, and J. Kubby, "High-speed scanning interferometric focusing by fast measurement of binary transmission matrix for channel demixing," *Optics Express,* vol. 23, no. 11, pp. 14168–14187, 2015.

[77] D. B. Conkey, A. M. Caravaca-Aguirre, E. Niv, and R. Piestun, "High-speed phase-control of wavefronts with binary amplitude DMD for light control through dynamic turbid media," in *SPIE MOEMS-MEMS,* 86170I-6. International Society for Optics and Photonics, 2013.

[78] I. M. Vellekoop and C. M. Aegerter, "Focusing light through living tissue," *Proceedings of SPIE,* vol. 7554, 755430, 2010.

[79] M. Jang, H. Ruan, I. M. Vellekoop, B. Judkewitz, E. Chung, and C. Yang, "Relation between speckle decorrelation and optical phase conjugation (OPC)-based turbidity suppression through dynamic scattering media: a study on *in vivo* mouse skin," *Biomedical Optics Express,* vol. 6, no. 1, pp. 72–85, 2015.

[80] Z. Yaqoob, D. Psaltis, M. S. Feld, and C. Yang, "Optical phase conjugation for turbidity suppression in biological samples," *Nature Photonics,* vol. 2, pp. 110–115, 2008.

[81] E. Larose *et al.,* "Correlation of random wavefields: an interdisciplinary review," *Geophysics,* vol. 71, no. 4, pp. SI11–SI21, 2006.

[82] F. van Beijnum, E. G. van Putten, A. Lagendijk, and A. P. Mosk, "Frequency bandwidth of light focused through turbid media," *Optics Letters,* vol. 36, no. 3, pp. 373–375, 2011.

[83] H. F. Arnoldus and T. F. George, "Phase-conjugated fluorescence," *Physical Review A,* vol. 43, no. 7, pp. 3675–3689, 1991.

[84] S. Feng, C. Kane, P. A. Lee, and A. D. Stone, "Correlations and fluctuations of coherent wave transmission through disordered media," *Physical Review Letters,* vol. 61, pp. 834–837, 1988.

[85] I. Freund, M. Rosenbluh, and S. Feng, "Memory effects in propagation of optical waves through disordered media," *Physical Review Letters,* vol. 61, pp. 2328–2331, 1988.

[86] J. H. Li and A. Z. Genack, "Correlation in laser speckle," *Physical Review E,* vol. 49, no. 5, p. 4530, 1994.

[87] I. M. Vellekoop and C. M. Aegerter, "Scattered light fluorescence microscopy: imaging through turbid layers," *Optics Letters,* vol. 35, pp. 1245–1247, 2010.

[88] J. Bertolotti, E. G. van Putten, C. Blum, A. Lagendijk, W. L. Vos, and A. P. Mosk, "Non-invasive imaging through opaque scattering layers," *Nature,* vol. 491, no. 7423, pp. 232–234, 2012.

[89] O. Katz, P. Heidmann, M. Fink, and S. Gigan, "Non-invasive single-shot imaging through scattering layers and around corners via speckle correlations," *Nature Photonics,* vol. 8, no. 10, pp. 784–790, 2014.

[90] S. Schott, J. Bertolotti, J.-F. Léger, L. Bourdieu, and S. Gigan, "Characterization of the angular memory effect of scattered light in biological tissues," *Optics Express*, vol. 23, no. 10, pp 13505–13516, 2015.

[91] G. Osnabrugge, R. Horstmeyer, I. N. Papadopoulos, B. Judkewitz, and I. M. Vellekoop, "Generalized optical memory effect," *Optica*, vol. 4, no. 8 pp. 886–892, 2017.

[92] R. K. Tyson, *Principles of Adaptive Optics*. Academic Press, 1998.

[93] J. F. de Boer, M. P. van Albada, and A. Lagendijk, "Transmission and intensity correlations in wave propagation through random media," *Physical Review B,* vol. 45, no. 2, p. 658, 1992.

[94] D. E. Milkie, E. Betzig, and N. Ji, "Pupil-segmentation-based adaptive optical microscopy with full-pupil illumination," *Optics Letters,* vol. 36, no. 21, pp. 4206–4208, 2011.

[95] J. Mertz, H. Paudel, and T. G. Bifano, "Field of view advantage of conjugate adaptive optics in microscopy applications," *Applied Optics*, vol. 54, no. 11, pp. 3498–3506, 2015.

[96] B. Y. Zel'dovich, V. I. Popovichev, V. V. Ragul'skii, and F. S. Faisullov, "Connection between the wave fronts of the reflected and exciting light in stimulated Mandel'shtam-Brillouin scattering," *JETP Letters,* vol. 15, p. 109, 1972.

[97] A. Yariv, "Phase conjugate optics and real-time holography [invited paper]," *IEEE Journal of Quantum Electronics,* vol. 14, no. 9, pp. 650–660, 1978.

[98] R. A. Fisher, *Optical Phase Conjugation*. Academic Press, 1983.

[99] M. Cui and C. Yang, "Implementation of a digital optical phase conjugation system and its application to study the robustness of turbidity suppression by phase conjugation," *Optics Express,* vol. 18, no. 4, pp. 3444–3455, 2010.

[100] C.-L. Hsieh, Y. Pu, R. Grange, and D. Psaltis, "Digital phase conjugation of second harmonic radiation emitted by nanoparticles in turbid media," *Optics Express,* vol. 18, no. 12, pp. 12283–12290, 2010.

[101] Y. Liu, P. Lai, C. Ma, X. Xu, A. A. Grabar, and L. V. Wang, "Optical focusing deep inside dynamic scattering media with near-infrared time-reversed ultrasonically encoded (TRUE) light," *Nature Communications,* vol. 6, 5904, 2015.

[102] D. Wang, E. H. Zhou, J. Brake, H. Ruan, M. Jang, and C. Yang, "Focusing through dynamic tissue with millisecond digital optical phase conjugation," *Optica,* vol. 2, no. 8, pp. 728–735, 2015.

[103] Y. Liu, C. Ma, Y. Shen, J. Shi, and L. V. Wang, "Focusing light inside dynamic scattering media with millisecond digital optical phase conjugation," *Optica,* vol. 4, no. 2, pp. 280–288, 2017.

[104] M. Fink, C. Prada, F. Wu, and D. Cassereau, "Self focusing in inhomogeneous media with time reversal acoustic mirrors," *IEEE Ultrasonics Symposium Proceedings,* vol. 2, pp. 681–686, 1989.

[105] C. Prada, J. L. Thomas, and M. Fink, "The iterative time reversal process: analysis of the convergence," *Journal of the Acoustical Society of America,* vol. 97, no. 1, pp. 62–71, 1995.

[106] X. Yang, Y. Pu, and D. Psaltis, "Imaging blood cells through scattering biological tissue using speckle scanning microscopy," *Optics Express,* vol. 22, no. 3, pp. 3405–3413, 2014.

[107] R. Barankov and J. Mertz, "High-throughput imaging of self-luminous objects through a single optical fibre," *Nature Communications,* vol. 5, 5581, 2014.

[108] A. M. Caravaca-Aguirre, E. Niv, D. B. Conkey, and R. Piestun, "Real-time resilient focusing through a bending multimode fiber," *Optics Express,* vol. 21, no. 10, pp. 12881–12887, 2013.

[109] I. M. Vellekoop, A. Lagendijk, and A. P. Mosk, "Exploiting disorder for perfect focusing," *Nature Photonics,* vol. 4, pp. 320–322, 2010.

[110] E. G. van Putten, A. Lagendijk, and A. P. Mosk, "Optimal concentration of light in turbid materials," *Journal of the Optical Society of America B,* vol. 28, pp. 1200–1203, 2011.

[111] B. Gjonaj, J. Aulbach, P. M. Johnson, A. P. Mosk, L. Kuipers, and A. Lagendijk, "Active spatial control of plasmonic fields," *Nature Photonics,* vol. 5, no. 6, pp. 360–363, 2011.

[112] B. Gjonaj, J. Aulbach, P. M. Johnson, A. P. Mosk, L. Kuipers, and A. Lagendijk, "Focusing and scanning microscopy with propagating surface plasmons," *Physical Review Letters,* vol. 110, no. 26, 266804, 2013.

[113] S. R. Huisman, T. J. Huisman, S. A. Goorden, A. P. Mosk, and P. W. Pinkse, "Programming balanced optical beam splitters in white paint," *Optics Express,* vol. 22, no. 7, pp. 8320–8332, 2014.

[114] J.-H. Park, C. Park, H. Yu, Y.-H. Cho, and Y. Park, "Active spectral filtering through turbid media," *Optics Letters,* vol. 37, no. 15, pp. 3261–3263, 2012.

[115] T. J. Huisman, S. R. Huisman, A. P. Mosk, and P. W. Pinkse, "Controlling single-photon Fock-state propagation through opaque scattering media," *Applied Physics B,* vol. 116, pp. 1–5, 2013.

[116] R. Horstmeyer, B. Judkewitz, I. Vellekoop, and C. Yang, "Secure storage of cryptographic keys within random volumetric materials," in *CLEO: Applications and Technology.* Optical Society of America, 2013.

[117] R. Pappu, B. Recht, J. Taylor, and N. Gershenfeld, "Physical one-way functions," *Science,* vol. 297, no. 5589, pp. 2026–2030, 2002.

[118] I. M. Vellekoop, "Feedback-based wavefront shaping," *Optics Express,* vol. 23, no. 9, pp. 12189–12206, 2015.

9 Focusing Light through Scattering Media Using a Microelectromechanical Systems Spatial Light Modulator

Yang Lu and Hari P. Paudel

9.1 Introduction

Optical microscopy has been used as a primary tool of biological subsurface imaging, where the features of interest are usually several hundreds of micrometers below the surface of tissue. However, the depth of imaging is greatly limited by the scattering property of the tissue. Confocal microscopy, invented by Minsky in 1961 [1], increased the imaging depth to a few hundreds of microns by selectively detecting the ballistic light from the scattered light component using a confocal pinhole at the detection plane of the scanning microscope. Two-photon microscopy, invented by Denk et al. in 1990 [2], further extended the imaging depth by utilizing the nonlinear property of two-photon process. However, these techniques can't prevent the scattering of light and rely on the ballistic light components. When imaging depth is more than 7 times of the scattering length, the intensity of scattered light inside the tissue will be greater than that of the ballistic light and these techniques began to fail. In 2007 Vellekoop showed that by shaping the wavefront of excitation beam, a speckle can be made several orders of magnitude brighter than background speckles [3]. This finding gave a new avenue to the deep tissue imaging.

Over the past decade, the development of Microelectromechanical Systems (MEMS) Spatial Light Modulator (SLM) has made enormous impact to the field of Adaptive Optics (AO) [4]. It has catalyzed rapid growth in astronomical imaging and other imaging applications, where fast aberration correction are critical, such as *in vivo* retinal imaging and subsurface imaging of scattering tissue. Its advantages have made this technology widely used: MEMS SLMs are inherently fast and compact as a result of its design and manufacturing process. More importantly, MEMS device cost almost an order of magnitude less than the technology they replace, which makes it affordable to be integrated into many experimental microscopy platforms that might benefit from AO-enhanced performance.

For subsurface imaging of biological tissue, image quality is usually compromised by both multiple-scattering events in the sample, and/or by an index of refraction mismatch between the tissue and the medium (e.g., water and oil). In theory, AO can be used to fix the aberration resulting from this index mismatch, i.e., mostly defocus and spherical aberration. However, multiple scattering makes it more challenging to measure, more difficult to fix with traditional closed-loop control, which involves a wavefront sensor. In order to overcome the deleterious effects from scattering, "sensorless" AO techniques are often required, in which an optimization approach iteratively reshapes its reflective surface and then measures the resulting image quality [5]–[10].

In the previous chapter, liquid crystal spatial light modulators (LC-SLMs) have been discussed extensively on their use for focus control in scattering media and biological samples. However, if the medium is dynamic, the feedback control needs to be fast enough to keep up with the wavefront [11]. For biological samples, the speckle decorrelation time is on the millisecond time scale [12], [13]. However, the LC-SLM's switching speed depends on the liquid crystal align rate, which is usually on the order of tens of Hz [14]. It is much slower than the kHz rate that a living biological sample requires. MEMS SLMs, on the other hand, have proved to be well suited for this type of control because they are fast (frame rates of >10 kHz), and can be shaped precisely with subnanometer precision and predictably with zero hysteresis [15]–[23].

In this chapter, we describe the use of MEMS device to enhance microscopic focusing in the scattering media. We begin the chapter with manufacturing and actuation mechanism of MEMS SLMs. Then we illustrate how focusing through scattering media is achieved using both monochromatic and chromatic light. In recent development, two- and three-photon microscopy is demonstrated with MEMS SLMs for scattered light control. Lastly, conjugate AO is provided as future direction to the readers to overcome field of view limitations most for subsurface imaging applications.

9.2   Background on MEMS SLM

In terms of manufacturing, the MEMS device production does not require exotic materials or much manufacturing tolerances: it can exploit a MEMS foundry and begins with an optically smooth, flat, and inexpensive substrate. Devices are batch-produced, although research and development costs are high, commercial production and replication expenses are low. In addition, hundreds of devices can be produced on each wafer, allowing broad parameter variations in a single batch-production cycle. The manufacturing of MEMS devices typically involves three sequential steps: depositing a thin film, patterning the film with a temporary mask, and etching the film through the mask. This cycle is repeated until the full multilayer structure is produced. Deposited films alternate between those that are structural (such as polycrystalline silicon) and those intended to be sacrificial (such as phosphosilicate glass). A final production step is to dissolve all of the sacrificial layers with a wet etch in hydrofluoric acid, yielding a released, fully assembled silicon device [24].

There are multiple actuation mechanisms for MEMS DMs, such as electrostatic actuation, magnetic actuation, thermal actuation, etc. We will only illustrate the first type of actuation (electrostatic) in this chapter because it is best candidate for fast feedback

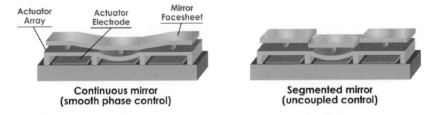

Continuous mirror
(smooth phase control)

Segmented mirror
(uncoupled control)

Figure 9.1 Schematic view of continuous face-sheet (left) and segmented face-sheet (right) mirror. Image reproduced with permission from Boston Micromachines Corp.

optimization control and has been widely used in a variety of applications, such as dynamic aberration control that requires high spatial and/or temporal frequencies. The architecture of such MEMS SLM is shown in Figure 9.1.

The success of the electromechanical design and manufacturing approach make MEMS device in high demand for scattering compensation: MEMS SLMs are mechanically stiff and lightweight, allowing their bandwidths to be controlled over tens of kilohertz [25]. The noncontact electrostatic actuation mechanism is repeatable to subnanometer precision, consumes almost no power, exhibits no hysteresis and is unaffected by trillions of operation cycles. More importantly, MEMS SLM devices are scalable; increasing the spatial resolution or the size of an SLM can be achieved by adding multiple lithographic masks.

Depending on the face-sheet, MEMS SLM can be categorized into two types: segmented and continuous. The segmented face-sheet type, or usually referred as SLM, comprises flat individual mirror segments. Each of the segments can move independently (piston mode). The continuous face-sheet type, on the other hand, has face-sheet mechanically connected to each other. It's often referred as a deformable mirror or DM. When one actuator is pulled down, the neighboring actuators are coupled. Generally, 10%–20% displacement by neighboring actuators is expected when using a DM.

9.3   MEMS SLM for Dynamic Scattering Media

The ability to focus light through highly scattering media has been studied by many groups. It has paramount significance of improving biological microscopy performance. A major challenge in many applications, however, is the dynamic nature of scattering in the media.

Previous studies have shown optical focusing through stationary scattering media using either coordinate descent optimization or measurement of the system's effective transmission matrix (as discussed in Chapter 5). Both techniques manipulate phase of an incident beam into scattering medium using a liquid crystal SLM, with a goal of optimizing the focus intensity at a point on the opposite side of the medium [3], [5]–[10]. In coordinate descent optimization, the intensity is measured and optimized iteratively for each of N input modes, where N is the number of spatial degrees of freedom in the SLM. In transmission matrix optimization, the relationship between optical input and output modes of the system is estimated from a group of N SLM input states and N corresponding output states. Using that relationship, one can optimize focus at any point in the measured field. In addition, one can use the estimated transmission matrix to predict the SLM input state that will optimize an arbitrary output state. We believe that coordinate descent optimization should be advantageous for focusing through a dynamic medium since it continuously adapts to the changing medium with each new measurement, whereas the transmission matrix optimization can only adapt after an ensemble of measurements [26].

In this section, we demonstrate steady focusing enhancement over a wide range of decorrelation time constants on the subsecond level. Moreover, the focusing enhancement as a function of number of degrees of freedom in the SLM is also quantified. The demonstration is done with speckle decorrelation time constants similar to those of typical biological tissue.

A high-speed segmented MEMS SLM was used (Boston Micromachines Corporation, Kilo-SLM) to maintain highly enhanced focus through media. The MEMS SLM comprises 1020 mirror segments in a 32×32 array where the four corner segments are inactive. Each segment measures 300 μm square and 3 μm thick. It can be translated in a surface-normal direction (i.e., piston mode) with a stroke of 1.5 μm. The translation is made through high precision voltage driver to an underlying electrostatic actuator. It is important to point out that the SLM was precalibrated using a surface mapping interferometer (Zygo NewView 6300), to allow subsequent direct phase control with ~0.01 wave accuracy at the laser wavelength of 532 nm. In the reported experiment, although the SLM has a mechanical response rate of >10k Hz, the 32×32 pixel CMOS sensor (USB 2 uEye LE) used for optimization feedback limited the overall system control frequency to ~330 Hz (control frequency is defined as the frequency at which the SLM can be updated and the sensor read) (Figure 9.2).

The goal of the optimization algorithm is to maximize the intensity of a single speckle grain in the camera sensor plane. In order to achieve maximum intensity, 1024 orthogonal 2D Walsh functions were utilized as a basis set for the optimization feedback control [27]–[30]. Each 2D Walsh function, W_k, has 32×32 (1024) terms that are bi-valued (+ 1 or −1). Each term corresponds to a mirror segment in the 32×32 SLM [31]. A collection of 1024 Walsh matrices scaled by corresponding coefficients c_k (in units of waves) represents an orthogonal basis set for the 1020 segment SLM. For simplicity, we ignore the fact that the four corner segments of the SLM are immobile.

To implement Walsh optimization, the modal coefficient c_k of each Walsh function W_k is adjusted sequentially to maximize the scalar optimization metric S defined in our experiments as the intensity of a selected pixel on the camera (in all of our experiments, pixel coordinate 16, 16 near the center of our 32×32 camera sensor was selected). For a particular Walsh mode, k, an initial measurement is made of the optimization metric, S_{k0} with a modal coefficient of 0. Next, the modal coefficient of the Walsh function is set to a prescribed value $+\alpha$, (typically, but not necessarily, $1/4$ of a wave), and a scalar measure (S_{k1}) of the optimization metric is made. Next, the modal coefficient of the Walsh function is set to a prescribed value $-\alpha$ and a third scalar measure (S_{k2}) of the desired optimization metric is made. Finally, the coordinate optimal value for this Walsh

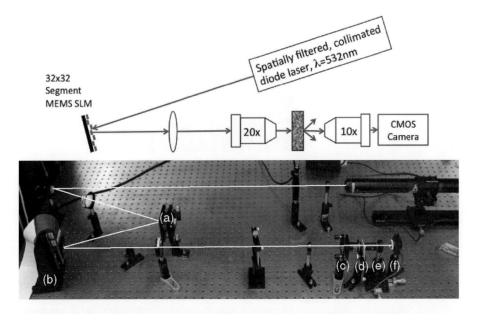

Figure 9.2 Schematic image of the experimental apparatus. A collimated, spatially filtered laser beam (**a**), reflects from the SLM surface (**b**), acquiring a spatially distributed phase shift. The wavefront at the SLM is reimaged by a 45× telescope comprising a 400 mm focal length lens and a 20× 0.4NA microscope objective (**c**), onto the near side of the scattering medium (**d**), a microscope objective on the far side of the scattering medium (**e**), projects a portion of the resulting speckle pattern onto the CMOS camera sensor (**f**). Reprinted with permission from [33], *Optics Express*.

function coefficient (in waves) is calculated using a standard technique for three-point phase-shifting interferometry [32]:

$$c_k = \tan^{-1}\left[\frac{(S_{k2} - S_{k1})\tan\left(\frac{\pi}{\alpha}\right)}{2S_{k0} - S_{k1} - S_{k2}}\right]$$

The process is repeated for all 1024 Walsh coordinate functions. Because each Walsh coordinate optimization requires three perturbations (three sequences of SLM output followed by camera input), the effective optimization frequency for Walsh coordinate optimization is one-third of the base control frequency, or ∼110 Hz.

In a qualitative demonstration of the technique, focus was enhanced through an 5mm thick section of chicken breast. Before optimization, the decorrelation time constant of the speckle at the camera was measured to be 3.4 s, based on an ensemble average auto-correlation for each pixel in the 32 × 32 pixel camera array decaying to the 3 dB point. Figure 9.3 illustrates the results. Focus enhancement, normalized by the mean camera intensity prior to optimization, reached approximately 160 with a standard deviation of 15, and did not diminish while the optimization controller continued to run for nearly two minutes. After the controller was stopped, with the SLM fixed at its final state, the enhancement dropped to a normalized value of ∼1 with an enhancement decay time

Figure 9.3 Experimental results for focus optimization though a 5 mm *ex vivo* sample of chicken breast tissue. Before each optimization, the mean intensity on the camera was measured. This value was then used as a normalization constant for subsequent enhancement measurements. (Left) Ensemble average autocorrelation for the fully developed speckle pattern on the camera prior to optimization. The decorrelation time constant was estimated to be ~3.4 s from these data. (Center) Optimized focus spot centered on the pixel at coordinate location (16, 16) within the 32-pixel square camera sensor area, with peak intensity ~160 times larger than the initial mean camera intensity measured within this area. (Right) Normalized peak intensity on the camera as a function of time before, during, and after optimization. After optimization is stopped, the focus enhancement decays with a first-order time constant of ~3.6 s to the background intensity level. Reprinted with permission from [33], *Optics Express*. A black and white version of this figure appears in some formats. For the color version, please refer to the plate section.

constant measured to be 3.6 s – approximately the same as the sample decorrelation time.

This result suggests, as expected, that the decay time after optimization corresponds to the speckle decorrelation time before optimization [33]. It is expected that the peak enhancement would be smaller for optimization through media characterized by shorter speckle decorrelation times, and vice versa, given a fixed controller update rate.

To further illustrate the technique, we quantify system performance as a function of media-dependent speckle decorrelation time. The experiment was conducted in which optimization was performed through a medium comprising two axially sequential adjacent cuvettes containing identical media. We move the first cuvette at constant speed while the second cuvette remained stationary. The medium comprised 0.75 μm diameter polystyrene spheres (Polybead® Microspheres) diluted to a concentration of 0.125% by volume and suspended in agarose gel. At this dilution the medium has a transport mean free path – the distance over which the direction of propagation becomes randomized of 4 mm. The first and second cuvettes were 4 mm and 1.2 mm thick respectively. The resulting speckle pattern was fully developed through these two samples. The decorrelation time was in proportion to the speed of the translating samples. Translation speeds were produced by a piezoelectric actuator. The speed was measured 75 nm/s to 3 μm/s by a linear variable differential transformer displacement gage. The speed range translates into speckle decorrelation times of 0.35 s to 18 s.

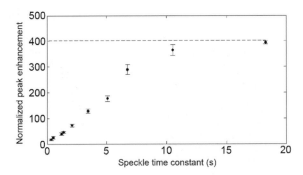

Figure 9.4 Peak optimized enhancement measured as a function of initial speckle decorrelation time for a dynamic scattering medium, with Walsh function coordinate optimization update rate of 110 Hz and 1020 phase-modulating segments. The dashed line corresponds to the mean peak enhancement level achieved when the medium was stationary. Error bars correspond to one standard deviation of enhancement fluctuation over a 10 s period at the end of the optimization trial. Reprinted with permission from [33], *Optics Express.*

Prior to the dynamic translation experiment, focus was set at the optimized SLM state at the combined media without translation. Peak normalized enhancement after optimization in this stationary case was approximately 380 with a standard deviation of $\sim$20. For all dynamic experiments, optimization duration was 90 s. Peak focus enhancement for translating media ranged from a low of 5 at a translation speed of 3 μm/s, corresponding to a speckle decorrelation time if 0.4 s, to a high of 380 at a translation speed of 75 nm/s, corresponding to a speckle decorrelation time of 20 s. Figure 9.4 illustrates the optimization results graphically for a number of dynamic conditions using 1020 independent SLM segments in the optimization.

If speckle evolution and optimization are considered to be competing first-order processes characterized by their respective time constants, we can estimate the expected peak enhancement with a function of the number of independent segments in the SLM, N, the time required to update each Walsh term in the optimization process T_i(0.085 s for the present system), and the speckle decorrelation time T_d, and a proportionality constant α. That relationship adapted from [3] is as follows:

$$\eta = \frac{\alpha N}{(1 + \frac{NT_i}{T_d})}$$

In the case of a static medium, this reduces to $\eta = \alpha N$. In our static medium experiments we found that $\alpha \sim$0.5, a constant of proportionality similar to that achieved experimentally by other researchers using liquid crystal SLMs. The theoretical maximum for α proposed in [3] is $\pi/4$.

To test the theory, we then conducted a series of experiments similar to those shown in Figure 9.5, but using fewer independent segments in the SLM and correspondingly fewer Walsh coordinate functions in the optimization algorithm. The optics of the system were not changed. The mirrors segments in the SLM were grouped into dependent clusters of 2×2, 4×4, and 8×8 "supersegments" corresponding to effective SLM array sizes

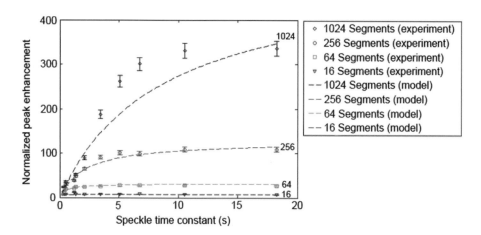

Figure 9.5 Normalized peak enhancement as a function of speckle decorrelation time and number of independent segments in the SLM. Error bars correspond to the span of mean enhancement for three trials at each experimental condition. Reprinted with permission from [33], *Optics Express.*

of 256, 64, and 16 respectively. The measured and modeled peak enhancement data are plotted as a function of speckle decorrelation time and number of segments in the SLM in Figure 9.5.

It can be seen from Figure 9.5 that shorter speckle decorrelation times associates with fast-changing dynamic media. Therefore the advantage of additional segments in the SLM becomes increasingly less significant in the overall task of enhancing focus intensity. This is expected: the additional degrees of freedom enabled by more segments require proportionately longer time to cycle through a full set of orthogonal Walsh coordinate optimizations. If the speckle decorrelation time is less than the time to complete that set of optimizations, the advantages of additional orthogonal coordinates is substantially diminished. The results obtained experimentally here correspond well both qualitatively and quantitatively to analytical model proposed by [34] for optimization through fluctuating media.

9.4   MEMS SLM for Scattered Light Control in Multiphoton Microscopy

Multiphoton microscopy has become an indispensable tool for subsurface imaging in scattering tissue because of its inherent selectivity to the ballistic photons. However, scattered light eventually limits the imaging depth to about five to six times of scattering length [35], [36]. If we could make a focus from the scattered light using coherent beam control technique, we can increase the imaging depth of multiphoton microscope. MEMS devices are orders of magnitude faster than LC-SLM and has been used for the wavefront correction in multiphoton microcopy [37]–[39]. The femtosecond laser source in multiphoton microscopes is polychromatic in nature. Therefore, the expected enhancement of optimized focus is lower than monochromatic illumination. Next two

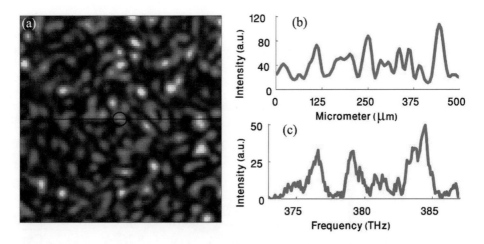

Figure 9.6 (**a**) Speckles of polychromatic light transmitted through a scattering media, (**b**) intensity profile of speckles along the black line, (**c**) spectral intensity profile of a single speckle.

subsections describe the model of focus enhancement with polychromatic light source and compares results with the two-photon florescence enhancement in multiphoton microcopy.

9.4.1   Focusing through Scattering Media with Polychromatic Light

The intensity of speckle focus by wavefront shaping is roughly equal to the number of active segments N in the SLM. If focus optimization is applied to M different spots simultaneously, the intensity of foci is reduced by the factor of M [8]. This is due to the fact that the SLM maps for full optimization are uncorrelated with each other. The focus formed by uncorrelated spectral modes applies the same argument that applied to the M independent spatially separated foci. When broadband light is scattered by random scattering media, different spectral bands lose their phase correlation due to the path length fluctuations and produces uncorrelated spectral modes. This section describes the relation of spectral bandwidth of polychromatic light with the speckle contrast and the coherent focus enhancement.

The pulse width of a typical femtosecond laser for two-photon microscopy is around 40 fs to 150 fs. The temporal pulse width Δt and spectral bandwidth Δv are Fourier transform pairs, given by an inverse relation $\Delta v \cong 1/\Delta t$. A polychromatic light source having source bandwidth Δv_l has finite coherent length $l_c \cong c/\Delta v_l$. In thick scattering sample, if the path length fluctuation is greater than temporal coherent length of source, it produces two or more uncorrelated frequency components. Therefore, the spectrum of a spatial single speckle has random fluctuation producing a spectral speckle. Figure 9.6a shows a speckle pattern produced by transmitting polychromatic light source through the scattering media. Figure 9.6b shows the intensity profile of speckles along the line drawn in Figure 9.6a while Figure 9.6b shows the spectral intensity profile of a single speckle enclosed in the circle shown in Figure 9.6a. The bandwidth of uncorrelated

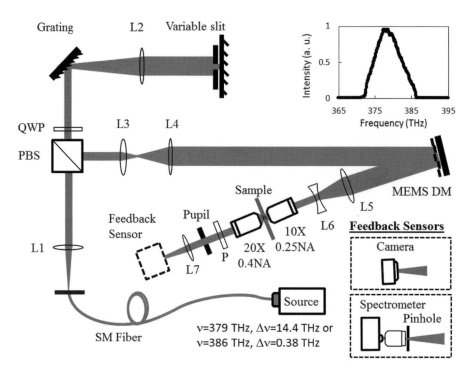

Figure 9.7 Schematic of the experimental setup. Reprinted with permission from [40], *Optics Express.*

frequency component is called sample bandwidth Δv_s. The number of uncorrelated frequency components M in a polychromatic speckle can be defined as $M = 1 + \Delta v_l / \Delta v_s$
[40]. Thicker sample has a narrower sample bandwidth, hence, more uncorrelated frequency components. The contrast of the speckle pattern is reduced by factor $1/\sqrt{M}$
[41]. The uncorrelated frequencies have independent phase pattern to produce a speckle
focus, so one might expect that the spectrally independent components may apply the
argument made for spatially M independent components, but the fact is that, this analogy
applies only after correction in bandwidth of uncorrelated frequency component that is
modified by the optimization process itself. This atypical relation of sample bandwidth
with optimization process is explained in detail in the subsequent paragraphs with the
experimental results.

A schematic of the experimental setup is illustrated in Figure 9.7. A 5 mW single-
mode fiber-coupled superluminescent diode (SLD) from Superlum (SLD-33-HP) is used
as a polychromatic light source. The bandwidth of SLD is 14.4 THz (29.8 nm) with central frequency 379 THz (790 nm). A setup consisting of a diffraction grating (1200
lines/mm), a 200 mm focal length lens (L2) and a variable-width slit in a standard
double-pass configuration control the spectral bandwidth of the illumination beam. The
illumination bandwidth ranges from 11.8 THz at the maximum (7 mm) slit width to
2.2 THz at the minimum (1mm) slit width. A 4.5 mW diode laser (Thorlabs CSP192)
with 0.35 THz bandwidth and 386 THz (778 nm) central frequency was used for the

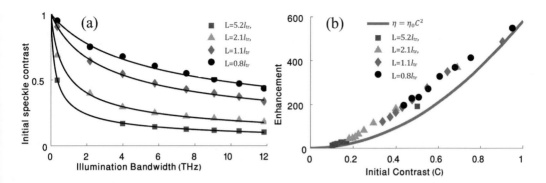

Figure 9.8 (a) Experimental results of initial speckle contrast C versus illumination bandwidth for four different thicknesses of samples. (b) Focus enhancement η versus speckle contrast C. Reprinted with permission from [40], *Optics Express.*

near monochromatic illumination source. The expanded illumination beams covers 900 segments (pixels) of SLM (Boston Micromachines Corporation, Kilo-SLM), each with surface area 0.09 mm^2. A linear polarizer and an iris are used produce a speckle image of controlled speckle size at the camera sensor (μEye USB 2LE). For wavelength selective optimization, a single speckle is passed through a 75 μm pinhole placed at the camera location. A 10× 0.25NA microscope objective is used to couple the speckle into a spectrometer (Thorlabs CCS175). Experiment was performed using four calibrated scattering samples having sample thicknesses L equals to 5.2 l_{tr}, 2.1 l_{tr}, 1.1 l_{tr} and 0.8 l_{tr}, where l_{tr} is the transport mean free path of samples. The size of speckle is made comparative larger than the pixel size of camera or the size of pinhole to prevent the speckle intensity from being averaging.

The spectral bandwidth of illumination source is calculated using equation $\Delta v_l = \left(\int S_0 (v) \, dv \right)^2 / \int S_0^2 (v) \, dv$, where S_0 is the measured spectral intensity profile. The spectral bandwidth (in terms of FWHM) is calculated from above formula is less sensitive to the exact shape of the spectral profile. Another advantage of this formula is that the temporal coherent time Δt_l is a direct Fourier transform pair of spectral bandwidth Δv_l. Initial speckle contrast C is measured at various of combination of illumination source bandwidths and sample bandwidths followed by measurement of speckle focus intensity enhancement.

The speckle contrast is defined as $C = \langle \sigma \rangle / \langle I \rangle$, where $\langle \sigma \rangle$ and $\langle I \rangle$ are the standard deviation and the average intensity of speckle respectively. For monochromatic illumination speckle contrast C is unity [41]. For polychromatic illumination, speckle contrast C is reduced a factor of square root of number of independent frequency components M. Therefore, the measured contrast of polychromatic illumination is given by relation $C = C_0 / \sqrt{M}$, where C_0 is the speckle contrast by monochromatic illumination. The measured C_0 is less than unity due to the experimental limitation caused by finite size of camera pixels and nonzero laser bandwidth. A near monochromatic 632.8 nm He-Ne laser is used to determine the experimental value of maximum speckle contrast.

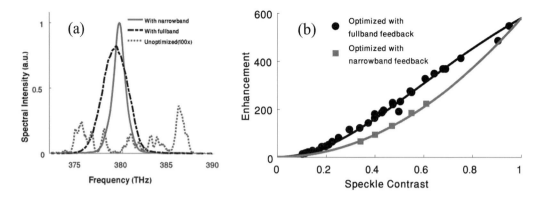

Figure 9.9 (**a**) Dot line: spectrum of a speckle from the sample $L = 1.1l_{tr}$, illumination source bandwidth 11.8 THz, solid line: spectral profile of optimized focus when feedback is 0.4 THz at the center of spectrum, dash line: spectral profile when feedback is full bandwidth. (**b**) Enhancement E versus speckle contrast C with broadening factors $\xi = 2$ (circle markers) and $\xi = 1$ (square markers) alongside the experimental data. Reprinted with permission from [40], *Optics Express*.

Figure 9.8a shows the experimental results of measured speckle contrast C using four scattering samples at various illumination source bandwidth. The contrast C with sample bandwidth Δv_s and source bandwidth Δv_l is given by a relation $C = \sqrt{\Delta v_s / (\Delta v_s + \Delta v_l)}$ [40]. This model provides the number of uncorrelated frequency components M in the speckle of contrast C at a given illumination bandwidth Δv_l. Therefore, the expected enhancement of polychromatic light having a contrast C can be written as $\eta = \eta_0/M = \eta_0 C^2$, where η_0 is the maximum possible enhancement with monochromatic illumination. Figure 9.8b shows the focus enhancement η versus speckle contrast C from experimental result and model. The experimental data points no longer fit the expected enhancement curve (red line), however, the enhancement from various scattering samples and illumination bandwidth appears to obey a relation with respect to the contrast. Interestingly, the experimental values were greater than expected values from the model. This unexpected enhancement gives a hint that the effective number of uncorrelated frequency components after optimization could be less than the initial number of uncorrelated frequency components. In order to confirm this hypothesis, a spectrally resolved feedback was set up, which is discussed in next paragraph.

To perform the spectrally resolved experiment, a single spatial speckle is passed through a pinhole and its spectral intensity profile is measured by a spectrometer. The dot line in Figure 9.9a shows the spectrum of a speckle from scattering sample of $L = 1.1l_{tr}$ at illumination source bandwidth 11.8 THz. The spectral width of each spectral speckle represents the sample bandwidth Δv_s, the frequency range over which the transmitted light remains monochromatic. The calculated value of Δv_s from the contrast model is 1.6 THz, therefore, the number of uncorrelated frequency modes M is about 8, which is consistent with spectrum profile. To optimized with spectrally resolved feedback a various spectral widths of speckle spectrum are selected for the feedback. The solid line in Figure 9.9a shows the spectral profile of optimized focus when optimization feedback is selected at a narrow spectral range (0.4 THz). The bandwidth of optimized focus is 2.56 THz, which is broader than sample bandwidth Δv_s. But it is consistent with report from tunable monochromatic source where the spectrum was found to coincide with the correlation of a spectral speckle, which has a larger bandwidth than spectral speckle itself [42]. A more interesting result was found when the optimization feedback was selected from the full spectrum (11.84 THz), which is shown by dash line in Figure 9.9a. Although the peak intensity of optimized focus is decreased slightly, there is a significant broadening of enhanced spectral width (5.12 THz) with net increase in focus intensity. This result proved the hypothesis that with polychromatic illumination the optimization process decreases the effective number uncorrelated frequency component in order to increase the overall intensity of focus. The focus enhancement η can be generalized in terms of effective number of uncorrelated frequency component M_{eff}. If spectral bandwidth is modified by factor ξ then the modified bandwidth will be $\xi \Delta v_s$. For full broadband feedback ξ is nearly 2 (Katz et al. report similar number in their temporal focusing experiment [43]). Therefore, the effective number of modes is

$$M_{eff} = 1 + \frac{\Delta v_l}{\xi \Delta v_s} = \frac{1}{\xi}(C^{-2} - 1) + 1$$

where C is the speckle contrast before optimization. For monochromatic source $C = 1$, hence, M_{eff} remains equal to unity. For polychromatic source, the focus enhancement is now modified by relation $\eta = \eta_0 / M_{eff}$. Figure 9.9b shows the enhancement η versus speckle contrast C with broadening factor $\xi = 2$ alongside the experimental data. To verify this model we further optimized the focus with narrow band (0.4 THz) feedback making $\xi = 1$. The new data points (square dots) in Figure 9.9b verified the model.

9.4.2   Focusing through Scattering Media Using Multiphoton Fluorescence Signal

In 2008, Vellekoop showed that with aid of a fluorescent bead as a guide star, the scattered light can be focused inside a fixed scattering medium [44]. In 2012, Tang et al. showed that the nonlinear two-photon processes could provide a guide star for focus intensity enhancement inside the scattering media [45]. This technique has been used to image fluorescent beads placed under the intact mouse skull [45], [46]. The detail of experiments and their results are described in Chapter 3, Section 3.1 and in Refs. [45], [46]. Experimental results in multiphoton microscopy show much less fluorescence enhancement than expected at a given degrees of freedom available in the SLM (about 20 times through the mouse skull for a thousand segments SLM [45], [46]). However, a closer analysis of difference of correction mechanisms between the single photon camera-based optimization and the two-photon fluorescence PMT-based optimization reveals the observed lower value of enhancement in two-photon fluorescence based optimization [47]. Figure 9.10 shows the optical layout of microscope with basic components shown. This microscope has a 3 W 140 fs 80 MHz repetition rate tunable

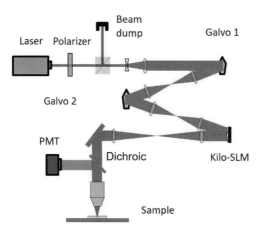

Figure 9.10 Optical layout of multiphoton adaptive optics microscope with its basic components. (Ref. [47])

Ti-Sapphire laser from Coherent. The laser is operated at 880 nm wavelength for two-photon excitation. To control the excitation power at sample the laser beam is passed through a motor-controlled polarizer (Thorlabs PRM1Z8E, AHWP05M-980) and polarization beam splitter (GT5-B). Two pairs of doublet achromatic lenses $f = 145$ mm and $f = 245$ mm are used to conjugate two scanning galvanometers (Thorlabs GVS011) with the SDM (Boston Micromachines Corporation, Kilo-SDM). The SDM is imaged onto the back pupil of an objective (Nikon N16XLWD-PF) by scanning lenses. The fluorescence signal is collected by the objective and detected by a photomultiplier tube (Hamamatsu H7422). The PMT signal is amplified by a preamplifier (Thorlabs TIA60) and digitized using a digitizer (Alazar Technologies ATS460). The digitizer is operated in external trigger mode for synchronized data acquisition. The feedback metric for the optimization algorithm is the total two-photon fluorescence signal detected by the PMT.

There are two major factors associated with the optimization in two-photon microscope, which make enhancement appeared to be less than the comparable experiments with coherent light optimization through scattering media. First, we must take into consideration that the signal in two-photon microscopy is the total integrated PMT response from all sources of two-photon fluorescence in the illumination field. When we park the beam (i.e., no scanning), the speckle field will cover many two-photon fluorescence sources. In general, this means the initial signal that we use as a baseline for optimization is much stronger (in above experiment it about $10\times$ stronger, the sketch in Figure 9.11a quantitatively illustrates the argument). The excitation speckles (shown by thin line) overlap with many fluorescent beads (shown in circles) to produce the fluorescence emission signal that is integrated over the sample area to produce the baseline signal (shown by thick line). This argument offers qualitative evidence that the initial fluorescence intensity collected by the objective is much higher than fluorescence intensity from a single bead, consequently, the enhancement will be lower than expected. Second, during the two-photon correction we parked the beam at an area where we observe

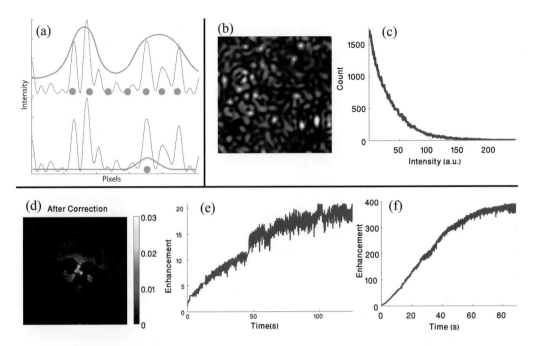

Figure 9.11 (**a**) Schematics of excitation speckles overlapping with fluorescent beads and fluorescence emission of by the beads, (**b**) typical speckle pattern, (**c**) camera pixels intensity count, (**d**) 1 μm fluorescent beads underneath the 280-μm-thick mouse skull after correction, (**e**) two-photon fluorescence intensity enhancement in SMPM microscope, (**f**) farfield speckle focus enhancement on camera through a 280-μm-thick mouse skull at 880 nm illumination. (Ref. [47])

maximum fluorescence. In other words, there has to be enough two-photon fluorescence to generate a signal, and that occurs when the brightest of the speckle overlaps with a bead. But the peak intensity of speckle is eight times brighter than the average speckle intensity.

To quantity this factor, we examined a camera image of typical speckle pattern as shown in Figure 9.11b. Figure 9.11c shows the camera pixel count as a function of intensity level for the same speckle pattern. The ratio of peak speckle intensity to average intensity is about 8.5. Above two arguments support that a factor of about 85 was hidden in the two-photon enhancement with respect camera based enhancement. If we include this factor with the two-photon PMT based enhancement the calculated excitation speckle focus enhancement will be $85 \times \sqrt{20} = 377$ times. Here, the enhancement factor 20 was taken from the enhancement we got in two-photon pupil AO correction when we imaged fluorescent bead through a 280-μm-thick mouse skull as shown in Figure 9.11d and Figure 9.11e . To further support this argument we optimized the far-field speckle focus at 880 nm femtosecond laser light through the same mouse skull. The maximum enhancement we got was 380 times (shown in Figure 9.11f). This number closely matched with the number we got from arguments.

The field of view of corrected image is much smaller than field of view of microscope, which is obviously one of the major limitations of imaging through or into the scattering media. The narrow field of view is due to presence of spatially variant aberration in the sample and corrected by a pupil SLM or a DM, which ideally can correct only the spatially invariant aberration. In next section, we have shown that the field of view of microscope can be significantly increase by placing the DM at a plane conjugate to the sample aberration plane.

9.5   Future Direction: Field of View Enhancement by Conjugate Adaptive Optics

In a scanning microscope, such as a two-photon microscope, a DM is inserted in the excitation beam path, most commonly in a plane conjugate to the back aperture of the objective [37], [38], [45]–[50], which is commonly called pupil AO. In principle, pupil AO is more effective for correcting spatially invariant aberrations. A common example of a spatially invariant aberration is an index mismatch at a planar interface, which introduces spherical aberration. An aberration introduced by the rough sample surface or localized index mismatch within the medium is a spatially variant aberration. The adaptive optics correction for the spatially variant aberration is effective only for the on-axis aberration, therefore, it corrects only for a limited field of view, which is called an aplanatic window. A more effective placement of the DM for spatially variant aberration is in a plane conjugate to the primary source of aberrations, called conjugate AO. The application of conjugate AO for *in vivo* multiphoton imaging has been demonstrated by Park et al. [39], which is described in detail in Chapter 3, Section 3.1 and 3.2. There are some studies by numerical simulations, which showed that benefits of conjugate AO persist even when only a single DM is employed [51]–[53]. In this section, we describe the conjugate AO in multiphoton microscope with a well-defined spatially variant interface aberrations located at a known plane with a wide-field AO correction technique using a continuous MEMS DM and compare its results with pupil AO correction with a segmented MEMS SLM [54].

In conjugate AO, the DM surface has one to one correspondence to the sample interface surface irrespective of incident beam scan angle. Therefore, when beam scans over the DM, a static compensation shape can be used to compensate wavefront errors corresponding to each beam scan position. In pupil AO, correction can be applied only for one beam scan angle position, which means that it cannot correct spatially varying aberration. Mertz et al. [55] showed that for an interface aberration characterized by wavefront surface normal standard deviation σ_ϕ and characteristic spatial period l_ϕ, the aberration correction field of view in pupil AO is limited to $2l_\phi/(1 + 2\sigma_\phi^2)^{1/2}$. This field of view becomes narrower with increasing phase variation, approaching the limit of $\sqrt{2}l_\phi/\sigma_\phi$ at higher phase deviation. On other hand, the conjugate AO can correct over the full microscope field of view by exactly matching the aberration surface to the DM surface.

The conjugate AO microscope was built by adding a scanning AO component to our existing pupil AO two-photon microscope shown in Figure 9.10. This component is added into the system by extending the optics in between second scan mirror and pupil deformable mirror (PDM). Figure 9.12 shows the detail of the optical layout for the

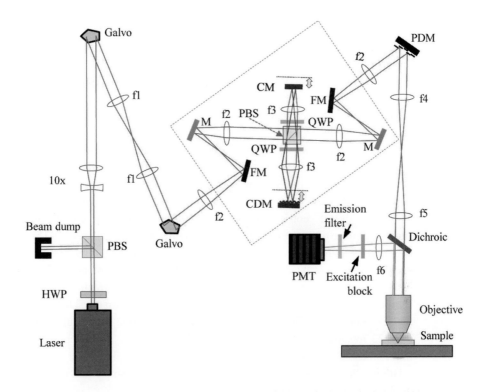

Figure 9.12 Optical layout of multiphoton pupil AO and conjugate AO. Conjugate AO component is shown by a dashed box. Reprinted with permission from [54], *Optics Express.*

conjugate AO components. A flip mirror diverts the laser beam toward the conjugate AO component. The center of the polarization beam splitter (PBS) (Thorlabs PBS252) and two galvos are in optically conjugate planes. A quarter wave plane, an $f = 40$ mm lens (Edmund Optics), and a mirror, one focal length distance from the lens, reflects the light back to the PBS. Another quarter wave plate, $f = 40$ mm lens and a continuous DM (1024 actuators, 3 μm linear stroke, >20 kHz, continuous face MEMS DM from BMC) at other end, reflects the light back to the PBS. The conjugate deformable mirror (CDM) and the conjugate mirror (CM) are in optically conjugate planes. The CDM and CM are mounted a translating platform, which shifts the CDM and CM together along the optical axis with respect to the rest of the components. The translatable carriage allows to adjust the position of CDM so that it can be conjugated to a range of axial positions between the microscope focal plane and the front pupil of the microscope objective. The number of active CDM actuators in conjugate AO compensation is determined by the size of beam on the CDM surface. This number changes as the CDM conjugates to planes close to or further from the sample. The pitch of CDM actuators is 0.4 mm. At a depth of 300 μm deep from the interface, 11 MEMS actuators were covered by the beam.

In order to demonstrate the performance of conjugate AO a known aberration (phase screen) is introduced in the system. A 2D sinusoidal pattern of peak-to-valley 3 μm and period 200 μm was patterned on a 30-μm-thick layer of AZ P4620 photoresist on top of a 300 μm thick glass substrate. A white light interferometer (Zygo NT6000) was used to measure the 3D surface profile of aberration sample, which is show in Figure 9.13d. Two-photon fluorescent signal was used as the feedback metric of aberration correction. The stochastic parallel gradient descent (SPGD) algorithm [56] was used to maximize the total fluorescence signal. Two-photon signal was sampled from the entire field of view by scanning the beam in a Lissajous pattern ($2f_x = 3f_y$). The signal integration time is matched with Lissajous scan period to remove the necessity of synchronization between scanning and signal integration.

A single layer of 1-μm-diameter fluorescent beads (Fluoresbrite, from Polysciences) attached to a microscope slide was imaged through the phase screen. The CDM is conjugated to the phase screen by translating $d \times M^2$ (12 mm) from image plane, where $M = 6.4$ is the telecentric magnification between CDM and phase screen. Figure 9.13a shows an image of fluorescent beads imaged through the phase screen. The phase screen distorted the light beam and blurred the image of beads. Figure 9.13b and Figure 9.13c show the images of beads taken after conjugate AO correction and pupil AO correction respectively. These results show that conjugate AO does correction over the entire FOV of microscope while pupil AO corrects only at center of image. The refractive index of phase screen is 1.63. For a 3 μm peak-to-valley variation the calculated phase variance σ_ϕ^2 is 22.76 square radians. The phase correlation length l_ϕ can be considered to be equal to the sinusoid period which is 200 μm. The correction window for pupil AO is given by relation $\sqrt{2}l_\phi/\sigma_\phi$ [55], which is 60 μm for the above sample and it is closely matched with result shown in Figure 9.13c.

The topography map of phase screen is shown in Figure 9.13d, which fairly matches with the topography map of corrected CDM surface shown in Figure 9.13e. In contrast, the phase map of PDM shown in Figure 9.13f has no resemblance to the topography of phase screen. The wavefront map $W(r, \theta)$ is twice the topography map in reflection mode and $(n - 1)$ times the topography map in transmission mode, where n is the refractive index of medium. Its phase map is related to wavefront map by $\Phi(r, \theta) = 2\pi/\lambda \times W(r, \theta)$.

In biological sample the aberration might be distributed along the axial thickness of sample. Therefore, the effective depth of correction would be the important parameter of conjugate AO correction. In order to test this range, an optimum correction is applied on the CDM when it was conjugated to the phase screen, then holding the corrected phase the CDM is translated and measured the effectiveness of correction over the axial ranges ± 100 μm [54].

Implementation of conjugate AO in scanning microscope is more complex and challenging than the counterpart, pupil AO. Two types of limitations were encountered. First, conjugate DM is located near the image plane in the optical system where laser beam diameter is smaller. In order to fill the DM pupil, or map the aberration to the DM surface, there has to be an enough magnification M_{mag} between the sample and DM. Displacement of DM from conjugate image plane to the conjugate aberration plane is proportional to the M_{mag}^2. As M_{mag} becomes larger, displacement increases rapidly, so does the beam diameter, and eventually is limited by the optics of the system. Instead, if magnification M_{mag} is designed to be smaller, assuming that DM has finer pixels

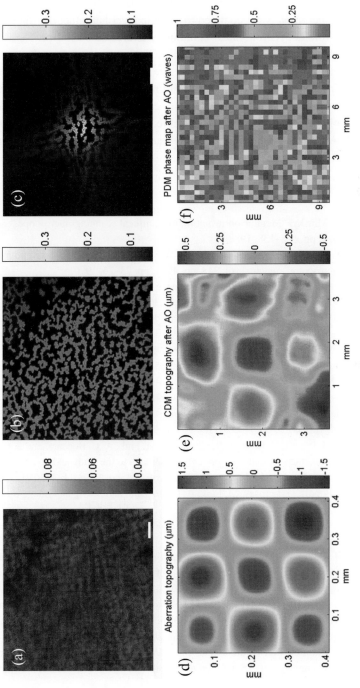

Figure 9.13 Images of fluorescent beads imaged through the phase screen (**a**) before correction, (**b**) after conjugate AO correction, and (**c**) after pupil AO correction. (**d**) Topography map of phase screen, (**e**) topography map of optimized CDM surface, and (**f**) phase map of pupil DM. Reprinted with permission from [54], *Optics Express*. A black and white version of this figure appears in some formats. For the color version, please refer to the plate section.

resolution, lenses in conjugate AO must have shorter focal length with larger diameter (i.e., thicker lenses), which might have detrimental effect on temporal dispersion in multiphoton microscopy. Apart from that, large cone angle of beam introduces spherical aberration, although fixable by pupil AO. The optics could be made simpler for transmission based correction elements. Another challenging problem in biological tissues is that the aberration is not confined to a single layer but rather distributed throughout the sample volume. Astronomical community overcame such a problems with multiconjugate AO, known by MCAO [57]–[60]. The experimental results presented in above experiments also shows that the conjugate AO correction is relatively long range in the axial direction, particularly in the case of modest to low NA. Such long range correction implies that a single DM correction can serve to compensate, at least partially, the axial range of volumetric aberrations. In future, conjunction of conjugate AO with pupil AO or MCAO may take the significant FOV advantage of conjugate AO in microcopy.

9.6   References

[1] M. Minsky, "Memoir on inventing the confocal scanning microscope," *Scanning*, vol. 10, no. 4, pp. 128–138, 1988.

[2] W. Denk, J. H. Strickler, and W. W. Webb, "Two-photon laser scanning fluorescence microscopy," *Science*, vol. 248, no. 4951, pp. 73–76, 1990.

[3] I. M. Vellekoop and P. Mosk, "Focusing coherent light through opaque strongly scattering media," *Opt. Lett.*, vol. 32, no. 16, pp. 2309–2311, 2007.

[4] T. Bifano, "Adaptive imaging: MEMS deformable mirrors," *Nat. Photonics*, vol. 5, no. 1, pp. 21–23, 2011.

[5] I. M. Vellekoop and C. M. Aegerter, "Scattered light fluorescence microscopy: imaging through turbid layers," *Opt. Lett.*, vol. 35, no. 8, pp. 1245–1247, 2010.

[6] I. M. Vellekoop, A. Lagendijk, and A. P. Mosk, "Exploiting disorder for perfect focusing," *Nat. Photonics*, vol. 4, no. 5, pp. 320–322, 2010.

[7] G. Montaldo, M. Tanter, and M. Fink, "Time reversal of speckle noise," *Phys. Rev. Lett.*, vol. 106, no. 5, pp. 1–4, 2011.

[8] S. M. Popoff, G. Lerosey, R. Carminati, M. Fink, A. C. Boccara, and S. Gigan, "Measuring the transmission matrix in optics: an approach to the study and control of light propagation in disordered media," *Phys. Rev. Lett.*, vol. 104, no. 10, pp. 1–4, 2010.

[9] S. Popoff, G. Lerosey, M. Fink, A. C. Boccara, and S. Gigan, "Image transmission through an opaque material," *Nat. Commun.*, vol. 1, p. 81, 2009.

[10] G. Lerosey, J. De Rosny, A. Tourin, and M. Fink, "Focusing beyond the diffraction limit with far-field time reversal," *Science*, vol. 315, no. 5815, pp. 1120–1122, 2007.

[11] M. Cui and C. Yang, "Implementation of a digital optical phase conjugation system and its application to study the robustness of turbidity suppression by phase conjugation," *Opt. Express*, vol. 18, no. 4, p. 3444, 2010.

[12] E. J. McDowell, M. Cui, I. M. Vellekoop, V. Senekerimyan, Z. Yaqoob, and C. Yang, "Turbidity suppression from the ballistic to the diffusive regime in biological tissues using optical phase conjugation," *J. Biomed. Opt.*, vol. 15, no. 2, 25004, 2010.

[13] M. Cui, E. J. McDowell, and C. Yang, "An *in vivo* study of turbidity suppression by optical phase conjugation (TSOPC) on rabbit ear.," *Opt. Express*, vol. 18, no. 1, pp. 25–30, 2010.

[14] M. Cui, "Parallel wavefront optimization method for focusing light through random scattering media," *Opt. Lett.*, vol. 36, no. 6, p. 870, 2011.

[15] E. A. Baker and Kevin L. Stappaerts, "Interferometric adaptive optics for high-power laser pointing and wavefront control and phasing," *J. Micro/Nanolithography, MEMS, MOEMS*, vol. 8, no. 3, 33040, 2009.

[16] K. L. Baker, E. a Stappaerts, D. Gavel, S. C. Wilks, J. Tucker, D. a Silva, J. Olsen, S. S. Olivier, P. E. Young, M. W. Kartz, L. M. Flath, P. Kruelevitch, J. Crawford, and O. Azucena, "High-speed horizontal-path atmospheric turbulence correction with a large-actuator-number microelectromechanical system spatial light modulator in an interferometric phase-conjugation engine," *Opt. Lett.*, vol. 29, no. 15, pp. 1781–1783, 2004.

[17] N. J. Kasdin, R. J. Vanderbei, and R. Belikov, "Shaped pupil coronagraphy," *Comptes Rendus Phys.*, vol. 8, no. 3–4, pp. 312–322, 2007.

[18] R. Belikov, E. Pluzhnik, M. S. Connelley, F. C. Witteborn, D. H. Lynch, K. L. Cahoy, O. Guyon, T. P. Greene, and M. E. McKelvey, "First results on a new PIAA coronagraph testbed at NASA Ames," *Tech. Instrum. Detect. Exopl. IV (San Diego, Calif., 2009) Proc. SPIE*, vol. 7440, 74400J, 2009.

[19] R. Belikov, E. Pluzhnik, F. C. Witteborn, T. P. Greene, D. H. Lynch, P. T. Zell, and O. Guyon, "Laboratory demonstration of high-contrast imaging at 2 lambda/D on a temperature-stabilized test bed in air," *SPIE 8151, Techniques and Instrumentation for Detection of Exoplanets*, vol. 8151, 815102, 2011.

[20] S. J. Thomas, R. Soummer, D. Dillon, B. Macintosh, D. Gavel, and A. Sivaramakrish-nan, "Testing the Apodized Pupil Lyot Coronagraph on the Laboratory for Adaptive Optics Extreme Adaptive Optics Testbed," *Astron. J.*, vol. 142, no. 4, p. 119, 2011.

[21] S. J. Thomas, R. Soummer, D. Dillon, M. Bruce, J. W. Evans, D. Gavel, A. Sivaramakrish-nan, C. Marois, and B. R. Oppenheimer, "Testing the APLC on the LAO ExAO testbed," *Adapt. Opt. Syst. Pts. 1–3*, vol. 7015, pp. 1–11, 2008.

[22] B. Macintosh, J. Graham, D. Palmer, R. Doyon, D. Gavel, J. Larkin, B. Oppenheimer, L. Saddlemyer, J. K. Wallace, B. Bauman, D. Erikson, L. Poyneer, A. Sivaramakrishnan, R. Soummer, and J. P. Veran, "Adaptive optics for direct detection of extrasolar planets: the Gemini Planet Imager," *Comptes Rendus Phys.*, vol. 8, no. 3–4, pp. 365–373, 2007.

[23] D. Gavel, S. Severson, B. Bauman, D. Dillon, M. Reinig, C. Lockwood, D. Palmer, K. Morzinski, M. Ammons, E. Gates, and B. Grigsby, "Villages: an on-sky visible wavelength astronomy AO experiment using a MEMS deformable mirror," *Proc. SPIE 6888, MEMS Adaptive Optics II*, vol. 6888, 688804, 2008.

[24] W. Piyawattanametha, "Review of optical MEMS devices," *Int. J. Appl. Biomed. Eng.*, vol. 8, no. 1, pp. 1–9, 2016.

[25] K. E. Petersen, "Silicon as a mechanical material," *Proc IEEE*, vol. 70, no. 5, pp. 1–50, 2005.

[26] D. B. Conkey, A. N. Brown, A. M. Caravaca-Aguirre, and R. Piestun, "Genetic algorithm optimization for focusing through turbid media in noisy environments," *Opt. Express*, vol. 20, no. 5, p. 4840, 2012.

[27] F. Wang, "Control of deformable mirror with light-intensity measurements through single-mode fiber," *Appl. Opt.*, vol. 49, no. 31, G60, 2010.

[28] F. Wang, "Wavefront sensing through measurements of binary aberration modes," *Appl. Opt.*, vol. 48, no. 15, pp. 2865–2870, 2009.

[29] F. Wang, "Utility transforms of optical fields employing deformable mirror," *Opt. Lett.*, vol. 36, no. 22, p. 4383, 2011.

[30] F. Wang, "High-contrast imaging via modal convergence of deformable mirror," *Astrophys. J.*, vol. 751, no. 2, p. 83, 2012.

[31] J. L. Walsh, "A closed set of normal orthogonal functions," *Am. J. Math.*, vol. 45, no. 1, pp. 5–24, 1923.

[32] H. Schreiber and J. H. Bruning, "Phase shifting interferometry." In *Optical Shop Testing* (ed. Malacara, D.) 547–666. John Wiley, 2007.

[33] C. Stockbridge, Y. Lu, J. Moore, S. Hoffman, R. Paxman, K. Toussaint, and T. Bifano, "Focusing through dynamic scattering media," *Opt. Express*, vol. 20, no. 14, pp. 15086–15092, 2012.

[34] I. M. Vellekoop and A. P. Mosk, "Phase control algorithms for focusing light through turbid media," *Opt. Commun.*, vol. 281, no. 11, pp. 3071–3080, 2008.

[35] P. Theer and W. Denk, "On the fundamental imaging-depth limit in two-photon microscopy," *J. Opt. Soc. Am. A*, vol. 23, no. 12, pp. 3139–3149, 2006.

[36] N. G. Horton, K. Wang, D. Kobat, C. G. Clark, F. W. Wise, C. B. Schaffer, and C. Xu, "*In vivo* three-photon microscopy of subcortical structures within an intact mouse brain," *Nat. Photonics*, vol. 7, no. 3, pp. 205–209, 2013.

[37] C. Wang, R. Liu, D. E. Milkie, W. Sun, Z. Tan, A. Kerlin, T.-W. Chen, D. S. Kim, and N. Ji, "Multiplexed aberration measurement for deep tissue imaging *in vivo*.," *Nat. Methods*, vol. 11, no. 10, pp. 1037–1040, 2014.

[38] L. Sherman, J. Y. Ye, O. Albert, and T. B. Norris, "Adaptive correction of depth-induced aberrations in multiphoton scanning microscopy using a deformable mirror," *J. Microsc.*, vol. 206, pp. 65–71, 2002.

[39] J.-H. Park, W. Sun, and M. Cui, "High-resolution *in vivo* imaging of mouse brain through the intact skull," *Proc. Natl. Acad. Sci.*, vol. 112, no. 30, p. 201505939, 2015.

[40] H. P. Paudel, C. Stockbridge, J. Mertz, and T. Bifano, "Focusing polychromatic light through strongly scattering media," *Opt. Express*, vol. 21, no. 14, pp. 17299–17308, 2013.

[41] J. W. Goodman, *Speckle phenomena in optics: theory and applications*. Roberts & Company, 2007.

[42] F. Van Beijnum, E. G. van Putten, A. Lagendijk, and A. P. Mosk, "Frequency bandwidth of light focused through turbid media," *Opt. Lett.*, vol. 36, no. 3, pp. 373–375, 2011.

[43] O. Katz, Y. Bromberg, E. Small, and Y. Silberberg, "Focusing and compression of ultrashort pulses through scattering media," *Nat. Photonics*, vol. 5, no. 6, pp. 372–377, 2010.

[44] I. M. Vellekoop, E. G. van Putten, A. Lagendijk, and A. P. Mosk, "Demixing light paths inside disordered metamaterials.," *Opt. Express*, vol. 16, no. 1, pp. 67–80, 2008.

[45] J. Y. Tang, R. N. Germain, and M. Cui, "Superpenetration optical microscopy by iterative multiphoton adaptive compensation technique," *Proc. Natl. Acad. Sci. U. S. A.*, vol. 109, no. 22, pp. 8434–8439, 2012.

[46] T. Bifano and H. Paudel, "Beam control in multiphoton microscopy using a MEMS spatial light modulator," *Proc. SPIE*, vol. 9083, p. 90830Q, 2014.

[47] H. P. Paudel, "Coherent beam control through inhomogeneous media in multi-photon microscopy," Boston University, 2015.

[48] K. Wang, D. E. Milkie, A. Saxena, P. Engerer, T. Misgeld, M. E. Bronner, J. Mumm, and E. Betzig, "Rapid adaptive optical recovery of optimal resolution over large volumes.," *Nat. Methods*, vol. 11, no. 6, pp. 625–628, 2014.

[49] N. Ji, D. E. Milkie, and E. Betzig, "Adaptive optics via pupil segmentation for high-resolution imaging in biological tissues," *Nat. Methods*, vol. 7, no. 2, p. 141, 2010.

[50] D. Sinefeld, H. P. Paudel, D. G. Ouzounov, T. G. Bifano, and C. Xu, "Adaptive optics in three-photon fluorescence microscopy," *Cleo 2015*, p. STu2K.8, 2015.

[51] Z. Kam, P. Kner, D. Agard, and J. W. Sedat, "Modelling the application of adaptive optics to wide-field microscope live imaging," *J. Microsc.*, vol. 226, no. 1, pp. 33–42, 2007.

[52] R. D. Simmonds, and M. J. Booth, "Modelling of multi-conjugate adaptive optics for spatially variant aberrations in microscopy," *J. Opt.*, vol. 15, 94010, 2013.

[53] T. Wu, and Meng Cui, "Numerical study of multi-conjugate large area wavefront correction for deep tissue microscopy," *Opt. Express*, vol. 23, no. 6, p. 7463, 2015.

[54] H. P. Paudel, J. Taranto, J. Mertz, and T. Bifano, "Axial range of conjugate adaptive optics in two-photon microscopy," *Opt. Express*, vol. 23, no. 16, pp. 20849–20857, 2015.

[55] J. Mertz, H. Paudel, and T. G. Bifano, "Field of view advantage of conjugate adaptive optics in microscopy applications," *Appl. Opt.*, vol. 54, no. 11, pp. 3498–3506, 2015.

[56] M. A. Vorontsov and V. P. Sivokon, "Stochastic parallel-gradient-descent technique for high-resolution wave-front phase-distortion correction," *J. Opt. Soc. Am. A – Optics Image Sci. Vis.*, vol. 15, no. 10, pp. 2745–2758, 1998.

[57] A. V. Goncharov, J. C. Dainty, S. Esposito, and A. Puglisi, "Laboratory MCAO test-bed for developing wavefront sensing concepts," *Opt. Express*, vol. 13, pp. 5580–5590, 2005.

[58] J. M. Beckers, "Increasing the size of the isoplanatic patch with multiconjugate adaptive optics," *Very Large Telesc. Their Instrument*, p. 693, 1988.

[59] D. C. Johnston and B. M. Welsh, "Analysis of multiconjugate adaptive optics," *J. Opt. Soc. Am. A*, vol. 11, no. 1, p. 394, 1994.

[60] R. Ragazzoni, E. Marchetti, and G. Valente, "Adaptive-optics corrections available for the whole sky," *Nature*, vol. 403, no. 6765, pp. 54–56, 2000.

10 Computer-Generated Holographic Techniques to Control Light Propagating through Scattering Media Using a Digital-Mirror-Device Spatial Light Modulator

Antonio M. Caravaca-Aguirre and Rafael Piestun

10.1   Introduction

Recent developments have shown the potential of wavefront shaping to control light propagating through a complex medium. While the key motivation is sensing, imaging, and energy deposition in biological tissue, there are significant potential applications in materials processing or the use of scatterers as optical components and in cryptography. Scattering media are often modeled as stochastic systems because of the overwhelming number of variables involved in its description, the difficulty of their determination, and the dynamic changes they undergo. However, once a specific medium is selected and observed over a relatively short time interval, it effectively becomes predictable (deterministic) and can be modeled as a specific realization of a stochastic process. Current state of the art spatial light modulation techniques, detectors, and computers have enabled the characterization of these scatterers or alternatively the optimization of light wavefronts to compensate or mitigate for the multiple-scattering events undergone inside the medium. Unfortunately, the typical time scales associated with live tissue are too short – in the order of milliseconds – for existing spatial modulation technologies [1]. Whereas limitation of blood flow [2] or the use of correlations [3] might help mitigate the need for fast modulation, novel mechanisms to spatially modulate, process the information, and optically encode it are required.

State of the art spatial light modulators (SLM) are based on liquid crystal on silicon devices, which have a limited refresh rate of (at best) a few hundred Hz. Ferroelectric LC devices can increase the speed up to a few kHz but only producing binary phase outputs [4]. Micromechanical devices composed of tiny mirror arrays are an attractive alternative because they can be modulated at tens of kHz, two orders of magnitude faster than LC-SLM [5]. Unfortunately, existing devices only enable a limited number ($\sim$1000) of degrees of freedom with continuous phase control [6], or a large number of degrees of freedom with two-state (on-off) switching [7, 8]. The digital light projection (DLP) device [9], which falls in the last category, is an attractive solution because of the high pixel count, high switching speed, and affordability. Each micromirror pixel of a DMD can rotate along the diagonal axis between two positions; to either deflect the light out or into the system. A schematic view of the elementary components and the geometry of each pixel is shown in Figure 10.1. Interestingly, the fact that DLPs can only generate binary on-off patterns does not prevent their application in phase wavefront shaping. One way to understand this capability is by invoking concepts of computer-generated holography.

In effect, computer-generated holograms (CGH) are optical components that generate arbitrary complex wavefronts encoded in (typically) binary amplitude masks [10]. There are multiple encoding schemes that can provide a given wavefront in a specific region of space [11]. One possible classification of CGHs is in terms of whether the desired wavefront is reconstructed along the axis normal to the mask [12] or off axis [11]. Another possible classification is in terms of the encoding technique; CGHs can be calculated by an optimization technique [13] or by an analytic equation directly derived from the target wavefront [11]. Regardless of the encoding process (e.g., analytic or optimization) or the reconstruction region (e.g., on-axis or off-axis), any binary mask can be considered as a realization of a CGH that shapes incoming wavefronts into complex propagating waves.

In this chapter we present techniques for wavefront shaping using binary masks and holographic concepts. We describe the encoding of phase off-axis using an analytic method to accelerate computation.

By controlling the experiment and computing with a field programmable phase array integrated with the DLP, it is possible to achieve the fastest adaptive wavefront modulation reported to date [14].

10.2 Binary Holographic Techniques for Phase Control

10.2.1 Lee Method

In this section we present a method for analytic calculation of the CGH based on the target phase. The main advantage of this method is the ease and speed of computation as well as the clean phase front obtained in the first diffraction order. The disadvantage is that the target wavefront is obtained off-axis and with relatively low diffraction efficiency as compared with iterative optimization techniques. In particular, we use an early method proposed by Lee [15], in which the encoding implements a binarization of the basic holographic equation obtained by interference of an object wave and a reference wave. The goal here is to encode the complex signal

$$s(x, y) = A(x, y) \exp\left(i\varphi(x, y)\right), \tag{10.1}$$

where $A(x, y)$ is the amplitude defined as positive, and $\varphi(x, y)$ is the phase which takes values in the domain $[-\pi, \pi]$, using a binary hologram. To find out the hologram, we can use a periodic binary amplitude grating, whose Fourier representation is given by:

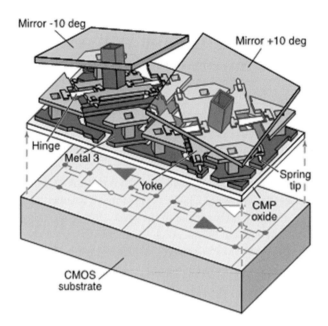

Figure 10.1 Schematic view of two DMD pixel showing the different layers composing it. Reproduced from www.nist.gov.

$$f(x, y) = \sum_n \frac{\sin(\pi nq)}{\pi n} \exp\left[in\left(2\pi(x+y)\alpha + \varphi(x, y)\right)\right] \tag{10.2}$$

where α is the carrier frequency of the periodic grating. This hologram will produce a first diffraction order proportional to $\frac{\sin(\pi q(x,y))}{\pi} \exp(\varphi(x, y))$ when illuminated with a plane wave. Therefore, the function $q(x, y)$ should be defined as

$$q(x, y) = \arcsin(A(x, y)) \tag{10.3}$$

to obtain the complex signal $s(x, y)$ at the first diffraction order. If the wavefront has constant amplitude, the parameter q can be tuned to determine the diffraction efficiency of the hologram. For instance, if q is 0.5, all the even terms vanish except the $m = 0$ term, and the diffraction efficiency of the first order is maximized to 10%.

To design the hologram, we can use the transmittance of the phase-only off-axis reference beam hologram, $t(x, y)$ given by

$$t(x, y) = 0.5\left\{1 + \cos\left[2\pi(x-y)\alpha - \varphi(x, y)\right]\right\}. \tag{10.4}$$

The binary amplitude hologram $h(x, y)$ is generated by thresholding the amplitude hologram, $t(x, y)$,

$$h(x, y) = \begin{cases} 0 & \text{if} \quad t(x, y) < 0.5 \\ 1 & \text{if} \quad t(x, y) > 0.5. \end{cases} \tag{10.5}$$

An example of four patterns that produce a set of discrete phase steps is shown in Figure 10.2. The proper selection of the carrier frequency α is important to avoid overlapping between the first and the second-order diffracted wave. If the spatial frequencies

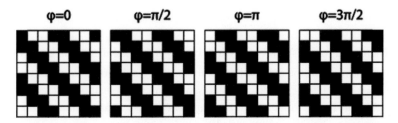

Figure 10.2 Four macro pixels from a Lee Hologram with an arbitrary carrier frequency that generate four different phases. In this example, $q = 0.5$. Reprinted with permission of [8], SPIE.

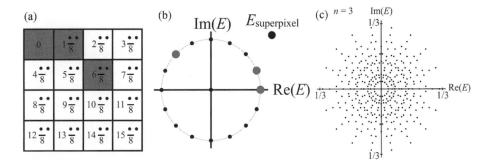

Figure 10.3 (a) Example of a 4×4 superpixel with the dark boxes indicating the mirrors that are on. (b) Complex field representation of each mirror (gray circles) and the total electric field of the superpixel (black circle). (c) Complex field that could be synthesized through combination of pixels. Reprinted with permission from [16], *Optics Express*.

in the x-direction are bounded by f_- and f_+, then the highest frequency in the first-order will be $\alpha + f_+$ and the lowest frequency in the second diffraction order will be $2\alpha + 2f_-$. To avoid the overlap of those two frequencies α needs to satisfy the following equation

$$\alpha > f_+ - 2f_-. \tag{10.6}$$

10.2.2 Superpixel Method

A different CGH approach that has been used to encode a phase distribution with the DMD was developed by Goorden et al. commonly referred as "superpixel" approach [16]. They used an imaging system composed of two lenses separated by a distance equal to the sum of the respective focal lengths ($f = f_1 + f_2$) and placed slightly off axis with respect to each other. This displacement introduces an extra phase factor in the target plane and therefore, the phase response of each micromirror is dependent on the position within the DMD. Combining groups of nxn pixels into a superpixel enables control of the desired amplitude and phase at the target plane. The lenses are placed in such a way that all the pixels inside the superpixel have a uniform response between 0 and 2π as shown in Figure 10.3. A pinhole placed in the Fourier plane between the lenses acts as a spatial filter blocking the high spatial frequencies. Therefore, the target plane response of a superpixel is the sum of the individual pixel responses as shown in Figure 10.3b.

The resolution or spatial bandwidth of the superpixel method is given by $\Delta k = \frac{2\pi r}{\lambda f_2}$ rad m^{-1} where r is the radius of the pinhole and f_2 is the focal length of the second lens. For n $= 4$, the number of discrete fields values that this method can reconstruct is 6561 as shown in Figure 10.3c. Therefore, arbitrary phase and amplitude light fields can be represented using this method [17–19].

10.3   Applications in Focusing through Complex Media

10.3.1   Focusing through Dynamic Turbid Media Using a DMD

One of the main applications of the high-speed modulation capability is focusing through scattering samples. Biological samples, which mostly scatter light in the visible and near-infrared spectrum, limit the penetration depth of current microscopy techniques. Furthermore, the dynamic nature of biological tissue due to blood flow or cell movements complicates the process of precalibration to image through; thus requiring to perform all measurements before the speckle pattern produced by the sample completely changes. The persistence time or decorrelation time of a turbid sample is the time over which the speckle pattern remains stable. Therefore, high-speed phase control techniques are necessary to compensate for the scattering produced by those dynamic samples.

Conkey et al. [20] introduced a high-speed phase mask optimization technique, which utilizes off-axis binary amplitude computer-generated holography implemented on a DMD. They demonstrated the ability to focus through dynamic samples more than one order of magnitude faster than the prior state of the art. A transmission matrix (TM) approach based on the projection of a Hadamard basis set (see Chapter 5) was used to reduce the timing required for feedback. This approach requires preloading a predefined set of phase masks in memory, thus reducing the data transfer time between the DMD and the computer during the experiment. Consequently, the DMD can display the different patterns at maximum frame rate during the experiment.

The TM is a subpart of the scattering matrix and describes the Green's function between the spatial input modes (or DMD pixels) and output modes (or CCD pixels) of the system. The knowledge of the TM brings fundamental insight into the medium and allows using its information, for example, to control the output light distribution. Obtaining the TM requires measurement of the complex optical field at the output CCD plane.

Previous implementation of the TM algorithms used phase-shifting interferometry with four different reference phases to determine the complex field [21]. To minimize the number of measurements, a three-phase reference method [22] can be implemented. Each Hadamard basis element projected is surrounded with a frame of a constant reference phase of 0, $\pi/2$ and π as shown in Figure 10.4a. Therefore, a fraction of the input light produces a static reference speckle through the sample that can be used as a

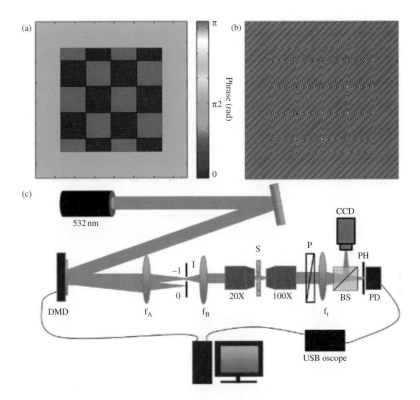

Figure 10.4 (a) An example of a Hadamard basis element surrounded by a phase reference and (b) the corresponding binary amplitude Lee hologram h(x,y) associated to the phase distribution shown in (a). (c) Experimental setup used to focus light through a scattering material using the binary amplitude Lee Hologram and the DMD. f_a, f_b, f_t: lenses; I: Iris; S: Scattering sample; BS: beam splitter; P: Polarizer; PH: Pinhole; PD: Photodetector. Reprinted with permission from [20], *Optics Express*. A black and white version of this figure appears in some formats. For the color version, please refer to the plate section.

reference to extract the amplitude and phase of the modulated light. The total number of measurements of this method is 3·N, corresponding to 25% reduction with respect to the 4·N measurements previously used [21]. The error obtained, taking into account the experimental noise level, is comparable to the error achieved with four-phase references. In this configuration, the so-called observed TM, K_{obs}, is obtained. The observed TM is related to the original TM by $K_{obs} = K \times S_{ref}$, where S_{ref} is the diagonal matrix representing the reference. For a more detailed discussion about the implications of using K_{obs} instead of K, see Chapter 5.

Because CCD cameras typically have a slower frame rate than the 22 kHz refresh rate corresponding to the DMD, a high-speed photodetector was utilized in the experiments. Hence, the TM measured was simplified into a N × 1 matrix, defined as the response of N input modes to a single output mode. The observed transmission matrix is calculated from the three intensity values recorded according to

$$k_{obs}^n = \frac{I^0 - I^{\pi/2}}{4} + i\frac{I^\pi - I^{\pi/2}}{4} \tag{10.7}$$

where I^φ is the intensity recorded for each reference phase φ. The appropriate phase mask that maximizes the intensity at the photodetector is obtained using $E_{in} = K_{obs}^t / |K_{obs}^t|$, where the superscript t indicates the transpose. The corresponding binary hologram calculated using Eq. 10.5 (shown in Figure 10.4b) is projected onto the DMD.

The experimental setup used is shown in Figure 10.4. A 532 nm CW laser (Millennia Spectra-Physics) illuminates the DMD (Texas Instruments DLP Discovery Kit D4100, 13.68 μm pixel size, 1024 × 768 micromirrors) where the binary Lee holograms are projected. A two lens system images the DMD onto the back aperture of a microscope objective (Olympus 20×, 0.5 NA) which focuses the light onto the scattering sample. An iris is place in the Fourier plane of the first lens to block all the diffraction orders except the first. The light scattered by the sample is collected by another microscope objective (Olympus 100×, 0.75 NA) and imaged into a photodetector. To limit the detection to one speckle grain, a 50 μm pinhole is place in front of it. The signal from the photodetector is digitalized by a digital USB oscilloscope and sent to a personal computer (PC), which computes the TM and calculates the optimal phase mask. The Lee hologram associated to the optimal phase mask is sent to the DMD via USB. A beam splitter and CCD camera is used to track the quality of the focus, but is not part of the TM measurement. The diffraction efficiency of the hologram achieved was 6%–10%.

The system described above is capable of measuring the transmission matrix composed of 256 Hadamard input modes in only 34 ms. However, transferring the data from the oscilloscope to the computer, calculating the optical phase mask and send it to the DMD requires an additional ∼270 ms (shown in Figure 10.5d). Thereafter, the optimal phase mask that maximizes the intensity at the photodetector is projected during a predetermined time. Therefore, this system is only suitable for samples with persistence times longer than 300 ms. Tissue phantoms with thickness of 1 mm, made from a mixture of gelatin and Intralipid added as a light scatterer were used to test the efficiency of the system. They reported average intensity enhancements at the focus ranging from 28 to 69 using samples with decorrelation times ranging from 350 ms to 850 ms (Figure 10.5). In this figure, half of the time the enhancement is close to zero and the other half, the enhancement is high. The enhancement is low while the system measures the TM and transfers the data from the USB oscilloscope to the computer, followed by the transmission of the Lee Hologram of the optimized phase mask to the DMD. After that, the optimal phase mask is projected on the DMD during 200 ms until a new TM is measured again. The images obtained with the CCD show the extent of the enhancement as a function of the persistence time and number of modes used to calculate the hologram.

10.3.2   FPGA Implementation

The data transfer and the computation required after the measurements of the TM described in the previous section is six times longer than the measurement time. To

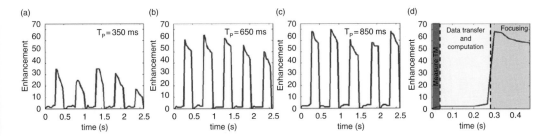

Figure 10.5 Graphs showing the enhancement of the focus spot versus the time with temporally dynamic turbid samples with a decorrelation speckle time of (**a**) 350 ms (**b**) 650 ms and (**c**) 850 ms. (**d**) The timing of the system corresponding to one cycle: 34 ms measuring TM, 270 ms of data transfer and 200 ms displaying the optimal phase mask. Reprinted with permission from [20], *Optics Express*.

improve the system timing and eliminate the bottleneck of data transfer with the computer, a hardware implementation of the TM has been developed [23]. The TM computation is implemented in a Virtex5 custom Field Programmable Gate Array (FPGA), part of the Texas Instruments Development Kit 4100. The analog signal generated by the photodetector is digitalized using a separate analog-to-digital converter with a buffered analog input, which generates the digital input for the FPGA board. The FPGA is responsible of controlling the DMD and triggering the analog signal conversion. Using the hardware implementation, the time required to calculate and project the Lee hologram of the optimized phase mask is reduced to 3 ms. Thus, a total time of only ~37 ms is required for the improved system to create a focus. The system is similar to the one shown in Figure 10.4, but substituting the oscilloscope and computer for the analog-to-digital converter and the FPGA. Tissue phantoms prepared with a mix of Gelatin, water and Intralipid with decorrelation times ranging from 10 ms to 85 ms were tested showing values of the SBR of the focus created ranging from 14 to 90 (Figure 10.6). As expected, the more dynamic samples generate lower SBR foci. In these situations the earliest input mode measurements become obsolete before all measurements are recorded, lowering the total SBR.

The implementation of the TM algorithm in an FPGA versus using a PC clearly shows an improvement in the total focus time. However, the hardware implementation lacks the flexibility to modify the system (replacing the photodetector for a high-speed camera, for instance) or to introduce new parts into it. To mitigate this problem, a look-up-table approach to compress the CGH data transfer between the DMD and the PC was proposed in Ref [25]. Several configurations with different number of input modes can be chosen where the DMD pixels are binned into square patches or macropixels. A custom C++ driver adapted for MATLAB for easy prototyping allows controlling the DMD at a few kilohertz frame rate with feedback. The reason for the improvement in the frame rate compared to the normal approach comes from the fact that only N bits of information (where N corresponds to the number of input modes selected in the configuration) have to be sent via USB, instead of the 1024 × 768 bits that would have been sent otherwise. This approach drastically reduces the required data transfer time, speeding the whole

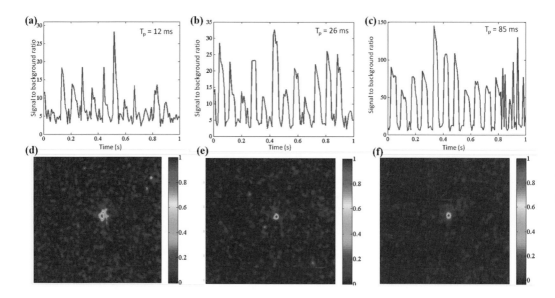

Figure 10.6 Signal-to-background ratio of the focus spot versus time with a temporally dynamic turbid sample with decorrelation speckle times of (**a**) 12 ms, (**b**) 26 ms, and (**c**) 85 ms. (**d–f**) The image of the focus created in (**a–c**) respectively recorded by the CMOS camera. Reprinted with permission of [24], SPIE. A black and white version of this figure appears in some formats. For the color version, please refer to the plate section.

process. This system can be used not only to focus light through scattering samples but also with MMFs [23] as described in the next section.

A different approach, [14] using digital optical phase conjugation (DOPC) (see Chapter 11) combined with hardware programing and a DMD is able to focus light in only 6 ms. In this case, a camera records the interference pattern between the optical beam exiting the scattering sample and a reference beam. The camera sends the image data to a host FPGA board, which is responsible for controlling the DMD as well. From the image acquired, using single-shot binary phase retrieval, [14] the FPGA computes the optical field and sends the conjugate field to the DMD. The DMD is placed in a conjugate plane with respect to the camera as shown in the experimental apparatus in Figure 10.7. In this approach, no Lee holograms are involved. Alternatively, the authors fit the binary amplitude modulation of the DMD into a phase-conjugation framework. When a pixel is turned on, the field produced can be decomposed into orthogonal phase vectors, as shown in Figure 10.7. If the phase difference between the optimal electric field calculated by phase retrieval and the one produced by the pixel, $\Delta\theta$, is less than $\pi/2$, the pixel is turned on. Otherwise, the pixel is turned off. This system was able to create a focus through a 2.3 mm thick unclamped *in vivo* mouse dorsal skin with a decorrelation time of less than 30 ms. The DOPC procedure was repeated 50 times per second maintaining the focus through the living sample. The intensity peaks recorded by an APD are shown in Figure 10.7c. The playback latency of the system is determined by the recording exposure, data transfer from the camera to the FPGA and from the FPGA to the DMD. In this system the latency is 5.3 ms, but the flexibility of the system could

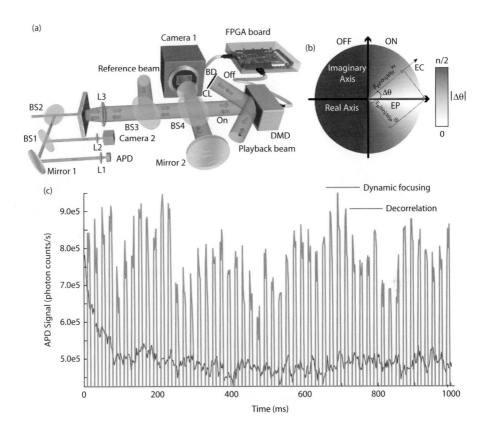

Figure 10.7 (**a**) DOPC experimental apparatus using the DMD. The scattered field emerging from the sample is interfered with a reference beam (clear arrows) and imaged by Camera 1 (pco.edge 5.5, PCO-TECH)). The FPGA (ViClaro IV GX Camera Link Development Kit, Microtronix) connected to the camera process the acquired image and compute the optimal pattern to be projected onto the DMD. The DMD is illuminated back (dark arrows) and the conjugated phase pattern is projected through the sample creating a focus at the opposite side. The conjugated focus is observed on Camera 2 (Prosilica GX 1920, Allied Vision) and the APD (SPCM-A1RH-14, Excelitas). (**b**) Representation of the complex field created by one DMD pixel. The field played back by the DMD E_p, can be decomposed in two orthonormal vectors: one parallel to the desired conjugated field E_c, and another one perpendicular that contributes to the background. (**c**) APD plot for sustainable foci and decorrelation focus. Reprinted with permission of [14], OSA.

reduce it below 1 ms using 1920×70 mirrors. The main disadvantage of the DOPC system is the need for precise alignment between the spatial light modulator and the camera sensor. Both sensor arrays need to have a pixel-to-pixel matching by aligning six degrees of freedom (x, y, z, tip, tilt, and rotation).

10.3.3 Focusing through Bending Multimode Fiber

In biological applications, an alternative method to overcome the diffusion of light due to scattering is to use thin optical endoscopes to both deliver and collect light from the

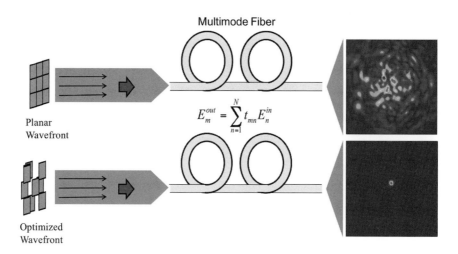

Figure 10.8 Sketch of how a MMF illuminated by a plane creates a speckle field at the distal tip. If the incident wavefront is optimized, it can compensate the different mode velocity to create a bright focus at the distal tip. Reprinted with permission of [26], SPIE.

sample. Typically, a micro-endoscope uses optical fibers with a high index of refraction core and a low index cladding to guide the light and visualize areas that are inaccessible by other means. Single-mode fibers [27], fiber bundles [28], graded-index lenses [29], and photonic crystal fibers [30] are some of the current alternatives. However, the low resolution (single-mode fibers), the large cross section (fiber bundles) or the rigidity and aberrations generated (graded-index) make none of these options optimal [31]. Since the 1970s [32, 33], researchers have studied the viability of using multimode optical fibers (MMFs) for imaging. However, the main disadvantage of MMFs, namely their modal dispersion remains an obstacle to implementation. When an image is sent through the MMF, it excites eigenmodes or principal modes of the fiber. The different mode velocity of each of these modes scrambles the phase creating a speckle field at the distal tip of the fiber (Figure 10.8). Recently, the advances in computation and wave control have made possible the applicability of the old idea of using multimode fibers for imaging. For example, the TM of the MMF can be measured (using wavefront shaping, DOPC, etc.), and used to calculate the optimal phase mask that creates a bright focus at the distal tip, as shown in Figure 10.8b. MMFs are an interesting alternative to fiber-bundle endoscopes because of their smaller diameter and larger number of propagating modes. The great challenge is still to characterize the MMF in real time because the mode coupling and velocity depends on the varying shape and the external environment of the fiber. Therefore, using DMDs to calibrate MMFs is an attractive approach.

The system shown in Figure 10.4 can be modified and adapted to focus light through a MMF obtaining a similar performance in terms of focusing time. The light is coupled into the MMF using a microscope objective with similar NA as the MMF. The ability to characterize the TM of the MMF in the millisecond time scale opens the possibility of focusing light through it even as it bends. Caravaca et al. [34] placed a step-index MMF (BFL22–365-Thorlabs, 365 μm diameter 0.22 NA) in a translation stage to modify the bending angle and demonstrated the capability of the system to create a focus at the

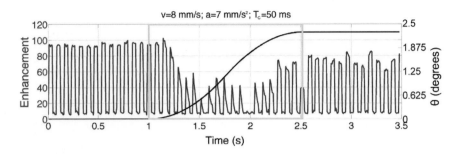

Figure 10.9 Enhancement of the focus spot at the distal tip of the MMF versus time (dark gray line). The light gray square delimits the period of time where the translation stage was moving. The black line shows the fiber bending angle respect to the origin. Reprinted with permission of [34], OSA.

distal tip of the fiber while the translation stage was on and the MMF was moving. Consistent focus enhancement values of ~100 were recorded using 256 input modes at a 13 Hz focusing rate (Figure 10.9).

Furthermore, the high-speed modulation of the DMD combined with a high-speed acquisition camera allows calibrating not just one output mode, but a few thousand different output modes at the distal tip of the fiber in only a few seconds. In addition, a EMCCD camera imaging the closer tip of the fiber in combination with a dichroic filter can collect the fluorescence signal that is emitted back through the fiber from a fluorescent sample placed at the distal tip, transforming the MMF into a florescence scanning endoscope.

Figure 10.10a shows the complete system proposed in Ref. [23]. In this case, the imaging procedure is composed of two steps. During the first step, the system measures the TM of the fiber and the hologram to create a focus in each output mode is calculated and stored in the memory of the DMD. After the fiber is calibrated, in a second step, the fluorescence sample is placed ~100 μm far from the distal tip (same output plane where the MMF was calibrated). Projecting all the optimal input phase patterns scans the focus at the distal tip. Measuring the integrated fluorescence signal that is emitted back through the MMF illuminating the EMCCD at the closer tip an image can be reconstructed.

A fluorescence sample made of 1 μm fluorescence beads in a cover slide was imaged using a 100 μm core diameter graded-index MMF (Newport F-MLD, 0.29NA). Figure 10.10 shows the results comparing the image of the beads under a custom wide-field fluorescence microscope (Figure 10.10b) and the image reconstructed using the MMF system (Figure 10.10c). Due to the high-speed, the system could be used to calibrate the fiber in multiple output planes enabling scanning of the focus in the axial direction by just modifying the input wavefront previously recorded.

10.4   Summary and Future Directions

This chapter reviewed the advances in light control through complex media using a DMD. Although these devices are known for their amplitude modulation capabilities,

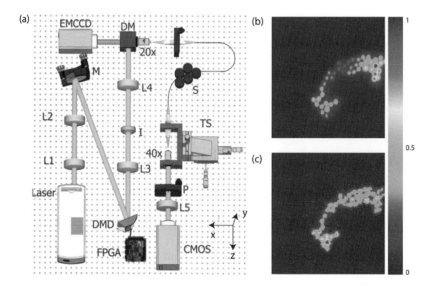

Figure 10.10 (a) Experimental apparatus to calibrate the MMF. (b) Image of 4 μm diameter fluorescence beads taken with a wide-field fluorescence microscope. (c) Same beads imaged through the MMF. Reprinted with permission of [26], SPIE. A black and white version of this figure appears in some formats. For the color version, please refer to the plate section.

the implementation of binary holographic techniques effectively converts them into phase-modulation devices. The fast decorrelation time of tissue, in the millisecond time scale, motivates the research and development of fast focusing and imaging methods. The main advantage of using these MEMS devices, is the high switching speed, up to more than 20 kHz, improving with respect to the widely used LC-SLMs. DMDs have allowed reducing the total focusing time through complex media to tens of milliseconds, close to the speckle decorrelation times of dynamic biological media (tissue). The signal-to-noise ratio (SNR) of the signal detected is critical in this type of high-speed experiments because SNR decrease with the integration time. Therefore, the implementation of algorithms designed for low SNR environments [35] will be required in the future.

Noninvasive feedback mechanisms are still required to be able to apply these techniques for *in vivo* imaging inside tissue. Furthermore, these high-speed systems can be adapted to maintain a focus through MMFs. As a result, the focus spot can be maintained during significant bending of the fiber, opening numerous opportunities for endoscopic imaging and energy delivery applications.

10.5   References

[1] Lazarev, G., Hermerschmidt, A., Krüger, S. & Osten, S. LCOS spatial light modulators: Trends and applications. In *Optical Imaging and Metrology* (eds. Osten, W. & Reingand, N.) 1–29 (Wiley, 2012).

[2] Liu, Y. *et al.* Optical focusing deep inside dynamic scattering media with near-infrared time-reversed ultrasonically encoded (TRUE) light. *Nat. Commun.* **6**, 5904 (2015).

[3] Judkewitz, Benjamin, *et al.* "Translation correlations in anisotropically scattering media." *Nature physics* **11.8**, 684. (2015).

[4] Walba, D. M. Fast ferroelectric liquid-crystal electrooptics. *Science* **270**, 250 (1995).

[5] Hornbeck, L. J. Deformable-mirror spatial light modulators. *Proc. SPIE* **1150**, 86–103 (1990).

[6] Perreault, J. A. Development and control of kilo-pixel MEMS deformable mirrors and spatial light modulators. *PhD Thesis* (2004).

[7] Dudley, D., Duncan, W. M. & Slaughter, J. Emerging digital micromirror device (DMD) applications. *Proc. SPIE* **4985**, 14–25 (2003).

[8] Chandrasekaran, S. N., Ligtenberg, H., Steenbergen, W. & Vellekoop, I. M. Using digital micromirror devices for focusing light through turbid media. *Proc. SPIE Emerging Digital Micromirror Device Based Systems and Applications VI* **8979**, 897905 (2014).

[9] Hornbeck, L. J. Digital light processing for high-brightness high-resolution applications. *Proc. SPIE* **3013**, 27–40 (1997).

[10] Lee, W.-H. Binary computer-generated holograms. *Appl. Opt.* **18**, 3661 (1979).

[11] Lee, W.-H., III. Computer-generated holograms: Techniques and applications. In *Progress in Optics* (ed. Wolf, E.) 119–232 (Elsevier, 1978).

[12] Piestun, R., Shamir, J., Weßkamp, B. & Bryngdahl, O. On-axis computer-generated holograms for three-dimensional display. *Opt. Lett.* **22**, 922 (1997).

[13] Fienup, J. R. Iterative method applied to image reconstruction and to computer-generated holograms. *Opt. Eng.* **19**, 193297 (1980).

[14] Wang, D. *et al.* Focusing through dynamic tissue with millisecond digital optical phase conjugation. *Optica* **2**, 728 (2015).

[15] Lee, W. H. Computer-generated holograms: Techniques and applications. In *Progress in Optics* (ed. Wolf, E.) (Elsevier, 1978).

[16] Goorden, S. A., Bertolotti, J. & Mosk, A. P. Superpixel-based spatial amplitude and phase modulation using a digital micromirror device. *Opt. Express* **22**, 17999 (2014).

[17] Mirhosseini, M. *et al.* Rapid generation of light beams carrying orbital angular momentum. *Opt. Express* **21**, 30196 (2013).

[18] Gong, L. *et al.* Observation of the asymmetric Bessel beams with arbitrary orientation using a digital micromirror device. *Opt. Express* **22**, 26763 (2014).

[19] Ren, Y.-X. *et al.* Dynamic generation of Ince-Gaussian modes with a digital micromirror device. *J. Appl. Phys.* **117**, 133106 (2015).

[20] Conkey, D. B., Caravaca-Aguirre, A. M. & Piestun, R. High-speed scattering medium characterization with application to focusing light through turbid media. *Opt. Express* **20**, 1733 (2012).

[21] Popoff, S. M. *et al.* Measuring the transmission matrix in optics: An approach to the study and control of light propagation in disordered media. *Phys. Rev. Lett.* **104**, 100601 (2010).

[22] Schreiber, H. B. Phase shifting interferometry. In *Optical Shop Testing* (ed. Malacara, D.) 547–666 (Wiley, 2007).

[23] Niv, E., Caravaca-Aguirre, A. M., Conkey, D. B. & Piestun, R. High-speed phase modulation using the DLP: Application in imaging through complex media. *Proc. SPIE 9376, Emerging Digital Micromirror Device Based Systems and Applications VII* **9376**, 937609 (2015).

[24] Conkey, D. B., Caravaca-Aguirre, A. M., Niv, E. & Piestun, R. High-speed, phase-control of wavefronts with binary amplitude DMD for light control through dynamic turbid media. *Invit. Pap. SPIE Photonics West* (2013).

[25] Caravaca-Aguirre, A. M., Niv, E. & Piestun, R. High-speed phase modulation for multimode fiber endoscope. In *Imaging and Applied Optics 2014*, ITh3C.1 (Optical Society of America, *2014*).

[26] Caravaca-Aguirre, A. M. & Piestun, R. Wavefront shaping for single fiber fluorescence endoscopy. *Proc. SPIE 9717, Adaptive Optics and Wavefront Control for Biological Systems II* **9717**, 97171B (2016).

[27] Kimura, S. & Wilson, T. Confocal scanning optical microscope using single-mode fiber for signal detection. *Appl. Opt.* **30**, 2143 (1991).

[28] Kim, D. *et al.* Toward a miniature endomicroscope: Pixelation-free and diffraction-limited imaging through a fiber bundle. *Opt. Lett.* **39**, 1921–1924 (2014).

[29] Jung, J. C. & Schnitzer, M. J. Multiphoton endoscopy. *Opt. Lett.* **28**, 902 (2003).

[30] Fu, L., Jain, A., Xie, H., Cranfield, C. & Gu, M. Nonlinear optical endoscopy based on a double-clad photonic crystal fiber and a MEMS mirror. *Opt. Express* **14**, 1027 (2006).

[31] Oh, G., Chung, E. & Yun, S. H. Optical fibers for high-resolution *in vivo* microendoscopic fluorescence imaging. *Opt. Fiber Technol.* **19**, 760 (2013).

[32] Gover, A., Lee, C. P. & Yariv, A. Direct transmission of pictorial information in multimode optical fibers. *J. Opt. Soc. Am.* **66**, 306 (1976).

[33] Yariv, A. Three-dimensional pictorial transmission in optical fibers. *Appl. Phys. Lett.* **28**, 88 (1976).

[34] Caravaca-Aguirre, A. M., Niv, E., Conkey, D. B. & Piestun, R. Real-time resilient focusing through a bending multimode fiber. *Opt. Express* **21**, 12881 (2013).

[35] Conkey, D. B., Brown, A. N., Caravaca-Aguirre, A. M. & Piestun, R. Genetic algorithm optimization for focusing through turbid media in noisy environments. *Opt Express* **20**, 4840 (2012).

11 Reflection Matrix Approaches in Scattering Media: From Detection to Imaging

Amaury Badon, Alexandre Aubry, and Mathias Fink

11.1 Introduction

Conventional imaging techniques are based on the first Born approximation and generally fail in the presence of aberrations or multiple scattering. In the reflection configuration, this approximation assumes a ballistic propagation of the incident wave between the source and a target inside a given medium, a reflection by the target, and once again a ballistic propagation between the target and the detector. Nevertheless, in the presence of inhomogeneities inside the surrounding medium, the emitted and/or the reflected wave can be distorted or scattered. Depending on the strength of the inhomogeneities, the resulting image can be slightly degraded or totally randomized. In order to restore the quality of the image, it is relevant to study the wave propagation in inhomogeneous media. This fundamental problem involves important applications, ranging from astronomical observations through a turbulent atmosphere to seismology or ultrasound imaging.

In this context, the use of array of sources and receivers offers a spatial diversity at the emission and the reception which is of great interest. For acoustics waves, an array of transducers offers the possibility to both shape the wavefront of the emitted pulse, and then record the reflected wavefield [1]. In seismology, the cross-correlation of codas recorded by a network of seismic stations directly yields the impulse response between each array element [2]. Computational techniques allow then to take advantage of the recorded information to image the Earth's crust [3]. Recently, in optics, the emergence of spatial light modulators (SLM) led to the development of numerous wavefront shaping techniques aiming to focus or image through scattering media [4]. Subsequently, a matrix approach of wave propagation through complex media was developed [5, 6]. In acoustics, for instance, it relies on the measurement of the Green's function – or impulse response – between each element of an array of transducers placed in front of a scattering sample. A so-called reflection matrix contains all the information available on the medium under investigation. Several techniques had then been developed to take advantage of this information for detection and imaging purposes.

We will first present these approaches in the context of biomedical ultrasound imaging and nondestructive testing where they were initially developed more than twenty years

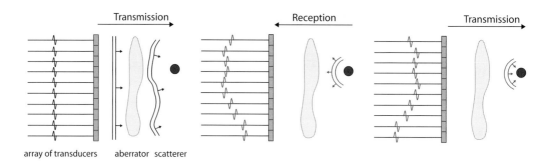

Figure 11.1 Time reversal through an aberrating medium. A plane wave is emitted by an array of transducers (transmit mode). The total backscattered field is recorded by the same elements (receive mode), time-reversed and reemitted (transmit mode). The time-reversed wavefield back-focuses on the target. Reprinted from [8], Copyright (1994), with permission from Elsevier.

ago. Using the wave nature of light, some of these approaches had recently been adapted to optics. In the second part, we will show, in particular, how the matrix formalism can help to push back the fundamental limit of multiple scattering for optical imaging in scattering media.

11.2   Reflection Matrix for Ultrasonic Imaging

11.2.1   Time-Reversal of Ultrasonic Waves

Ultrasonic waves provide a powerful modality to probe inhomogeneous media such as biological tissues due to its versatility, its noninvasive aspect and the already developed technology. In particular, piezoelectric transducers enables both a precise control of an emitted wave and a precise measurement of a reflected wavefield. Indeed, the electronic capability of transducers allows not only to emit an ultrasound pulse of typically some MHz, but also to record the temporal dependence of the reflected wavefield. In the early 1990's, Fink demonstrated the possibility to take advantage of the temporal information of a scattered wavefield to focus through an heterogeneous medium on a single scatterer [7]. Such an experiment first consists in emitting a pulse from an element or from an array of sources placed on one side of a complex medium (here a medium containing an aberrating layer; see Figure 11.1). The emitted wave propagates through the medium and then illuminates the target that acts as a secondary source. Then, the same array of transducers records the backscattered field, corresponding here to the impulse-responses or the Green functions between the target and the array of transducers. Finally, these impulse-responses are time-reversed and sent back from the array of sensors. From a measured signal largely spread in both time and space, a tight focus is obtained.

   This founding experiment showed the possibility to use scattered waves, to focus or transmit coherent information through a simple inhomogeneous medium. However, this experiment is limited due to the necessity to have in the zone of interest a unique scatterer acting as a coherent source. If one aims to focus on several targets or image such

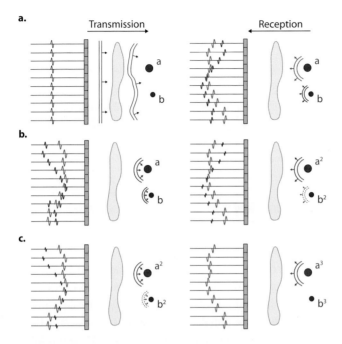

Figure 11.2 Iterative time reversal experiment in a medium made of an aberrating layer and two indivudal scatterers. (**a**) A plane wave is sent from the array of transducers (transmit mode). The backscattered wave field is recorded by the transducers (receive mode). (**b**) This field is time-reversed and reemited. (**c**) This procedure is repeated until a selective focusing is reached. Reprinted from [8], Copyright (1994), with permission from Elsevier.

a medium through an aberrating layer, things are more complicated. A matrix approach in a reflection configuration is needed and can be introduced by means of the concept of iterative time reversal.

11.2.2   Iterative Time-Reversal in Multi-target Media

We now consider an array of N transducers, acting both as transmitters and receivers, insonifying an inhomogeneous medium made of an aberrating layer and two individual scatterers A and B. We suppose that these two scatterers are separated by a distance $d > \lambda$ and display reflectivities equal to a and b respectively, with $a > b$. Let us consider, for example, the emission of a plane wave from the array of transducers. The incident wavefront is first distorted by the aberrating layer and then partially reflected by the two scatterers. This reflected wave is then distorted again by the aberrator and recorded by the transducers (receive mode). Here we consider that we are in a single scattering regime.

The reflected wavefield is then time reversed and reemitted (transmit mode). Due to the proportionality between the amplitude of the recorded signal and the reflectivities of the scatterers, the new incident wave focuses on the two scatterers with an amplitude given by their respective reflectivities, a and b. By iterating this procedure n times, the amplitudes of the waves focusing on the two reflectors are equal to a^n and b^n , respectively [9]. With $a^n >> b^n$, the incident wave will ultimately focus on the target with the stronger reflectivity, here a.

By subtracting the emitted signal from the first recorded signal, it is then possible to focus selectively on the second scatterer, and so on. However, this method is demanding in time as the number of scatterer increases. The combination of iteration procedures and subtraction operations converges toward a set of wavefronts that are the invariants of the time reversal process. These invariants are the eigenvectors of the time reversal operator [8] that is defined in the next paragraph.

11.2.3   Decomposition of the Time-Reversal Operator in a Multi-target Medium

The analysis of the time reversal operator allows to mimic in the computer all the steps of an iterative time reversal experiment. Experimentally, this operator can be deduced from the reflection matrix associated with the array of transducers. In a sequential acquisition procedure, a transducer at position $\mathbf{u_{in}}$ emits a short pulse toward the medium under investigation. The backscattered wavefield is then recorded by all the transducers at position $\mathbf{u_{out}}$. This yields the N impulse responses $K(\mathbf{u_{out}}, \mathbf{u_{in}}, t)$ between the source at $\mathbf{u_{in}}$ and each receiver. This procedure is then repeated for each element acting as a source. The N^2 measured impulses responses form a time-dependent reflection matrix $\mathbf{K_{u,u}}$ of dimension $N \times N$:

$$\mathbf{K_{u,u}}(t) = [K(\mathbf{u_{out}}, \mathbf{u_{in}}, t)] \tag{11.1}$$

This reflection matrix fully characterizes the propagation of a wave inside the medium. For a given incident wavefield $\varphi_{in}(\mathbf{u_{in}}, t)$ emitted by the transducers, the received signals are given by:

$$\varphi_{out}(\mathbf{u_{out}}, t) = \sum_{\mathbf{u_{in}}} K(\mathbf{u_{out}}, \mathbf{u_{in}}, t) \otimes \varphi_{in}(\mathbf{u_{in}}, t) \tag{11.2}$$

where the symbol $\otimes$ denotes a convolution over time. The last equation can be rewritten in the frequency domain in a matrix form:

$$\Psi_{out}(\omega) = \mathbf{K_{u,u}}(\omega).\Psi_{in}(\omega) \tag{11.3}$$

where $\Psi_{out}(\omega)$ and $\Psi_{in}(\omega)$ are the received and transmitted vector signals of N components. For a monochromatic wave, we can simply reinterpret the iterative time-reversal experiment previously described with the matrix formalism. We have seen that this method consists in repeating a procedure made of an emission/reception/phase conjugation/reemission step. This procedure can be modeled with the time reversal operator $\mathbf{T_{u,u}} = \mathbf{K_{u,u}}.\mathbf{K_{u,u}}^\dagger$, product of the two operators $\mathbf{K_{u,u}}$ and $\mathbf{K_{u,u}}^\dagger$, each one being associated with one step of the time reversal operation. Looking for the wavefront that will focus selectively on each scatterer of a multi-target medium is equivalent to find the invariants of this operator. Mathematically, the time reversal invariants can be deduced from the eigenvalue decomposition of the time reversal operator $\mathbf{T_{u,u}}$ or, equivalently, from the singular value decomposition of $\mathbf{K_{u,u}}$. It consists in writing the reflection matrix as follows [8]:

$$\mathbf{K_{u,u}} = \mathbf{U}.\Sigma.\mathbf{V}^{\dagger} = \sum_{i=1}^{N} \sigma_i \mathbf{U_i}(\mathbf{u_{out}}).\mathbf{V_i}(\mathbf{u_{in}})^{\dagger} \qquad (11.4)$$

with Σ a diagonal matrix containing the singular values σ_i in decreasing order. Here, $\mathbf{U}$ and $\mathbf{V}$ are unitary matrices whose columns correspond to the output and input singular vectors $\mathbf{U_i}$ and $\mathbf{V_i}$, respectively. Once the matrix is measured, the processing is then purely numerical. The singular value distribution provides interesting information: in the single scattering regime and in the case of point like targets, the number of significant singular values is exactly the number of targets provided they are resolved by the system. In our experiment, we expect to observe two singular values significantly higher than the others. On the one hand, a physical backpropagation of the associated singular vectors $\mathbf{V_1}$ and $\mathbf{V_2}$ selectively focuses on each scatterer (Figure 11.3c) and the phase law associated with each singular vectors automatically compensates for the distortion induced by the aberrating layer. On the other hand, a numerical backpropagation of $\mathbf{V_1}$ and $\mathbf{V_2}$ yields of an image of each scatterer. The quality of the image depends on the backpropagation model. If the medium between the array of transducers and the target is known, the resolution of the image is only limited by diffraction. If an effective homogeneous medium model is used, the images of the two targets may still suffer from aberrations.

The DORT method (French acronym for Decomposition of the Time Reversal Operator) takes advantage of the reflection matrix to separate the contributions of a limited number of well resolved scatterers in homogeneous or heterogeneous media as long as the first Born approximation is valid [8]. Note that this assumption of a *sparse* object is not always valid and that it exists many situations where it fails. What happens if the object of interest is made of a random distribution of unresolved scatterers such as, for instance, biological tissues in ultrasound imaging? In this configuration, even if the first Born approximation is still valid, the reflection matrix becomes random as the various impulse responses result from the interference between the backscattered echoes. In this situation, we will show (see Section 11.2.5) that a statistical approach based on the Van Cittert–Zernike theorem allows to extract the distortion induced by the aberrating layer.

Another interesting situation occurs in ultrasonic nondestructive testing of coarse grains materials where multiple scattering becomes important [10]. In presence of multiple scattering, the one-to-one association between the time of flight of an echo and the position of a scatterer is no longer valid and the interpretation of the time gated reflection matrix becomes more difficult. A separation between the single and multiple scattering contributions becomes necessary (see Section 11.2.4).

11.2.4   Reflection Matrix in the Single Scattering and Multiple Scattering Regimes

Let us now consider an array of N transducers facing a random scattering medium (Figure 11.3a). When an incident wave is emitted by one of the transducers, the backscattered wave contains two contributions:

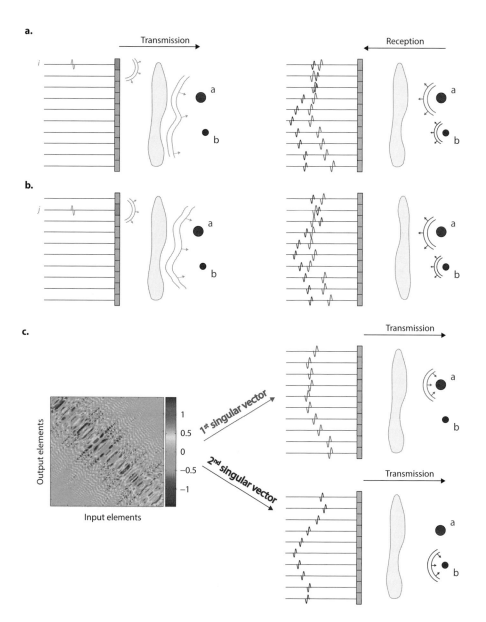

Figure 11.3 DORT method. (**a**) A short pulse is sent by the transducer i. The backscattered field is recorded by the transducers (receive mode). The field measured by the transducer j directly gives the impulse response $(\mathbf{u_j},\mathbf{u_i},t)$. (**b**) The procedure is repeated for each input element of the array. (**c**) A singular value decomposition of the matrix $\mathbf{K_{u,u}}$ at one frequency enables to focus selectively on the two scatterers.

- A single scattering (SS) contribution: the incident wave undergoes only one scattering event before coming back to the sensor(s). This is the contribution which is taken advantage of, because there is a direct relation between the arrival time t of the echo and the distance d between the sensor and the scatterer, $t = 2d/c$ (c is the sound velocity). Hence an image of the medium's reflectivity can be built from the measured signals.

- A multiple scattering (MS) contribution: the wave undergoes several scattering events before reaching the sensor. Multiple scattering is expected to take place when scatterers are strong and/or concentrated. In this case there is no more equivalence between the arrival time t and the depth of a scatterer. Thus, classical imaging fails when multiple scattering dominates.

In this part, we want to show how a matrix approach is relevant for the separation of these two contributions. As previously explained, a time-dependent reflection matrix is obtained from the measurement of the N^2 impulse responses of the medium between the elements of the array. In practice, the time-dependent signals are time-gated (by a window with sufficient duration) and a short-time Fourier transform is applied. Hence, for each value of time of arrival t, we can define a frequency dependent matrix $K_{u,u}(t, \omega)$ and use Eq. 11.3 in the frequency domain. In these conditions, for each arrival time t, the reflection matrix can be written as a sum of two matrices containing the SS and the MS contributions such that

$$\mathbf{K_{u,u}}(t, \omega) = \mathbf{K_{u,u}^S}(t, \omega) + \mathbf{K_{u,u}^M}(t, \omega) \tag{11.5}$$

In an imaging context, the information of interest is contained in the SS matrix but is unfortunately often drowned into a predominant MS background. Even for arrival times close to the expected time of the echo from the target, the MS contribution can prevent from a clear observation of the target echo (Figure 11.4c). Aubry and Derode proposed to separate these two components using the input-output correlations of the reflection matrix [11, 12].

Interestingly, it was shown that though $\mathbf{K_{u,u}^S}$ and $\mathbf{K_{u,u}^M}$ are both random matrices, they do not have the same statistical behavior at all. In particular, while $\mathbf{K_{u,u}^M}$ only exhibits short-range spatial correlations (Figure 11.4b), $\mathbf{K_{u,u}^S}$ displays a long-range correlation along its antidiagonals (i.e., for array elements i and j such that $i + j$ is constant), whatever the realization of disorder (Figure 11.3c). Physically, this can be understood as the equivalent, in a backscattering configuration, of the well-known memory effect in optics [13, 14]. When an incident plane wave is rotated by an angle θ, the far field speckle image measured in transmission is shifted by the same angle θ (or $-\theta$ if the measurement is carried out in reflection), as long as θ does not exceed the angular correlation width $\Delta\theta$ (Figure 11.5a). In the single scattering regime, the memory effect spreads over the whole angular spectrum ($\Delta\theta = \pi/2$). This accounts for the fact that the matrix coefficients are coherent along a given antidiagonal when only single scattering takes place. Indeed if two pairs of array elements $(i_1;j_1)$ and $(i_2;j_2)$ are on the same antidiagonal then we have $i_1+j_1=i_2 + j_2$. Changing the direction of emission amounts to changing i_1 into i_2. As a result, in reflection the speckle image will be tilted so that the signal that was received in j_1 will be coherent with the new signal in $j_2 = j_1 - (i_2 - i_1)$ (see Figure 11.5b). On the contrary when MS dominates, the correlation width $\Delta\theta$ is inversely proportional to the characteristic size of the diffusive halo W inside the medium. $\Delta\theta$ thus

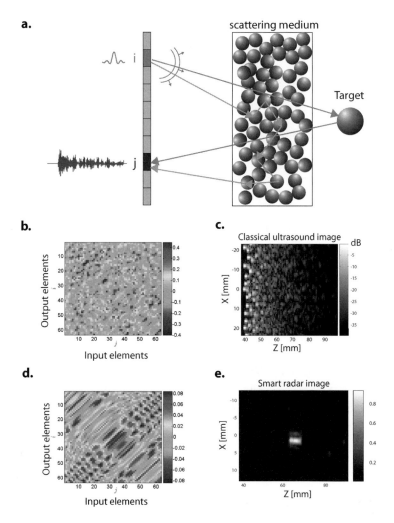

Figure 11.4 Reflection matrix in the presence of a scattering medium. (**a**) The impulse responses $K(\mathbf{u_i}, \mathbf{u_j}, t)$ between transducers i and j form the reflection matrix $\mathbf{K_{u,u}}$. (**b**) Example of the real part of a reflection matrix $\mathbf{K_{u,u}}$ when the multiple scattering contribution is predominant. (**c**) Conventional ultrasound image built from $\mathbf{K_{u,u}}$. (**d**) Real part of the extracted single scattering contribution $\mathbf{K_{u,u}^S}$. (**e**) Image of the target deduced from the DORT method applied to $\mathbf{K_{u,u}^S}$. Reprinted from [12], with the permission of AIP Publishing. A black and white version of this figure appears in some formats. For the color version, please refer to the plate section.

diminishes as $t^{-1/2}$ and, as soon as $\Delta\theta$ becomes smaller than the angular aperture of one array element, the matrix coefficients become random and uncorrelated (Figure 11.4b).

The memory effect can thus be taken advantage of to separate the single-scattering and the multiple-scattering contributions. This is achieved via rotations and projection of the reflection matrix [12]. These manipulations amount to selecting, within $\mathbf{K_{u,u}}$, the part of it that presents the aforementioned antidiagonal correlation. The resulting filtered matrix corresponds to the single scattering contribution $\mathbf{K_{u,u}^S}$, albeit weak (see

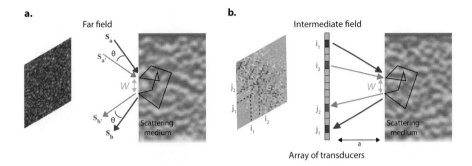

Figure 11.5 Memory effect. (**a**) Memory effect in optics (illumination and recording of the reflected wavefield from the far-field): When an incident plane wave is tilted by an angle θ, the speckle wavefield reflected by the scattering layer is tilted by the opposite angle $-\theta$. The correlation width $\Delta\theta$ scales as λ/W, with W the mean distance between the first and last scattering events in the scattering layer. (**b**) Memory effect in acoustics (array of transducers placed at a finite distance a from the sample): The memory effect manifests itself along the antidiagonals of the reflection matrix. Matrix elements, $K(u_{j_1}, u_{i_1}, t)$ and $K(u_{j_2}, u_{i_2}, t)$, such that $i_1 + j_1 = i_2 + j_2$, are strongly correlated. The correlation length scales as $\lambda a/W$.

Figure 11.4d). It is thus a way to get rid of multiple scattering. On the one hand, applying the DORT method to the single scattering matrix allows to image a target (Figure 11.4e) whose presence was initially completely hidden behind a strongly scattering layer (Figure 11.4c). This approach has been successfully applied to the nondestructive testing of coarse grain steels [10]. On the other hand, the multiple scattering contribution can be isolated and investigated for characterization purposes [15].

11.2.5 Overcoming Aberrations in Single Scattering Regime with the Focusing Matrix Approach (FDORT)

In ultrasonic biomedical imaging, tissues behaves mainly as a random distribution of nonresolved scatterers. As the ultrasonic mean free path in soft tissue is larger than the body thickness, the first-order Born approximation is valid and single scattering processes dominate. However, classical beam forming techniques made also the assumption of uniform speed, which fails in presence of subcutaneous layers of fat that behave like aberrating layers. Is it possible from the measurement of the reflection matrix of such a random medium to extract the exact Green's functions of the medium and therefore to compensate for the aberrations? As we may assume a ballistic propagation between the transmit/receive transducers and the scatterers, we will show that the canonical matrix $\mathbf{K}^S_{u,u}$ that relates the transducer positions $\mathbf{u}_{in}$ to $\mathbf{u}_{out}$ is not always the best choice. Introducing the concept of an array of virtual transducers located inside the medium can greatly simplify the interpretation of the time reversal operator. It allows to express the time-reversal operator and the reflection matrix in a different basis and to give a physical meaning to the singular vectors associated to a random medium.

11.2.5.1 The Virtual Transducers

The focused DORT (FDORT) method [16, 17] is an implementation of the DORT method using focused transmits instead of single elements transmits (see Figure 11.6). In FDORT, a focused beam is transmitted by a group of transducers inside the medium and the received signals are recorded for each element. A time-gated window selects the echoes coming from any depth and a Fourier transform is then taken and the focused beam is translated step by step. Each focused beam can be conceptually replaced by a virtual transducer located at the focus. It is like you have a virtual array of transmitted sources located at the focal depth z (see Figure 11.6c). The position of these virtual sources is then given by $\mathbf{r} = (x, z)$ and the new reflection matrix can be written as

$$\mathbf{K_{u,r}} = [K(\mathbf{u}, \mathbf{r})]. \tag{11.6}$$

$\mathbf{K_{u,r}}$ can be seen as the interelement response matrix between a virtual transducer located at $\mathbf{r}$ and a real transducer located at $\mathbf{u}$. Therefore the singular decomposition of this matrix can be written as

$$\mathbf{K_{u,u'}} = \mathbf{U(u)}.\Sigma.\mathbf{V(r)}^{\dagger} \tag{11.7}$$

with $\mathbf{U}$ containing the invariants of the time-reversal operator in the physical array plane and $\mathbf{V}$ containing the invariants expressed in the virtual array plane. In the case of well-resolved scatterers, $\mathbf{U}$ contains the Green's functions between each scatterer and the physical array while $\mathbf{V}$ contains the Green's functions between the same scatterer and the virtual array. Note that, if the scatterer is located shallower than the virtual array plane, the associated singular vector $\mathbf{V_i}$ then corresponds to the conjugate of the causal Green's function, i.e., the advanced Green's function, with a reversed curvature of the recorded wavefront (see Figure 11.6c).

11.2.5.2 FDORT in Speckle Noise

The main interest of the FDORT method arises for a random scattering medium with singly scattered echoes [17]. The possibility to focus in transmit allows to select a small zone of interest inside the scattering medium. The reflected signals, recorded by the array elements, are random signals resulting from the superposition of all the individual echoes from the scatterers located at the focus. However, the backscattered wavefield observed in the array plane is not completely random as it can be seen in Figure 11.7c. There is a strong correlation between the echoes recorded on adjacent array elements. This spatial correlation extends along the whole array aperture and is related to well-known Van Cittert–Zernike theorem that described the propagation (ballistic) of statistical property of a wavefield radiated by a spatially incoherent source.

Here the incoherent source is the random distribution of scatterers located in the focal spot at depth z. As the incident focused beam is described at the focal depth by a focused wavefield $p(x)$, this incoherent source is modulated by the directivity pattern associated with $p(x)$. The Van Cittert–Zernike theorem predicts that the spatial correlation of the wavefield radiated by an incoherent source and observed for a pair of elements (m, n) only depends on the distance between the elements, $|\mathbf{u_n} - \mathbf{u_m}|$. It is actually proportional to the Fourier transform of the square of $p(x)$. For example, for a homogeneous medium and a rectangular transmit aperture containing a limited number of transducer,

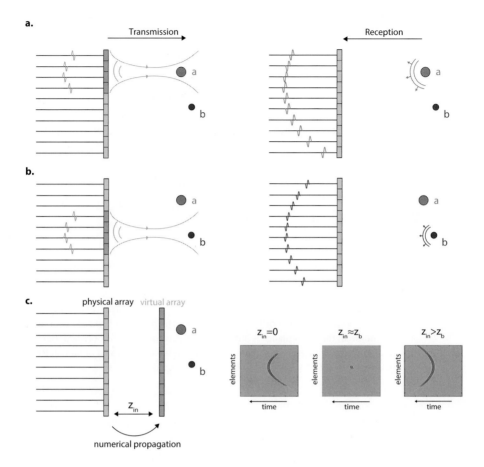

Figure 11.6 FDORT method. (**a,b**) Acquisition of the focusing matrix. One element $K(u_j, r_i, t)$ of this matrix is the signal received by element j after the ith focused beam has been transmitted. (**c**) The focusing matrix describes the wave propagation between the physical array of transducers and a virtual array. Its singular vectors present different curvatures in the time domain according to the distance between the two arrays. Reprinted with permission from [16]. Copyright 2008, Acoustic Society of America.

$p(x)$ is a sinc function. Therefore the amplitude of the cross correlation for pair of elements decreases as a triangle function (see Figure 11.7d) when the distance between elements increases. The time-reversal operator $\mathbf{T}_{\mathbf{u},\mathbf{u}}$ associated with FDORT directly yields a good estimate of the spatial correlation matrix. As the transmitted focused beams scan different portions of the scattering medium, the columns of $\mathbf{K}_{\mathbf{u},\mathbf{r}}$ represents different realizations of the backscattered signal received by the array. The time reversal operator elements can be expressed as

$$\mathbf{T}_{\mathbf{u},\mathbf{u}} = [T(\mathbf{u_m}, \mathbf{u_n})] = \sum_{\mathbf{r}} \sum_{\mathbf{r'}} K(\mathbf{u_m}, \mathbf{r}).K(\mathbf{u_n}, \mathbf{r'})^{\dagger} \tag{11.8}$$

The sum over the focused transmits at $\mathbf{r}$ and $\mathbf{r'}$ can be seen as an average over realizations of disorder of the signal cross correlation received on elements m and n. If the number

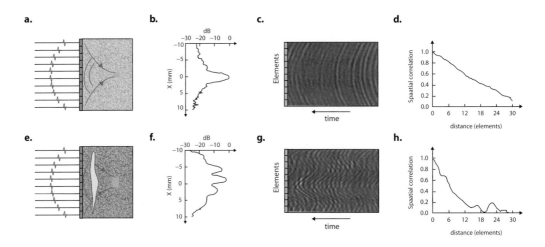

Figure 11.7 Spatial coherence of backscattered field. (**a**) Focusing of an incident field in a random distribution of scatterers. (**b**) Spatial distribution of the amplitude in the focal plane: A diffraction-limited focal spot is obtained. (**c**) B-scan of the backscattered field. (**d**) Spatial correlation of the scattered wavefield: A triangle shape is obtained. (**e**) Focusing of an incident field in a random distribution of scatterers through an aberrating layer. (**f**) Spatial distribution of the amplitude in the focal plane: A strongly degraded focal spot is obtained. (**d**) Spatial correlation of the scattered wavefield: The triangle shape is far to be obtained. Reprinted with permission from [18]. Copyright 1994, Acoustic Society of America.

of realizations is sufficient (enough transmit beams), the triangle law appears along the antidiagonals of the time reversal operator as it can be seen in Figures 11.8b and 11.8c. This remarkable property means that, in the single scattering regime, the FDORT time-reversal operator $\mathbf{T_{u,u}}$ is not random but deterministic.

On a mathematical standpoint, everything happens as if we were performing a time reversal experiment on a deterministic reflector, located at the focal depth, whose reflectivity is proportional to $|p(x)|^2$, the incident beam intensity. By averaging over many realizations a random mirror made of many scatterers, we have built a deterministic virtual reflector. As the DORT method is performed on this virtual reflector, the first eigenvector exactly focused at the focal point and the phase law of the eigenvector allows a perfect focusing. Note that as the reflectance of the virtual reflector is not a Dirac distribution, there is not a single eigenvector but several as it is observed for extended objects [19, 20]. In the case of perfect focusing with a sinc function, it is possible to show that C, the ratio of the first eigenvalue to the sum of all the eigenvalues, is exactly the focusing criterion (or the coherence factor) introduced by Mallart and Fink [18].

This technique becomes really interesting in presence of aberrating layers that distort the transmit beams used in the focusing reflection matrix acquisition. As shown in Figure 11.7, the beam distortion due to the aberrating layer introduces a defocusing and the presence of strong sidelobes at the focal depth. Therefore the spatial correlation of the backscattered field decreases more quickly with the distance between transducers. It

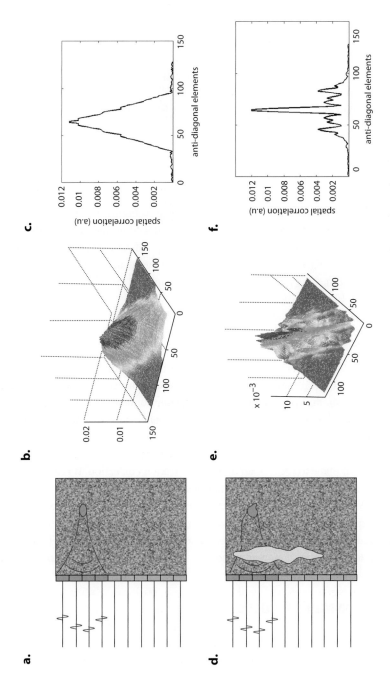

Figure 11.8 Relation between the Van Cittert–Zernike theorem and the focusing matrix. (**a**) Recording of the focusing matrix in a random medium. (**b**) Spatial correlation matrix and its antidiagonal elements. (**c**) A triangle shape is obtained. (**d**) Recording of the focusing matrix in the presence of an aberrating layer. (**e**) Spatial correlation matrix and its antidiagonal elements (**f**) Due to the aberrating layer, a narrower spatial correlation curve is obtained. Reprinted with permission from [17]. Copyright 2008, Acoustic Society of America.

becomes sharper than a triangle function as the focused beam is wider (see Figures 11.7e–11.7h). This can also be seen in Figures 11.8d–11.8f.

In order to keep the virtual reflector interpretation, in presence of an aberrating layer, we have to guarantee that the different transmit beams used to scan a portion of the scattering medium, conserve the same directivity patterns. This occurs only if the different incident beams are steered in the same isoplanetic patch. Therefore by measuring only a restricted matrix with a sufficient number of columns in an isoplanetic patch (typically 25) we can extract, from the singular value decomposition of the focusing matrix, the Green's function of a single virtual scatterer located in the main lobe. We are thus able to compensate for the distortion induced by the aberrating layer. However, as the virtual reflector is now wider in presence of aberrations and can display different side lobes (a collection of points that are not well resolved; see Figure 11.9a), the distribution of the eigenvalues is different and the factor C diminishes. The first eigenvector is now mainly the Green's function of the brightest spot but, as it is not well separated from the other points, it is perturbed by these contributions. A first estimate of the aberration profile is obtained from the phase of the first eigenvector and can be used to improve the focusing. However, the result is still perfectible. The process can therefore be iterated and the transmits are partially corrected. As shown in Figure 11.9, after four iterations, the iterative focusing process converges and the procedure has selected the brightest spot. The phase of the first eigenvector gives a very good estimation of the distortion induced by the aberrating layer. This technique can be used in medical ultrasound imaging to correct by zones the ultrasonic image of a biological phantom with aberrations as it can be seen in Figures 11.9d–11.9f.

11.2.6   Conclusion

In this first part, several matrix approaches of ultrasound propagation through complex media have been presented. Depending on the nature of the inhomogeneities and their concentration, the wave can undergo aberrations and/or multiple scattering events. Different strategies have been proposed to cope with these fundamental issues whether it be for focusing, detection or imaging purposes. The matrix analysis of the backscattered wavefield can be made either in the transducers' plane to get rid of multiple scattering or in the focal plane to provide an optimal aberration correction. It is also worth noting that the use of plane waves enables attainment of ultrafast frame rates over a large field of view. With more than 1000 frames/s, ultrafast ultrasonic imaging appears as a promising technique for functional imaging [21]. These results illustrate the potential and the flexibility of a matrix approach for the study of complex media. The transposition of these ideas to optics is now presented in the second part of this chapter.

11.3   Reflection Matrix for Optical Imaging

Optical microscopes have for a long time played a crucial role in biomedical research. Their ability to provide micrometer scale information about a specimen in a noninvasive

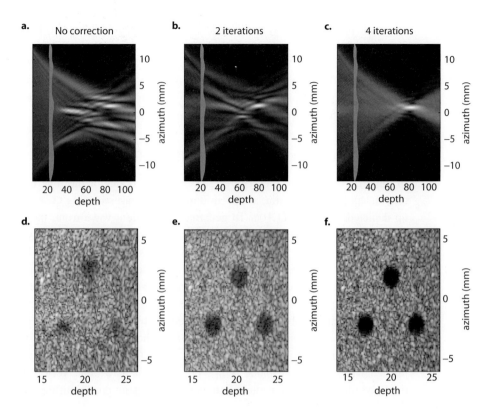

Figure 11.9 Iterative focusing and imaging through an aberrating layer. (**a–c**) Simulated transmit fields in the presence of a far-field phase aberattor located at 20 mm depth for a different number of iteration. (**d–f**) Image of simulated speckle phantom for the same conditions. Reprinted with permission from [17]. Copyright 2008, Acoustic Society of America.

manner has permitted numerous scientific advances. However, the performance of a microscope is notably degraded by the presence of inhomogeneities, due to the specimen itself or to the presence of an immersion liquid for instance. Restoring the resolution of a microscope, even in the presence of an aberrating layer, yields crucial applications for biomedical observations [22]. In that context, the recording of a reflection matrix enables such an achievement.

As previously seen, the reflection matrix approach is experimentally based on the measurement of impulse responses between elements of the same array. For ultrasound or microwaves, such a formalism is particularly appropriate as the wavefield can be simultaneously generated and recorded by the same multi-element array. In contrast, it has always been a challenge to emit and measure time varying wavefields in optics. Indeed, light oscillates at extremely high frequencies that cannot be sampled by any electronic equipment. This has pushed the optics community to develop coherent sources such as lasers, and to probe light fields using interferometric methods. Therefore, two different schemes are necessary for the control of the incident light and for the measurement of the backscattered wavefield. The coherent control of the incident wavefield was made possible recently with the development of spatial light modulators (SLM). In 2007, Vellekoop and Mosk showed how such a device can shape an incident wavefront in order to focus light through an opaque material [23]. This pioneering work was followed by the development of a transmission matrix approach dedicated to the control of light through complex media [5]. This approach was later extended to a reflection configuration for detection and imaging purposes [24, 25].

11.3.1 Selective Focusing in Multitarget Media: The DORT Method in Optics

In this first section, we report on the extension to optics of the time reversal operator concept [5]. To that aim, a reflection matrix is measured using the experimental set up depicted in Figure 11.10a. To generate the incident wave front, the beam from a single mode laser (frequency ω) is expanded and spatially modulated by a SLM. The surface of the SLM is imaged on the pupil of a MO, thus a pixel of the SLM ($\mathbf{u_{in}}$) matches the transverse component of the wave vector in the focal plane. The modulated input beam is focused onto the surface of a sample. The latter one is here made of gold nanoparticles deposited on a glass slide. A weakly aberrating slab is positioned between the object and the microscope objective. As seen in Figure 11.10b, this slab prevents from a clear observation of the nanoparticles. The backscattered wavefield interferes with a reference beam on a CCD camera in an image plane conjugated with the sample plane ($\mathbf{r_{out}}$). For all Hadamard input vectors displayed by the SLM, the complex wavefield is then obtained using phase-shifting interferometry.

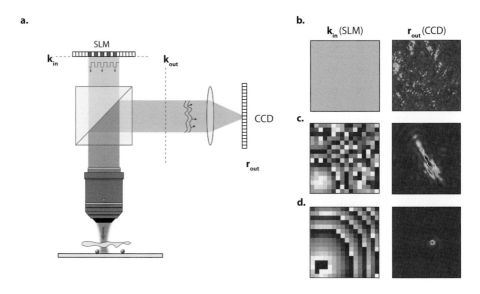

Figure 11.10 Focusing light through an aberrating layer. (**a**) Experimental setup dedicated to the measurement of the focusing matrix between each pixel of an SLM in the pupil plane and each pixel of a CCD in the object plane. (**b**) Phase mask displayed by the SLM (left) and image acquired by the CCD. (**c**) Phase mask calculated from the focusing matrix (left) and image acquired by the CCD. (**d**) Control experiment without the aberrating layer. Reprinted figure with permission from [25]. Copyright (2011) by the American Physical Society.

A projection from the Hadamard to the canonical basis, gives us the reflection matrix between the SLM pixels ($\mathbf{u_{in}}$) and the CCD pixels ($\mathbf{r_{out}}$). We adopt the following notation:

$$\mathbf{K_{r,u}} = [K(\mathbf{r_{out}}, \mathbf{u_{in}}, \omega)] \tag{11.9}$$

Note that in optics, the arrays of sources and receivers are bidimensional, thus $\mathbf{K_{r,u}}$ is in principle a four-dimensional matrix. For the sake of simplicity, the 2D arrays are concatenated into 1D vectors. In that representation, $\mathbf{K_{r,u}}$ is a matrix of dimension $N \times M$, with M and N the number of pixels contained by the CCD and the SLM, respectively. These dimensions can be up to several millions. However the acquisition and the manipulation of such matrices would be impossible in practice. In a typical experiment, we rather use macro-pixels such that $N = M \sim 1000$. Once the reflection matrix is recorded, a SVD is applied. It consists in writing:

$$\mathbf{K_{r,u}} = \mathbf{U(r_{out})} . \Sigma . \mathbf{V(u_{in})}^\dagger \tag{11.10}$$

with Σ a diagonal matrix and $\mathbf{U}$ and $\mathbf{V}$ are unitary matrices whose columns correspond to the output and input singular vectors $\mathbf{U_i}$ and $\mathbf{V_i}$, respectively. Under the single scattering approximation and for point-like scatterers, each scatterer is associated mainly to one significant eigenstate. While the lowest singular values are associated to the noise (residual specular reflections, laser fluctuations, CCD readout noise, etc.), few singular values are associated with the gold nanoparticles.

The input singular vector $\mathbf{V_i}$ corresponds to the wavefront focusing on the ith particle and the output singular vector $\mathbf{U_i}$ is the corresponding scattering pattern measured by the CCD in reflection. Figure 11.10 shows explicitly the effect of aberrations by comparing the singular vectors $\mathbf{U_1}$ and $\mathbf{V_1}$ (associated to the brightest scatterer) with and without the aberrating layer. In free space, the wavefront focusing on a particle corresponds to a Fresnel phase zone plate (Figure 11.10d). In contrast, in the presence of an aberrating layer (Figure 11.10c), $\mathbf{U_1}$ is strongly modified to compensate for the wavefront distortions. Note that even if the aberrations are compensated for the focusing, the image provided by $\mathbf{V_1}$ still suffers from the aberrations.

In this experiment, the success of the DORT method is due to the fact we are in the single scattering regime. In the presence of multiply scattered light, the one-to-one association between each singular state and each scatterer is no longer guaranteed. Therefore, a removal of MS light becomes necessary.

11.3.2   The Double Focusing Matrix in the Single Scattering Regime

In turbid media, MS starts to predominate beyond a few scattering mean free paths. To cope with this fundamental issue, several approaches have been proposed to enhance the SS contribution. The first option consists of separating SS from MS photons by means of time gating. However, electronic devices (CCD, photodiodes ...) are not sufficiently fast to select directly the photons of interest. One can use coherence time gated techniques which exploits the temporal coherence of the single scattered wave combined with a reference beam to achieve such a discrimination [26]. The second option is to spatially discriminate SS and MS. In acoustics, we have seen the possibility to take advantage of the memory effect to perform this procedure directly in the transducers plane $(\mathbf{u}_{in}, \mathbf{u}_{out})$. In optics, thanks to lenses and microscope objectives, similar discrimination can be achieved either in the sample/image plane $(\mathbf{r}_{in}, \mathbf{r}_{out})$ or in the far-field $(\mathbf{k}_{in}, \mathbf{k}_{out})$. While confocal microscopy or two-photon microscopy are based on the first possibility, Wonshik Choi and his colleagues recently demonstrated the possibility to take advantage of input-output correlations of reflection matrix elements to discriminates SS and MS photons in the far-field (k-space) [27].

Here, inspired by the FDORT method developed in acoustics (see 11.2.5), the reflection matrix is directly investigated in the focal plane (point-to-point basis). A set of time-dependent reflection coefficients $K(\mathbf{r}_{in}, \mathbf{r}_{out}, t)$ are measured between each point of the focal plane identified by the vectors $\mathbf{r}_{in}$ at the input and $\mathbf{r}_{out}$ at the output (see Figure 11.11a). In practice, the CCD is placed in the output pupil plane of the MO $(\mathbf{r}_{in})$ but a numerical Fourier transform propagates the recorded field into the image plane $(\mathbf{r}_{out})$. This procedure is similar to the numerical focusing performed in the FDORT method to create a virtual array of pixels/transducers inside the medium. The coefficients $K(\mathbf{r}_{out}, \mathbf{r}_{in}, t)$ form the double focusing reflection matrix $\mathbf{K}_{\mathbf{r},\mathbf{r}}$. The temporal resolution is provided by the temporal coherence of the light source, here a femtosecond laser. The time of flight t is chosen to match with the ballistic time. In the following, we will no longer mention the time dependence of the measured reflection matrix since the time t is now fixed. The experimental setup and procedure are presented in a simplified manner in Figure 11.11a.

In this section, we first investigate the so-called double focusing matrix in free space to characterize the signature of the SS contribution. Figure 11.11b displays the reference reflection matrix $\mathbf{K}_{\mathbf{r},\mathbf{r}}$ measured for a ZnO bead deposited on a microscope. The reflection matrix is measured over a field of view of 40×40 mm^2, mapped with 289 input wavefronts. In this case, $\mathbf{K}_{\mathbf{r},\mathbf{r}}$ contains two main contributions:

- The specular echo from the glass slide that emerges throughout the diagonal of the matrix
- The strong bead echo that arises for position $\mathbf{r}_{in} \sim \mathbf{r}_b$ with $\mathbf{r}_b$ being the position of the bead in the focal plane.

Thus, the SS contribution only emerges along the diagonal and closed diagonal elements of $\mathbf{K}_{\mathbf{r},\mathbf{r}}$. This is accounted for by the fact that a singly scattered wavefield can only come from points illuminated by the incident focal spot. A time-gated confocal image can be deduced from the reflection matrix by only considering its diagonal elements, such that

$$I_0(\mathbf{r}) = |K(\mathbf{r}, \mathbf{r})| \tag{11.11}$$

The corresponding image is displayed in Figure 11.11c. It is equivalent to an en face OCT image. Not surprisingly, it shows a clear image of the target on the microscope glass slide. The same observation can be deduced from an experiment where a US Air Force (USAF 1951) resolution target is placed in the focal plane of the MO whose image is nicely recovered from the diagonal elements of the reflection matrix (see Figures 11.11d and 11.11e).

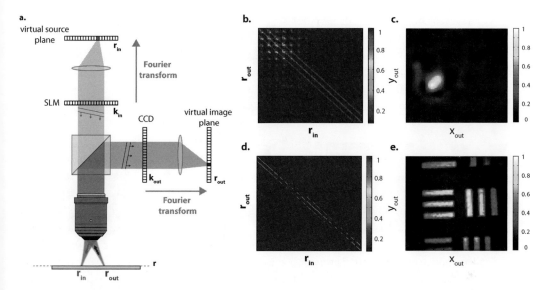

Figure 11.11 Imaging with the double focusing matrix in the single scattering regime. (**a**) Experimental setup. A femtosecond laser beam is shaped by an SLM acting as a dynamic diffraction grating. A set of incident plane waves is thus emitted from the SLM and focused at a different position in the focal plane of an MO. The backscattered wavefield is collected through the same MO and measured by a CCD camera in the pupil plane of the MO. A numerical Fourier transform propagates the recorded field into the image plane. (**b**) Reflection matrix $\mathbf{K_{r,r}}$ measured for a ZnO bead deposited at the surface of a glass slide. (**c**) En-face OCT image deduced from $\mathbf{K_{r,r}}$. (**d**) Reflection matrix measured for a USAF 1951 resolution target. (**e**) Corresponding en-face OCT. Figure adapted from [28], with permission from AAAS. A black and white version of this figure appears in some formats. For the color version, please refer to the plate section.

11.3.3 The Double Focusing Matrix in the Multiple Scattering Regime

Let us now consider the experiment with the bead on the glass slide but with two paper sheets placed between the MO and the object. The optical thickness of such a scattering medium is equal to $12.25\ell_s$. The ballistic wave has to go through $24.5\ell_s$ back and forth, thus undergoing an attenuation of $\exp(-24.5) \sim 2 \times 10^{-11}$ in intensity. The SS-to-MS ratio (SMR) of the reflected wavefield is estimated to be 10^{-11} in the MO back focal plane. For an incident plane wave, this means that only 1 scattered photon of out 1000 billion is associated to an SS event from the target. Theoretically, the target is far to be detectable and imaged in this experimental configuration, whether it be by conventional microscopy (SMR, $\sim 10^{-10}$), confocal microscopy (SMR, $\sim 10^{-8}$), or OCT (SMR, $\sim 10^{-5}$). This experimental situation is thus particularly extreme, even almost desperate, for a successful imaging of the target. Figure 11.12b displays the reflection matrix $\mathbf{K_{r,r}}$ measured in the presence of the scattering layer. Contrary to the SS contribution that emerges along the diagonal and closed-diagonal elements of $\mathbf{K_{r,r}}$ (Figures 11.12b and 11.12d), MS randomizes the directions of light propagation and gives rise to a random reflection matrix. Nevertheless, one can try to image the target by considering

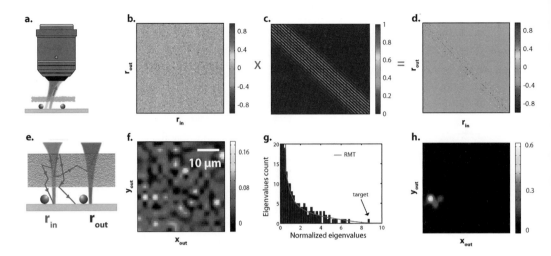

Figure 11.12 Target detection in the deep MS regime. (**a**) Two paper sheets are placed between the object, here a ZnO bead, and the MO. (**b**) Time-gated reflection matrix measured in the presence of the scattering medium. (**c**) Characteristic SS matrix. (**d**) SS matrix deduced from the projection of $\mathbf{K}_{\mathbf{r},\mathbf{r}}$ onto S. (**e**) Principle of the double focusing matrix. An input beam is focused at position $\mathbf{r}_{\mathbf{in}}$ (left beam). A numerical focusing is performed at reception to distinguish the light that comes from $\mathbf{r}_{\mathbf{out}} = \mathbf{r}_{\mathbf{in}}$ and the scattered light. (**f**) En face OCT image deduced from $\mathbf{K}_{\mathbf{r},\mathbf{r}}$. (**g**) Eigenvalue histogram of the time reversal operator $\mathbf{T}_{\mathbf{r},\mathbf{r}}^{S}$ compared to the random matrix prediction (solid line): the largest eigenvalue emerges from the MS noise. (**h**) Smart-OCT image deduced from the first eigenstate of $\mathbf{K}_{\mathbf{r},\mathbf{r}}^{S}$. Figure adapted from [28], with permission from AAAS. A black and white version of this figure appears in some formats. For the color version, please refer to the plate section.

the diagonal of $\mathbf{K}_{\mathbf{r},\mathbf{r}}$. The corresponding en face OCT image is shown in Figure 11.12f. As theoretically expected, MS still predominates despite confocal filtering and coherence time gating. An image of speckle is thus obtained without any enhancement of the intensity at the expected target location.

An alternative route is now proposed to image and detect the target behind the scattering layer. The smart-OCT approach first consists in filtering the MS contribution in the measured reflection matrix $\mathbf{K}_{\mathbf{r},\mathbf{r}}$. To that aim, the $\mathbf{K}_{\mathbf{r},\mathbf{r}}$ matrix is projected on a characteristic SS matrix **S**, whose elements are given by

$$S(\mathbf{r}_{\mathbf{out}}, \mathbf{r}_{\mathbf{in}}) = \exp\left(-\frac{|\mathbf{r}_{\mathbf{out}} - \mathbf{r}_{\mathbf{in}}|^2}{l_c^2}\right) \tag{11.12}$$

l_c is a tunable parametric length that accounts for the fact that the ballistic signal does emerge not only along the diagonal of the reflection matrix but also along off-diagonal elements. l_c is governed by two factors:

- The coherence length of the ballistic wavefield in the focal plane: In addition to ballistic attenuation and MS, the scattering layer also induces aberrations that degrade the focusing quality of the ballistic wavefront and enlarge the point spread function of the imaging system.

- The size of the target: the target signal does not only emerge along the diagonal elements of R in the absence of the scattering layer. This is accounted for by the size of the target that is larger than the resolution cell.

Mathematically, the projection of $\mathbf{K_{r,r}}$ can be expressed as a Hadamard product with $\mathbf{S}$:

$$\mathbf{K^S_{r,r}} = \mathbf{K_{r,r}} \circ \mathbf{S} \qquad (11.13)$$

This mathematical operation thus consists in keeping the diagonal and closed-diagonal coefficients of $\mathbf{K_{r,r}}$, where the SS contribution arises, and filtering the off-diagonal elements of $\mathbf{K_{r,r}}$ mainly associated with the MS contribution. It can be seen as a digital confocal operation with a virtual pinhole mask of size l_c [29]. In the present experiment, the SS matrix $\mathbf{K^S_{r,r}}$ is deduced from $\mathbf{K_{r,r}}$ by considering that $l_c = 5$ µm. The resulting matrix is displayed in Figure 11.12d. $\mathbf{K^S_{r,r}}$ contains the SS contribution as wanted, plus a residual MS. This term persists because MS signals also arise along and close to the diagonal of $\mathbf{K_{r,r}}$.

Once this SS matrix is obtained, one can apply the DORT method to extract the target signal among the residual MS. Indeed, the one-to-one association between an eigenstate of is only valid under an SS approximation. Hence, the DORT method cannot be applied to the raw reflection matrix because it contains an extremely predominant MS contribution. The trick here is to take advantage of the SS matrix $\mathbf{K^S_{r,r}}$. A singular value decomposition of $\mathbf{K^S_{r,r}}$ is performed. In the case of the double focusing matrix, the input and output singular vectors $\mathbf{U_i}$ and $\mathbf{V_i}$ contains directly the contribution in the sample plane of the scatterer associated to the singular value σ_i. Figure 11.12g displays the histogram of the eigenvalues σ_i^2 normalized by their average. This is compared to the distribution that would be obtained in a full MS regime [28]. The histogram of $\sigma_i^2/ < \sigma_i^2 >$ follows this distribution except for the largest eigenvalue, σ_1^2, which is actually beyond the superior bound of the MS continuum of eigenvalues. This means that the first eigenspace is associated to the target [10, 12]. The combination of the first input and output singular vectors, $|\mathbf{U_1} \circ \mathbf{V_1}|$, forms the smart-OCT image displayed in Figure 11.12h. The image of the target is nicely recovered. The comparison with the en face OCT image displayed in Figure 11.12d unambiguously demonstrates the benefit of smart- OCT in detecting a target in the deep MS regime (L = 12.25 ℓ_s). Note that the target image does not exactly match with the reference image (Figure 11.12c). This difference can be accounted for by the residual aberration effects induced by the scattering layer itself.

11.3.4 Imaging through Thick Biological Tissues

Following this experimental proof of concept, we now apply our approach to the imaging of an extended object through biological tissues. A positive U.S. Air Force (USAF) 1951 resolution target placed behind an 800-mm-thick layer of rat intestine tissues is imaged through an immersion objective (see Figure 11.13a). The reflection matrix is measured over a field of view of 60 × 60 µm^2 (see the square in Figure 11.13a) with 961 input wavefronts. The diagonal of $\mathbf{K_{r,r}}$ yields the en-face OCT image displayed in

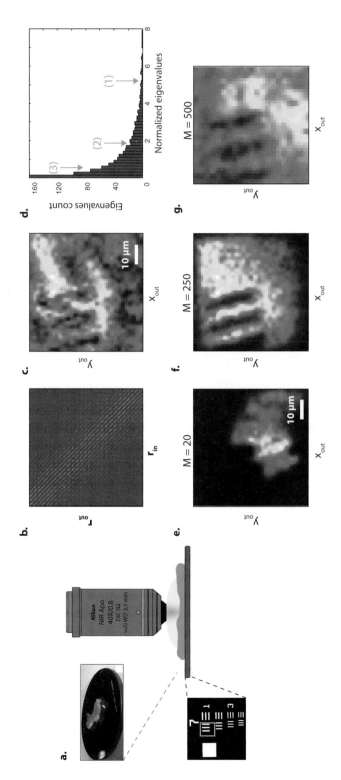

Figure 11.13 Imaging through thick biological tissues with the double focusing matrix. (**a**) Schematic of the experimental conditions. A resolution target is placed in the focal plane of an immersion MO with an 800-μm thick layer of rat intestine on top of it. (**b**) Reflection matrix $\mathbf{K_{r,r}}$. (**c**) Corresponding en face OCT image. (**d**) Eigenvalue histogram of the time reversal operator $\mathbf{T_{r,r}^S}$. (**e–g**) Smart OCT images obtained from the 20,250 and 500 first eigenstates of the filtered matrix $\mathbf{K_{r,r}^S}$. From [28], reprinted with permission from AAAS.

Figure 11.13b. Because of the aberration effects and MS events induced by the biological tissues, the three bars of the USAF target cannot be recovered. To overcome these detrimental effects, the MS filter is applied to the raw matrix $\mathbf{K_{r,r}}$ ($l_c = 8$ μm), yielding the SS matrix $\mathbf{K^S_{r,r}}$.

Iterative time reversal is then performed. In previous experiments, the object being imaged consisted of one bead. This sparsity implied that only a one eigenstate was needed to recover the image of the bead. In the present case, the USAF target is an extended object. It is thus associated with a large number M of eigenstates, with M scaling as the number of resolution cells contained in the object. To estimate the rank M of the object, one can observe the contrast of the image as a function of the number n of eigenstates considered for the imaging process. The images obtained for $n = 20, 250$, and 500 are displayed in Figures 11.13e–11.13g. For $n = 250$, the three bars of the USAF target are recovered nicely, and the comparison with the en face OCT image (Figure 11.13c) is striking. This experimental result demonstrates the benefit of our approach for deep tissue imaging. The importance of a correct determination of M can be highlighted by the images built from the first 20 and 500 eigenstates of $\mathbf{K^S_{r,r}}$ (see Figures 11.13e and 11.13g, respectively). On the one hand, considering too few eigenstates only provides a partial imaging of the field of view (Figure 11.13e). On the other hand, considering too many eigenstates blurs the image because the weakest eigenvalues are mainly associated with the MS background (Figure 11.13g).

11.3.5   Conclusion

In this second part, we presented several matrix approaches developed in optics to focus, detect or image in complex media. Similarly to acoustics, the analysis of the reflection matrix can be made preferentially in the focal plane or in the far-field depending on the experimental conditions. A double focusing matrix here appeared to offer the best performances for a discrimination between SS and MS photons. Notably, the high number of degrees of control offered by an SLM on optical waves enables ultra-deep imaging through scattering media. In particular, an imaging-depth limit of $22\ell_s$ is predicted for smart-OCT in biological tissues, hence drastically pushing back the fundamental MS limit in optical imaging (see Figure 11.14).

However, the use of a femtosecond laser for the recording of the reflection matrix is invasive and therefore not suitable for biomedical applications. Moreover, the scanning of the whole field-of-view at the input is time consuming. To circumvent these experimental issues, recent studies have paved the way toward a more efficient, full-field and noninvasive measurement of the reflection matrix. The point-to-point time-dependent Green's functions of a scattering medium can actually be deduced from the correlation of the wavefield reflected by the sample when it is placed under an incoherent white light illumination [30, 31].

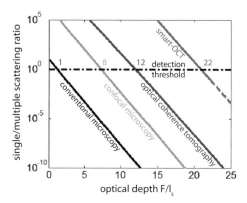

Figure 11.14 Imaging-depth limit in human soft tissues. This graph compares the SMRs expected for conventional microscopy, confocal microscopy, OCT, and smart-OCT as a function of the optical depth. These curves have been computed with parameters considered in [28]. From [28], reprinted with permission from AAAS.

11.4   References

[1] Fink M. Time reversal of ultrasonic fields. I. Basic principles. IEEE Transactions on Ultrasonics, Ferroelectrics, and Frequency Control. 1992;39(5):555–566.

[2] Campillo M, Paul A. Long-range correlations in the diffuse seismic coda. Science. 2003;299(5606):547–549.

[3] Shapiro NM, Campillo M, Stehly L, Ritzwoller MH. High-resolution surface-wave tomography from ambient seismic noise. Science. 2005;307(5715):1615–1618.

[4] Mosk AP, Lagendijk A, Lerosey G, Fink M. Controlling waves in space and time for imaging and focusing in complex media. Nature Photonics. 2012;6:283.

[5] Popoff SM, Lerosey G, Carminati R, Fink M, Boccara AC, Gigan S. Measuring the transmission matrix in optics: An approach to the study and control of light propagation in disordered media. Physical Review Letters. 2010;104(10):100601.

[6] Kim M, Choi Y, Yoon C, Choi W, Kim J, Park QH, et al. Maximal energy transport through disordered media with the implementation of transmission eigenchannels. Nature Photonics. 2012;6(9):581–585.

[7] Fink M, et al. Time-reversed acoustics. Scientific American. 1999;281(5):91–97.

[8] Prada C, Fink M. Eigenmodes of the time reversal operator: A solution to selective focusing in multiple-target media. Wave Motion. 1994;20(2):151–163.

[9] Prada C, Wu F, Fink M. The iterative time reversal mirror: A solution to self-focusing in the pulse echo mode. Journal of the Acoustical Society of America. 1991;90(2):1119–1129.

[10] Shahjahan S, Aubry A, Rupin F, Chassignole B, Derode A. A random matrix approach to detect defects in a strongly scattering polycrystal: How the memory effect can help overcome multiple scattering. Applied Physics Letters. 2014;104(23):234105.

[11] Aubry A, Derode A. Random matrix theory applied to acoustic backscattering and imaging in complex media. Physical Review Letters. 2009;102(8):084301.

[12] Aubry A, Derode A. Detection and imaging in a random medium: A matrix method to overcome multiple scattering and aberration. Journal of Applied Physics. 2009;106(4):044903.

[13] Freund I, Rosenbluh M, Feng S. Memory effects in propagation of optical waves through disordered media. Physical Review Letters. 1988;61(20):2328.

[14] Feng S, Kane C, Lee PA, Stone AD. Correlations and fluctuations of coherent wave transmission through disordered media. Physical Review Letters. 1988;61(7):834.

[15] Aubry A, Derode A. Multiple scattering of ultrasound in weakly inhomogeneous media: Application to human soft tissues. Journal of the Acoustical Society of America. 2011;129(1):225–233.

[16] Robert JL, Fink M. The time-reversal operator with virtual transducers: Application to far-field aberration correction. Journal of the Acoustical Society of America. 2008;124(6):3659–3668.

[17] Robert JL, Fink M. Green's function estimation in speckle using the decomposition of the time reversal operator: Application to aberration correction in medical imaging. Journal of the Acoustical Society of America. 2008;123(2):866–877.

[18] Mallart R, Fink M. Adaptive focusing in scattering media through sound-speed inhomogeneities: The van Cittert Zernike approach and focusing criterion. Journal of the Acoustical Society of America. 1994;96(6):3721–3732.

[19] Aubry A, de Rosny J, Minonzio JG, Prada C, Fink M. Gaussian beams and Legendre polynomials as invariants of the time reversal operator for a large rigid cylinder. Journal of the Acoustical Society of America. 2006;120(5):2746–2754.

[20] Robert JL, Fink M. The prolate spheroidal wave functions as invariants of the time reversal operator for an extended scatterer in the Fraunhofer approximation. Journal of the Acoustical Society of America. 2009;125(1):218–226.

[21] Montaldo G, Tanter M, Bercoff J, Benech N, Fink M. Coherent plane-wave compounding for very high frame rate ultrasonography and transient elastography. IEEE Transactions on Ultrasonics, Ferroelectrics, and Frequency Control. 2009;56(3):489–506.

[22] Booth MJ. Adaptive optical microscopy: The ongoing quest for a perfect image. Light: Science & Applications. 2014;3(4):e165.

[23] Vellekoop IM, Mosk A. Focusing coherent light through opaque strongly scattering media. Optics Letters. 2007;32(16):2309–2311.

[24] Popoff SM, Lerosey G, Fink M, Boccara AC, Gigan S. Image transmission through an opaque material. Nature Communications. 2010;1(6):1–5.

[25] Popoff SM, Aubry A, Lerosey G, Fink M, Boccara AC, Gigan S. Exploiting the time-reversal operator for adaptive optics, selective focusing, and scattering pattern analysis. Physical Review Letters. 2011;107:263901.

[26] Fujimoto JG, Puliafito C, Margolis R, Oseroff A, De Silvestri S, Ippen E. Femtosecond optical ranging in biological systems. Optics Letters. 1986;11(3):150–152.

[27] Kang S, Jeong S, Choi W, Ko H, Yang TD, Joo JH, et al. Imaging deep within a scattering medium using collective accumulation of single-scattered waves. Nature Photonics. 2015;9(4):253–258.

[28] Badon A, Li D, Lerosey G, Boccara AC, Fink M, Aubry A. Smart optical coherence tomography for ultra-deep imaging through highly scattering media. Science Advances. 2016;2(11):e1600370.

[29] Goy AS, Psaltis D. Digital confocal microscope. Optics Express. 2012;20(20):22720–22727.

[30] Badon A, Lerosey G, Boccara AC, Fink M, Aubry A. Retrieving time-dependent Green's functions in optics with low-coherence interferometry. Physical Review Letters. 2015;114(2):023901.

[31] Badon A, Li D, Lerosey G, Boccara AC, Fink M, Aubry A. Spatio-temporal imaging of light transport in highly scattering media under white light illumination. Optica. 2016;3(11):1160–1166.

Part V

Time Reversal, Optical Phase Conjugation

12 Wavefront-Engineered Optical Focusing into Scattering Media Using Ultrasound- or Perturbation-Based Guide Stars: TRUE, TRAP, SEWS, and PAWS

Xiao Xu, Cheng Ma, Puxiang Lai, and Lihong V. Wang

12.1   Introduction

Focusing light is essential in most optical imaging modalities. In (nearly) transparent media such as clear air or water, optical focusing is simple: light propagation is predictable by geometric and diffractive optics, and controllable by beam forming with lenses and other optical elements. In scattering media such as biological tissues, simple geometric focusing is distorted by elastic scattering that dominates the interaction between light and matter. Beyond one transport mean free path ($\sim$1 mm in human skin), light propagation is so randomized by multiple scattering that direct focusing becomes infeasible. This makes it difficult to "look" into or through a thick scattering medium.

For biological applications, focusing light inside instead of through a scattering medium is more useful. To this end, a variety of approaches have been explored. One approach is optical clearing, which artificially reduces scattering by introducing into a scattering medium optical clearing agents that match the refractive indices of the scatterers with their background [1]. This approach has limited effectiveness in suppressing scattering, and can undesirably alter the optical and physiological properties of the medium. Another approach is optical phase conjugation (OPC) with embedded light-emitting particles [2, 3], which suppresses the turbidity of a scattering medium in a two-pass procedure [4]. In the first pass, a laser beam is delivered into the turbid medium. The diffused light passing through the embedded particles is absorbed and reemitted at different wavelengths, and the wavefront of the frequency-shifted light is recorded after being multiply scattered inside the medium; in the second pass, the recorded wavefront is phase conjugated at the shifted frequency and sent back into the medium. In the ideal case of lossless propagation and full-aperture detection, the same multiple-scattering process is time reversed, and eventually the wavefront converges at the particle locations to form foci. A third approach, wavefront shaping (WFS), can actively adjust the wavefront of a beam to achieve light focusing inside [5] a turbid medium with an embedded or external guide star. The wavefront adjustment relies on a feedback mechanism that depends on the presence of a light-emitting guide star (e.g., a fluorescence bead) for the incident light to focus on.

In this chapter, we introduce several methods capable of focusing light deep inside scattering media noninvasively, without invasively implanting guide stars. The first method – time-reversed ultrasonically encoded (TRUE) light focusing – uses focused ultrasound as the guide star, and selectively phase-conjugates light whose frequency is acousto-optically shifted. The focal spot size is acoustic-diffraction-limited, and can be further reduced ([6], [7], and [8]). The second method – time-reversed adapted-perturbation (TRAP) light focusing – detects perturbation in the scattered light due to dynamic objects, and focuses by subsequent OPC. This method can reach optical-diffraction-limited focus because the focus tracks the dynamic object. The third method – ultrasonically encoded wavefront shaping (SEWS) – uses focused ultrasound as a guide star, which tags light via the acousto-optic effect as WFS feedback to reach an acoustic-diffraction-limited focal spot size. The last method – photoacoustically guided wavefront shaping (PAWS) – employs nonlinear photoacoustic signals as the feedback for WFS to achieve optical-diffraction-limited focusing.

12.2   Time-Reversed Ultrasonically Encoded (TRUE) Light Focusing

12.2.1   Principle

Time-reversed ultrasonically encoded (TRUE) optical focusing [9] uses ultrasonic modulation and optical phase conjugation of multiply scattered coherent light to dynamically focus light into a scattering medium in a two-pass configuration. In the first pass, a laser beam and a focused ultrasound beam are incident on a medium which is optically highly scattering and acoustically transparent, e.g., soft biological tissue. The ultrasound beam modulates or encodes the diffused photons inside the medium. The ultrasonically encoded light emanates from the ultrasonic focus inside the turbid medium, propagates out of the medium, where its wavefront is recorded on a phase-conjugate mirror (PCM). In the second pass, the wavefront is time reversed by OPC and converges back to the ultrasonic focus.

Since optical phase conjugation and the "guide star" concept are the two main building blocks of TRUE focusing, their principles are introduced first. Optical phase conjugation is a well-established method, whose validity is based on the validity of time-reversal in wave propagation. In a source-free lossless medium, the wave equation of the electric field vector $\nabla^2 \vec{E} - \frac{1}{c^2}\frac{\partial^2}{\partial t^2}\vec{E} = 0$ (where c is the wave speed in the medium and t is time) conserves its form under time reversal operation $\mathcal{T}$: $t \rightarrow -t$. Therefore, after phase-conjugate reflection, the wavefront retraces its previous evolution so exactly that distortions introduced by a scattering medium are reversed. In practice, the time-reversal operation is implemented through a process which includes recording the wavefront of the scattered light from a medium, replicating it, and sending it back. Unlike mirror reflection, which takes place at the speed of light, the phase-conjugate reflection typically takes much longer (usually >1 ms) to implement. For time-reversal to work effectively, the scattering medium must remain static during this period. Any movement of the scattering medium, such as the internal motion induced by blood vessels in biological tissue, can potentially degrade the quality of time reversal. Further discussions on this point will be presented later in the chapter.

The guide star can be either a light focus outside the scattering medium or an embedded light-emitting particle at the desired location inside the medium. In TRUE focusing, the guide star concept is extended to a virtual source [10], where an ultrasound beam modulates or encodes the diffused coherent light such that the ultrasound focus becomes the virtual source of the modulated photons. The mechanism of the ultrasonic encoding of the diffuse light has been modeled in the framework of diffuse wave spectroscopy [11, 12]. After a coherent laser beam is transmitted through a turbid medium, the electric field $E(t)$ of the multiply scattered light can be described as a superposition of the component plane waves $E_\rho(t)$ traveling through the different scattering paths ρ, i.e., $E_s(t) \sim \exp\left[i \sum_j \vec{K}_j \cdot \vec{r}_j(t)\right]$, and the autocorrelation function of the electric field $E(t)$ can be expressed as $G_1(\tau) = \int_0^\infty p(\rho)\langle E_\rho(t) E_\rho^*(t+\tau)\rangle d\rho$, where $\vec{K}_j$ and $\vec{r}_j(t)$ are the wave vector and position vector after the jth scattering event, and $p(\rho)$ is the probability of light traveling along the scattering path s. The ultrasonic encoding adds a phasemodulation term for the scattering paths that cross the ultrasound beam, $E_s(t) \sim \exp\left[im_s \sin(2\pi f_{US}t + \theta_s) + \varphi_s\right]$, which contains spectral components at the ultrasound frequency f_{US} and its harmonics nf_{US}, since $\exp\left[im_s \sin(2\pi f_{US}t + \theta_s) + \varphi_s\right] = \sum_{n=-\infty}^{+\infty} J_n(m_s) \exp\left[i(2\pi n f_{US}t + n\theta_s + \varphi_s)\right]$, where m_s is the modulation strength, φ_s and θ_s are the relative phase angles for the specific scattering path and modulation respectively, and $J_n(m_s)$ is the nth-order Bessel functions of m_s.

The mechanism of TRUE focusing and its effect inside a scattering medium can be illustrated with Monte Carlo simulation of the propagation of the diffused coherent light $E(f_0 - f_{US})$, the ultrasonically encoded light $E(f_0)$, and their conjugations $E^*(f_0 - f_{US})$ and $E^*(f_0)$, where f_0 denotes the laser frequency. In scattering medium such as soft biological tissues, the light–medium interaction is dominantly elastic scattering, and can be characterized by the scattering mean free path l and anisotropy g. For example, in human breast, $l \approx 0.1$ mm, $g \approx 0.9$ [13] Optical absorption is much less than scattering and is neglected here. At depths beyond one transport mean free path $l' = l/(1 - g)$, light propagation is sufficiently randomized. A photon is scattered ~ 70 times on average before exiting a scattering layer of thickness $L = 40l$. With increasing optical thickness, the light intensity of $E(f_0 - f_{US})$ and $E(f_0)$ decreased much more slowly than the ballistic light, as observed experimentally. The light that can be recorded and time-reversed by PCM is therefore predominantly multiple-scattered in thick media.

The trajectories of $E(f_0 - f_{US})$, $E(f_0)$, $E^*(f_0 - f_{US})$ and $E^*(f_0)$, shown in Figure 12.1, appeared to be random walks. However, owing to the deterministic nature of the medium at any instant, time reversal led to the convergence of $E^*(f_0 - f_{US})$

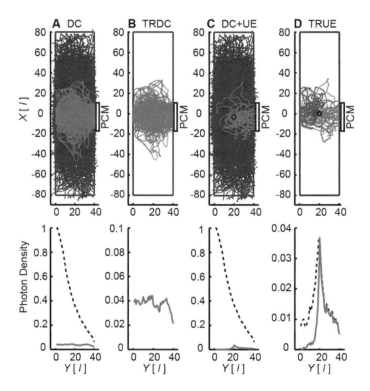

Figure 12.1 Monte Carlo simulated light propagation through tissue. Adapted by permission from Macmillan Publishers Ltd: Nature Photonics [9], copyright 2011.

and E^* (f_0). Without ultrasonic encoding, E^* ($f_0 - f_{US}$) converged to the incident location of E ($f_0 - f_{US}$). With ultrasonic encoding, E^* (f_0) converged to the ultrasonic focus instead, which was the source of E (f_0)

12.2.2   Experimental Implementation

To implement TRUE focusing experimentally (as shown schematically in Figure 12.2), the light from a long-coherence length laser is split into a sample beam E_S and two mutually conjugated plane wave reference beams E_R and E_R^*. Before propagating diffusively through the medium, E_S is spectrally tuned to $f_S = f_0 - f_{US}$ by acousto-optic modulators (AOM), where f_0 is the laser frequency and f_{US} is the frequency shift. A focused ultrasonic wave of frequency f_{US} traverses the medium and modulates the diffused light. The ultrasonically modulated light, having spectral components $f_+ = f_0$ and $f_- = f_0 - 2f_{US}$, can be regarded as emanating from a virtual source defined by the ultrasonic focus. Outside the medium, the diffused light E_S is holographically recorded in the presence of E_R onto a PCM, here a photorefractive $Bi_{12}SiO_{20}$ (BSO) crystal capable of dynamic recording. The only stationary hologram that can be recorded on the

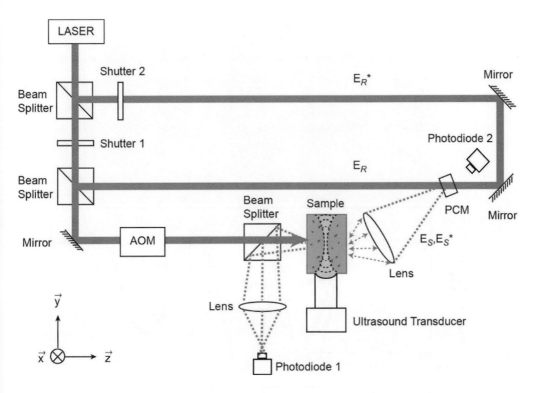

Figure 12.2 Schematic of the experimental setup for TRUE optical focusing. Adapted by permission from Macmillan Publishers Ltd: Nature Photonics [9], copyright 2011.

PCM is from the interference between E_R (f_0) and E_S (f_0), which have identical frequencies. Then, the hologram is read by E_R^* to generate a time-reversed copy of E_S (f_0), denoted as E_S^* (f_0). By reversibility, E_S^* (f_0) back-traces the trajectory of E_S (f_0) and converges to its virtual source, thereby achieving optical focusing into the scattering medium.

Direct visualization of TRUE focusing is not possible when the focus is located inside a turbid medium; instead, the following imaging experiment is conducted as a proof-of-concept demonstration. A phantom sample (Figure 12.3a) is laterally scanned along the x-axis, and four 1D images of the sample are acquired (Figure 12.3b). The first two are acquired without either AOM tuning or ultrasonic modulation. To form the first image – a "DC" image, E_S ($f_0 - f_{US}$) is detected by a photodiode at the PCM position. To form the second image – a "TRDC" image, E_S^* ($f_0 - f_{US}$) is transmitted back through the sample and detected by photodiode 1. To form the third image – a "UOT" image based on conventional ultrasound-modulated optical tomography (UOT) [14, 15], E_S (f_0) is spectrally filtered by the BSO and then detected by photodiode 2. To form the fourth image – a "TRUE" image, E_S^* (f_0) is transmitted back through the sample and detected by photodiode 1.

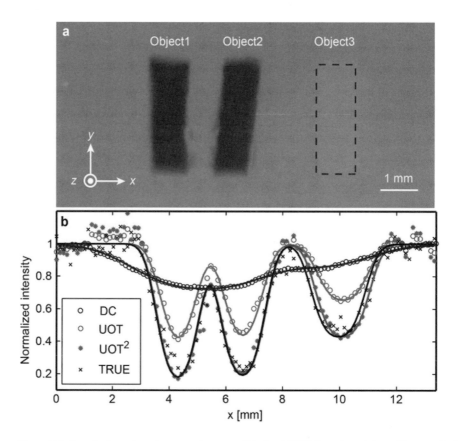

Figure 12.3 Results from imaging experiments validating TRUE focusing. Adapted by permission from Macmillan Publishers Ltd: Nature Photonics [9], copyright 2011.

The sample is a 10-mm thick tissue-mimicking phantom slab with three 1-mm wide blocks embedded in its mid-plane – two absorbing and one strongly scattering. Using TRUE to focus light at the mid-plane and photodiode 1 to detect $E_S^*(f_0)$ after it transmits through the phantom, one can reconstruct the image of the three targets with submillimeter resolution. Moreover, if light indeed converges to the ultrasonic focus, then the theoretical resolution should be $1/\sqrt{2}$ of the ultrasound focal width – a result of the light passing through the virtual source twice [9]. This prediction can be confirmed by comparing the TRUE image with the UOT image. In UOT, the light $E_S(f_0)$ passes through the virtual source just once, and the imaging resolution is determined by the ultrasound focal width. Therefore, in theory, the TRUE signal should be proportional to roughly the square of the UOT signal. The experimental results in Figure 12.3 also support this quadratic behavior.

An immediate application of TRUE focusing is fluorescence imaging at greater tissue depth than previously achievable [16]. Figure 12.4 shows a tomographic image of an 8-mm-thick turbid layer having a fluorescence labeled structure at its mid-plane. The two dyed objects cannot be distinguished by conventional fluorescence imaging because of the poor imaging resolution as a result of the diffusion of the excitation light. In

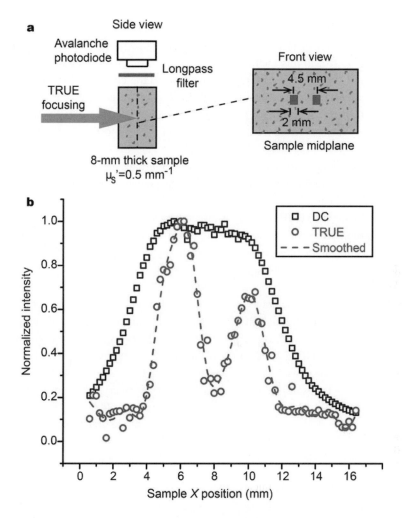

Figure 12.4 Fluorescence imaging with TRUE focusing. Modified from Ref. [16]. © Astro Ltd. Reproduced by permission of IOP Publishing. All rights reserved.

contrast, TRUE can focus the excitation light at depths greater than the diffusion limit, and elucidate the fine structure that is otherwise inaccessible.

12.2.3   Discussion

It is important to note that there are a number of different ways to implement TRUE focusing experimentally. Most notable are two methods for optical phase conjugation. In the above example, an analog device – a photorefractive BSO crystal – is used as the phase-conjugate mirror. The wavefront of $E_S(f_0)$ is holographically engraved in the BSO when $E_S(f_0)$ and $E_R(f_0)$ illuminate the crystal. The conjugate wavefront $E_S^*(f_0)$ is holographically generated when the BSO crystal is illuminated by $E_R^*(f_0)$. Alternatively, OPC can be performed digitally with the combination of a digital camera (e.g., an sCMOS camera), a computer, and a spatial light modulator (SLM) such as a liquid crystal on silicon spatial light modulator (LCOS-SLM) or a digital micromirror device (DMD) [17, 18]. In this method, the camera records the interference patterns between $E_S(f_0)$ and $E_R(f_0)$, and transfers the patterns to the computer for computation of the wavefronts of $E_S(f_0)$ and $E_S^*(f_0)$. Then the computer controls the spatial light modulator to display the desired wavefront mask (e.g., a phase delay pixel map), which transforms the wavefront of the incident light $E_R^*(f_0)$ to that of the phase conjugate $E_S^*(f_0)$

Because of their different technical implementations, analog and digital TRUE focusing methods have distinct advantages and disadvantages, which make each suitable for different applications.

The efficacy of TRUE focusing is affected by several factors – most important are the number of controls N for generating the wavefront $E_S^*(f_0)$, the speed of the OPC procedure, and the attainable optical power or energy in $E_S^*(f_0)$. The most significant figure of merit to characterize the quality of focusing in practice is the peak-to-background ratio (PBR). Theoretically, $PBR \propto N/M$, where N is the number of independently controlled segments on the wavefront, and M is the number of speckle grains inside the focal region [5]. In digital TRUE focusing, N is limited by the available pixels on the SLM and digital camera, e.g., $N = 1920 \times 1080$ when both operate in high definition mode. In analog TRUE focusing, N is the number of speckles recorded onto the PCM. When the speckle density is kept constant on the PCM, N scales with the recording area of the PCM – for example, a large area photorefractive crystal or polymer. So N ranges from 1.25×10^7 to 2.5×10^7 for a recording density of 5000–10,000 pixels/mm^2 and a recording area of 2500 mm^2 [19], an order of magnitude higher than the N of the SLM in high definition mode.

The speed of TRUE focusing is another important parameter, especially for *in vivo* biomedical imaging, when optical time reversal has to be executed within the speckle decorrelation time, which is usually on the order of milliseconds [20, 21]. A photorefractive PCM can record $E_S(f_0)$ and generate $E_S^*(f_0)$ with a response time as fast as 1 ms [22, 23]. In a digital system, time reversal comprises a cascade of processes, including wavefront capture by the camera, image transfer from the camera to the computer and from the computer to the SLM, and image processing for solving the wavefront of $E_S(f_0)$ and $E_R^*(f_0)$. Therefore, digital TRUE focusing is significantly slower than its analog counterpart [24].

The third parameter is the time reversal gain. To measure how much light is brought to the focus in $E_R^*(f_0)$ compared to the original ultrasonically encoded light $E_S(f_0)$, the time reversal gain can be defined as $G = \left|E_S^*(f_0)\right|^2/|E_S(f_0)|^2$. Obviously, G can be computed either in terms of the attainable optical power or optical energy. When a photorefractive material is used as the PCM, the dynamic holographic readout simultaneously erases the existing hologram on the PCM, so that G is usually much less than unity in terms of optical energy. It is, however, possible to achieve a gain greater than 1 in terms of optical power. A $33,000\times$ optical power gain has been obtained when the photorefractive PCM is read out by a high-power short-pulsed laser whose temporal width is below the hologram decay time in the PCM [25]. On the other hand, in a digital TRUE focusing system, the generation of the phase-conjugated light by an SLM is physically decoupled from the holographic recording device (CCD or CMOS camera). Thus the power (energy) of the phase-conjugated light is proportional to the power (energy) of the reference beam illuminating the SLM, and is therefore restricted only by the damage threshold of the SLM. Optical gains of as high as 5×10^4 can be achieved in a digital system [24].

12.3 Time-Reversed Adapted-Perturbation (TRAP) Light Focusing

12.3.1 Introduction

To focus light into a scattering medium, one can first illuminate the medium using light with a regular wavefront. The diffuse light passing through the desired focus is then tagged and selectively detected. Time reversing the tagged light forms a focus at the target location.

The focusing problem is illustrated in Figure 12.5. For simplicity, we represent light as a scalar field by ignoring polarization. Suppose light is to be focused inside a scattering medium, and we denote $\mathbf{E}_F$ (a column vector whose elements correspond to speckle grains within the focus) to be the field at that focal point. The field on an outside detection plane is denoted as $\mathbf{E}_D$ (a column vector whose elements correspond to speckle grains on the detector). Field propagation from the focal point to the detection plane is characterized by a transmission matrix $\mathbf{T}$ [26, 27]. Due to reciprocity, the transmission

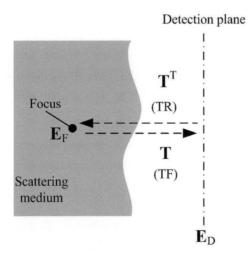

Figure 12.5 Definition of parameters in TRAP focusing. The fields at the focus and on the detection plane are denoted as $\mathbf{E}_F$ and $\mathbf{E}_D$, respectively. Transmission through the scattering medium is described by matrices $\mathbf{T}$ and $\mathbf{T}^T$ for time-forward (TF) and time-reversed (TR) propagation.

matrix describing light propagation from the detection plane to the focus is $\mathbf{T}^{\mathrm{T}}$, where T in the superscript denotes matrix transposition.

In TRUE focusing (see the preceding Section), light-tagging is realized by ultrasonic modulation at the target location, using focused ultrasound with a frequency f_{US}. The ultrasonic modulation imposes frequency shifts of $\pm f_{\mathrm{US}}$ on the light passing through the focus. On the detection plane, lock-in detection (such as phase-stepping holography [28]) is performed to selectively measure in parallel the phases of the frequency-shifted light at all pixels.

If a tissue's optical properties undergo nonperiodic changes, such as those caused by blood flow, neural activity, and heartbeat, time-reversed adapted-perturbation (TRAP) optical focusing [29] can be used to convert the nonperiodic and nonemissive changes to guide stars. TRAP focusing proceeds as follows: First, the scattered fields before and after the refractive index change are measured. Second, the two fields are digitally subtracted to obtain a differential field. Finally, the differential field is phase conjugated. As long as the background medium is stable during the procedure, the phase-conjugated differential field will focus to the index-perturbed location.

Mathematically, the field E_{F} at the point of perturbation, and the field distribution $\mathbf{E}_{\mathrm{D}}$ on the detection plane are linked by the transmission matrix $\mathbf{T}$ (dimensions: N by M, where N and M are the number of speckle grains on the spatial light modulator and within the focus, respectively) by

$$\mathbf{E}_{\mathrm{D}}^{(1)} = \mathbf{T}\mathbf{E}_{\mathrm{F}}^{(1)} + \mathbf{B}, \tag{12.1}$$

$$\mathbf{E}_{\mathrm{D}}^{(2)} = \mathbf{T}\mathbf{E}_{\mathrm{F}}^{(2)} + \mathbf{B}. \tag{12.2}$$

In the above equations, the superscripts denote different times at which the fields are measured, and $\mathbf{B}$ is a time-invariant background. Subtracting Eq. (12.1) from (12.2), we obtain a similar equation for the differential field:

$$\Delta\mathbf{E}_{\mathrm{D}} = \mathbf{T}\Delta\mathbf{E}_{\mathrm{F}}, \tag{12.3}$$

where $\Delta\mathbf{E}_{\mathrm{D}} = \mathbf{E}_{\mathrm{D}}^{(2)} - \mathbf{E}_{\mathrm{D}}^{(1)}$, $\Delta\mathbf{E}_{\mathrm{F}} = \mathbf{E}_{\mathrm{F}}^{(2)} - \mathbf{E}_{\mathrm{F}}^{(1)}$. The field at the point of perturbation resulting from phase conjugation of $\Delta\mathbf{E}_{\mathrm{D}}$ can be found by

$$\mathbf{E}_{\mathrm{F}}^{(\mathrm{TR})} = \mathbf{T}^{\mathrm{T}}(\Delta\mathbf{E}_{\mathrm{D}})^{*}. \tag{12.4}$$

Here, taking the complex conjugate of $\Delta\mathbf{E}_{\mathrm{D}}$ is equivalent to time reversal [4]. By inserting Eq. (12.3) into (12.4) and assuming $\mathbf{T}^{\dagger}\mathbf{T} \approx 1$ († denotes conjugate transpose), one gets

$$\mathbf{E}_{\mathrm{F}}^{(\mathrm{TR})} \approx (\Delta\mathbf{E}_{\mathrm{F}})^{*}. \tag{12.5}$$

Note that in the Born approximation, the perturbation affects the local field only, thus $\Delta\mathbf{E}_{\mathrm{F}}$ is zero everywhere except at the index-perturbed location.

The mechanism of TRAP focusing can be understood as follows. On the detection plane, the field is composed of two parts due to linearity of the scattering process in the Born approximation: a time variant part originating from the perturbation, and a stable background due to the rest of the medium. Thus, subtracting the fields recorded at two instants cancels the stable background and recovers a differential field that originates

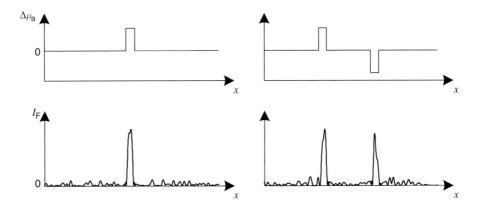

Figure 12.6 Plots of the absorption perturbations $\Delta\mu_a$ and the corresponding time-reversed light intensity distributions along a line passing through the focus (left column) or foci (right column). The position along the line is denoted by x. Left column: a single, time-variant absorber. Right: a moving absorber.

from the perturbed region. As a result, time reversing this differential field converges energy exclusively to the perturbed region. Note that, in practice, a portion of the total energy will be lost to the stationary medium as a background due to partial phase conjugation [18].

In Figure 12.6, two scenarios of TRAP focusing are shown. Here, it is assumed that the absorption (characterized by absorption coefficient μ_a) of the perturbed region is time variant. The first row of Figure 12.6 illustrates the change in attenuation coefficient $\Delta\mu_a$ near the perturbation. The left column shows the situation for a stationary absorber with varying absorption. Because the absorption increases over time, $\Delta\mu_a$ is positive on the perturbed spot and zero elsewhere. The right column shows an alternative situation where a light-absorbing object moves from right to left. In this case $\Delta\mu_a$ is positive (negative) at the new (old) object location, and is zero elsewhere. The panels on the bottom are the simulated light intensity distributions near the focus. Notice that backgrounds exist in these plots due to partial phase conjugation, which is accounted for by assuming the elements of the transmission matrix to be independent and identically distributed circular Gaussian random variables.

12.3.2   Implementation

For *in vivo* applications, TRAP focusing should meet the following three criteria:

(1) High energy gain: Defined by the energy ratio between the focusing and probing light in the focus. Obviously, an energy gain greater than unity is desired.
(2) High speed: The focusing procedure must be sufficiently fast to accommodate the background medium change. Mathematically, the transmission matrix **T** should not change during the focusing procedure, including field measurements and time reversal.

(3) Large number of control elements: As shown previously, a background exists if the phase conjugation is partial. The focus would vanish if the PBR is reduced to unity. Since the focusing PBR is proportional to the total number of control elements [18], having a spatial light modulator (SLM) with as many control elements as possible is key to achieving high focusing quality.

For criterion (1), digital SLMs are preferred to analog crystals. To date, despite a power gain exceeding unity [25], no analog scheme has achieved a truly high energy gain. However, using a digital SLM, achieving an energy gain much greater than unity is easy [18].

Criterion (2) deserves more scrutiny. First, the field measurement must be accomplished within a timescale much shorter than the field decorrelation time associated with the perturbation. Second, the background scattering medium is generally time variant as well. According to the mathematical derivation in the preceding section, the transmission matrix $\mathbf{T}$ must stay almost unchanged for TRAP to properly focus. But, if $\mathbf{T}$ varies, the field changes contributed by the desired perturbation and the host medium will couple and become inseparable by subtraction. The decorrelation time of biological tissue depends on the specific application, and, due to blood flow, can be less than a millisecond [23]. Accordingly, the entire focusing procedure should take less than a millisecond to adapt to the tissue decorrelation.

From the discussion of criterion (1), we conclude that the digital implementation of TRAP focusing is more practical. For digital measurement of fields, phase-stepping holography is commonly used. Despite its simplicity and robustness, this method has an evident drawback – low speed. As shown in the first row of Figure 12.7, each field measurement requires four camera exposures. In most cases, the camera's frame rate becomes the ultimate speed-limiting factor. For example, if the camera's frame rate is 50 Hz, a field measurement may take 20 ms × 3 = 60 ms. The long measurement time

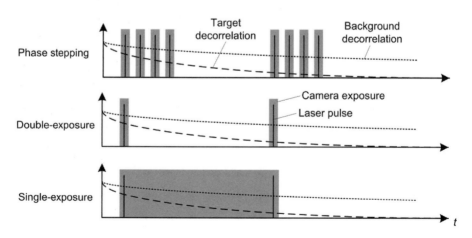

Figure 12.7 Temporal sampling schemes for TRAP focusing. Top row: phase-stepping holography. Middle row: double-exposure TRAP focusing. Bottom row: single-exposure TRAP focusing.

poses great challenges for applications where perturbations are fast. This consideration justifies the necessity of single-camera-exposure field measurement. When each camera exposure encompasses a single laser shot, then the effective exposure time is defined by the laser pulse width if the duration of the laser pulse is much shorter than the camera exposure time (a condition easily met by Q-switched lasers with nanosecond pulse widths). Accordingly, if one of the single-exposure field measurement techniques introduced below is employed, the duration of the measurement will be reduced to a time scale defined by the laser pulse width.

Before discussing single-exposure schemes, we must emphasize the role of binary light modulation (either amplitude or phase) in accelerating the focusing process. First, binary modulation needs only 1 bit per pixel, as opposed to multiple bits in grayscale modulation. Such efficient bit usage implies a minimized data transfer time between the image processing unit and the SLM. Second, binary modulation is compatible with fast SLMs, such as digital micromirror devices (DMDs) or ferroelectric liquid crystal SLMs; thus the actuation time can be significantly shorter than that afforded by nematic liquid-crystals. Binary modulation schemes do not excessively sacrifice the focal PBR, contrary to what one might expect. The PBRs of binary amplitude and binary phase modulations are 20% and 40% of that of phase-only modulation, respectively, sufficient for many applications [30]. Given these advantages of binary modulation, the following discussion will concentrate on single-exposure field measurement methods specifically tailored for it.

To understand the principles of single-exposure field measurements, we begin by expressing the total light intensity impinging on a camera pixel (which represents a single spatial mode and is in *italic* to represent a scalar quantity) as

$$I_{\text{tot}}^{(1)} = \left| E_{\text{D}}^{(1)} + E_{\text{R}} e^{-ik_\theta x} \right|^2, \tag{12.6}$$

$$I_{\text{tot}}^{(2)} = \left| E_{\text{D}}^{(2)} + E_{\text{R}} e^{-i(k_\theta x + \varphi)} \right|^2. \tag{12.7}$$

In Eqs. (12.6) and (12.7), the superscripts "(1)" and "(2)" label the laser pulse number. Here we consider a general case where the sample and reference beams subtend an angle $\theta = \sin^{-1}(k_\theta/k_0)$ (the sample beam propagates along $+z$, k_0 is the wavenumber, and k_θ is the wave vector of the reference beam along x). A phase shift, φ, whose value is either 0 or π, is applied to the reference beam in the second exposure. Our goal is to reconstruct the conjugate differential field $\Delta E_{\text{D}}^* = (E_{\text{D}}^{(2)} - E_{\text{D}}^{(1)})^*$ from $I_{\text{tot}}^{(1)}$ and $I_{\text{tot}}^{(2)}$.

If we assume $\varphi = 0$ and expand the right-hand side (R.H.S.) of Eqs. (12.6) and (12.7) and subtract the results, we obtain (without loss of generality, we set the phase of the reference to 0):

$$\Delta I_{\text{tot}} = \left| E_{\text{D}}^{(2)} \right|^2 - \left| E_{\text{D}}^{(1)} \right|^2 + \Delta E_{\text{D}}^* |E_{\text{R}}| e^{-ik_\theta x} + \Delta E_{\text{D}} |E_{\text{R}}| e^{ik_\theta x}. \tag{12.8}$$

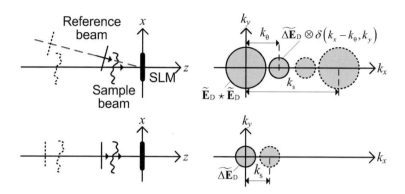

Figure 12.8 Spatial sampling schemes (left column) and their corresponding frequency spectra (right column). Upper row: off-axis sampling. Lower row: on-axis sampling.

The two-dimensional (2D) Fourier transform of Eq. (12.8) comprises multiple terms:

$$\widetilde{\Delta I}_{tot}(k_x, k_y) = \widetilde{E_D^{(2)}}(k_x, k_y) \star \widetilde{E_D^{(2)}}(k_x, k_y) - \widetilde{E_D^{(1)}}(k_x, k_y) \star \widetilde{E_D^{(1)}}(k_x, k_y)$$
$$+ |E_R| \widetilde{\Delta E_D^*}(-k_x, -k_y) \otimes \delta(k_x + k_\theta, k_y) + |E_R| \widetilde{\Delta E_D}(k_x, k_y) \otimes \delta(k_x - k_\theta, k_y)$$

$$(12.9)$$

In Eq. (12.9), the tilde sign ($\sim$) denotes the Fourier transform of the corresponding quantity, the symbol $\star$ represents the (2D) autocorrelation integral, and the symbol $\otimes$ denotes the (2D) convolution operation. $\delta(k_x, k_y)$ is the 2D Dirac delta function in the spatial-frequency domain. From Eq. (12.9), one concludes that if the angle θ is sufficiently large, the last two terms (in which the phase of ΔE_D is encoded) can be separated from the autocorrelation terms in the spatial-frequency domain. This separation is shown in Figure 12.8, upper row. The left panel shows the experimental arrangement, where the reference beam interferes with the sample beam on the surface of the SLM at an oblique angle. The right panel illustrates the frequency domain signal in Eq. (12.9). Assuming that the signal ΔE_D has a highest frequency B_S (the radius of the small circle), the big circle representing the autocorrelation term ($\tilde{E}_D \star \tilde{E}_D$) occupies twice the signal bandwidth. To separate the cross terms from the autocorrelation terms, $k_\theta > 3B_S$ is required. Consequently, to avoid aliasing due to sampling by the camera pixels, the sampling frequency k_S should satisfy $k_S > 8B_S$ (shown as the circles with dashed boundaries).

From the above analysis, the phase map of the desired signal ΔE_D^* can be directly inferred from a single-shot, off-axis interferogram. At a fixed spatial sampling frequency, k_s, the maximum signal frequency of the sample beam is $B_S = k_s/8$. Following the phase retrieval, a binary modulation pattern can be readily obtained from

$$p = \begin{cases} 1, & \text{for } -\frac{\pi}{2} \leq \arg\left(\Delta E_D^*\right) < \frac{\pi}{2} \\ 0, & \text{for } -\pi \leq \arg\left(\Delta E_D^*\right) < -\frac{\pi}{2} \text{ or } \frac{\pi}{2} \leq \arg\left(\Delta E_D^*\right) < \pi \end{cases} \quad (12.10)$$

When applying the pixel assignment scheme in Eq. (12.10) to an amplitude-modulation-based device (such as a DMD), binary amplitude modulation is achieved. In practice, binary phase modulation can be derived from the binary amplitude modulation by adding a zero-order block [30].

Aside from the off-axis scheme, one can employ an on-axis scheme for multiple benefits, as discussed below. In the case of on-axis sampling, Eq. (12.8) reduces to

$$\Delta I_{\text{tot}} = \left| E_{\text{D}}^{(2)} \right|^2 - \left| E_{\text{D}}^{(1)} \right|^2 + 2\,|\Delta E_{\text{D}}|\,|E_{\text{R}}|\cos\Phi, \tag{12.11}$$

where we use Φ for the phase of ΔE_{D}. Under the condition $2\,|\Delta E_{\text{D}}|\,|E_{\text{R}}| \gg \left| E_{\text{D}}^{(2)} \right|^2 - \left| E_{\text{D}}^{(1)} \right|^2$, one readily gets:

$$\Delta I_{\text{tot}} \approx 2\,|\Delta E_{\text{D}}|\,|E_{\text{R}}|\cos\Phi. \tag{12.12}$$

According to Eq. (12.12), a binarization scheme equivalent to Eq. (12.10) is to set a pixel value to unity if $\Delta I_{\text{tot}} \geq 0$, or to zero otherwise. The advantages of on-axis sampling are now evident: First, compared to the off-axis scheme, it is computationally simpler, which can effectively accelerate the focusing process. Second, it significantly increases the signal spatial bandwidth, as shown in Figure 12.8, lower row. Since no spectral separation is required as in the off-axis scheme, the maximum frequency of the signal is $B_S = k_s/2$, limited only by the Nyquist–Shannon sampling theorem. Compared to its off-axis counterpart ($B_S = k_s/8$), the on-axis scheme has $4\times$ more bandwidth, thus $16\times$ more spatial mode sampling capacity. This advantage is important because the focusing PBR is proportional to the total number of spatial modes that one system can handle.

However, a major concern of employing the on-axis scheme is contamination from the squared terms on the R.H.S. of Eq. (12.11). More specifically, if the condition $2\,|\Delta E_{\text{D}}|\,|E_{\text{R}}| \gg \left| E_{\text{D}}^{(2)} \right|^2 - \left| E_{\text{D}}^{(1)} \right|^2$ is violated, the terms on the R.H.S. of the above inequity act as noise in the binarization criterion in Eq. (12.12), which ultimately results in errors in the binary modulation pattern and consequently a reduced PBR. We will come back to this problem later.

The focusing procedure described above, whether employing off-axis or on-axis spatial sampling, involves two camera exposures, whose temporal diagram is depicted in Figure 12.7, second row. Consequently, the speed of the entire focusing cycle is ultimately limited by the camera frame rate. Ideally, a TRAP hologram can be formed within a single camera exposure, decoupling the focusing speed from the camera's frame rate. This is possible when the temporal sampling method shown in the bottom row of Figure 12.7 is applied. In this paradigm, a phase shift of π radians is introduced to every other pulse in the pulse train. As a result, one single camera exposure covering two such sign-flipped pulses automatically subtracts the sample field before digitalization. Mathematically, instead of subtracting Eqs. (12.7) and (12.6), we add them and set $\varphi = \pi$ to get

$$I_{\text{tot}}^{(1)} + I_{\text{tot}}^{(2)} = \left| E_{\text{D}}^{(1)} \right|^2 + \left| E_{\text{D}}^{(2)} \right|^2 + 2\,|E_{\text{R}}|^2 + \Delta E_{\text{D}}^*\,|E_{\text{R}}|\,e^{-ik_\theta x} + \Delta E_{\text{D}}\,|E_{\text{R}}|\,e^{ik_\theta x}. \tag{12.13}$$

Notice that the above calculation is performed automatically on each camera pixel, rather than digitally processed in the computer. If k_θ is sufficiently large, we can separate ΔE_{D}^* from the rest of the terms on the R.H.S. of Eq. (12.13) in the spatial-frequency domain, as discussed previously. Alternatively, one may choose to employ the on-axis scheme for simpler computation. To do so, a constant value of $2|E_R|^2 + \left\langle \left|E_D^{(1)}\right|^2 + \left|E_D^{(2)}\right|^2 \right\rangle = 2\left(|E_R|^2 + \langle|E_D|^2\rangle\right)$, where $\langle a \rangle$ represents the time average of a, is first digitally subtracted from Eq. (12.13). Then, if we assume $2|\Delta E_D||E_R| \gg \left|E_D^{(1)}\right|^2 + \left|E_D^{(2)}\right|^2 - \left\langle \left|E_D^{(1)}\right|^2 + \left|E_D^{(2)}\right|^2 \right\rangle$, the binarization scheme described by Eqs. (12.12) and (12.10) can be subsequently applied. In both on-axis and off-axis cases, the binary hologram can be derived from a single camera exposure. The hologram formation time is determined by the laser pulse repetition rate, rather than the camera frame rate. Since laser repetition rates can be made much higher than camera frame rates, the single-exposure sampling scheme endows TRAP focusing with the highest speed.

In conclusion, double-exposure TRAP focusing enables fast field measurement, and single-exposure focusing can further achieve a hologram synthesis time defined by the laser pulse interval. When used in conjunction with on-axis spatial sampling, binary light modulation can be achieved with simpler data processing, less data to transfer, and consequently faster response. In real practice, one has to ensure that the terms encoding ΔE_D^* in Eqs. (12.8) and (12.13) dominate the rest of the terms, which can be achieved by amplification by a strong reference field E_R. In general, this condition is satisfied when the intensity ratio between the sample and reference beams is much less than unity. Violating this condition results in binarization error and consequently reduced focal PBR. Computer simulations were performed to show the dependence of the focal PBR on the beam intensity ratio I_s/I_R, with results shown in Figure 12.9. In the simulation, there are 100 spatial modes on the focal and detection planes, and the elements of the transmission matrix are independent and identically distributed circular Gaussian random variables. For verification, the curves from analytical models are co-plotted. Detailed theoretical analysis can be found in [30].

12.3.3   Application

One key feature of TRAP is its ability to focus light onto endogenous optical contrast agents, such as red blood cells (RBCs). Since moving RBCs are intrinsic guide stars and are exclusively confined within blood vessels, TRAP focuses light onto the entire vasculature. This feature is potentially useful in photoacoustic tomography of blood vessels [31] and treatment of port wine stains [32]. Living tissues exhibit fast decorrelation due to dynamic objects (red blood cells in this case), and slow decorrelation contributed by the relatively stationary background. TRAP focusing must accurately detect the former, while robustly adapting to the latter.

In the tissue-mimicking phantom experiment shown in Figure 12.10, a phantom was constructed from two ground-glass diffusers as scattering media with a silicone tube (300 μm inner diameter) containing flowing blood placed in between. The mimic blood vessel was completely invisible outside of the ground glass. To mimic tissue decorrelation with controlled correlation time, the scattering layers were mounted onto a motorized stage to allow translation at controlled speeds. The speckle correlation time of the scattered light was measured as a function of the media's movement speed

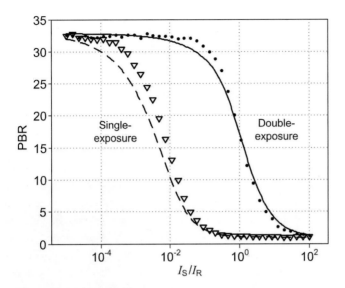

Figure 12.9 Focal PBR plotted against the beam ratio I_S/I_R for single- and double-exposure TRAP focusing. Points and lines are from simulation and analytical models, respectively. Adapted with permission from [30], Optica.

and subsequently fitted as an exponential decay, allowing precise control over the media's correlation time. Bovine blood diluted in a 1:8 volume ratio with phosphate-buffered saline solution was pumped through the silicone tube at a constant flow speed of ~1 mm/s. As the blood flowed, the far-field speckle pattern kept flickering, indicating perturbation by the moving scatterers. While double-exposure, on-axis TRAP focusing was performed at a repetition rate of 7 Hz, the scattering layers were translated, superimposing another dynamic speckle pattern, though varying less rapidly. The cross-sectional intensity profile along the x-axis is plotted in the left panel on the bottom row of Figure 12.10. For comparison, the figure shows the focal light intensity distributions corresponding to stationary and moving scattering layers. Despite an intensity drop, the focus is clearly visible when the media decorrelation time is 0.28 s. The focal light intensity pattern for stationary diffusers is shown on the right.

Another similar technology, termed Time Reversal by Analysis of Changing wavefronts from Kinetic targets (TRACK) [33], was used for flow cytometry in a reflective configuration. In the demonstration shown in Figure 12.11, a microfluidic channel was hidden behind a ground-glass diffuser. Before the cytometric measurement, polystyrene beads flowing inside the channel were used to generate dynamic backscattering of light. Using an off-axis double-exposure sampling scheme, a time-reversed focus was created inside the microfluidic channel, and the position and illumination intensity of the focus were maintained during the successive flow measurements (assuming stationary background medium). Fluorescence beads were then introduced into the same channel, passing through the focus at a constant flow rate. The resulting time-varying fluorescence signal was recorded by a photomultiplier tube (PMT) positioned outside the

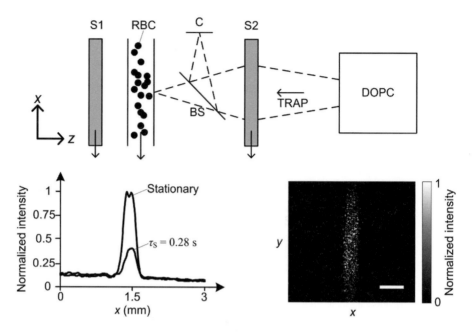

Figure 12.10 TRAP focusing into blood-flowing tube. Upper row: schematic. BS, non-polarizing beam splitter; C, camera for focus inspection; DOPC, digital optical phase conjugation; RBC, red blood cell; S, scattering layer. Lower row: cross-sectional intensity profiles for focusing into stationary and moving scattering media (left) and 2D focal light intensity distribution for stationary scattering media (right). Scale bar: 0.5 mm.

scattering medium on the digital optical phase conjugation side. The fluorescence signal peaked whenever a particle passed through the focus, enabling the cytometric count shown on the bottom row of Figure 12.11.

12.3.4   Discussion

In essence, TRAP and TRACK focusing are time reversal technologies that employ novelties, i.e., transient changes in a pattern from one time frame to the next, in a scattering medium as guide stars. Novelty detection is accomplished by differentiating the scattered field. Differentiation was proven effective in novelty detection in the past. For example, an optical novelty filter installed in an imaging system ensures that only dynamic scenes are imaged [34], which is achieved by an interferometer or a two-beam-coupling photorefractive crystal. The device is balanced and switched to an "off" state by quasi-static scenes, and can be immediately turned on when novelty in the tranquil background disturbs the balance. In addition, subtraction of the scattered light intensity or field was used to improve laser speckle imaging of blood perfusion [35] and assist optical particle tracking [36]. In the microwave regime, an imaging method was proposed for tracking moving objects in clutter by subtracting multistatic data matrices and performing subsequent time reversal [37]. Another relevant

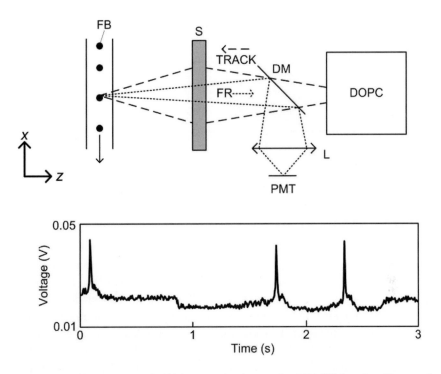

Figure 12.11 Flow cytometry behind a scattering layer using TRACK focusing. Upper: schematic. DM, dichroic mirror; DOPC, digital optical phase conjugation; FB, Fluorescent beads; L, lens; PMT, photomultiplier tube; S, scattering layer. Lower: time trace of the PMT output as fluorescent beads flow though the microfluidic channel. Adapted with permission from [33], Optica.

technology, holographic interferometry, reveals novelty in a scene by reconstructing linearly superimposed holograms recorded at different times [38].

TRAP and TRACK are fully compatible with the fast DOPC module developed by Wang *et al.* [39]. The module implements a 1920×1080 resolution DMD interfaced with an FPGA board to achieve optical time reversal with a latency of 5.3 ms. The marriage of the two technologies for *in vivo* deep tissue applications is under way. The focusing technology discussed in this Section is not restricted to endogenous contrast agents. Exogenous agents with controllable optical properties, such as magnetomotive particles [40], voltage-sensitive dyes [41], and photo-switchable dyes and proteins [42, 43], can be readily incorporated to make TRAP and TRACK focusing more powerful.

12.4   Ultrasonically Encoded Wavefront Shaping (SEWS) and Photoacoustically Guided Wavefront Shaping (PAWS)

12.4.1   Introduction

Another way to correct scattering-induced phase distortion and focus light beyond one transport mean free path in scattering media is through wavefront shaping techniques

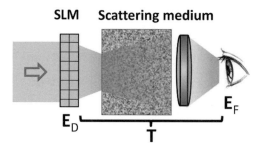

Figure 12.12 The concept of wavefront compensation using a spatial light modulator (SLM). The incident and perceived electric fields are denoted as matrices $\mathbf{E}_D$ and $\mathbf{E}_F$, respectively. Transmission through the scattering medium is described by matrix $\mathbf{T}$.

[5, 26, 44–46]. While related to optical phase conjugation (OPC)-based time-reversal methods, wavefront shaping has a different principle of operation. In optical time-reversal, focusing is achieved by holographically recording and then phase conjugating the diffused light wavefront originating from an internal guide star. This guide star can be provided in various ways, including radiation from embedded nanoparticles [47] or fluorescent beads [2], or light encoded by focused ultrasound (TRUE [9, 16, 18, 24, 48, 49] and TROVE [6]) or adapted-perturbation (TRAP [29, 30]). In wavefront shaping, focusing is achieved by optimizing the phase distribution of the illuminating light using spatial light modulators (SLMs), which are phase-programmable arrays.

Here we briefly review the principle of WFS. A detailed account of WFS can be found in earlier chapters (Chapters 8–10). The concept is illustrated in Figure 12.12. As introduced in the preceding section, the optical field E_F at a targeted position behind or inside a scattering medium can be expressed by

$$E_F = TE_D \tag{12.14}$$

where E_D is the light field on the SLM plane external to the scattering medium, T is the transmission matrix of the medium and the optical system, and M and N are the degrees of freedom of E_F and E_D, respectively. According to Eq. (12.14), the field at one point (labeled by index m) within the focus, $E_{F,m}$, can be expressed as a linear combination of the fields coming from different segments of the SLM:

$$E_{F,m} = \sum_{n=1}^{N} |t_{mn}| \, e^{i\varphi_{mn}} \, |E_{D,m}| \, e^{i\varphi_n}. \tag{12.15}$$

Here $|E_{D,m}|$ and φ_n are respectively the amplitude and phase modulation by one segment n on the SLM, while $|t_{mn}|$ and φ_{mn} are respectively the amplitude and phase scrambling due to scattering. For thick tissue or tissue-like scattering media, φ_{mn} is randomly distributed with phase wrapping between 0 and 2π. If there is no phase modulation or only random phase modulation by the SLM, fields from different segments add up randomly, forming a speckle pattern at the targeted position. But if the SLM is programmed with a modulation that cancels out the scattering-induced phase scrambling, i.e., $\varphi_n = -\varphi_{mn}$, light fields from different segments add up in phase at the targeted position, forming a bright optical focus [26].

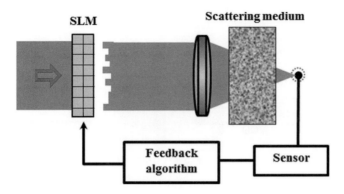

Figure 12.13 Implementation of iterative wavefront shaping.

Such optimized SLM phase compensation can be achieved by using iterative search algorithms [50, 51] or by directly measuring the phase distortion for each SLM element and applying the reverse [27]. Figure 12.13 illustrates how iterative wavefront shaping can be implemented. First, we define a targeted position, creating a guide star. A signal from the guide star is used as feedback to an iterative optimization algorithm, which searches for the phase pattern that maximizes the guide star signal, and hence the light delivered to the targeted position, forming an optical focus. Theoretically, the intensity at the focus can be improved by [5]

$$R = \frac{\pi}{4}\frac{N-1}{M} + 1, \tag{12.16}$$

where N is the number of independently controlled segments on the SLM, and M is the number of speckle grains within the focal region.

To obtain the feedback signal from the guide star, physical access to the targeted position is usually required, whether directly by using a photodiode or CCD camera pixel to detect *in situ* speckles [26, 50–52], or indirectly by using fluorescent probes [5, 45, 46]. In biomedical applications, however, using such probes within tissue is usually undesirably invasive and sometimes potentially toxic. Furthermore, in both cases, optical focusing is restricted to fixed positions defined by the physical probes. To overcome these issues, researchers have recently proposed focused ultrasound mediation, including ultrasonically encoded (UE) light [53] or photoacoustic (PA) signals [54–57], as the feedback. Both UE light and PA sensing are noninvasive and nontoxic, and they allow for dynamic focusing by simply translating the ultrasonic focus. Moreover, in biological tissue, ultrasound has orders of magnitude weaker scattering than light does [58, 59], enabling ultrasonically limited resolution at great depths in biological tissue. If nonlinearity is exploited, the resolution can be even finer.

12.4.2 Ultrasonically Encoded Wavefront Shaping (SEWS)

Using UE light as the guide star, Tay *et al.* proposed a technique called ultrasonically encoded wavefront shaping (SEWS) [53] to focus light within scattering media noninvasively.

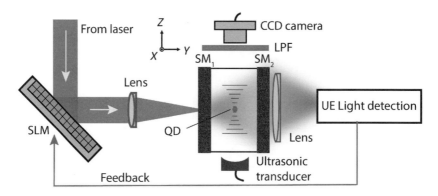

Figure 12.14 Schematic of the optical setup of SEWS. For a proof-of-principle demonstration, two ground glass diffusers (SM_{1-2}) simulate a diffuse sample. A clear gelatin layer inserted between the diffusers enables clear optical access from the top for a CCD camera and acoustic coupling for the ultrasound field. The ultrasound (5 cycles at 6 MHz in the pilot study) propagates and encodes the scattered light in the clear layer. To visualize the optical focus, we inserted a quantum dot (QD) bar in the ultrasound focal volume. A CCD camera positioned at the top acquired the fluorescent emissions. A long-pass filter (LPF) prevented non-fluorescent light from entering the CCD camera. *XYZ* are the system coordinates (*Y* is the optical illumination direction, and *Z* is the ultrasound propagation axis).

Figure 12.14 is a schematic of this technique, where the incident signal beam from a laser, initially planar, is spatially tailored by a phase-only liquid crystal on silicon (LCoS) SLM. Within the sample, the signal beam is multiply scattered and phase encoded [60] by the applied focused short-pulsed ultrasound field (6 MHz central frequency, five cycles). The resulting diffuse signal beam, including both ultrasonically encoded and nonencoded portions, is collected outside the sample by lenses or a fiber bundle, and directed into a detection scheme, such as a photorefractive interferometer [14, 61–64]. With this scheme, the ultrasound-induced phase encoding can be converted into a light intensity offset that is proportional to the intensity of the scattered light traversing the ultrasonic focus. This offset signal, denoted as UE light, is preamplified, digitized, and subsequently fed back to the optimization algorithm.

Figure 12.15 shows results from an optimization procedure when the SLM was divided into 20 × 20 independently controlled segments for practical reasons, and a genetic algorithm [51] was used to obtain the optimum SLM phase pattern that maximized the UE light intensity. For each iteration, the highest UE signal value was recorded (Figure 12.15a). After 600 iterations, it was found that the UE light intensity was increased by ~11 times over the uniform or randomized value (Figure 12.15b), indicating a corresponding increase of the *in situ* light intensity within the ultrasound focus. Such light focusing can also be visualized directly, as shown in Figure 12.15c and d, by using a CCD camera mounted above the target to capture the fluorescence excitations from the quantum dot (QD) bar in real time during the optimization procedure. As seen, when the uniform and the randomized patterns were displayed, the fluorescence emissions were relatively evenly distributed, suggesting that light had been fully scrambled by the first diffuser. With the optimized pattern, however, the peak fluorescence

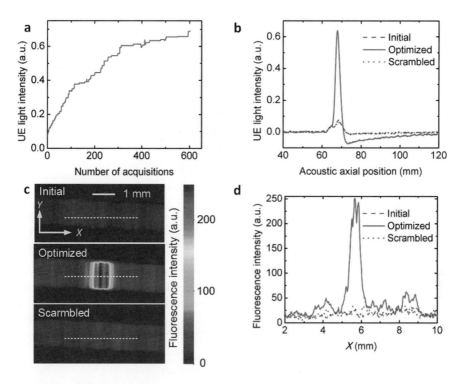

Figure 12.15 SEWS optimization results. (**a**) Improvement of the UE light intensity as the genetic algorithm ran. Key algorithm parameters included a population size of 30, an initial mutation rate of 0.12, and a final mutation rate of 0.01. (**b**) Measured UE light signals when a uniform phase pattern (dashed curve), an optimized phase pattern (solid curve), and a randomized phase pattern (dotted curve) were displayed on the SLM, respectively. (**c**) Fluorescence images of the QD bar captured by the CCD camera when the uniform, optimized, and randomized phase patterns used in (**b**) were displayed on the SLM. (**d**) Cross-sectional fluorescence intensity profiles along the three white dotted lines in (**c**). Adapted by permission from Macmillan Publishers Ltd: Scientific Reports [53], copyright 2014. A black and white version of this figure appears in some formats. For the color version, please refer to the plate section.

intensity was increased by ∼10 times, in good agreement with the increase of the UE light intensity. Moreover, the full width at half maximum (FWHM) size of the focal spot was ∼700 μm, reasonably close to the transverse FWHM of the ultrasound focus (400 μm). These results clearly demonstrate that by using SEWS one can noninvasively focus light down to an ultrasonically determined focal spot within a scattering medium. It must be noted that fluorescence from embedded molecules in this study was used only to visualize the optimization procedure, but not as feedback to the iterative algorithm as in Ref. [2].

12.4.3 Photoacoustically Guided Wavefront Shaping (PAWS)

Another way to remotely and noninvasively guide the wavefront shaping optimization process uses PA signals as the feedback. This technique, called photoacoustically guided wavefront shaping (PAWS), was first proposed by Kong *et al.* [54], and then further improved and extended by different groups, such as Chaigne *et al.* [55, 65], Conkey *et al.* [66], and Lai *et al.* [56, 57].

The photoacoustic effect describes the conversion of photon energy into acoustic waves [31]. First, transient photon energy is absorbed by molecules, causing a rapid temperature rise. Due to thermal expansion, the heat is converted to ultrasonic waves, which can be detected by an ultrasound transducer. Generally, the detected PA signal amplitude can be expressed by

$$PA = k \iint A\,(x,y)\,\Gamma(T)\mu_a F\,(x,y)\,dxdy, \qquad (12.17)$$

where k is a constant, $A(x, y)$ is the acoustic detection sensitivity distribution normalized as $\iint A(x, y)dxdy = 1$, Γ is the Grueneisen parameter (usually a function of the medium's temperature), μ_a is the optical absorption coefficient of the medium, and $F(x, y)$ is the optical energy density (or fluence) within the ultrasonic focus. Eq. (12.17) shows clearly that the PA signal amplitude is linearly proportional to the optical fluence F, as long as the temperature and the absorption coefficient of the targeted region remain relatively constant. Therefore, if we use PA signals as the feedback for wavefront shaping (Figure 12.16), maximizing the PA signals' amplitude leads to maximized optical fluence within the ultrasonic focus [54] and accordingly an optical focus within a highly scattering medium.

Using PA signals as the internal guide star has many advantages. For example, it allows deep penetration; it can pinpoint the targeted position accurately. Moreover, it is noninvasive and label-free. However, since the energy is maximized within the ultrasonic focus, the optical focal spot size is typically acoustic diffraction-limited. For example, it is ~50 um for a 50 MHz focused transducer. Within this region, many optical modes may exist, resulting in a large M and a limited peak intensity enhancement ratio, according to Eq. (12.16).

Many microscopic applications benefit from tighter focusing, such as an optical diffraction-limited focal spot with an intense peak intensity. Researchers have recently shown this tighter focus is feasible if nonlinear, instead of linear, PA signals are used as the feedback [56].

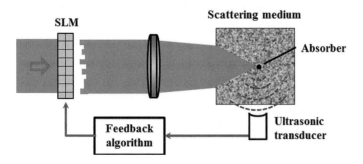

Figure 12.16 The principle of PAWS.

Recall that the PA signal is proportional to the Grueneisen parameter Γ, which increases with local temperature for many materials, including water and blood [67, 68]. Within the thermal confinement time, the temperature rise due to the absorption of light lingers and changes Γ accordingly, creating what is called the Grueneisen relaxation effect. A dual-pulse excitation approach has been developed to exploit this phenomenon: The first laser pulse heats the sample, generating a PA signal denoted as PA_1, whose amplitude can be written as

$$PA_1 = k \iint A(x, y) \Gamma_0 \mu_a F(x, y) \, dxdy, \tag{12.18}$$

where Γ_0 is the Grueneisen parameter at the base temperature. Note that the absorption of the first pulse also increases the value of the Grueneisen parameter. Before the heat dissipates completely, a second laser pulse is fired when the medium temperature is

$$\Gamma = \Gamma_0 + \Delta\Gamma = \Gamma_0 + \eta\Gamma_0'\mu_a F, \tag{12.19}$$

where η is a constant coefficient that converts the absorbed optical energy density into temperature rise, and Γ_0' is the first-order derivative of the Grueneisen parameter with respect to the temperature at T_0. Therefore, the amplitude of the PA signal generated by the second laser pulse is

$$PA_2 = k \iint A(x, y) [\Gamma_0 + \eta\Gamma_0'\mu_a F]\mu_a F(x, y) \, dxdy. \tag{12.20}$$

The difference between these two PA signal amplitudes is then

$$\Delta PA = PA_2 - PA_1 = k\eta\Gamma_0'\mu_a^2 \iint A(x, y) F^2(x, y) \, dxdy. \tag{12.21}$$

Since ΔPA is proportional to the square of the optical fluence, it is referred to as the "nonlinear PA signal amplitude." Let us consider two simple cases showing how such nonlinearity can achieve narrower focusing. In the first case, there are multiple speckle grains within the ultrasound focus, and in the second case there is only a single speckle grain. In both cases, the total energy within the ultrasound focus is the same. As we know, these two cases will generate equal linear PA signals. But if we consider the square of the fluence, the single speckle grain in the second case will generate higher nonlinear PA signals. Theoretically, all the optical energy will be concentrated on a single speckle grain, maximizing the nonlinear PA signal amplitude that can be used as feedback for optimization.

Experiments with the setup illustrated in Figure 12.17 were performed to validate the theory. A 532 nm pulsed laser (10 ns; 0–30 kHz; pulse energy <0.2 mJ) was the illumination source; a portion of its output was sampled by a photodiode to compensate for fluctuations in the laser output energy. The other beam was then reflected off the SLM, which was divided into 192 × 108 independent blocks. The experimental sample was a ground-glass diffuser and a thin layer of blood. The diffuser scattered light completely, and the blood absorbed light and generated PA signals. The optical fluence on the blood layer was about 0.1 mJ/cm^2 before optimization, and the speckle grain size was estimated to be around 5 μm. The blood was covered by a layer of plastic film

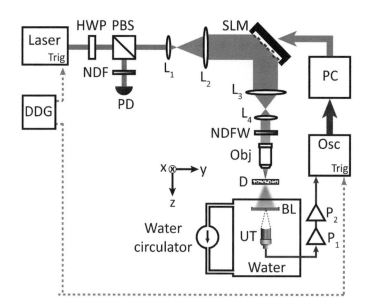

Figure 12.17 Schematic diagram of an experimental PAWS setup. BL, blood layer; D, ground glass diffuser; DDG, digital delay generator; HWP, half-wave plate; L_{1-4}, lenses; NDF, neutral density filter; NDFW, neutral density filter wheel; Obj, microscopic objective ($10\times$, 0.25 NA); Osc, oscilloscope; P_{1-2}, signal amplifiers; PBS, polarizing beam splitter; PC, computer; PD, photodiode; SLM, spatial light modulator; Trig, trigger signal; UT, ultrasonic transducer; *XYZ*, coordinate system Adapted by permission from Macmillan Publishers Ltd: Nature Photonics [56], copyright 2015.

and immersed in water to sonically couple it with the generated acoustic waves, which were detected by a 50 MHz focused ultrasonic transducer. The signal detected by the transducer was amplified, digitized, and sent to a computer, where a genetic algorithm controlled the updating of the phase pattern on the SLM. Initially it was planned to directly take the difference of two PA signal amplitudes as the feedback. But in a pilot study it was found that the laser source, even at maximum output, still could not provide sufficient intensity to generate reliable nonlinear PA signals when the blood was positioned far behind the diffuser. Accordingly, the researchers carried out the experiment in two stages (Figure 12.18). In the first stage (linear PAWS), regular single-pulse PA signals were used as the feedback to increase the optical fluence within the ultrasonic focal region. In the second stage (nonlinear PAWS), two optical pulses separated by 40 µs (well within the thermal confinement time of 189 µs under the experimental conditions) were fired every 20 ms. The difference between the two PA signals was then used as the feedback for optimization, which eventually led to single-speckle-scale optical focusing.

Experimental results from the both optimization stages are shown in Figure 12.19. Initially, a random phase pattern was displayed on the SLM. Under an optical fluence of 0.1 mJ/cm^2, a very small PA signal was recorded, taking an average over eight traces (Figure 12.19a). As the genetic algorithm-based optimization proceeded, the PA signal

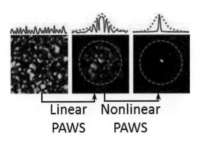

Figure 12.18 The two-stage optimization procedure. The dashed circles represent the acoustic focal region. Adapted by permission from Macmillan Publishers Ltd: Nature Photonics [56], copyright 2015.

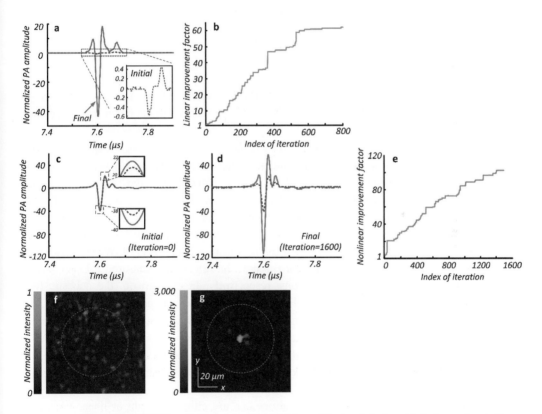

Figure 12.19 PAWS optimization results. (**a**) Linear PA signals before (dotted curve) and after (solid curve) linear optimization. (**b**) Linear PA signal improvement factor versus iteration index. (**c–d**) PA signal pairs (dotted curve for the first, and solid curve for the second) before (**c**) and after (**d**) nonlinear optimization. (**e**) Nonlinear PA signal improvement factor versus iteration index. (**f–g**) Speckle patterns at the ultrasonic focal plane before (**f**) and after (**g**) the two PAWS optimization procedures. The dashed circles represent the measured ultrasonic focal region (50 MHz, −6 dB). Adapted by permission from Macmillan Publishers Ltd: Nature Photonics [56], copyright 2015.

amplitude increased with the index of iteration (Figure 12.19b), until after 800 iterations the feedback almost plateaued. At that point, a much stronger PA signal was obtained (Figures 12.19a and 12.19b), one whose amplitude was about 60 times higher than that of the initial PA signal. This result suggests that the fluence within the ultrasonic focal region had been increased to 6 mJ/cm^2, which was sufficient to generate reliable nonlinear PA signals and allowed moving onto the second (nonlinear) optimization stage.

In the second stage, initially the two PA signals did not differ much (Figure 12.19c). The amplitude of the second PA signal was only slightly higher than that of the first PA signal, suggesting low nonlinearity. Nevertheless, their difference was used as the feedback for optimization, and Figure 12.19e shows how the nonlinear signal increases as the algorithm proceeds. After ~1500 iterations, the nonlinear PA signal amplitude was increased by ~100 times. Accordingly, the amplitude of the second PA signal became ~160% higher than that of the first PA signal (Figure 12.19d). Note that the increases in nonlinearity and peak fluence are proportional. Therefore, the total peak optical fluence improvement from two optimization stages was $60 \times 100 = 6000$ times, which is a really exciting improvement. For example, a 6000 times improvement in the SNR would make an imaging system much more sensitive and potentially able to image much deeper. Or, if you are doing laser surgery, you could damage targeted tissue much more efficiently, even in deep regions.

The effect of optical focusing and the peak intensity enhancement factor can be confirmed from the speckle pattern images (Figures 12.19f and 12.19g) at the ultrasonic focal plane, which were captured by a CCD camera after the optimization procedures with the blood layer moved away. Before optimization or when a random phase pattern was displayed on the SLM, a random speckle pattern was seen, with many speckle grains (~169 through estimation) within the ultrasonic focus (Figure 12.19f, profiled by the dashed circle). After the linear and nonlinear PAWS optimizations, only a single bright spot was seen (Figure 12.19g), which measured 5.1 μm and 7.1 μm along the X and Y directions, respectively. In comparison, the acoustic focal region width was 65 μm (FWHM), about 10 times larger than the optical focal spot size. Clearly, single speckle grain optical focusing in scattering media has been achieved.

12.4.4   Summary

We have shown in this Section that one can use either UE light or linear PA signals to noninvasively guide optical wavefront shaping to achieve an efficient acoustically diffraction-limited optical focus inside scattering media. To break through the acoustic resolution limit, Lai *et al.* [56] recently used nonlinear photoacoustic signals, based on the Grueneisen relaxation effect, as the feedback for iterative optimization, achieving optically diffraction-limited optical focusing with a superior peak fluence gain (~6000 times) in scattering media. Such an intense and highly confined optical focus in scattering media can benefit many micrometer-scale optical applications, especially if the optimization speed can be improved significantly. Such an improvement would enable biomedical applications in tissue, where optical speckles decorrelate within milliseconds due to physiological motions such as blood flow and aspiration.

12.5   References

[1] Tuchin, V. V., *A clear vision for laser diagnostics (review)*. IEEE Journal of Selected Topics in Quantum Electronics, 2007. **6**(13): pp. 1621–1628.

[2] Vellekoop, I. M., M. Cui, and C. Yang, *Digital optical phase conjugation of fluorescence in turbid tissue*. Applied Physics Letters, 2012. **101**(8): 081108.

[3] Hsieh, C.-L., et al., *Imaging through turbid layers by scanning the phase conjugated second harmonic radiation from a nanoparticle*. Optics Express, 2010. **18**(20): pp. 20723–20731.

[4] Yaqoob, Z., et al., *Optical phase conjugation for turbidity suppression in biological samples*. Nature Photonics, 2008. **2**(2): pp. 110–115.

[5] Vellekoop, I. M., et al., *Demixing light paths inside disordered metamaterials*. Optics Express, 2008. **16**: pp. 67–80.

[6] Judkewitz, B., et al., *Speckle-scale focusing in the diffusive regime with time reversal of variance-encoded light (TROVE)*. Nature Photonics, 2013. **7**(4): pp. 300–305.

[7] Yang, Q., et al., *Time-reversed ultrasonically encoded optical focusing using two ultrasonic transducers for improved ultrasonic axial resolution*. Journal of Biomedical Optics, 2013. **18**(11): 110502.

[8] Si, K., R. Fiolka, and M. Cui, *Breaking the spatial resolution barrier via iterative sound-light interaction in deep tissue microscopy*. Scientific Reports, 2012. **2**: 748.

[9] Xu, X., H. Liu, and L. V. Wang, *Time-reversed ultrasonically encoded optical focusing into scattering media*. Nature Photonics, 2011. **5**(3): pp. 154–157.

[10] Atlan, M., et al., *Evidence of a millimeter size virtual light source for acousto-optical tomography in scattering media using a chirp modulation of the ultrasound*. 2003.

[11] Pine, D. J., et al., *Dynamical correlations of multiply scattered light*. Scattering and Localization of Classical Waves in Random Media, 1990. **8**: p. 312.

[12] Wang, L. V., *Mechanisms of ultrasonic modulation of multiply scattered coherent light: an analytic model*. Physical Review Letters, 2001. **87**(4): 043903.

[13] Srinivasan, S., et al., *Interpreting hemoglobin and water concentration, oxygen saturation, and scattering measured in vivo by near-infrared breast tomography*. Proceedings of the National Academy of Sciences, 2003. **100**(21): pp. 12349–12354.

[14] Ramaz, F., et al., *Photorefractive detection of tagged photons in ultrasound modulated optical tomography of thick biological tissues*. Optics Express, 2004. **12**(22): pp. 5469–5474.

[15] Xu, X., et al., *Photorefractive detection of tissue optical and mechanical properties by ultrasound modulated optical tomography*. Optics Letters, 2007. **32**(6): pp. 656–658.

[16] Lai, P., et al., *Focused fluorescence excitation with time-reversed ultrasonically encoded light and imaging in thick scattering media*. Laser Physics Letters, 2013. **10**(7): 075604.

[17] Cui, M., and C. Yang, *Implementation of a digital optical phase conjugation system and its application to study the robustness of turbidity suppression by phase conjugation*. Optics Express, 2010. **18**(4): pp. 3444–3455.

[18] Wang, Y. M., et al., *Deep-tissue focal fluorescence imaging with digitally time-reversed ultrasound-encoded light*. Nature Communications, 2012. **3**: p. 928.

[19] Hariharan, P., *Basics of Holography*. 2002: Cambridge University Press.

[20] Lev, A. and B. Sfez, In vivo *demonstration of the ultrasound-modulated light technique.* JOSA A, 2003. **20**(12): pp. 2347–2354.

[21] Draijer, M., et al., *Review of laser speckle contrast techniques for visualizing tissue perfusion.* Lasers in Medical Science, 2009. **24**(4): pp. 639–651.

[22] Solymar, L., D. J. Webb, and A. Grunnet-Jepsen, *The Physics and Applications of Photorefractive Materials.* 1996: Clarendon Press.

[23] Liu, Y., et al., *Optical focusing deep inside dynamic scattering media with near-infrared time-reversed ultrasonically encoded (TRUE) light.* Nature Communications, 2015. **6**: 5904.

[24] Si, K., R. Fiolka, and M. Cui, *Fluorescence imaging beyond the ballistic regime by ultrasound-pulse-guided digital phase conjugation.* Nature Photonics, 2012. **6**(10): pp. 657–661.

[25] Ma, C., X. Xu, and L. V. Wang, *Analog time-reversed ultrasonically encoded light focusing inside scattering media with a 33,000× optical power gain.* Scientific Reports, 2015. **5**: 8896

[26] Vellekoop, I. M., and A. P. Mosk, *Focusing coherent light through opaque strongly scattering media.* Optics Letters, 2007. **32**(16): pp. 2309–2311.

[27] Popoff, S. M., et al., *Measuring the transmission matrix in optics*: An approach to the *study and control of light propagation in disordered media.* Physical Review Letters, 2010. **104**(10): 100601.

[28] Yamaguchi, I., and T. Zhang, *Phase-shifting digital holography.* Optics Letters, 1997. **22**(16): pp. 1268–1270.

[29] Ma, C., et al., *Time-reversed adapted-perturbation (TRAP) optical focusing onto dynamic objects inside scattering media.* Nature Photonics, 2014. **8**(12): pp. 931–936.

[30] Ma, C., et al., *Single-exposure optical focusing inside scattering media using binarized time-reversed adapted perturbation.* Optica, 2015. **2**(10): pp. 869–876.

[31] Wang, L. V. and S. Hu, *Photoacoustic tomography*: In vivo *imaging from organelles to organs.* Science, 2012. **335**(6075): pp. 1458–1462.

[32] Chen, J. K., et al., *An overview of clinical and experimental treatment modalities for port wine stains.* Journal of the American Academy of Dermatology, 2012. **67**(2): pp. 289–304.

[33] Zhou, E. H., et al., *Focusing on moving targets through scattering samples.* Optica, 2014. **1**(4): pp. 227–232.

[34] Anderson, D. Z., J. Feinberg, and D. M. Lininger, *Optical tracking novelty filter.* Optics Letters, 1987. **12**(2): pp. 123–125.

[35] Liu, R., J. Qin, and R. K. Wang, *Motion-contrast laser speckle imaging of microcirculation within tissue beds* in vivo. Journal of Biomedical Optics, 2013. **18**(6): 060508.

[36] Miccio, L., et al., *Particle tracking by full-field complex wavefront subtraction in digital holography microscopy.* Lab Chip, 2014. **14**(6): pp. 1129–1134.

[37] Fouda, A. E., and F. L. Teixeira, *Imaging and tracking of targets in clutter using differential time-reversal techniques.* Waves in Random and Complex Media, 2012. **22**(1): pp. 66–108.

[38] Brooks, R., L. Heflinger, and R. Wuerker, *9A9-Pulsed laser holograms.* Quantum Electronics, 1966. **2**(8): pp. 275–279.

[39] Wang, D., et al., *Focusing through dynamic tissue with millisecond digital optical phase conjugation.* Optica, 2015. **2**(8): pp. 728–735.

[40] Jin, Y., et al., *Multifunctional nanoparticles as coupled contrast agents.* Nature Communications, 2010. **1**: p. 41.

[41] Peterka, D. S., H. Takahashi, and R. Yuste, *Imaging voltage in neurons.* Neuron, 2011. **69**(1): pp. 9–21.

[42] Rust, M. J., M. Bates, and X. Zhuang, *Sub-diffraction-limit imaging by stochastic optical reconstruction microscopy (STORM)*. Nature Methods, 2006. **3**(10): pp. 793–796.

[43] Patterson, G. H., and J. Lippincott-Schwartz, *A photoactivatable GFP for selective photo-labeling of proteins and cells*. Science, 2002. **297**(5588): pp. 1873–1877.

[44] Vellekoop, I. M., and A. P. Mosk, *Universal optimal transmission of light through disordered materials*. Physical Review Letters, 2008. **101**(12): 120601.

[45] van Putten, E. G., A. Lagendijk, and A. P. Mosk, *Optimal concentration of light in turbid materials*. Journal of the Optical Society of America B, 2011. **28**(5): pp. 1200–1203.

[46] Aulbach, J., et al., *Spatiotemporal focusing in opaque scattering media by wave front shaping with nonlinear feedback*. Optics Express, 2012. **20**(28): pp. 29237–29251.

[47] Hsieh, C.-L., et al., *Digital phase conjugation of second harmonic radiation emitted by nanoparticles in turbid media*. Optics Express, 2010. **18**(12): pp. 12283–12290.

[48] Lai, P., et al., *Reflection-mode time-reversed ultrasonically encoded (TRUE) optical focusing into turbid media*. Journal of Biomedical Optics, 2011. **16**(8): 080505.

[49] Lai, P., et al., *Time-reversed ultrasonically encoded (TRUE) optical focusing in biological tissue*. Journal of Biomedical Optics, 2012. **17**(3): 030506.

[50] Vellekoop, I. M., and A. P. Mosk, *Phase control algorithms for focusing light through turbid media*. Optics Communications, 2008. **281**(11): pp. 3071–3080.

[51] Conkey, D. B., et al., *Genetic algorithm optimization for focusing through turbid media in noisy environments*. Optics Express, 2012. **20**(5): pp. 4840–4849.

[52] Cui, M., *Parallel wavefront optimization method for focusing light through random scattering media*. Optics Letters, 2011. **36**(6): pp. 870–872.

[53] Tay, J. W., et al., *Ultrasonically encoded wavefront shaping for focusing into random media*. Scientific Reports, 2014. **4**: p. 3918.

[54] Kong, F., et al., *Photoacoustic-guided convergence of light through optically diffusive media*. Optics Letters, 2011. **36**(11): pp. 2053–2055.

[55] Chaigne, T., et al., *Controlling light in scattering media noninvasively using the photo-acoustic transmission-matrix*. Nature Photonics, 2014. **8**(1): pp. 58–64.

[56] Lai, P., et al., *Photoacoustically guided wavefront shaping for enhanced optical focusing in scattering media*. Nature Photonics, 2015. **9**(2): pp. 126–132.

[57] Lai, P., et al., *Optical focusing in scattering media with photoacoustic wavefront shaping (PAWS)*. Proc. SPIE, 2014. **8943**: 894318.

[58] Yao, G., and L. V. Wang, *Theoretical and experimental studies of ultrasound-modulated optical tomography in biological tissue*. Applied Optics, 2000. **39**(4): pp. 659–664.

[59] Prince, J. L., and J. M. Links, *Medical Imaging: Systems and Signals*. 2006: Pearson Prentice Hall.

[60] Wang, L. V., *Mechanisms of ultrasound modulation of multiply scattered coherent light: An analytic model*. Physical Review Letters, 2001. **87**(4): 043903.

[61] Murray, T. W., et al., *Detection of ultrasound-modulated photons in diffuse media using the photorefractive effect*. Optics Letters, 2004. **29**(21): pp. 2509–2511.

[62] Lai, P., R. A. Roy, and T. W. Murray, *Quantitative characterization of turbid medium using pressure contrast acousto-optic imaging*. Optics Letters, 2009. **34**(18): pp. 2850–2852.

[63] Lai, P., et al., *Real time monitoring of high-intensity focused ultrasound lesion formation using acousto-optic sensing*. Ultrasound in Medicine and Biology, 2011. **37**(2): pp. 239–252.

[64] Suzuki, Y., et al., *High-sensitivity ultrasound-modulated optical tomography with a photorefractive polymer*. Optics Letters, 2013. **38**(6): pp. 899–901.

[65] Chaigne, T., et al., *Light focusing and two-dimensional imaging through scattering media using the photoacoustic transmission matrix with an ultrasound array.* Optics Letters, 2014. **39**(9): pp. 2664–2667.

[66] Conkey, D. B., et al., *Super-resolution photoacoustic imaging through a scattering wall.* Nature Communications, 2015. **6**: 8902.

[67] Wang, L. V., and H.-I. Wu, *Biomedical Optics: Principles and Imaging.* 2007: John Wiley.

[68] Yao, J., et al., *Absolute photoacoustic thermometry in deep tissue.* Optics Letters, 2013. **38**(24): pp. 5228–5231.

13 Transmission Matrix Correlations

Roarke Horstmeyer, Ivo M. Vellekoop, and Benjamin Judkewitz

13.1 Introduction

Focusing light through strongly scattering media was long believed to be impossible. However, even though scattering is extremely complex, the scattering process is linear and deterministic – meaning that it can be controlled and utilized. To focus light into any point within or across biological tissue, we need to know the correct input wavefront (Chapter 5). Finding that wavefront is nontrivial, but a number of approaches addressing this challenge have been demonstrated over the past few years. When the goal is to focus across scattering media, the correct wavefront can be obtained in three ways: iterative optimization [1], phase conjugation [2], and measuring the scattering transformation [3, 4, 5]. In many cases, however, the goal is to image inside a scattering sample, e.g. during *in vivo* imaging experiments, and there is no direct optical access to the target plane. In such cases, nonlinear [6], fluorescent, kinematic [7, 8], acousto-optic [9, 10, 11] and photo-acoustic [12, 13] guide stars can be used to find the right input wavefront [14] (also see Chapter 8).

However, the above list of guidestar techniques determine the optimal wavefront to focus at just a single target location at a time. For imaging, we need to be able to scan the focus to a number of different locations, and thus need to know the correct wavefront for each of these targeted spots. One way to summarize this collection of wavefronts is through what is commonly referred to as the transmission matrix (TM). When light is viewed as a wave, scattering can be described as a linear transformation of an input wavefront (e.g. at the tissue surface) to an output wavefront (at the target plane within the tissue). Because this input-output process is linear, it can be represented by a matrix equation (see Chapter 5). The input (plane A) and output (plane B) wavefronts can each be represented by vectors $U(x_a)$ and $U(x_b)$, and the so-called transmission matrix $T_x = T(x_b, x_a)$ defines the transformation between the two planes: $U(x_b) = T(x_b, x_a)U(x_a)$. Here, we will limit our attention to one transverse spatial dimension x for simplicity, but U and T are generally a function of both x and y. Owing to its discrete nature, the transmission matrix is especially amenable to experimental observation. If the 2D input and output planes were sampled at, for example, 1000-by-1000 = 10^6 pixels (this could be a 500 μm field-of-view at an optical resolution of 0.5 μm), the transmission matrix would have 10^{12} elements. This example illustrates the overwhelming scale of the challenge faced by scattering media imaging approaches: If all 10^{12} TM elements had to be measured, imaging through scattering media in real time would be practically impossible.

Of course, any statistical a priori information about the TM would help reduce its complexity and make it easier to measure within a finite time window. Luckily, the TM for biological media (e.g., tissue) can often exhibit a large amount of redundancy and inter-element correlation. In this chapter, we examine three specific forms of correlation. We discuss strategies to both measure and exploit these correlations in an attempt to achieve real-time imaging through scattering material.

Before proceeding, it is helpful to introduce an alternative representation of the TM. We've already described how the (spatial) transmission matrix $T(x_b, x_a)$ connects the optical field at an input plane A and output plane B along spatial coordinates. It is also possible to connect the optical field at each plane as a function of wavevector, k. To do so, we use the well-known Fourier relationship connecting an optical field to its wavevector decomposition: $FU(x) = U(k)$, where F denotes a discrete Fourier transform (DFT) matrix applied to the vector $U(x)$. Inserting this matrix equation into our above equation connecting $U(x_a)$ and $U(x_b)$ leads to the relation, $T(k_b, k_a) = F_{2D}T(x_b, x_a)$, where $T(k_b, k_a)$ is now our "k-space" transmission matrix and F_{2D} represents a two-dimensional Fourier transform that includes a horizontal flip. In summary, we will be working with both spatial and k-space TMs throughout this chapter, and the two are connected by a linear transform.

13.2   Angular (Traditional) Memory Effect

As noted above, the transmission matrix is often not completely random. The first source of nonrandomness that we will examine concerns the average magnitude of the elements of the spatial transmission matrix. Specifically, when a small point source is placed on one side of a scattering medium, it is common to observe that the transmitted speckle is spatially localized to a finite "envelope" along the output surface.

Mathematically, this envelope in space also manifests itself as an envelope along the diagonal of the transmission matrix, where entries far from the diagonal vanish. Since this effect is present in the intensity of the light, it affects the square of the spatial transmission matrix elements, $|T_x(x_b, x_a)|^2$. Furthermore it is an average effect, so we thus define what is known as the "intensity propagator" as

$$P_x(x_b, x_a) = \left\langle |T_x(x_b, x_a)|^2 \right\rangle, \tag{13.1}$$

where $\langle \cdot \rangle$ denotes an ensemble average over many possible configurations of the same scattering material. The intensity propagator connecting point x_a along the input surface to point x_b along the output surface typically encompasses an enveloping effect with larger values closer to its diagonal (i.e., a larger average intensity when x_a and x_b are closer in value). In a strongly scattering sample, this propagator may be calculated using diffusion theory. It will have an extent on the order of the sample thickness. In other types of samples, $P(x_b, x_a)$ may be found through different means, such as Monte Carlo ray tracing. For the sake of our arguments, we do not need to make any assumptions about $P_x(x_b, x_a)$, other than that it is known.

A second source of nonrandomness in the transmission matrix are interelement correlations. If two elements are correlated, measuring one will give some information about the other, thereby reducing the amount of information that is required to reconstruct the full TM. As we will see later, interelement correlations and the average intensity envelope described above are closely related. In the seminal work by Feng et al. [15], the correlations within the T_k transmission matrix were studied. It was found that the intensity correlation function of the matrix elements can be described by three terms:

$$C_I(k_b, k_a; k_{b'}, k_{a'}) = \langle \delta T_k(k_b, k_a) \delta T_k(k_{b'}, k_{a'}) \rangle = C_I^{(1)} + C_I^{(2)} + C_I^{(3)} \tag{13.2}$$

where $\delta T_k(k_b, k_a) = T_k(k_b, k_a) - \langle T_k(k_b, k_a) \rangle$. Usually, the dominant contribution is $C_I^{(1)}$, which is given by

$$C_I^{(1)}(k_b, k_a; k_{b'}, k_{a'}) = D_1 \langle T_k(k_b, k_a) \rangle \langle T_k(k_{b'}, k_{a'}) \rangle \delta (\Delta k_a - \Delta k_b) F_1 (\Delta k_a), \tag{13.3}$$

with $\Delta k_a = k_{a'} - k_a$ and $\Delta k_b = k_{b'} - k_b$, and δ is the Dirac delta function for a continuous field description or the Kronecker delta for a discretized TM. D_1 is a constant factor of order unity and $F_1(x) = x^2 / \sinh^2(x)$. Traditionally, the correlation described by $C_I^{(1)}$ is known as the "optical memory effect." To distinguish between a different memory effect that we will introduce shortly, we will refer to this correlation as the "angular memory effect." This effect states that a small tilt in the incident wave results in an equal tilt in the transmitted wave at the output surface. As the tilt angle approaches the memory effect angle (determined by the width of F_1), the transmitted wave will start to decorrelate.

In the context of imaging through scattering media, the angular memory effect can prove to be quite useful when the scattering surface and the object of interest are separated by a finite distance. Let's say, for example, that we already know the k-space transmission matrix and would like to use this knowledge to form a focus through the scatterer and onto the object at the distant plane. Then, we can display an optimized input wavefront $U_{opt}(k_a)$ on the SLM such that we may form a spherical wave along the output surface via the matrix-vector product, $U_{sphere}(k_b) = T(k_b, k_a)U_{opt}(k_a)$. This spherical wave will propagate to form a focus at the distant object plane. Now, if we tilt $U_{opt}(k_a)$ to form $U_{opt}(k_a - d)$, the memory effect tells us that $U_{sphere}(k_b)$ will also tilt to form $U_{sphere}(k_b - d)$ (to a first approximation). This tilted spherical wave will then propagate to a shifted point on the object located at the distant plane, just like the focal point of a tilted lens will shift away from the optical axis. We diagram this idea in Figure 13.1. Thus, we can tilt a single optimized wavefront around to scan out an image along the object plane. Note that this analysis requires a finite distance between the scatterer and object plane to convert the tilting effect into the spatial shift required for point scanning.

Although Feng et al. [15] used a diagrammatic approach to calculate the memory effect, it was later pointed out in Ref. [16] that the function F_1 is equal to Fourier transform of the average intensity propagator of light, P_x, connecting the input and output surface of the scatter: $|F(P_x(x_b - x_a))|^2$, where F denotes a Fourier transform. This fact was demonstrated experimentally a few years later [17]. This elegant relation shows that

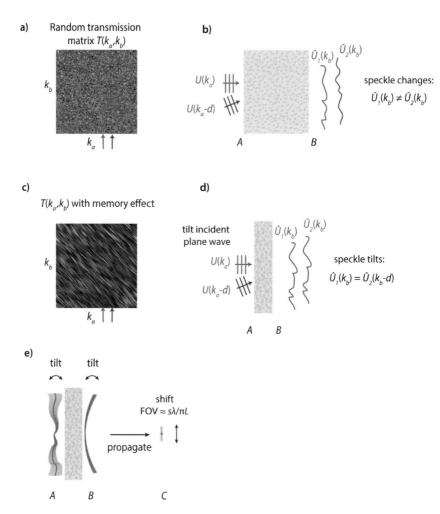

Figure 13.1 Angular memory effect. (**a**) Hypothetical random transmission matrix and (**b**) corresponding scattering medium. Incident wave $U(k_a)$ results in a speckle pattern $\hat{U}_1(k_b)$. When the incident wave is tilted, the result is a completely different, uncorrelated speckle pattern $\hat{U}_2(k_b)$. (**c**) If there is a memory effect, the angular transmission matrix is striped along the diagonal, and a tilt of the input wavefront causes the output wavefront to tilt but otherwise remain approximately the same. (**d**) At a distance behind the scattering layer, this tilt results in a shift of the wavefront. (**e**) With the appropriate input wavefront, this effect can be used to translate a focus. Adapted with permission from Macmillan Publishers Ltd: Nature Physics [20], copyright 2015.

any a priori information about the intensity propagator P_x (i.e., the enveloping effect discussed above) will give rise to a correlation in the elements of T_k.

As can be seen in Eq. 13.2 and Eq. 13.3, it is assumed that F_1 only depends on the difference between the angle of incident light, Δk_a, and not on the values of k_a or $k_{a'}$ by themselves. In other words, F_1 is assumed to be tilt invariant. In addition, it is assumed that the intensity propagator is translation invariant: $P(x_b, x_a) = P(x_b - x_a)$. This assumption is valid for a slab with constant thickness and homogeneous scattering properties. However, as we will see next, the memory effect also occurs under other far more relaxed conditions.

We will now proceed to derive the optical memory effect for the most general case where the intensity propagator $P(x_b, x_a)$ is known, but not necessarily translation invariant (see Figure 13.2). We start with our original definition of the intensity propagator in Eq. 13.1, $P_x(x_b, x_a) = \langle |T_x(x_b, x_a)|^2 \rangle$. This function expresses the finite average intensity envelope along the output surface, caused by an input point at a specific location $x_a(i)$, within its ith column. In an attempt to examine the effect of this envelope on the k-space transmission matrix, we may apply our two-dimensional Fourier transform operator from the previous section to both sides of this equation:

$$F_{2D}P_x(x_b, x_a) = \sum_{k_a, k_b} \langle T_k(k_b, k_a)T_k(k_b - \Delta k_b, k_a - \Delta k_a) \rangle \propto C(\Delta k_b, \Delta k_a) \quad (13.4)$$

We have applied the cross-correlation theorem to arrive at the right-hand side of Eq. 13.4 (i.e., the Fourier transform of the product of two functions is the cross-correlation of their Fourier transforms). The resulting cross-correlation function $C(\Delta k_b, \Delta k_a)$ is the angular correlation function of interest, where again $\Delta k_a = k_{a'} - k_a$ and $\Delta k_b = k_{b'} - k_b$. In the special case that P_x is translation invariant (i.e., $P_x(x_b, x_a) = P_x(x_b - x_a)$), the correlation function reduces to

$$C(\Delta k_a, \Delta k_b) \propto \delta_{\Delta k_a, \Delta k_b} F^{\Delta x \to \Delta k_b} [P_x(\Delta x)], \quad (13.5)$$

which matches Eq. 13.3.

In the context of bio-imaging, the question arises whether the optical memory effect also occurs inside scattering media. Schott et al. [18] recently measured a memory effect across biological samples. The above derivation makes clear that there is no restriction whatsoever on the position of the input and output plane locations, z_a and z_b (they can be anywhere inside or outside of a scattering medium). Therefore, the optical memory effect must occur inside scattering media exactly as it does in transmission: tilting the incident field results in an equal tilt at plane z_b. However, this effect may be hard to observe in isolation, since a tilt in the wavefront at z_b has no effect on the intensity distribution of the speckle field at that plane. Further research may identify a particular use for this memory effect inside scattering media.

13.3   Translation Memory Effect

Our derivation of the angular memory effect in the previous section relied upon the intensity propagator. We began by observing how an input point source of light will create a diffuse spot with limited spatial extent on the output surface. This, in effect, specifies an "enveloped" or banded structure along the spatial transmission matrix

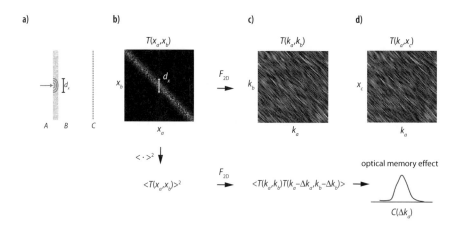

Figure 13.2 Origin of the angular memory effect. (**a**) A light source illuminates a point at the input surface (A) of a thin diffuser. As light scatters to the other side (B) it spreads to a limited diameter d_x before propagating through free space to plane C. (**b**) A corresponding TM that maps input space at plane A to output space at plane B. (**c**) The corresponding angular TM can be obtained by a 2D Fourier transform (and flipping horizontally). The stripes indicate angular correlations. (**d**) The angle (at the input plane A) to space (at the target plane C) TM indicates that a tilt of the input wavefront causes a shift at the distant target plane.

$T_x(x_a, x_b)$, where matrix entries closer to the main diagonal tend to be larger in magnitude than entries further away, which leads to correlations within the wavevector transmission matrix $T_k(k_b, k_a)$.

The above derivation presents a natural question. Can the same banded structure that we sometimes observe along T_x also appear within T_k? And if so, what sort of correlations might this structure cause? To answer this question, we may retrace our analytical steps above, but now start our analysis in k-space. Doing so will unveil a new type of memory effect.

The Fourier dual of a point source is a plane wave. Thus, we begin our new analysis by considering how an incident plane wave is transformed on average after it passes through a scattering slab. In some materials that are especially anisotropic (i.e., highly forward scattering, like many biological samples), a certain amount of directionality will be preserved on the output surface. This directionality may be maintained even after multiple scattering events. Instead of spreading across all possible wavevectors, the incident plane wave will, on average, spread into a finite output "cone" of light.

Exactly like the finite intensity envelope that masks T_x, an average wavevector envelope will mask T_k. Mathematically, we define a k-space intensity envelope to encompass this finite spread:

$$P_k(k_b, k_a) = \langle |T(k_b, k_a)| \rangle^2 \tag{13.6}$$

A diagram of an example enveloped k-space transmission matrix is in Figure 13.3. One can approximate the k-space intensity envelope with a simple experiment: illuminate a sample of interest with a plane wave, measure the resulting scattered

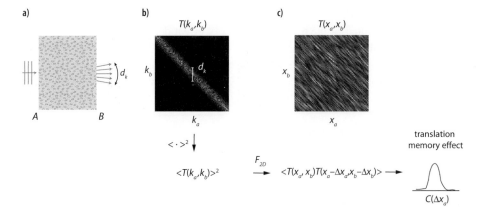

Figure 13.3 Translation memory effect. (**a**) A plane wave illuminates the input surface (A) of a sample. As light scatters to the other side (B) rays spread to a limited angular range d_k. (**b**) A corresponding TM that maps input angle at plane A to output angle at plane B. (**c**) The corresponding spatial TM can be obtained by a 2D Fourier transform (and flipping horizontally). The stripes indicate spatial shift correlations. The translation memory effect was first introduced in Ref. [20].

field, and compute its spectrum. A simple theoretical model is also available to predict the envelope as a function of slab thickness L, albedo ω, scattering cross-section a, and single-scattering phase function (i.e., anisotropy parameter) h, as derived by Kokhanovsky [19]:

$$P_k(k_b - k_a) = P_k(\Delta k) = \frac{1}{\lambda} F^{x \to \Delta k} \exp\left[-aL\left(1 - \hat{h}(x)\right)\right]. \qquad (13.7)$$

Here we have assumed a shift-invariant envelope for simplicity (this latter assumption is not required in general), and $\hat{h}(x)$ is the Fourier transform of the single-scattering phase function (e.g., the Fourier transform of the Henyey-Greenstein phase function). Unlike the spatial intensity envelope predicted with the diffusion approximation, this model holds under the small angle approximation and thus applies within the regime of highly forward scattering [20].

Just like the traditional memory effect, we may now derive the translation memory effect by taking the Fourier transform of the k-space intensity propagator P_k and applying the cross-correlation theorem:

$$F_{2D}^{k \to \Delta x} P_k = \sum_{x_a, x_b} \langle T_x(x_a, x_b) T_x^*(x_a - \Delta x_a, x_b - \Delta x_b) \rangle \propto C(\Delta x_a, \Delta x_b). \qquad (13.8)$$

Here, $C(\Delta x_a, \Delta x_b)$ measures the amount of correlation remaining between the input and output optical fields if the former is shifted by $\Delta x_a = x_a - x_{a'}$, and/or if the latter by $\Delta x_b = x_b - x_{b'}$. If the k-space intensity propagator only depends upon difference coordinates, then we may reexpress Eq. 13.8 as

$$C(\Delta x_a, \Delta x_b) \propto \delta_{\Delta x_a, \Delta x_b} F^{\Delta k \to \Delta x_b} [P_k(\Delta k)], \qquad (13.9)$$

where $\Delta k = k_b - k_a$ defines the difference coordinate. We see that Eq. 13.9 is the Fourier dual of the traditional memory effect defined via Eq. 13.5. It predicts the presence of translation correlations as a result of a finite wavevector intensity envelope. Experimental measurements of $C(x)$ for forward-scattering slabs of different thickness are presented in [20].

13.4   Ray Transmission Tensor

So far, we have mathematically defined the input-output behavior of a scatterer across both space ($T_{xx} = T(x_b, x_a)$) and wavevector ($T_{kk} = T(k_b, k_a)$) and examined the resulting correlations within each domain. These two transmission matrix models offer different representations of the same underlying scattering phenomena. Given the response of the scattering process in space (T_{xx}), we can always determine its response in wavevector (T_{kk}) by applying a discrete Fourier transform (DFT) to the matrix rows and columns. We could also easily invert this operation, or compute T_{xk} or T_{kx} with just one DFT.

In this section, we develop a third possible representation for scattering. As we've seen in the previous sections, sometimes it is useful to work with T_{xx} (e.g., when examining the spatial memory effect), and sometimes is helpful to use T_{kk} (e.g., during experimental measurement of T). Assuming perfect measurement, both matrices fully characterize the linear scattering process. But, neither form always offers an ideal model. For example, it might not be direct to connect the values within T_{xx} or T_{kk} to the refractive index distribution of the object of interest. Alternatively, examining just the properties of T_{xx} or T_{kk} alone might not be the most direct way to understand the above memory effects.

To address these shortcomings, we now model the input-output behavior of a scattering object as both a function of x and k, simultaneously. Instead of probing T_{xx} with a shifting point source, or T_{kk} with a tilting plane, we may move a single "ray" across each position and direction on the input side while examining the material response. Through this new "ray transmission" model, we hope to draw new connections between the transmission matrix, the scattering object of interest, and also concepts like scattering complexity and transform sparsity, which we will return in the next section.

13.4.1   Ray Input-Output Matrix

Here, we first introduce a "ray" input-output model of scattering based upon geometric optics [21]. We will connect this model with wave fields in the following subsection. We again work within a 2D (x, z) geometry for simplicity, and define our input and output planes, A and B, along two axial locations of interest, z_a and z_b, on either side of our scattering material. Along the input side, we may define the radiance of one ray entering the system at spatial location x_a and angle θ_a as $L(x_a, \theta_a)$. We may do the same for rays along the output plane with $L(x_b, \theta_b)$. We note that type of parameterization is utilized often in ray transfer analysis [21] and is sometimes referred to as a light field [22].

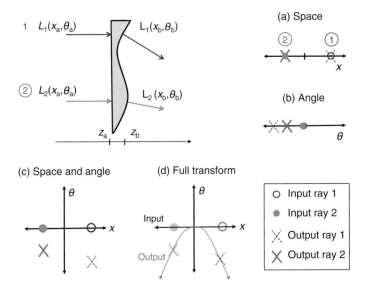

Figure 13.4 Ray input-output. Two rays, L_1 and L_2, are transformed uniquely by a thin warped sheet of glass from input side A to output side B. We denote Ray 1 as an empty circle at the input and a dashed "x" at the output, while ray 2 is a filled circle and "x," respectively. (**a**) The transformation plotted only as a function of ray position. Here, position does not change. (**b**) Similar plot only as a function of ray angle, from which it is challenging to ascertain the surface shape of the warped glass. (**c**) Plot of the transformation in ray space (as a joint function of space and angle, where, if we plot the transformation of all straight input rays across the output surface, we will form a parabolic curve, which informs us of the surface profile of the warped glass sheet (via a derivative relationship; see text).

To understand the possible benefits of this ray space formalism, it is helpful to consider the example of light passing through a thin piece of warped glass, which obeys the approximations required of geometric optics and will uniquely alter the angle of an input ray as a function spatial location (see Figure 13.4). This type of optical disturbance is often encountered in setups that use adaptive optics. Since we assume the sheet of glass is thin, the spatial location of each input ray does not significantly change between the input and output planes. Thus, just plotting how this glass sheet transforms each ray as a function of space only, as shown in Figure 13.4a, does not reveal very much about the curved glass structure. If we instead plot the transformation of each ray as a function of angle, as in Figure 13.4b, a variation between input and output is clear but the map is not very instructive (i.e., with overlapping incident rays at $\theta = 0$).

Third, we plot both the position and direction of each incident ray in Figure 13.4c. In this "ray space" diagram, the full behavior of the warped glass sheet becomes quite clear: the different angular deviation of each ray is connected to its distinct spatial location. It is instructive to go through the exercise of mapping how a straight input ray (traveling at $\theta = 0$) along all input locations transforms into an output ray for this particular piece of warped glass, whose thickness follows a cubic profile along space ($t(x) \propto x^3$). All straight input rays in this ray-space diagram form a straight line (purple). We may write the input light field of parallel rays (i.e., a plane wave) as $L(x_a, \theta_a) = \delta(\theta_a)$. After passing through the warped glass, the output rays form a parabola in ray space (blue): $L(x_b, \theta_b) = \delta(\theta_b - cx_b^2)$, where c is a constant relating to the degree glass of curvature. This parabolic output follows from the observation that the deviation of rays passing through a thin plate follows the gradient of the imparted phase delay [22] (i.e., the parabolic shape follows from a cubically varying thickness).

We may use a generalized four-dimensional ray transformation to express the input-output response of an arbitrary object of interest. If we assume the input and output coordinates are continuous both in space and angle, then we may write

$$L(x_b, \theta_b) = \iint R(x_b, \theta_b, x_a, \theta_a) L(x_a, \theta_a) dx_a d\theta_a, \tag{13.10}$$

where $R(x_b, \theta_b, x_a, \theta_a)$ is the ray transformation tensor mapping input rays to output rays. For the simple example of the cubic glass plate in Figure 13.4, this specific ray transformation will take the following form:

$$R(x_b, \theta_b, x_a, \theta_a) = \delta \left((\theta_b - \theta_a) - c(x_b - x_a)^2 \right). \tag{13.11}$$

Note that the form of R here directly reflects the surface profile of the curved glass. We also see that this ray transformation tensor is quite sparse (i.e., most of its entries are zero). We will soon return to this observation.

13.4.2   The Wigner Distribution Function

An accurate description of the interaction of light with objects at the microscopic scale must account for scattering and diffraction. To connect the concept of ray space with wave fields, we now introduce a useful function known as the Wigner distribution function (WDF), which will eventually help us express the transmission matrix as a four-dimensional mapping, much like the ray transformation tensor above.

Similar to the Fourier transform, the Wigner distribution function (WDF) transfers the representation of a complex optical field from one domain to another. Unlike the Fourier transform, which can transform a function of space to a function of spatial frequency ($U(x)$ to $U(k)$), the Wigner distribution maps a function of one spatial variable to a joint function of two Fourier conjugate variables: $U(x)$ to $W_U(x, k)$. This joint (x, k) space is often referred to as "phase space." Although W_U is a function of both conjugate variables, it maintains a one-to-one relationship with the one-dimensional function U (up to a constant phase shift).

By representing the original field $U(x)$ in phase space, the WDF effectively offers a picture of localized spatial frequency content. Specifically, the value of $W_U(x_c, k_c)$ informs us about the amount of optical power of the field U at location x_c propagating in the direction k_c. This is quite similar to how the light field value $L(x_c, \theta_c)$ specifies a particular ray in geometric optics. While a helpful connection, we note that this interpretation must be taken loosely, as the WDF is not necessarily a non-negative function. We refer the interested reader to [23] for additional details about connecting the WDF to radiance and ray space.

Mathematically, the WDF of the coherent optical field $U(x)$ is given as

$$W_U(x,k) = \int U\left(x + \frac{x'}{2}\right) U^*\left(x - \frac{x'}{2}\right) e^{-ikx'} dx'. \tag{13.12}$$

It is helpful to think about this function $W_U(x, k)$, applied to the field $U(x)$, as a reversible transformation, much like the Fourier transform.

13.4.3   Wigner Input-Output Relationship

We may now define the input-output behavior of a scattering system in phase space. To map the WDF of the optical field on the input side, $W_U(x_a, k_a)$, to that on the output side, $W_U(x_b, k_b)$, we may re-express the original transmission matrix equation, $U_b(x_b) = T(x_b, x_a)U_a(x_a)$, as the following:

$$W_U(x_b, k_b) = \iint V(x_b, k_b, x_a, k_a) W_U(x_a, k_a) dx_a dk_a. \tag{13.13}$$

Here, the input and output WDFs, $W_U(x_a, k_a)$ and $W_U(x_b, k_b)$, are WDF transforms of the input and output fields, $U_a(x_a)$ and $U_b(x_b)$, following Eq. 13.12. The function V connects these input and output phase space distributions (Figure 13.5). We refer to it as

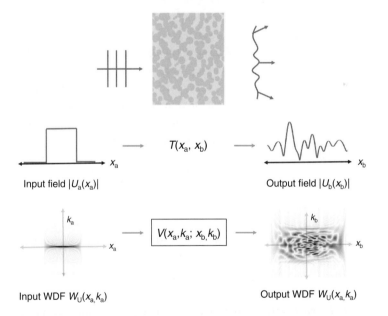

Figure 13.5 Wigner input-output relationship. Similar to the concept of ray space plot presented in Figure 13.4, we may also represent a coherent field as a joint function of space and spatial frequency (i.e., wavevector) using the Wigner distribution function (WDF). (top) A spatially confined input plane wave (left) will scatter into an output speckle field as defined by the transmission matrix. (bottom) Analogously, The WDF of the incident plane wave will transform through a scatterer into its corresponding output WDF as defined by the ray transmission tensor (see Eq. 13.13 in the text). See Ref. [24] for details. A black and white version of this figure appears in some formats. For the color version, please refer to the plate section.

a light-field transmission tensor (LTT). It is a similar transformation as the WDF, but is applied to both the rows and columns of the transmission matrix:

$$
V(x_b, k_b, x_a, k_a)
$$
$$
= \iint T\left(x_a + \frac{x_{a'}}{2}, x_b + \frac{x_{b'}}{2}\right) T * \left(x_a - \frac{x_{a'}}{2}, x_b - \frac{x_{b'}}{2}\right) e^{-ik_a x_{a'} + ik_b x_{b'}} \, dx_{a'} dx_{b'}.
$$

$$(13.14)$$

Unlike the 2D transmission matrix T, V is a 4D real function. V may also be formed from the k-space transmission matrix T_{kk} with a nearly analogous equation.

The LTT is a particular example of a ray spread function [23] that mathematically defines the output behavior of the scattering system to an input delta function in both space and wavevector. For example, if we probe the scatterer with the input field $W_U(x_a, k_a) = \delta(x_a - x_c, k_a - k_c)$, which conceptually represents one "ray" entering the scattering system at location x_c with wavevector k_c, then we will obtain one 2D slice of V. Due to the uncertainty principle, it is not possible to form this type of single-ray input WDF, but such an interpretation draws a helpful analogy between the light-field transmission tensor here and the ray transformation tensor in Eq. 13.10.

Briefly returning to our warped sheet of glass example, if one were to measure its transmission matrix (either T_{xx} or T_{kk}), then one could compute V (using Eq. 13.14) to find its LTT. The resulting function would closely resemble the quadratic form of Eq. 13.11 (equaling this form at the small wavelength limit). Just like we noted about Eq. 13.10, the LTT will also be a relatively sparse function – specifically, it is a 2D surface defined within a 4D space at the small wavelength limit. Furthermore, its quadratic shape will directly reflect the cubic profile of the index of refraction of this thin transparent example object. Thus, it appears that the LTT space may at times be well-suited for connecting the transmission matrix to the spatial structure of samples. Additional details linking the concept of the LTT to the memory effect are presented in Ref. [24].

13.5   Outlook

Often, optical scattering is not quite as random as the transmission matrix implies. One might not have to measure all of its values for a complete picture of light's input-output behavior. We have examined three examples of nonrandomness: the optical memory effect, the translation memory effect, and sparsity introduced within a four-dimensional ray space. Figure 13.6 uses a simple simulation to put these memory effects into context.

There are of course other sources of nonrandomness that one might also take advantage of when imaging or focusing through scattering media. Following insights from the field of signal processing, any a-priori information about an experiment is often helpful when interpreting data. For example, knowledge of the average material absorption or any boundary effects might better inform an imaging/focusing attempt. In addition, other correlations might exist within the material's response to light, for example along the temporal and/or spectral dimensions, and also potentially across polarization, to further reduce the required number of measurements. Furthermore, we only considered

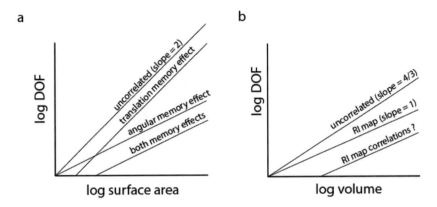

Figure 13.6 The effect of correlations on TM information content. (**a**) Hypothetical experiment of varying the input area of a scattering slab with constant thickness. Degrees of freedom (DOF) as a function of surface area for a slab with constant thickness. (**b**) (Schematic) degrees of freedom (DOF) as a function of volume for an equilateral cube.

first- and second-order correlations of the optical field in the above analysis. Higher order correlations and their connection to the C_2 and C_3 correlations for scattered intensities may also point toward additional strategies to minimize measurements.

Memory effects are dominant in the ballistic and quasi-ballistic regime. Do we expect all correlations to vanish in the diffuse regime? We can only speculate about as yet undiscovered TM correlations, but there are good reasons to be optimistic if we consider the following example of an equilateral scattering cube with side length L. The TM from one surface to the opposing surface will have $\left(\frac{2}{\lambda}L\right)^2$ input modes and the same number of output modes and the number of TM elements will scale with L^4. On the other hand, we know that the TM is the result of wave propagation through a scattering medium and could, in principle, be numerically derived from the 3D refractive index map $n(x, y, z)$ of the medium. In contrast to the TM, which scales with L^4, the number of wavelength-sized elements within the 3D refractive index map scales with L^3. This scaling is independent of the size of the cube and whether scattering is diffuse or quasi-ballistic. Thus, there is a description of the TM, which is shorter than the TM itself, even in the diffuse regime. This description could be further compressed if there were correlations within 3D refractive index map. Future work aimed at "inverting" wave propagation and deducing the refractive index map from the TM may uncover a range of scientifically interesting and practically useful correlations.

We hope these considerations illustrate the importance of searching for non-randomness in scattered light.

13.6   References

[1] I. M. Vellekoop and A. Mosk, "Focusing coherent light through opaque strongly scattering media," Opt. Lett. 32(16), 2309–2311 (2007).

[2] Z. Yaqoob, D. Psaltis, M. S. Feld and C. Yang, "Optical phase conjugation for turbidity suppression in biological samples," Nature Photon. 2, 110–115 (2008).

[3] S. Popoff, G. Lerosey, M. Fink, A. C. Boccara and S. Gigan, "Image transmission through an opaque material," Nature Comm. 1(81), 1–5 (2010).

[4] T. Cizmar and K. Dholakia, "Exploiting multimode waveguides for pure fibre-based imaging," Nature Comm. 3(1027), 1–9 (2012).

[5] M. Kim, Y. Choi, C. Yoon, W. Choi, J. Kim, Q. H. Park and W. Choi, "Maximal energy transport through disordered media with the implementation of transmission eigenchannels," Nature Photon. 6, 581–585 (2012).

[6] C. L. Hsieh, Y. Pu, R. Grange and D. Psaltis, "Digital phase conjugation of second harmonic radiation emitted by nanoparticles in turbid media," Opt. Express 18, 12283–12290 (2010).

[7] E. H. Zhou et al., "Focusing on moving targets through scattering samples," Optica 1(4), 227–232 (2014).

[8] C. Ma et al., "Time-reversed adapted-perturbation (TRAP) optical focusing onto dynamic objects inside scattering media," Nature Photon. 8(12), 931–936 (2014).

[9] X. Xu, H. Liu and L.V. Wang, "Time-reversed ultrasonically encoded optical focusing into scattering media," Nature Photon. 5(3), 154–157 (2011).

[10] Y. M. Wang et al., "Deep-tissue focal fluorescence imaging with digitally time-reversed ultrasound-encoded light," Nat. Commun. 3, 928 (2012).

[11] B. Judkewitz et al., "Speckle-scale focusing in the diffusive regime with time reversal of variance-encoded light (TROVE)," Nature Photon. 7(4), 300–305 (2013).

[12] T. Chaigne et al., "Controlling light in scattering media non-invasively using the photoacoustic transmission matrix," Nature Photon. 8(1), 58–64 (2014).

[13] P. Lai et al., "Photoacoustically guided wavefront shaping for enhanced optical focusing in scattering media," Nature Photon. 9(2), 126–132 (2015).

[14] R. Horstmeyer, H. Ruan and C. Yang, "Guidestar-assisted wavefront-shaping methods for focusing light into biological tissue," Nat Photon. 9(9), 563–571 (2015).

[15] S. Feng, C. Kane, P. A. Lee and A. D. Stone, "Correlations and fluctuations of coherent wave transmission through disordered media," Phys. Rev. Lett. 61(7), 834–837 (1988).

[16] R. Berkovits, M. Kaveh and S. Feng, "Memory effect of waves in disordered systems: A real-space approach," Phys. Rev. B 40(1), 737–740 (1989).

[17] J. H. Li and A. Z. Genack, "Correlation in laser speckle," Phys. Rev. E 49, 45304533 (1994).

[18] S. Schott, J. Bertolotti, J. F. Léger, L. Bourdieu and S. Gigan, "Characterization of the angular memory effect of scattered light in biological tissues," Opt. Express 23, 13505–13516 (2015).

[19] A. A. Kokhanovsky, "Small-angle approximations of the radiative transfer theory," J. Phys. D: Appl. Phys. 30, 2837–2840 (1997).

[20] B. Judkewitz, R. Horstmeyer, I. M. Vellekoop, I. N. Papadopoulos and C. Yang, "Translation correlations in anisotropically scattering media," Nature Phys. 11(8) (2015).

[21] J. Goodman, Introduction to Fourier Optics (McGraw-Hill, 1996)

[22] Z. Zhang and M. Levoy, "Wigner distributions and how they relate to the light field," IEEE International Conference on Computational Photography (ICCP), 1–10 (2009).

[23] M. J. Bastiaans, J., "Wigner distribution function and its application to first-order optics", Opt. Soc. Am. 69, 1710 (1979).

[24] G. Osnabrugge, R. Horstmeyer, I. N. Papadopoulos, B. Judkewitz and I. M. Vellekoop, "Generalized optical memory effect," Optica 4, 886–892 (2017).

Part VI

Shaped Beams for Light Sheet Microscopy

14 Light Sheet Microscopy with Wavefront-Shaped Beams: Looking Deeper into Objects and Increasing Image Contrast

Alexander Rohrbach

14.1 Introduction

Light propagation through inhomogeneous, disordered materials is still an enigmatic problem with unpredictable output, since complex multiparticle light scattering results in uncountable phase delays from scattered or absorbed photons. In coherent optics, strong intensity modulations arise from the interference of ballistic and diffusive photons and thus generate deterministic chaotic intensity distributions after some dozens of microns of propagation through scattering materials such as biological tissue. This circumstance is detrimental to the quality of an image p(x,y,z) in light sheet–based microscopy (LSBM), where a thin plane within the sample is illuminated by a sheet of light, as shown on the right side of Figure 14.1. In the ideal, but unrealistic case the light sheet consists of purely ballistic photons, which do not interact with the various scatterers inside the sample to be imaged. However, only recently it has been shown [1, 2]that the relative number of ballistic photons could be increased by holographically shaping the phase of the incident laser beam. This effect leads not only to enhanced penetration depths, but consequently also reduces diffusive photons or beam deflections by scattering objects.

In LSBM, the light sheet is usually launched by separate illumination optics, which are oriented perpendicularly to the detection optical system, such that the propagation of illumination light through the sample can be observed. LSBM has been successfully used especially in modern developmental biology [3] or in neurology [4, 5]. LSBM exhibits large advantages in the observation of highly dynamic (living) samples [69–], since objects are scanned planewise or linewise (DSLM, digital scanned light sheet microscopy), but not pointwise as in confocal microscopy. LSBM illuminates only the part of the object that is in the plane of focus of the detection objective as illustrated in Figure 14.2. Therefore, it makes more efficient use of the illumination light (dose) than confocal microscopy[10], which illuminates the whole sample for each plane that is imaged.

Many technical improvements have been achieved in LSM within the last decade – on the detection side, but especially on the illumination side. It turned out that scanning

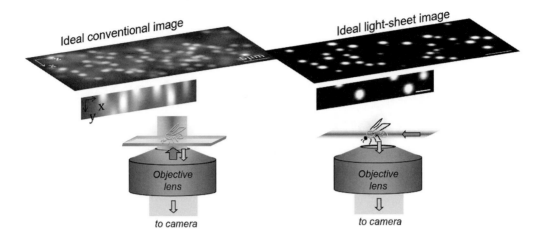

Figure 14.1 Fluorescent imaging with an epi-fluorescence and an ideal light sheet microscope. Since the whole probe (the fly) is illuminated in conventional microscopy, blurry background structures from out-of-focus objects become visible in the image sections p(x,z) and p(x,y). However, if the object is illuminated with a thin sheet of light, only the part of the object within the focal plane is imaged and blurry background structures remain invisible, leading to a strongly improved image contrast.

a beam [6] through the focal plane results in a more homogeneous light sheet [11] than by forming a static light sheet generated by a cylindrical lens, known as selective plane illumination microscopy (SPIM). Image quality decreases along the propagation axis of the illumination beam due to scattering and absorption. A direct way to reduce this problem consists in two-sided illumination [12]. First studies on Bessel beams in light sheet microscopy in large scattering samples have [2, 8, 11, 12] demonstrated improved image quality whereas later studies [8] achieved higher resolution in small samples like single cells.

In the scanning type, a typically rotationally symmetric beam $h_{ill}(x, y, z)$ is moved in x-direction across the object by scan mirrors during the integration time of a camera. Linewise illuminated parts of the object are imaged and composed to a 2D image by scanning the illumination beam laterally (SB $=$ *scan beam*). Then the object is shifted by y_s to address the third dimension.

For a single illumination beam $h_{ill}(\mathbf{r} - \mathbf{x_s})$ at position $\mathbf{x_s}$, the image on the camera is

$$b_{SB}(\mathbf{r} - \mathbf{x_s} - \mathbf{y_s}) = (f(\mathbf{r} - \mathbf{y_s}) \cdot h_{ill}(\mathbf{r} - \mathbf{x_s})) * h_{det}(\mathbf{r}) \qquad (14.1)$$

where $*$ denotes convolution. For an illumination beam scanned laterally within $[-x_m; x_m]$, the image reads:

$$b_{scan}(\mathbf{r} - \mathbf{y_s}) = \left(\int_{-x_m}^{x_m} f(\mathbf{r} - \mathbf{y_s}) \cdot h_{ill}(\mathbf{r} - \mathbf{x_s}) \, dx_s \right) * h_{det}(\mathbf{r}) \qquad (14.2)$$

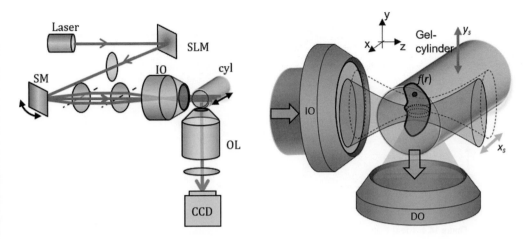

Figure 14.2 Principles of a light sheet microscope. A more advanced type of light sheet microscope uses a spatial light modulator (SLM), which produces the desired beam shape right behind it. The holographically shaped beam is imaged by two 4f-systems into the focal plane of the detection objective (DO). A scan mirror displaces the illumination beam laterally within the integration time of the camera (CCD) to generate a thin sheet of light. Right: The illumination objective (IO) and the detection objective lens (DO) are aligned to each other in a rectangular configuration. A gel cylinder embedding the fluorescently labeled specimen is moved through the light sheet along y-direction to generate an image stack.

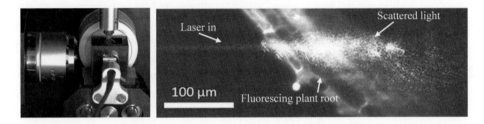

Figure 14.3 Sample chamber with fluorescing beam (left) and scattering of illumination light at specimen (right). A collimated laser beam impinges on a plant root, which is fluorescently labeled (individual cells in grayscale are visible). Light is coherently scattered and redirected at refractive index changes induced by the biological object, leading to a strong broadening of the beam.

14.1.1   Beam Spreading and Scattering

When the illumination light propagates through a medium with an inhomogeneous refractive index distribution, scattering leads to a redistribution of light momentum and energy in the imaged plane, which causes prominent dark and bright stripes. Hence, both the light sheet quality and the image quality are deteriorated. Furthermore, spreading of the beam (Figure 14.3) along the propagation direction decreases axial resolution. Moreover, contrast for higher penetration depths is decreased by the augmented illumination of out-of-focus objects, which generate unwanted ghost images (Figure 14.4).

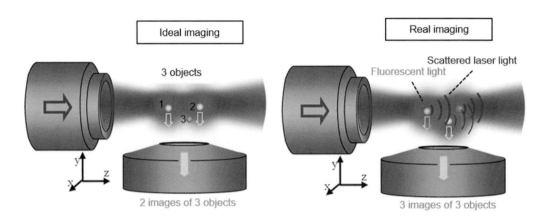

Figure 14.4 Ideal (left) and real (right) fluorescence excitation of three beads. The upper two beads are within the extent of ideal light sheet and within the focal plane of the detection objective. As a result of coherent scattering of the illumination beam (from left), also bead #3 outside the light sheet is excited to fluorescence. This unwanted image of the bead is a ghost image.

Approaches to remove background blur by postprocessing and to increase image contrast include structured illumination [8, 13, 14] and variations thereof [15, 16]. As they require several full-frame images for each final image the acquisition speed is reduced. Alternatively, larger samples can be illuminated more homogeneously by impinging light from opposite sides [4] and scattering artifacts like stripes can be reduced by fast pivoting of the light sheet in the image plane [12]. However, these approaches to increase image quality do not reduce light scattering and beam spreading.

The total (the nonideal) intensity of the illumination beam is obtained by a superposition of the incident and the multiply scattered beam:

$$h_{ill}(\mathbf{r}) = |\mathbf{E}_{inc}(\mathbf{r}) + \mathbf{E}_{sca}(\mathbf{r})|^2$$
$$= |E_{inc}|^2 + \left(|E_{sca}|^2 + 2Re\left\{E_{inc} \cdot E_{sca}^*\right\}\right) = h_{inc}(\mathbf{r}) + h_{sca}(\mathbf{r}) \qquad (14.3)$$

The resulting intensities can be split into an unscattered and scattered illumination intensity $h_{inc}(\mathbf{r}) + h_{sca}(\mathbf{r})$, hence resulting in two images after convolution with $h_{det}(\mathbf{r})$: the ideal image and the superimposed ghost image [11].

$$p(\mathbf{r}) = ((h_{inc}(\mathbf{r}) + h_{sca}(\mathbf{r})) \cdot f(\mathbf{r})) * h_{det}(\mathbf{r}) = p_{ideal}(\mathbf{r}) + p_{ghost}(\mathbf{r}) \qquad (14.4)$$

The ghost image, however, is hardly correlated with the ideal image.

14.1.2   Control of Beam Spreading through Holographic Illumination

To generate an image, which is close to the object, it is an obvious goal to minimize the ghost image intensity. Besides a change of the wavelength (using other lasers), the only control over the complex imaging process is to change the phase of the incident illumination light – typically by using an SLM. The intensity right behind the computer hologram is:

$$h_{inc}(x, y) = |E_{inc}(x, y) \cdot \exp{(i\phi_{holo}(x, y))}|^2 = |E_{holo}(x, y)|^2 \tag{14.5}$$

Thus, the 3D ghost image changes with the field $E_{holo}(x, y)$ behind the hologram

$$p_{ghost}(\mathbf{r}) = \left(\left(|E_{sca}|^2 + 2Re\left\{ E_{holo} \cdot E_{sca}^* \right\} \right) \cdot f(\mathbf{r}) \right) * h_{det}(\mathbf{r}) \tag{14.6}$$

The incident and the scattered field is separated into an amplitude and the phase term. Here, the unscattered field (the ballistic photons) can be approximated by a mean phase $\phi_{holo}(\mathbf{r}) = \bar{\phi}(\mathbf{r}) + \phi_0$ and the multiply scattered photons (the diffusive photons) are characterized by their phase variations $\delta\phi(\mathbf{r})$. Using $E_{tot}(\mathbf{r}) = E_{holo}(\mathbf{r}) + E_{sca}(\mathbf{r}) = A_{holo}(\mathbf{r}) \cdot e^{i\bar{\phi}(\mathbf{r})} + A_{sca}(\mathbf{r}) \cdot e^{i\delta\phi(\mathbf{r})}$, we find for the ghost image:

$$p_{ghost}(\mathbf{r}) = \left(\left(|E_{sca}|^2 + 2|E_{holo}| \cdot |E_{sca}| \cdot \cos{(\phi_{holo}(\mathbf{r}) - \delta\phi(\mathbf{r}))} \right) \cdot f(\mathbf{r}) \right) * h_{det}(\mathbf{r}) \tag{14.7}$$

By choosing the right phase profile $\phi_{holo}(x, y)$ on the SLM, the phase variations $\delta\phi(\mathbf{r})$ can be minimized. In a first step the hologram phase $\phi_{holo}(x, y)$ can be chosen such that selfreconstructing Bessel beams are generated. Bessel beams do not change their initial phase profile very much such that beam spreading and phase variations $\delta\phi(\mathbf{r})$ are small.

14.1.3   Bessel Beams Can Self-Reconstruct in Inhomogeneous Media

A light sheet microscope using self-reconstructing beams (MISERB) allows better control for the propagation of the illumination beams through the scattering sample. In particular, the usage of a laterally scanned Bessel beams enables a smooth illumination of the sample within the focal plane of the detection lens. Bessel beams can be created using annular apertures [17] or axicons [18] and show an increased depth of focus, i.e., the transverse intensity profile of Bessel beams changes only very little over a considerable distance. This propagation invariance was explained by strongly reduced diffraction in free space [19]. Moreover, Bessel beams are able to self-reconstruct, i.e., regain their initial profile behind isolated obstacles [20, 21].

Bessel beams, which have a δ-ring as their angular spectrum, can be generated by adding a conical phase profile $\phi(x, y) = \phi_0 - q(x^2 + y^2)^{1/2}$, a phase axicon, e.g., through an SLM. The parameter q is proportional to the focusing numerical aperture NA of the illumination objective. Further details are explained in section 14.4.

Bessel beams are able to self-reconstruct even inside inhomogeneous, scattering media and thereby open a series of applications in light sheet microscopy [1]. By exploiting this property, MISERB addresses the following three factors, which have been limiting the image quality in light sheet microscopy. First, due to light's natural beam spreading, there is a trade-off between the illumination beam's depth of field and its beam waist. An improvement of the axial resolution, i.e., in optical sectioning, can only be achieved by a stronger axial focusing of the light sheet, thereby reducing the depth of field of the illumination beam [22]. Second, as the illumination light path lies in the plane to be imaged, scattering and absorption result in strongly visible artifacts in

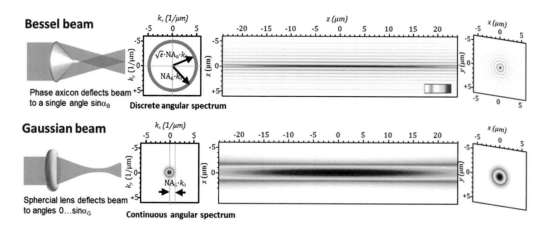

Figure 14.5 A phase axicon with a deflection $\sin\alpha_B$ generates a Bessel beam with a δ-ring like angular spectrum. Spherical lens with the relatively low focusing angle $\sin\alpha_G$ generates a Gaussian beam, with the same z-axial extent as the Bessel beam but with a much stronger extent in lateral xy-direction.

the image such as dark and bright stripes behind scatterers (ghost images). These artifacts can, so far, only be suppressed in visibility by tilting the light sheet in the image plane [12]. Third, scattering causes additional nonuniform spreading of the illumination beam along the propagation into the sample. This leads to a limited penetration depth, i.e., a decrease in image quality and signal strength along the illumination axis.

The amount and distribution of the energy carried in beams with identical depths of focus are important properties of illumination beams. Ideal zero-order Bessel beams have a radial intensity profile given by $I(r) = I_0 \cdot |J_0(k_r r)|^2$, J_0 being the Bessel function of the zeroth order. In this idealized description the intensity $I(r)$ does not depend on the propagation distance z and the energy in the rings is infinite. The angular spectrum is infinitely thin and can be written as $\tilde{E}(k_r) = E_0 \cdot \delta\,(k_r - k_0 NA)$, where $k = |\mathbf{k}| = 2\pi/\lambda$ is the wave number in vacuum, k_r its radial component of the wave vector and $NA = n \cdot \sin\alpha$ the focusing numerical aperture (IO). Multiplying a propagator onto the ideal annular spectrum $E_0 \cdot \delta\,(k_r - NA \cdot k_0) \cdot \exp\left(-i\Delta z\sqrt{k_n^2 - k_r^2}\right)$ to obtain the spectrum in a distance Δz, has no effect on the beam profile. The strongly reduced beam spreading can be attributed to the suppression of diffraction in free space, which is why Bessel beams are also called nondiffracting beams [19]. However, the realistic case of a (zero-order) Bessel beam that has an angular spectrum with a finite ring width can be described by

$$\tilde{E}(k_r) = \tilde{E}_0\left(step(k_0 NA - k_r) - step(\sqrt{\varepsilon}\,k_0 NA - k_r),\right) \tag{14.8}$$

where step(x) denotes the Heaviside step function. $\varepsilon < 1$ is the ring width parameter, i.e., the area ratio of the inner to the outer disk. For $\varepsilon = 0$, the back aperture is homogeneously illuminated and a conventional beam with a limited depth of field $\Delta z = 2\lambda/(1 - \cos\alpha)$ is created. The intensity profile $I(r) = |HT(f)|^2$ is obtained from equation (14.8) by a Hankel-Transform (HT). Normalization of the beam intensity to one results in

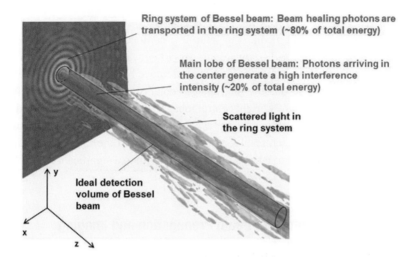

Figure 14.6 Simulated light scattering of a Bessel beam with electric field profile $J_0(r)$. Both the Bessel beam's ring system and the central main lobe transfer momentum to the many scatterers (not visible). However, the main maximum inside the ideal detection volume hardly change its propagation direction in contrast to conventional Gaussian beams. Reprinted by permission from Macmillan Publishers Ltd: Nature Communications [24], copyright 2012.

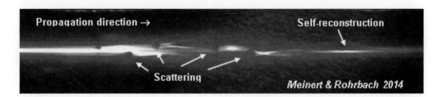

Figure 14.7 A Bessel beam hits 7 μm large glass beads, scatters and self-reconstructs its profile after a specific distance. Image width: 400 μm. Experiment: 2-photon excitation inside a fluorescing gel.

$$I_{Bes}(r) = \left[\frac{2}{k_0 NA\,(\varepsilon - 1) \cdot r\pi^2} \left(J_1\left(k_0 NA\, r \right) \quad \sqrt{\varepsilon} J_1\left(\sqrt{\varepsilon}\, k_0 NA\, r \right) \right) \right]^2, \qquad (14.9)$$

The depth of field (DOF) of beams with an angular spectrum described by eq. (14.8) can be estimated using the width of the spectrum's projection on the k_z-axis [23], defined by $k_z = (k_0^2 - k_r^2)^{1/2}$ via the Ewald sphere:

$$\Delta k_z = k_0 \left(\sqrt{n^2 - \varepsilon NA^2} - \sqrt{n^2 - NA^2} \right) \qquad (14.10)$$

Although Bessel beams are advantageous for imaging with respect to penetration depth and directional propagation stability [11], the accompanying ring system around their central lobe illuminates out-of-focus regions and strongly decreases contrast. In [8] it was shown that the negative effect of the rings can be avoided by using structured illumination, and/or two-photon fluorescence excitation by pulsed lasers.

An efficient way to block the photons emitted from the Bessel beam's ring system even in large scattering media is to setup line-confocal detection and exploit the propagation stability of Bessel beams along one straight line [24]. By using a slit moving parallel to the Bessel beam, it is possible to remove background light generated by the Bessel ring system, so that the final image contains only that part of the object illuminated by the Bessel beam's bright central lobe. This principle can be realized technically quite easily by using a rolling shutter available in many modern sCMOS cameras [25].

This detection approach results in a multiplication of the illumination and detection probability functions in axial direction and therewith in a significant loss of background photons.

14.2   Computer Simulations of Beam Propagation and Imaging

The great advantage of computer simulations is the ability to compare 3D object distribution as input with 3D image distribution as output. In this way it is not only possible to assess the theoretical feasibility or quality of an imaging method (see e.g., a comparison between confocal-line detection and structured illumination at the end of this chapter), it is also well possible to compare different imaging modes for different object configurations or set up conditions.

14.2.1   Propagation of Single Beams

In a first step, it is useful to investigate the scattering behavior of different beam types propagating through isolated dielectric spheres, which are typically of glass or polystyrene, and have typical diameters of about 0.5 μm to 5 μm. In contrast to the experiments, the computer simulations allow to analyze the intensity, the amplitude and the phase of the electric field, both in position space $E(x, y, z_0)$ and frequency space $\tilde{E}(k_x, k_y, z_0)$, at each plane z_0 within the 3D object distribution. This is shown in Figure 14.8, where the intensities of a Bessel beam at the start and at the end of the propagation volume, consisting of three dielectric spheres, are displayed. Disturbances of the distribution in space domain and in frequency domain of both phase and intensity can be well compared and are of special interest, since the planes $I(x, y, z_0)$ and $\tilde{I}(k_x, k_y, z_0)$ (power spectrum) before and behind the object are accessible in the experiment.

To obtain the intensities inside and behind the inhomogeneous object, we numerically calculate the electric field by using a scalar Beam Propagation Method (BPM) [26]. Here the angular spectrum of the electric field at a distance $z + dz$ propagating through the sample can be described by

$$\tilde{E}(k_x, k_y, z + dz) = FT\left[E(x, y, z) \cdot e^{-ik_0 \cdot \delta n(x,y,z) \cdot dz}\right] \cdot e^{-i \cdot dz \cdot \sqrt{(k_0 n_m)^2 - k_x^2 - k_y^2}} \qquad (14.11)$$

$FT[\ldots]$ denotes the Fourier Transform in x and y. The space dependent refractive index $n(\mathbf{r}) = n_m + \delta n(\mathbf{r})$ changes by $\delta n(\mathbf{r}) < \delta n_{max}$ around the mean value n_m, typically defined by the index of the scatterer's environment.

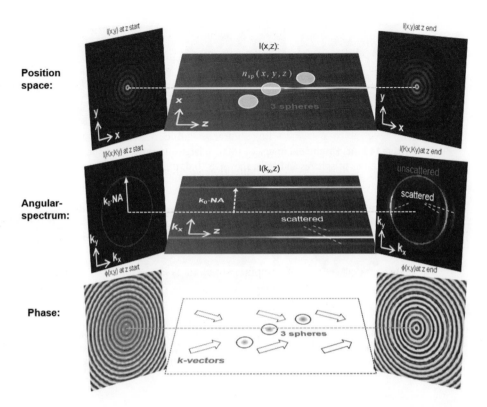

Figure 14.8 Numerical beam propagation using the BPM of a Bessel beam through three dielectric spheres along the x-direction. Scattering at the spheres has only little impact on the intensity cross-sections both in position space $|E(x, y, z_0)|^2$ and in the angular spectrum representation $|\tilde{E}(k_x, k_y, z_0)|^2$. The phase distribution at the end is mainly disturbed in the negative x-direction.

A distribution of spheres or cells can be described by different refractive index distributions with $n_m = 1.33$ and $\delta n = 0.08$. The first index distribution is that of a large centered sphere (sp) of radius R_{sp} such that

$$n_{sp}(x, y, z) = n_m + \delta n \cdot step\left(R_{sp} - \sqrt{x^2 + y^2 + z^2}\right) \qquad (14.12)$$

where $step(r) = 1$ if $r > 0$ (and 0 otherwise) designates the Heavyside step function. The other index distribution is that of a cell cluster (CC), which also has a radius of R_{sp}, but contains $N \approx 2 \cdot 10^4$ single cells (sc). These cells consist of a shell with an outer radius $R_{sc} \approx 1$ μm and inner radius $0.85 \cdot R_{sc}$ reflecting the increased index $n = 1.33 + 0.08$ of the cell membrane and actin cortex relative to a mean index $n = 1.33 + \frac{2}{3} \cdot 0.08$ of the cell plasma (the nucleus is disregarded).

The cells are arranged in arbitrary order inside a spherical volume.

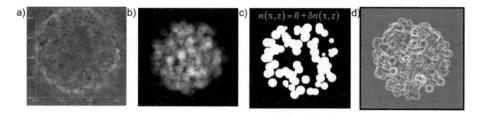

Figure 14.9 (a) Brightfield image of 180 μm large cell cluster and simulated refractive index distributions of spheres and cells, shown as (**b,d**) projections and (**c**) slice.

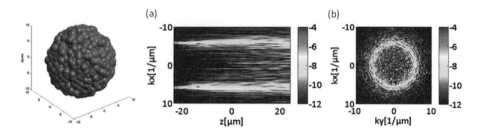

Figure 14.10 Volume rendered model of the refractive index distribution of a cell cluster (left). Power spectrum of Bessel beam incident on a spherical cell cluster. (**a**) Along the propagation direction, $\log(I(k_x, z))$, (**b**) behind the spherical cell cluster, $\log(I(k_x, k_y))$.

$$n_{cc}(\mathbf{r}) = n_m + \partial n \cdot \sum_{j=1}^{N} \left[step(R_{sc} - |\mathbf{r} - \mathbf{r}_j|) - \frac{1}{3} step(0.85 R_{sc} - |\mathbf{r} - \mathbf{r}_j|) \right] \quad (14.13)$$

The volume fraction of the cells inside the spherical volume cluster is 85%, leaving small volume gaps between the cells with index $n_m = 1.33$. Figure 14.9 displays the microscopy image of a real cell cluster and simulated index distributions $n_{sp}(\mathbf{r})$ and $n_{cc}(\mathbf{r})$ as defined in the equations above.

14.2.2   Imaging Beam Propagation through a Cluster of Spheres

Our imaging simulation is based on a standard light sheet–based microscope using laterally scanned illumination beams, where a spatial light modulator (SLM) modulates the phase of the incident illumination beam to $\phi_{holo}(x, y)$, which would be e.g., conical in the case for a Bessel beam.

It is now possible to vary the SLM phase hologram $\phi_{holo}(x, y)$ to generate different types of Bessel beams (see eq. (14.5)), which have different energies (photon densities) in the ring system and different ring frequencies, mainly expressed by the parameters ε and NA (see eq. (14.9)). While the Bessel beam propagates through a cluster of spheres, the thin, ring like power spectrum of the Bessel beams starts to broaden. Thereby, both the amount and the direction of scattered light is estimated by analyzing the redistribution of k-vectors at the end of the propagation volume as shown exemplarily in Figure 14.10.

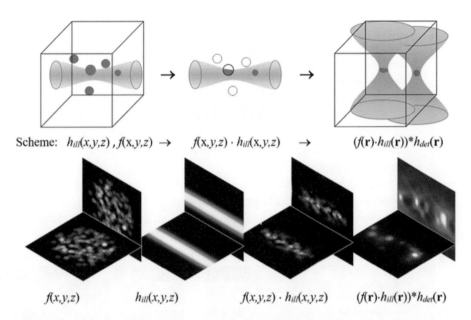

Scheme: $h_{ill}(x,y,z)$, $f(x,y,z)$ → $f(x,y,z) \cdot h_{ill}(x,y,z)$ → $(f(\mathbf{r}) \cdot h_{ill}(\mathbf{r}))*h_{det}(\mathbf{r})$

$f(x,y,z)$ $h_{ill}(x,y,z)$ $f(x,y,z) \cdot h_{ill}(x,y,z)$ $(f(\mathbf{r}) \cdot h_{ill}(\mathbf{r}))*h_{det}(\mathbf{r})$

Figure 14.11 Schematic image formation (top row) and wave optical simulation (bottom row). A cluster of spheres $f(x, y, z)$ is illuminated by an increasingly distorted beam/light sheet $h_{ill}(x, y, z)$, such that the product of both is convolved with $h_{det}(x, y, z)$).

14.2.3 Influence of the Detection Pointspread Function

It is of extraordinary importance to distinguish between the aberrations caused by i) the coherent illumination light propagating, scattering and interfering inside the medium, and ii) by the imaging process. Although the image formation is characterized by fluorescence photons propagating through the object to the camera, we simplify this process by a convolution of the inhomogeneously illuminated part of the object, with a specific detection pointspread function $h_{det}(\mathbf{r})$, as described by eqs. (14.3) and (14.4). This principle is further illustrated in Figure 14.11. Through this simulation procedure, it is well possible to identify image aberrations or backgrounds, which are further masked by the anisotropic low-pass filtering in the imaging process (convolution with 3D detection PSF). Only by a thorough understanding of the unwanted image contributions due to scattering, it is possible to compensate them in a certain way.

14.2.4 Line-Confocal Detection versus Structured Illumination for Bessel Beam Illumination

Line-confocal detection and structured illumination are different imaging modes on the detection side to remove background blur. Comparing different imaging modes with simulated data is advantageous, since on the one hand, exactly the same object is imaged and postprocessed. On the other hand, such simulations allow the assessment of the effect of scattering and beam broadening on the image contrast enhancement. This works especially well, if the realistic illumination – where scattering is accounted for

– is compared to an ideal illumination where scattering is disregarded. We have performed advanced computer simulations, where we simulated the propagation of Bessel beams through a volume ($n_{bg} = 1.33$) with the dimensions $24 \times 24 \times 48 \ \mu m^3$ that contains ≈ 4000 silica spheres with a diameter of $d = 0.75 \ \mu m$. To account for scattering by the spheres, we used the beam propagation method (BPM) to calculate $h_{sca}(x, y, z)$.

If scattering is taken into account, beam spreading occurs and the modulation amplitude and phase of the structured illumination patterns deteriorate. In consequence, more background blur is subtracted and the reconstructed images become darker at higher penetration depths.

The following comparison of imaging techniques is displayed in Figure 14.12. Images obtained for line-confocal detection and structured illumination (SI3) with two grid periods (3 μm and 6 μm) and a sample illumination by an ideal nonscattered light sheet are

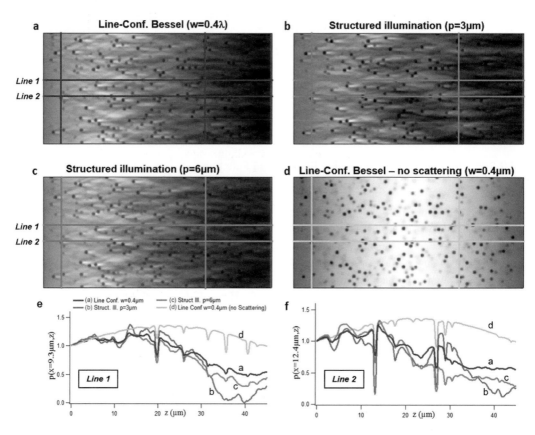

Figure 14.12 Line-confocal and structured illumination images of a scattering sample (Simulation). A Line-confocal detection image (w = 0.4 μm) is shown in (**a**). Reconstructed structured illumination images (SI3 mode) using a 3.0 μm period and a 6.0 μm period are shown in (**b**) and (**c**), respectively. For comparison, the corresponding line-confocal detection image for ideal illumination by a nonscattered Bessel beam is shown in (**d**). (**e, f**) Image line scans along the propagation direction of Bessel beams through a scattering medium for all illumination modes. Adapted by permission from Macmillan Publishers Ltd: Nature Communications [24], copyright 2012.

shown. The reconstructed images for structured illumination reveal a stronger decline in signal along z as the in-focus image information has been subtracted by the postprocessing due to decreased modulation contrast, which is a result of the broadening and scattering of the illumination beams. In the line-confocal, case a loss in intensity from left to right due to scattering is also visible, but the light sheet remains smoother while still offering good sectioning at higher signal strength. The decline of p(z) can be seen in the intensity profiles along the beam propagation axis that are shown in Figures 14.12e and 14.12f.

14.3 Holographic Beam Shaping: Experimental Results

14.3.1 The Right Phase Profile Can Enhance the Penetration Depth

Bessel beams can self-reconstruct their profile even in the presence of many obstacles such as light scattering biological cells. This reconstruction works because the photons scattered in the center of the beam are replaced continuously by new photons coming in from the conical angle defined by the initial phase on the SLM. These photons from the side, forming the Bessel beam's ring system, arrive nearly in phase in the beam center despite strong phase delays generated by the scatterers. Through an unexpectedly constructive interference of many photons, the high intensity maximum in the beam center is formed, which decreases only slightly with increasing penetration through a cluster of spheres (Figure 14.13).

The somewhat unexpected conclusion is that wavefront shaping of a beam influences the penetration depth into scattering medium.

14.3.2 Beam Self-Reconstruction in Human Skin

Another comparison between conventional Gaussian beams and Bessel beams regarding the penetration depth was performed at a several hundred μm large piece of fresh human skin embedded in agarose gel. The outer part of the skin consists of the epidermis and the dermis, separated by the Basal membrane. The first stiff layer of dead cells in the epidermis, the stratum corneum can be distinguished from a layer of living cells (keratinocytes). The laser beam propagates first through the stratum corneum and then through the epidermis (Figure 14.14). The layers all have different, nonhomogenous refractive index distributions $n(\mathbf{r})$ and different fluorophore concentrations $C(\mathbf{r})$.

A comparison of the self-reconstruction ability of a Gaussian and a Bessel beam, both laterally scanned is shown in Figure 14.14 where the propagation image of a single beam $h_{ill}(x,z,b_0)$ is overlaid in orange-hot colors with the gray-scaled image of the human skin. Again, the Gaussian beam generates more illumination artifacts in the form of stripes, whereas the Bessel beam enables reduced scattering and a more homogenous illumination without stripes.

In addition, the reduced scattering of the Bessel beam at the densely packed skin cells results in a 55% increase of the average penetration depth d as indicated by the fluorescence intensity linescans $\bar{F}(Z)$ inset into the image. We find $d_{Gauss} = 50\ \mu m$ and

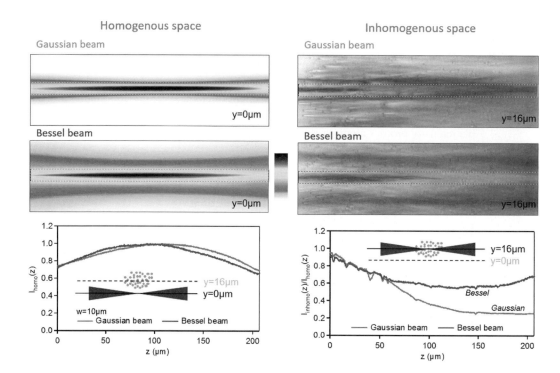

Figure 14.13 Measured penetration depths of illumination beams in a cluster of silica spheres. Left column: Images of the fluorescence intensity $I_{hom}(x, z)$ in a homogeneous medium for a single Gaussian beam and Bessel beam with the same depth of field. Right column: Images of the fluorescence intensity $I_{inh}(x, z)$ of both beams propagation through a cluster of spheres. The pseudo-color image of the intensity of the single beam is overlaid with the grayscale image resulting from a scanned beam. Bottom row: Comparison of normalized axial intensity linescans $I_{hom}(0, z)$ inside homogeneous medium and axial intensities $I_{inh}(0, z)$ inside an inhomogeneous medium. Adapted by permission from Macmillan Publishers Ltd: Nature Photonics [1], copyright 2010. A black and white version of this figure appears in some formats. For the color version, please refer to the plate section.

$d_{Bessel} = 77$ μm in the epidermis ($z = 138$–225 μm). The axial change of fluorescence intensities can be separated in a region for the stratum corneum (left) and epidermis (middle), whereas the dermis (right) is hardly visible. Assuming a mean constant fluorophore concentration $C(x, z)$, we average $\bar{F}(z) = \bar{C}(z) \cdot \bar{h}_{ill}(z)$ over the image width $2x_m = 175$ μm. The axial decay of fluorescence can be written as

$$\bar{F}(z, \phi_{holo}) = \frac{1}{2x_m} \int_{-xm}^{xm} C(x, z) \cdot |\mathbf{E}_{holo}(x, z) + \mathbf{E}_{sca}(x, z)|^2 \, dx$$

$$= \frac{1}{2x_m} \int_{-xm}^{xm} C(x, z) \cdot (I_{holo} + I_{sca} + 2 |E_{holo}| |E_{sca}| \cos (\Delta\phi_{holo})) \, dx \qquad (14.14)$$

$$\approx \bar{F}(z_0) \cdot \exp (-z/d(\phi_{holo}))$$

Gaussian beam

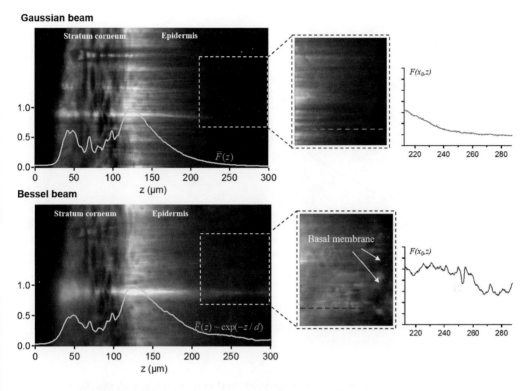

Bessel beam

Figure 14.14 Maximum-selection images of human skin. Illumination by a conventional beam (top) and a Bessel beam (bottom). Both types of beams illuminate the skin from left to right. An image from a single Gaussian beam and single Bessel beam is overlayed in orange-hot colors. Averaged intensity linescans $\bar{F}(z) \sim \bar{h}_{ill}(z)$ show an exponential decay through the epidermis. Inset area: A part of the epidermis close to the Basal membrane is magnified and autoscaled (boxes with dashed border) revealing single cells only for Bessel beam illumination. Right: line scans $F(x,z)$ normalized to $F(x,z=0)$ for $x = x1, x2$ (indicated by dashed lines) show the strong increase in contrast for the Bessel beam illumination: The penetration depth is increased by 50% relative to conventional Gaussian beams. Adapted with permission from [2], *Optics Express*.

where the illumination intensity $h_{ill}(\mathbf{r}) = |\mathbf{E}_{tot}(\mathbf{r})|^2$ separates into a field $\mathbf{E}_{tot}(\mathbf{r})$ with an average and a fluctuating phase $\Delta\phi_{holo}(\mathbf{r}) = \phi_{holo}(\mathbf{r}) - \delta\phi(\mathbf{r})$. With increasing z the number of diffusive photons increase and the correlation of their phases decreases in the epidermis. In other words, the root-mean square deviation of the phase $\delta\phi(z) = \sqrt{-\ln\left(\bar{F}(z)/\bar{F}_0\right)}$ grows, while the intensity $\bar{F}(z) \sim \bar{h}_{ill}(z)$ along the beam axis, i.e., the number of nonscattered, ballistic photons falls off approximately exponentially, as indicated in eq. (14.14).

In other words, the incident holographically shaped field $\mathbf{E}_{holo}$ determines the effective scattering cross section and the penetration depth $d(\phi_{holo})$ in dense media – which is a remarkable result. Although pointed out by earlier studies [27], the strength of this effect has been unexpected. This effect is further manifested by the magnified and autoscaled image areas and linescans of Figure 14.14, which reveal structures on a single

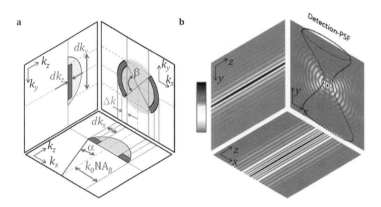

Figure 14.15 A sectioned Bessel beam in k-space and in real space. (**a**) The extents dk_x, dk_y, dk_z, of the sectioned angular spectrum (red areas) as a projection of the Ewald spherical cap define the dimensions of the beam (**b**) The intensity cross-sections $I(x, y, z = 0)$, $I(x, y = 0, z)$ and $I(x = 0, y, z)$ illustrate the section angle β, but also show the propagation invariance. Adapted with permission from [28], *Optics Express*.

cell level with the self-reconstructing Bessel beam – structures that are invisible with a Gaussian illumination beam.

14.3.3   Section Bessel Beams Have a Cross Section Complementarity to the Detection PSF

For all illumination techniques presented so far, the thickness of the illuminated volume (the light sheet) increases with the depth of field. Therefore, optical sectioning and image contrast decrease for large samples.

A novel type of beam is the sectioned Bessel beam [28] that can be generated by blocking opposite sections of the beam's angular spectrum. Sectioned Bessel beams are similar to Mathieu beams, a solution to the Helmholtz-equation in parabolic coordinates, of which the propagation behavior has only been tested in homogeneous space [10, 11].

Other than Bessel beams with a complete ring-shaped angular spectrum, the spectrum of a sectioned Bessel beam consists of two opposed sections of the ring – each spanning over an angle β, as pointed out in Figure 14.15 in the $k_x k_y$-plane. The red-shaded areas are projections of the Ewald sphere onto the three orthogonal planes in k-space. In the $k_x k_y$-plane the angular spectrum representation $\tilde{E}(k_x, k_y)$ exhibits a point symmetry with respect to the optical z-axis. As an extension to eq. (14.8), the spectrum can be sufficiently well described by

$$\tilde{E}_{SeB}(k_x, k_y)$$

$$= \tilde{E}_0 \left(step(k_0 NA_B - k_r) - step(\sqrt{\varepsilon} \, k_0 NA_B - k_r) \right) \cdot rect \left(\frac{k_y}{2k_0 NA_B \cdot \sin(\beta/2)} \right) \tag{14.15}$$

where β is the angular width and rect is defined as $rect(x/b) = 1$ if $|x| \leq b/2$ and 0 otherwise. $\beta = 180°$ results in a conventional Bessel ring spectrum, since the rect-function

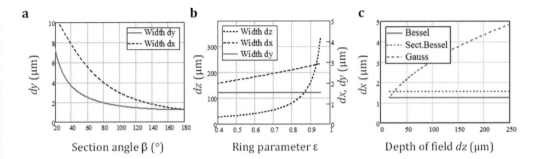

Figure 14.16 Change of the effective beam widths. (**a**) The lateral beams widths of a sectioned Bessel beam change slowly along the detection axis y, but fall off strongly in beam scanning x-direction for angles $\beta > 50°$. (**b**) The depth of field dz can be controlled by the ring thickness ε, which has only a minor effect on the lateral width dx and no effect on dy. The curves shown are for $\beta = 0.6\pi \sim 110°$. (**c**) The width dy of the Gaussian beam increases with increasing depth of field dz, which is not the case for both types of Bessel beams according to the angular spectrum estimation. Adapted with permission from [28], *Optics Express*.

does not cut-off any frequencies. Correspondingly, the field of a sectioned Bessel beam in real space can be obtained by an inverse Fourier transform:

$$E_{\text{SeB}}(x, y, z) = \frac{A}{r} \left(J_1 \left(k_0 \text{NA}_\text{B}\, r \right) - \sqrt{\varepsilon} J_1 \left(\sqrt{\varepsilon}\, k_0 \text{NA}_\text{B}\, r \right) \right) * \text{sinc} \left(k_0 \text{NA}_\text{B} \cdot \sin(\beta/2) \cdot y \right)$$

$$(14.16)$$

where $r = \sqrt{x^2 + y^2 + z^2}$ is the distance to the beam center, $A = \left(2\pi^2 k_0 \text{NA}\, (\varepsilon - 1) \right)^{-1}$ is a factor and the symbol (*) denotes the convolution operation. The intensity cross-sections $|E_{\text{SeB}}(x, y)|^2$, $|E_{\text{SeB}}(x, z)|^2$ and $|E_{\text{SeB}}(y, z)|^2$ are shown in Figure 14.16b. Here x is the beam scanning direction parallel to the light sheet, y is the direction of the detection optical axis perpendicular to the light sheet.

From the uncertainty relation $dj \cdot dk_j = 2\pi$ $(j = x, y, z)$ we obtain the following extents of a sectioned Bessel beam with section angle β:

$$dx = \frac{\lambda}{\text{NA}_\text{B} \cdot \left(1 - \sqrt{\varepsilon} \cos(\beta/2) \right)} \qquad dy = \frac{\lambda}{\text{NA}_\text{B} \cdot \sin(\beta/2)} \qquad dz = \frac{4 \cdot \lambda \cdot n}{\text{NA}_\text{B}^2 \cdot (1 - \varepsilon)}$$

$$(14.17)$$

These beam extents dx, dy, and dz represent effective widths and correspond to real space standard deviations obtained by the inverse of the standard deviations in k-space [12].

As shown Figure 14.16, a Bessel beam's depth of field dz can be steered efficiently by varying the ring parameter $\varepsilon = 0.7 \ldots 0.95$. This has a small effect on the lateral width dx and no effect on dy. Since also the ring parameter is adjusted by wavefront shaping (see Section 14.4), this sectioned conical phase of the Sectioned Bessel beam (SBB) allows to generate illumination beams with enhanced depth of field but without becoming remarkably thicker.

14.3.4   Optical Transfer of Sectioned Bessel Beams in Conjunction with Orthogonal Imaging

The profile of a sectioned Bessel Beam is characterized by the fact that the central lobe of the beam is mainly fed by well interfering, "self-healing" photons from the sides along the scan axis x. If sectioned Bessel beams are used to illuminate a thin line and fluorescence along this line is detected using the confocal-line detection scheme, it is expected that less background will be recorded. As visible in Figure 14.15b, the outline of the detection PSF overlaps with the section Bessel beam only in the beam center. This kind of beam-engineering [28] bares an interesting application potential and will be investigated in detail in the following.

The confocal system point spread function can be assumed to be a product of intensities:

$$h_{sys}(x, y) = h_{ill}(x, y) \cdot h_{det}(x, y). \tag{14.18}$$

An image p_{CL} from a fluorophore distribution $c_F(\mathbf{r})$ obtained by confocalline detection (CL) can be approximated by

$$p_{CL}(\mathbf{r}) = [h_{ill}(\mathbf{r}) \cdot h_{det}(\mathbf{r})] * c_F(\mathbf{r}) \tag{14.19}$$

In Fourier space this is a convolution of the corresponding transfer functions:

$$H_{sys}(\mathbf{k}) = H_{ill}(\mathbf{k}) * H_{det}(\mathbf{k}) = AC\left[\tilde{E}_{ill}(\mathbf{k})\right] * AC\left[\tilde{E}_{det}(\mathbf{k})\right] \tag{14.20}$$

where $\tilde{E}_{ill}(\mathbf{k})$ and $\tilde{E}_{det}(\mathbf{k})$ are the Fourier transforms of the illuminating field (see eq. (14.16)) and of the coherent detection PSF. AC[...] designates the autocorrelation.

This confocal transfer function $H_{sys}(\mathbf{k}, \beta)$ is displayed for various section angles Figure 14.17. The first row illustrate projections of the angular spectrum $\tilde{E}_{ill}(k_x, k_y, \beta)$ along the beam propagation axis z. The second row shows $H_{ill}(k_x, k_y, \beta) = AC[\tilde{E}_{ill}(k_x, k_y, \beta)]$, whereas $H_{sys}(k_x, k_y, \beta)$ is shown in the third row. From the shape of $H_{sys}(k_x, k_y, \beta)$, the beam's sectioning capability can be estimated. It is roughly expressed by the ratio of high spatial frequencies (HSF) to medium/low spatial frequencies (LSF) of the MTF (see also dotted ellipse in bottom of Figure 14.17 and see Figure 14.18). The LSF components lie to the center of the axis region, which appears in yellow and white for all beams indicating the largest amplitudes. Medium amplitudes are shown in red and extend to the highest frequencies (i.e., largest distances to the center) for the conventional Bessel beam. The line-profiles shown to the right of Figure 14.17 confirm that the Bessel beam MTF contains the highest spatial frequencies. However, the support for medium frequencies (see extent of yellow areas in $H_{sys}(k_x, k_y)$ is smaller in comparison to the sectioned Bessel beam for $\beta = 0.6 \ldots 0.8\pi$. This finding indicates the different sectioning capability for sectioned Bessel Beams as analyzed in detail below.

14.3.5   Contrast Analysis through Low-Pass and High-Pass-Filtered Images

Figure 14.18a shows an image of a fluorescently labeled cell cluster without any visible decrease in contrast along the illumination direction (left to right) using Bessel beams,

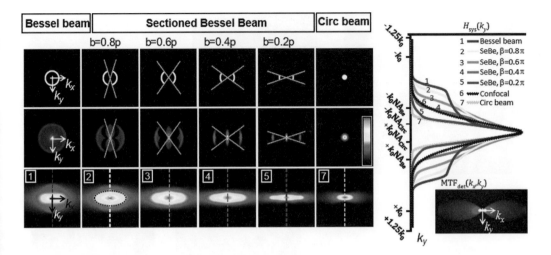

Figure 14.17 Transfer functions of fields and intensities for (sectioned) Bessel beams and a conventional (flat-top) beam. (**a**) The table shows the coherent transfer functions $\tilde{E}(k_x, k_y)$ (angular spectrum) in the upper rows, the intensity transfer functions $AC(\tilde{E})$ in the second row, and the system transfer function $H_{sys}(k_x, k_y)$ in the bottom row. The red and yellow areas in $H_{sys}(k_x, k_y)$ give an estimate for the ratio of high-frequency and medium-frequency information of the beam. Vertical line scans $H_{sys}(k_y)$ along the detection axis are plotted on the right and reveal the ratios of high- and low-frequency contributions. The detection $MTF_{det}(k_x, k_y)$ is shown as an inset (bottom right). Adapted with permission from [28], *Optics Express.*

Figure 14.18b reveals a decay in contrast with Gaussian beam illumination (both with confocal-line detection). Among a variety of existing definitions to analyze the image contrast Q, an elegant and well applicable method is that presented by Truong et al. [29]. Based on this concept, a 2D image can be decomposed into a low-pass (LP) and high-pass (HP) filtered image as shown in Figure 14.18b. The image decomposition depends sensitively on the corner frequency $k_C = 2\pi/d_{max}$, which can be estimated by the extent d_{max} of the largest structure inside the object.

The images p(x,z) are Fourier transformed to $\tilde{p}(k_x, k_z) = FT\left[p(x,z)\right]$ and separated into images with high spatial frequencies (HSF, where $k_r = (k_x^2 + k_y^2)^{1/2} \geq k_C$) and low spatial frequencies (LSF, where $k_r = (k_x^2 + k_y^2)^{1/2} < k_C$) defined by the corner frequency $k_C = 1/\mu m$. The ratio of the average high-pass and low-pass filtered intensities $|\tilde{p}(k_x, k_z)|$ provides a quality parameter for image contrast Q, which is defined by

$$Q(y) = \frac{HSF}{LSF} = \frac{\iint_{k_r \geq k_C} |\tilde{p}(k_x, k_z, y)| \, dk_x dk_z}{\iint_{k_r < k_C} |\tilde{p}(k_x, k_z, y)| \, dk_x dk_z} \qquad (14.21)$$

This expression represents an alternative signal-to-background ratio and effectively describes the usable in-focus intensity signals HSF relative to the image out-of-focus intensity LSF. Since the contrast decreases with the detection depth y, it is useful to analyze $Q(y_i)$ for different image planes $p(x, y_i, z)$ along the detection direction y.

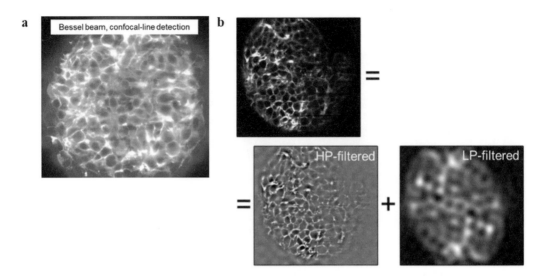

Figure 14.18 Principles of image contrast analysis – here for a tumor multicellular spheroid imaged with confocal-line detection. (**a**) Projection of image slices $\int p(x, y, z)\, dy$ obtained by a Bessel beam illumination. (**b**) Image p(x,y0,z) illuminated with a Gaussian beam and the decomposition into high-pass and low-pass filtered image,
$$p(x, y_0, z) = p_{HP}(x, y_0, z) + p_{LP}(x, y_0, z).$$

14.3.6   Propagation Properties of Sectioned Bessel Beams inside Tumor Cell Clusters

In addition to the analytical concepts and numerical simulations, experimental results obtained from the imaging of cancer cell clusters are presented and analyzed in this section. To test the performance of sectioned Bessel beams we acquired 3D stacks of 250 μm large Tumor Multicellular Spheroids (Mouse colon carcinoma CT26 spheroids cultured during 4 days, fixed in 4% paraformaldehyde, stained with Alexa488).

The performance of sectioned Bessel beam illumination combined with confocal-line detection is compared to three other imaging modes: scanned Gaussian beam illumination as well as confocal-line detection with Gaussian beams and conventional Bessel beams. For each mode, a stack of 13 images with a spacing of 10 μm was recorded. Figures 14.19a–14.19d show images p(x,y$_i$,z) for y$_i$ = 100 μm, located approximately in the center of the spheroid. The images reveal that confocal-line detection is able to improve the visually perceived contrast well above the level of scanned Gaussian beams. However, in the case of Gaussian beam illumination, the signal for large penetration depth (right side) becomes very weak and the structure of the cell walls is hardly visible anymore. The penetration depth was analyzed by integrating the images shown in Figures 14.19b–14.19d, along the x-axis to obtain p$_{avg}$(z) (Figure 14.19e). This measurement quantifies the beam's on axis power. This result confirms the simulation results:

In the images of tumor multicellular spheroids, sectioned Bessel beams enable the same penetration depth as conventional Bessel beams, but at the same time offer a

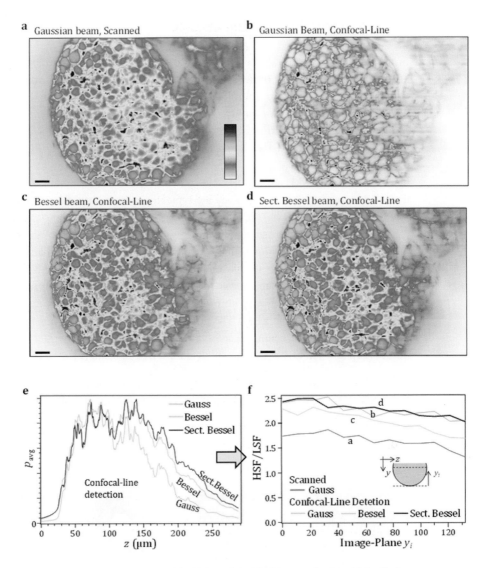

Figure 14.19 Images of tumor multicellular spheroids. Images of spheroidal cell clusters were imaged with Gaussian beam illumination using wide-field detection (**a**), and using confocal-line detection (**b**). For Bessel beams (**c**) and sectioned Bessel beams (**d**) images are shown only with confocal-line detection. The average image intensity along the propagation axis for b,c,d is shown in (**e**). (**f**) Shows the image contrast measured by the ratio of high and low spatial frequencies for the whole image stack. Layers deeper within the spheroid (larger i) show less contrast. The scale bar is 20 μm. Adapted with permission from [28], *Optics Express*.

similar signal-to-background (Q = HSF/LSF) as Gaussian beams with confocal-line detection. Especially in the deeper parts of the sample, sectioned Bessel beams offer significantly higher signal strength. We measured an increase in contrast over conventional Bessel beams of approximately 15%.

14.4   Technical Realization of Beam Shaping

In the previous sections we have outlined and documented the enormous advantages of shaping the wavefronts of the illumination beams in light sheet microscopy. On the one hand different types of illumination beams can be compared directly at the same object and, even more interestingly, in combination with different detection modes. A further advantage of wavefront shaping the illumination beam is that the dimensions of the beam can be attempted to the spatial extent and structural composition (scattering properties) of the 3D specimen. Among the different beam shaping elements, spatial light modulators (SLM) allow a flexible and sufficiently fast switching of the illumination mode (around 30–100 Hz). They are typically operated in reflection mode and are often based on liquid crystal on silicon, LCOS).

Typical beam modes are generated by a two-dimensional profile of the beam phase $\phi_{bm}(x,y)$, which is superimposed by a linear wedge $\phi_{wt}(x)$, to separate the useful first diffraction order from the unwanted zero th diffraction order. As illustrated in Figure 14.20, the sum of the beam phase and the phase grading is multiplied with an annular aperture function (within a radius r_{min} and outer radius r_{max}), to obtain the final hologram for the SLM. The aperture function $A(x,y)$ defines the lateral and the axial extent of the beam. Hence the final hologram can be written as $T(x, y) = A(x, y) \cdot \exp(i\phi_{bm}(x, y) + i\phi_{wt}(x))$, which is for a Bessel beam [2]

$$A(x, y) = step\,(r_{max} - r) \cdot step\,(r - r_{min}), \quad \phi(x, y) = k_r r + k_{x0} x \qquad (14.22)$$

where $r = \sqrt{x^2 + y^2}$. The inner and outer radii of the hologram r_{min} and r_{max}, define the lateral and axial extent of the beam focus, z_{start}-z_{end}, as well as its axial position

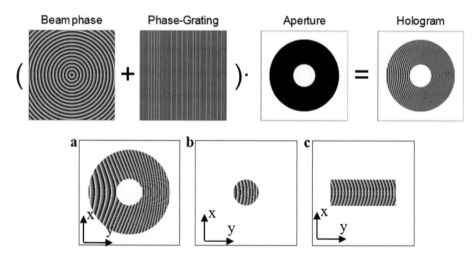

Figure 14.20 Top row: Scheme for the composition of holograms. Bottom row: Phase holograms for different light sheet illumination beams. The hologram for Bessel (**a**) and Gaussian beam (**b**) and a cylindrical light sheet (**c**) is shown in grayscale where a phase shift of $\delta\phi = 0$ is indicated in white and $\delta\phi = 2\pi$ in black. Adapted with permission from [2], *Optics Express*.

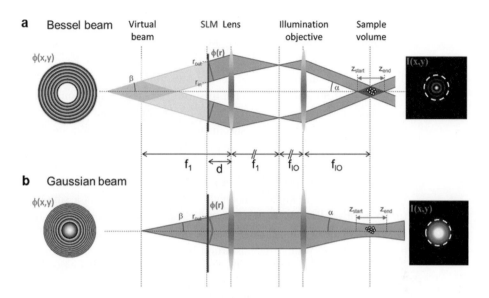

Figure 14.21 The optical path for holographic beam shaping and positioning. The figure shows the generation of Bessel (**a**) and Gaussian beams (**b**) by applying adequate phases on the beam incident on the SLM. A telecentric 4f lens system images a virtual beam created by the SLM (placed in a distance d < f_1 to the lens L1) into the sample volume. A conical phase with ring-shaped aperture (shown by $\phi(r)$) is used for Bessel beams. For Gaussian beams, a spherical phase with a circular aperture is applied. Adapted with permission from [2], *Optics Express.*

relative to the focal length of the illumination objective lens. This is further explained in Figure 14.21. The frequency k_r describes the deflection angle of the phase axicon, the ring thickness $r_{max} - r_{min}$ controls the beam's depth of field and the amount of energy in the Bessel ring system.

The bottom row of Figure 14.20 displays three different holograms for (a) a Bessel beam, (b) a Gaussian beam, and (c) a cylindrical wave, which all have the same depth of field z_{start}-z_{end}, but reveal a completely different propagation behavior through the specimen, resulting in a completely different scattering an interference pattern and thereby also in a different image.

As shown in Figure 14.21, the SLM is not positioned in a conjugate plane to the sample volume, but in a distance d to the next lens generating the spectrum of the field in a distance f_1. This arrangement opens the opportunity to shift the beam position relative to the sample volume, and improves the pixel sampling on the hologram because the beam diameter increases with increasing distance to the conjugate plane (virtual beam).

14.5   Future Work

Computer-generated holograms as the basic beam shaping elements in light sheet microscopy turn out to be the most flexible and precise optical devices to adjust the illumination beam to the object. Both penetration depth and image contrast can be maximized for each object by choosing the right hologram. Only recently, fascinating new approaches have been realized, by a parallelized illumination of weakly scattering objects, such as the holographic generation a lattice of Bessel beams [30], to improve the speeds of beam scanning and image acquisition.

However, if imaging speed is not the main issue, but rather a minimization of image artifacts and thereby an optimization of image contrast and revelation of small structures within a large scattering specimen, one has to think about iterative illumination concepts in light sheet microscopy [31, 32]. To speed up the rate of optimization and iteration, coherent scattering at the object might be better than fluorescence imaging, because sufficiently many coherent photons reach the detector thus allowing iterative image acquisition at higher rates and little photodamage. As a simple demonstration for an iterative illumination correction for phase distortions caused in the object plane, Figure 14.22 depicts a principle setup of how to analyze and correct the angular spectrum of a Bessel beam after distortion by a displaced glass capillary [31]. Both the forward scattering and sideward scattering can be analyzed to feed this information back to the spatial light modulator (SLM), which generates the optimal illumination hologram within a few iterations, either for each single position of a scanned beam or for a confined volume area within the specimen (e.g., the head region between the eyes of a zebrafish embryo). Figure 14.23 shows that it is possible to illuminate different zones of an extended object with different beam positions, beam intensities, and beam profiles, which can be realized by the large interference cross section of Bessel beams.

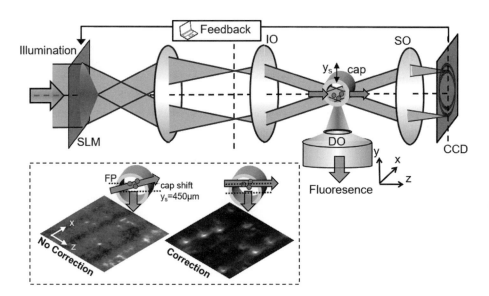

Figure 14.22 Light sheet setup with feedback holographic aberration correction. In the illumination path a SLM enables feedback phase correction resulting from phase distortion of the shifted capillary (cap). The beam distortion is read out with a camera (CCD) in Fourier space. Bottom inset: Images of fluorescent beads inside a capillary shifted by y_s without and with illumination correction. Adapted with permission from [31], Optics Letters.

Figure 14.23 Experimentally obtained image of a holographically shaped Bessel beam propagating through a fluorescence labeled gel. The beam propagates from left to right and is characterized by parallel propagation axes and three axially separated illumination zones 1, 2, and 3.

Advanced computer simulations for beam propagation, faster computer holograms for wavefront shaping and faster cameras in combination with novel detection schemes and analysis procedures, will ensure this field of research a bright and exciting future.

Acknowledgments

This chapter would not have been possible without the great contributions from Dr. Florian Fahrbach and Tobias Meinert, who also did a careful proofreading together with Luis Koebele. I am thankful for their help.

14.6   References

[1] F. O. Fahrbach, P. Simon, and A. Rohrbach, "Microscopy with self-reconstructing beams," *Nature Photonics,* vol. 4, pp. 780–785, 2010.

[2] F. O. Fahrbach and A. Rohrbach, "A line scanned light-sheet microscope with phase shaped self-reconstructing beams," *Optics Express,* vol. 18, pp. 24229–24244, 2010.

[3] J. Huisken and D. Y. R. Stainier, "Selective plane illumination microscopy techniques in developmental biology," *Development (Cambridge, England),* vol. 136, pp. 1963–1975, 2009.

[4] H. U. Dodt, U. Leischner, A. Schierloh, N. Jahrling, C. P. Mauch, K. Deininger, J. M. Deussing, M. Eder, W. Zieglgansberger, and K. Becker, "Ultramicroscopy: three-dimensional visualization of neuronal networks in the whole mouse brain," *Nature Methods,* vol. 4, pp. 331–336, 2007.

[5] T. F. Holekamp, D. Turaga, and T. E. Holy, "Fast three-dimensional fluorescence imaging of activity in neural populations by objective-coupled planar illumination microscopy," *Neuron,* vol. 57, p. 661, 2008.

[6] P. J. Keller, A. D. Schmidt, J. Wittbrodt, and E. H. K. Stelzer, "Reconstruction of zebrafish early embryonic development by scanned light sheet microscopy," *Science,* vol. 322, pp. 1065–1069, 2008.

[7] A. B. Arrenberg, D. Y. R. Stainier, H. Baier, and J. Huisken, "Optogenetic control of cardiac function," *Science,* vol. 330, pp. 971–974, 2010.

[8] T. A. Planchon, L. Gao, D. E. Milkie, M. W. Davidson, J. A. Galbraith, C. G. Galbraith, and E. Betzig, "Rapid three-dimensional isotropic imaging of living cells using Bessel beam plane illumination," *Nature Methods,* vol. 8, pp. 417–423, 2011.

[9] J. G. Ritter, R. Veith, A. Veenendaal, J. P. Siebrasse, and U. Kubitscheck, "Light sheet microscopy for single molecule tracking in living tissue," *PloS One,* vol. 5, e11639, 2010.

[10] E. G. Reynaud, U. Krzic, K. Greger, and E. H. K. Stelzer, "Light sheet-based fluorescence microscopy: more dimensions, more photons, and less photodamage," *HFSP Journal*, vol. 2, pp. 266–275, 2008.

[11] A. Rohrbach, "Artifacts resulting from imaging in scattering media: a theoretical prediction," *Optics Letters*, vol. 34, pp. 3041–3043, 2009.

[12] J. Huisken and D. Y. R. Stainier, "Even fluorescence excitation by multidirectional selective plane illumination microscopy (mSPIM)," *Optics Letters*, vol. 32, pp. 2608–2610, 2007.

[13] P. J. Keller, A. D. Schmidt, A. Santella, K. Khairy, Z. Bao, J. Wittbrodt, and E. H. K. Stelzer, "Fast, high-contrast imaging of animal development with scanned light sheet-based structured-illumination microscopy," *Nature Methods*, vol. 7, pp. 637–642, 2010.

[14] T. Breuninger, K. Greger, and E. H. K. Stelzer, "Lateral modulation boosts image quality in single plane illumination fluorescence microscopy," *Optics Letters*, vol. 32, pp. 1938–1940, 2007.

[15] S. Kalchmair, N. Jahrling, K. Becker, and H. U. Dodt, "Image contrast enhancement in confocal ultramicroscopy," *Optics Letters*, vol. 35, pp. 79–81, 2010.

[16] J. Mertz and J. Kim, "Scanning light-sheet microscopy in the whole mouse brain with HiLo background rejection," *Journal of Biomedical Optics*, vol. 15, 016027, 2010.

[17] W. T. Welford, "Use of annular apertures to increase focal depth," *Journal of the Optical Society of America*, vol. 50, pp. 749–753, 1960.

[18] J. W. Y. Lit and R. Tremblay, "Focal Depth of a Transmitting Axicon," *Journal of the Optical Society of America*, vol. 63, pp. 445–449, 1973.

[19] J. Durnin, "Exact-solutions for nondiffracting beams. 1. The scalar theory," *Journal of the Optical Society of America A*, vol. 4, pp. 651–654, 1987.

[20] Z. Bouchal, J. Wagner, and M. Chlup, "Self-reconstruction of a distorted nondiffracting beam," *Optics Communications*, vol. 151, pp. 207–211, 1998.

[21] V. Garces-Chavez, D. McGloin, H. Melville, W. Sibbett, and K. Dholakia, "Simultaneous micromanipulation in multiple planes using a self-reconstructing light beam," *Nature*, vol. 419, pp. 145–147, 2002.

[22] J. A. N. Buytaert and J. J. J. Dirckx, "Design and quantitative resolution measurements of an optical virtual sectioning three-dimensional imaging technique for biomedical specimens, featuring two-micrometer slicing resolution," *Journal of Biomedical Optics*, vol. 12, 2007.

[23] C. W. McCutchen, "Generalized aperture and the three-dimensional diffraction image," *Journal of the Optical Society of America*, vol. 54, pp. 240–244, 1964.

[24] F. O. Fahrbach and A. Rohrbach, "Propagation stability of self-reconstructing Bessel beams enables contrast-enhanced imaging in thick media," *Nature Communications*, vol. 3, p. 632, 2012.

[25] E. Baumgart and U. Kubitscheck, "Scanned light sheet microscopy with confocal slit detection," *Optics Express*, vol. 20, pp. 21805–21814, 2012.

[26] M. D. Feit and J. A. Fleck, "Light propagation in graded index optical fibers," *Applied Optics*, vol. 17, pp. 3990–3998, 1978.

[27] S. Colak, C. Yeh, and L. W. Casperson, "Scattering of focused beams by tenuous particles," *Applied Optics*, vol. 18, pp. 294–302, 1979.

[28] F. O. Fahrbach, V. Gurchenkov, K. Alessandri, P. Nassoy, and A. Rohrbach, "Self-reconstructing sectioned Bessel beams offer submicron optical sectioning for large fields of view in light-sheet microscopy," *Optics Express*, vol. 21, pp. 11425–11440, 2013.

[29] T. V. Truong, W. Supatto, D. S. Koos, J. M. Choi, and S. E. Fraser, "Deep and fast live imaging with two-photon scanned light-sheet microscopy," *Nature Methods,* vol. 8, pp. 757–760, 2011.

[30] B.-C. Chen, W. R. Legant, K. Wang, L. Shao, D. E. Milkie, M. W. Davidson, C. Janetopoulos, X. S. Wu, J. A. Hammer, Z. Liu, B. P. English, Y. Mimori-Kiyosue, D. P. Romero, A. T. Ritter, J. Lippincott-Schwartz, L. Fritz-Laylin, R. D. Mullins, D. M. Mitchell, J. N. Bembenek, A.-C. Reymann, R. Böhme, S. W. Grill, J. T. Wang, G. Seydoux, U. S. Tulu, D. P. Kiehart, and E. Betzig, "Lattice light-sheet microscopy: Imaging molecules to embryos at high spatiotemporal resolution," *Science,* vol. 346, 1257998, 2014.

[31] T. Meinert, B.-A. Gutwein, and A. Rohrbach, "Light-sheet microscopy in a glass capillary – Feedback holographic control for illumination beam correction," *Optics Letters,* vol. 42, pp. 350–353, 2016.

[32] L. A. Royer, W. C. Lemon, R. K. Chhetri, Y. N. Wan, M. Coleman, E. W. Myers, and P. J. Keller, "Adaptive light-sheet microscopy for long-term, high-resolution imaging in living organisms," *Nature Biotechnology,* vol. 34, pp. 1267–1278, 2016.

15 Shaped Beams for Light Sheet Imaging and Optical Manipulation

Tom Vettenburg and Kishan Dholakia

15.1 Introduction

Optical approaches to imaging the natural world have taken center stage in many applications, particularly in the biomedical arena. The surprisingly powerful approach of using photonics coupled with the use of both endogenous and exogenous fluorophores has led to unprecedented insights in numerous biological systems and disease progression. Methods such as confocal and multiphoton imaging have revolutionized our understanding of the biomedical world in the course of the last decades; however, the nature of such point scanning techniques is associated with a high potential for photodamage when performing live cell imaging. Minimizing irradiance while simultaneously imaging a large field-of-view in thick samples required a radically different approach. Light sheet fluorescence microscopy (LSFM), or selective plane illumination microscopy (SPIM), has come to the fore as an "optical sectioning" method that provides high contrast, fast, and with minimal sample exposure. This simple, yet elegant, modification of a conventional microscope consists in going from an epi- or transmission geometry to one where the illumination and detection are orthogonal to one another (see Figure 15.1). The origins of this method date back more than one hundred years, with R. A. Zsigmondy (Nobel Prize winner in 1925), who performed a key demonstration of the heterogeneous nature of colloid solutions by studying the light scattering from cranberry glass. The orthogonal illumination geometry has been rediscovered for fluorescence microscopy studies. Confining the illumination to an orthogonal "light sheet" yields the sought-after combination of high contrast with several orders of magnitude lower sample exposure than with point-by-point confocal or multiphoton imaging [1, 2], and correspondingly lower photo-bleaching [3]. This opens the door to live cell imaging for developmental biology and neuroscience.

This chapter covers three main advanced themes in light sheet imaging. Firstly, we discuss the widely recognized issues of sample induced optical aberrations. In itself this a broad and topical area given the importance of probing deeper into biological systems. This requirement arises as we wish to gain functional imaging which often can only be gleamed at depths currently difficult to attain (e.g., >1 mm). The scattering and absorptive properties of tissue compromise contrast in thick tissue. This issues is exacerbated in fluorescence light sheet microscopy for which the fluorescence excitation and detection paths do not coincide. In this chapter we discuss different approaches to adaptive aberration measurement and correction of both the illumination and the detection paths for deep tissue light sheet microscopy.

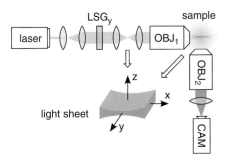

Figure 15.1 Simplified schematic of a light sheet microscope. The illumination path propagates horizontally along the x-axis, and the fluorescence detection path propagates vertically along the z-axis. The laser beam is expanded by a telescope and its focus is spread out along the y-axis by a light sheet generating element (LSG$_y$), such as a scanning mirror or a cylindrical lens. The light sheet is reimaged into the sample by the excitation objective (OBJ$_1$). The light emitted by fluorescence is collected along the z-axis by the detection objective (OBJ$_2$) and reimaged onto a camera (CAM), similar as in a wide-field microscope.

Phase-front modulating elements are typically used in adaptive optics to achieve light fields that closely resemble the truncated Gaussian laser beam outputs that are ubiquitously used in microscopy. In Section 15.3, we discuss how deliberately introducing a controlled phase modulation can offer important advantages. It is well known that the depth of focus of a Gaussian beam is tightly coupled to its wavelength and the spot size: the tighter we focus a beam the shorter its depth of focus. In terms of resolution for light sheet imaging, this has key limitations: the broadening of the Gaussian beam beyond the Rayleigh range reduces resolution. Light sheet illumination created by propagation-invariant light fields has been demonstrated to successfully address this issue. Despite the presence of side lobes for all propagation-invariant fields, important advances have been achieved with their judicious use in both single and two-photon fluorescence excitation. In other chapters the reader will have seen the advantages offered by Bessel light fields to form the light sheet. More complex embodiments include structured light fields and lattices which modulate the Fourier plane annular structure of the Bessel mode. In this chapter we concentrate on a powerful alternative for single photon light sheet microscopy, the Airy light field.

Finally, in Section 15.4, we conclude by looking more broadly at combining all forms of light sheet imaging with optical trapping. This is an example of the growing recognition that multimodal approaches can bring unique advantages for more advanced biomedical studies.

15.2 Wavefront Correction

15.2.1 Introduction

To a certain degree, optical aberrations are present in every imaging system. An example from every day experience is the visible twinkling of star light, a time-dependent deviation from the point-like image one would expect. One may expect the star-light to arrive with approximately planar wavefronts that are focused to a single point at the back of our eye. However, the constantly changing atmosphere induces deviations in the wavefront before it reaches the ground-based telescope, thus causing optical aberrations. Although the wavefront deviations are relatively small compared to the scale of the instrument, variations on the scale of a wavelength result in a visible degradation of the point image. A similar problem arises when imaging biological specimen with a fluorescence microscope. The wavefronts emitted by a fluorophore will deviate from the perfect sphere due to the inhomogeneities in the optical properties of biological tissue, thus preventing the microscope objective and detection optics to form a well-defined focus and sharp image. Adaptive optics have been used both in astronomy and in microscopy to achieve a sharp image in the presence of such aberrations.

15.2.2   Aberration Correction in Light Sheet Microscopy

Light sheet microscopy stands out for its ability to image intact specimen or organs at depths beyond what can typically be achieved using alternative optical microscopy techniques. Although the light sheet illumination improves the contrast, the optical properties of most biological tissues preclude meeting the diffraction-limited resolution at imaging depths of a few hundred of microns. As in wide-field microscopy, scattering, or merely refractive index inhomogeneities, distort the spherical wavefront emitted from a point emitter in the sample, thus preventing the microscope objective and tube-lens from forming a perfect focus on the detector array surface. Unlike most wide-field modalities, the light sheet microscope's illumination path is focused through a different part of the sample. To approach optimal imaging conditions, both the detection and the illumination paths require aberration correction independently.

The light sheet can be produced either by introducing a cylindrical lens or by rapidly swiping the beam focus across the field-of-view one or more times while capturing the image. Any deviation from the optical cylindrical or spherical wavefront will result in a wider than desired light sheet, in principle deteriorating the optical sectioning capabilities of the light sheet microscope. Adaptive optics has a dual purpose in light sheet microscopy: ensuring optimal lateral resolution, as well as optimal axial confinement.

15.2.2.1   Correction of the Fluorescence Detection Path

Unlike a wide-field microscope, the light sheet microscope can excite fluorescence deep into the tissue without the background fluorescence that limits contrast in wide-field microscopy. The ability to image at depth can only be fully exploited if image sharpness is maintained through multiple layers of inhomogeneous tissue. Scattering invariably causes resolution loss, prompting the use of optical clearing methods to homogenize the optical properties of the tissue *ex vivo*. Yet, light sheet microscopy is a very promising method for *in vivo* imaging where optical inhomogeneities cannot be disregarded. The resolution loss due to optical aberrations can however be mitigated using adaptive optics as was demonstrated for light sheet microscopy by Bourgenot et al. [4]. In this work the authors use a deformable mirror as the correcting device, placed in a plane conjugate

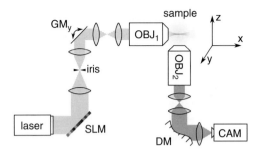

Figure 15.2 Simplified schematic of a digitally scanned light sheet microscope with adaptive correction of both the illumination path and the fluorescence detection path. An expanded laser beam is collimated onto the active surface of a reflective spatial light modulator (SLM). An iris is used as a spatial filter to remove light outside of the region of interest. The active surface of the SLM is reimaged onto a beam scanning device such as a galvanometer-mounted mirror (GM_y). The scanning device is conjugated to the back aperture of the excitation objective (OBJ_1) so that the focus in the sample is rapidly swiped along the y-axis to form a virtual light sheet. The emitted fluorescence is collected by the fluorescence objective (OBJ_2) which has its back aperture reimaged onto a deformable mirror (DM) before an image is formed on the detector surface of the camera (CAM).

to the back aperture of the detection objective as shown in Figure 15.2. To measure the aberration, the deformable mirror is first calibrated without the sample so that the mirror's membrane can be steered in 12 orthogonal Zernike modes. As low-order aberrations are expected to be dominant, the first Zernike modes, excluding piston, tip, and tilt, are corrected for. Piston only affects the phase, not the intensity, while tip and tilt cause a shift in the image and therefore do not alter image sharpness. By capturing images with a series of Zernike mode mirror shapes, the coefficients are determined for each mode to maximize an image sharpness metric. A well-chosen metric must lead to the diffraction-limited image without any remaining aberrations. In this state the mirror shape exactly cancels the wavefront error and inverts the optical aberrations.

15.2.2.2 Correction of the Illumination Path

In a light sheet microscope, optical aberrations are not only present in the orthogonal detection path, but also in the optical path illuminating the focal plane. The distinct advantage of light sheet microscopy is the confinement of the illumination to a single plane. Yet optical aberrations may preclude the formation of a thin homogeneous light sheet that has a diffraction-limited width. The broadened illumination leads to a loss in contrast and resolution whilst unnecessarily irradiating the sample. Precise control over the light sheet becomes more important when using higher NA objectives that confine the light sheet width to sub-cellular features. In this section we show that adaptive optics needs not be restricted to the detection path, aberrations can also be overcome in the illumination path. In principle a second deformable mirror could be used to correct the light sheet illumination; however, a spatial light modulator has significant advantages when used in the illumination path of a light sheet microscope. Although the wavelength-independent operation of the deformable mirror is an important advantage when considering light with a significant spectral width such as that of fluorescence, it should be noted that the number of degrees of freedom is typically limited, e.g., to 52 in the aforementioned study [4]. The main advantages of the spatial light modulator are its high number of independent degrees of freedom, on the order of one million, and its relatively low crosstalk and cost. As a diffractive element, the spatial light modulator is more suitable for the modulation of light with a single wavelength and polarization [5], such as the laser light typically used in light sheet microscopes. Furthermore, diffraction losses in the illumination path are less important since commercially available laser powers are several times higher than what is typically required for light sheet microscopy. In contrast, fluorescence light emitted by the sample is scarce, has a broad spectrum, and is typically randomly polarized. The deformable mirror is therefore a better choice for the detection path, while the spatial light modulator is more flexible and cost effective for use in the illumination path.

To achieve a diffraction-limited light sheet width, Dalgarno et al. [6] applied the *in situ* aberration correction approach described by Cizmar et al. [7], to a digitally scanned light sheet microscope. In a digitally scanned light sheet microscope [8], a focused beam is rapidly scanned to form a sheet of light in the focal plane. Although rapidly scanning a beam adds significant complexity to the instrument, it offers more control over the homogeneity of the illumination. It should also be noted that some of the more recent light sheet microscopy variants rely on a scanned beam implementation and would not work with the original cylindrical lens implementation [9–12]. To determine the aberration, the digital scanning is stopped temporarily so that the laser beam is focused to a fixed position in the sample. Initially sample-induced aberrations will prevent the formation of a tight focus; however, by measuring the intensity of a reference embedded in the sample while modulating the laser beam, the intensity can be maximized at the reference position. As a reference, one can use a small endogenous fluorescent feature of the sample, or a bright fluorescent microsphere introduced during biological development or sample preparation. The intensity of the reference can be monitored by tracking the value of one or a few pixels in the captured images.

Several algorithms exist for determining the optical aberration and therefore the correction pattern to be imposed by the spatial light modulator. Iterative approaches are discussed in detail in Chapter 8. Here we opt for a non-iterative measurement of orthogonal modes from which the correction can be readily determined in a single step. Orthogonal mode measurements are efficient when the light propagation is not yet in the diffusive regime and all contributing modes can be measured within the required time frame. As the orthogonal Zernike modes of the deformable mirror, independent modes can be chosen on the spatial light modulator. Dalgarno et al. [6] choose the modes as non-overlapping identical rectangles in a grid on the spatial light modulator surface. As depicted in Figure 15.3a, at any one time light is transmitted only through two rectangles, one at the center of the spatial light modulator and a second one that is scanned across its surface. The two light fields are combined by the objective to interfere in the focal region. The spatial light modulator can control the relative phase of both fields. The phase difference maximizing the constructive interference at a specific position can be readily found by monitoring the intensity of a bright point scatterer or a small fluorescent

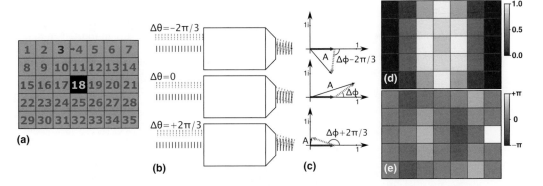

Figure 15.3 The *in situ* measurement of aberrations by interference from two rectangular sub-apertures. (**a**) Indication of the areas used on a spatial light modulator. One sub-aperture at the center is opened, while a second aperture is scanned through all positions of the active surface of the spatial light modulator. At each position, the phase difference between the light passing the central rectangle and the scanned rectangle is cycled through the phases $-2\pi/3$, 0, and $2\pi/3$. (**b**) The waves entering the objective and the interference in the sample. (**c**) The interference at the sample probe position as the coherent sum of two components. The measured intensity is an indication of the magnitude of interference (thin arrow), of the static reference field (solid thick arrow), and the scanned field (dotted thick arrow). Three measurements allow the determination of the relative amplitude and phase of the scanned field (**d**, **e**). An example measurement result showing the presence of coma. The complex amplitude and phase are shown separately in panels (**d**) and (**e**).

emitter that acts as an intensity probe in the sample. The relative phase that maximizes the intensity, also minimizes the residual wavefront error for the corresponding rectangular section of the aperture. By repeating the phase optimization for all rectangular sub-apertures on a grid covering the back aperture of the objective, a tiled aberration correction pattern can be constructed and encoded on the spatial light modulator. This modulation pattern reverses the phase-errors induced by the sample and maximizes the focal intensity at the reference point in the sample.

To determine the optimum phase for each back aperture section, a minimum of three intensity measurements are required while changing the phase of the scanned sections with respect to the central section. Light wave transmitted through an individual section, s, of the back aperture is a harmonically oscillating electric field proportional to $e^{-i\omega t}$, where ω is the angular velocity of the illumination light. Propagation to the reference point in the sample causes an amplitude reduction and phase delay that can be represented succinctly by an *a priori* unknown complex constant $A_s e^{i\phi_s}$, where A_s is the relative amplitude of the electric field that reaches the reference point from aperture section s, and ϕ_s is phase difference incurred by the propagation through the sample. Furthermore, the spatial light modulator can add an additional phase delay, θ_s, to the light traversing the back aperture section s. The real electric field oscillation, received from aperture section s at the reference point in the sample can be understood as the real part of the complex function $Es(t)$:

$$E_s(t) = e^{-i\omega t} A_s e^{i(\phi_s + \theta_s)}. \tag{15.1}$$

The intensity measured at this point will be constant and proportional to $|A_s|^2$, independent of any other variable. However, when the laser light is transmitted simultaneously through two aperture sections, r and s, the field at the reference point in the sample is

$$E_{rs}(t) = E_r(t) + E_s(t) \tag{15.2}$$

$$= e^{-i\omega t} A_r e^{i(\phi_r + \theta_r)} + e^{-i\omega t} A_s e^{i(\phi_s + \theta_s)} \tag{15.3}$$

$$= e^{-i\omega t} A_r e^{i(\phi_r + \theta_r)} \left[1 + A e^{i(\Delta\phi + \Delta\theta)} \right], \tag{15.4}$$

where $A = A_s/A_r$ is the relative amplitude of the two interfering waves, while $\Delta\phi$ and $\Delta\theta$ are the phase differences between the two light propagation paths incurred by the sample and the spatial light modulator, respectively. The corresponding intensity is proportional to

$$I_{rs,\Delta\theta} \propto \left| 1 + A e^{i(\Delta\phi + \Delta\theta)} \right|^2 = 1 + |A|^2 + 2A \cos(\Delta\phi + \Delta\theta), \tag{15.5}$$

where $\Delta\theta$ can be controlled during the measurement.

The two unknown variables, A and $\Delta\phi$, can be determined from three intensity measurements, e.g., for $\Delta\theta$ equal to 0, $\pi/2$, and $3\pi/2$, as shown in Figures 15.3b and 15.3c. Regular spacing of the phases improves the conditioning of the system of equations and simplifies the calculation of A and $\Delta\theta$. Note that only the final term of Eqn. 15.5 is dependent on $\Delta\theta$ and that this term equals the sum $Ae^{i(\Delta\phi + \Delta\theta)} + Ae^{i(-\Delta\phi - \Delta\theta)}$. It follows that multiplying the intensities $I_{rs,\Delta\theta}$ by $e^{-i\Delta\theta}$ and calculating their mean yields $Ae^{i\Delta\phi}$, for uniformly distributed phases, $\Delta\theta$. This immediately reveals both the relative amplitude A and phase $\Delta\phi$ between the light waves traversing the sample from sections s and r. A complete map of the aberration can be obtained by repeating this process for all possible back aperture sections s with respect to the reference section, r, for instance at the center of the back aperture as shown in Figures 15.3d and 15.3e.

Although correction of the phase assures constructive inteference at the position of the sample probe, a non-uniform amplitude may also cause a deviation from the theoretical intensity distribution at focus. In principle the spatial light modulator should invert the measured aberration $Ae^{i\Delta\phi}$ both in phase and amplitude. The phase can readily be inverted by the spatial light modulator; however, amplitude modulation is only possible by selectively reducing the deflection efficiency to the lowest measured transmission. The resulting transmission efficiency will often be too low to be of practical value. A more efficient approach could be to restrict the amplitude reduction to the areas of the spatial light modulator that produce the highest field amplitudes, e.g., those larger than 25% of the maximum transmission, thereby limiting the reduction in efficiency while still modulating the dominant components correctly.

The rectangles used during the aberration measurement process correspond to an orthogonal basis for the measurements. This prevents redundancy in the measurements and allows the combination of per-mode optimizations by simple summation. It should be noted that alternative orthogonal bases exist. Albeit less intuitive, a practical choice of orthogonal modes are the complex harmonic functions $e^{i(k_x x + k_y y)}$, where (k_x, k_y) and

(x, y) are the spatial angular frequency and location in the spatial light modulator plane. For a spatial light modulator with a rectangular active area $\Delta x \times \Delta y$, the orthogonal modes are the components of the discrete Fourier transform with k_x and k_y any integer multiple of $2\pi/\Delta x$ and $2\pi/\Delta y$, respectively. Instead of deflecting only a small rectangular fraction of the light, each of these modes will deflect all the light reaching the spatial light modulator surface as a planar wave in the direction (k_x, k_y, k_z), where $k_z^2 = (2\pi/\lambda)^2 - \left(k_x^2 + k_y^2\right)$ for the wavelength λ. To determine the aberration, instead of using two sub-apertures, two modes are combined by addition of their complex harmonic functions (Figures 15.4a and 15.4b). This produces two beams that travel at a slightly different angle toward the back aperture of the excitation objective as shown in Figure 15.4c. Note that before addition, the modes must first be scaled by a factor of 0.5 to ensure that the modulation has a maximum magnitude of 1 and can be reproduced accurately with the spatial light modulator. Although the interference patterns in the sample will be very different from those created by rectangular sub-apertures, the variation of the probe intensity with incident phase difference, $\Delta\theta$, can be used in the same way to sense how the sample affects different light paths (Figure 15.4d). The maximum

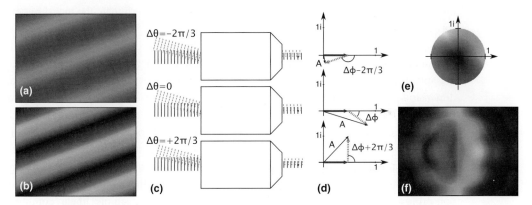

Figure 15.4 The *in situ* measurement of aberrations by interference of two plane waves at the back aperture of the objective. (**a**) Depiction of the two interfering waves at a given time. The axially incident reference wave has a uniform electric field amplitude across the back aperture plane. The second plane wave, propagating at an angle with the optical axis, has a linear phase dependency across the back aperture, resulting in an interference pattern with periodic amplitude. (**b**) Depiction of the combined complex-valued field at the back aperture plane where brightness corresponds to amplitude and hue to phase, encoded as indicated by the color-legend in panel (**e**). (**c**) The two waves entering the objective and the interference in the sample for incident phase differences $\Delta\theta \in \{-2\pi/3, 0, 2\pi/3\}$. (**d**) The interference at the sample probe position as the coherent sum of two components. The measured intensity is proportional to the squared magnitude of the interference (thin arrow), of the static reference field (solid thick arrow), and the non-axial wave (dotted thick arrow). Three measurements allow the determination of the relative amplitude and phase of the light path corresponding to the non-axial direction. By repeating the measurements for different incident directions, a Fourier component of the aberration is measured per direction. (**e**) The complex value depiction used in (**b**) and (**f**). (**f**) An example measurement result for the same aberration as in Figures 15.3d and 15.3e.

intensity is obtained when $\Delta\theta$ cancels out the sample induced phase. Instead of determining the amplitude and phase of the light transmission from a specific sub-aperture to the sample probe, the measured values now correspond to the transmission of a plane wave incident on the back aperture from a specific direction. As with scanning the sub-aperture position, scanning the angle of the incident wave allows us to determine the complete sample induced aberration. The aberration is now calculated as a linear combination of the harmonic functions, weighted by the measured amplitude and shifted by the measured phase (Figures 15.4e and 15.4f). Note that the tiling of rectangular sub-apertures can similarly be understood as a linear combination of functions with a finite support.

Although the use of harmonic functions as a basis may appear somewhat contrived, it has the important practical advantage that the measurement is more light efficient. Instead of blocking all but two rectangular areas, approximately half the light is used when combining two harmonic modes, thus reducing the required illumination power, or measurement time, with respect to that of the method using rectangular apertures. A second practical advantage is that by measuring the smallest propagation angle first, one can obtain rapidly a low-resolution approximation of the aberration and correction pattern. By measuring for increasing propagation angles a continuously improved aberration correction is obtained. This often makes it possible to terminate the measurement early when low order aberrations dominate.

In general the measurement process and the final correction require both amplitude and phase modulation, either to block all but two rectangles or to produce the smoothly varying amplitude of two interfering planar waves. The measured two-dimensional function will be complex, with an amplitude and a phase, its complex argument. While correction of the phase alone may be sufficient to obtain high intensity at the focus, approaching the diffraction-limited focal spot requires the correction of both the phase and the amplitude. However, most spatial light modulators are either amplitude-only or phase-only modulators. This is particularly pertinent when measuring using harmonic functions. By ignoring the amplitude modulation, non-orthogonal, suboptimal, basis functions are used and a simple linear combination may lead to a bias in the measurement. Section 15.2.3 therefore discusses how both phase and amplitude modulation can be readily achieved using only a phase modulator.

15.2.3   Complex Modulation Using a Phase-Only Spatial Light Modulator

The aberration measurement protocols assume the ability of complex modulation, i.e., modulating both the phase and the amplitude of the light field. However, in general, spatial light modulating devices don't offer independent control over both the phase and the amplitude. Although a phase modulating device can be combined with an amplitude modulating device using a re-imaging telescope, it has been demonstrated that complex modulation can also be achieved using a single phase-only modulator. Several schemes have been proposed [13–19]. In what follows we discuss an efficient and straightforward implementation.

Although a phase-only spatial light modulator cannot alter the total intensity of the light, the intensity distribution after a lens can be controlled. An appropriate complex modulation scheme could project excess power outside the region-of-interest. If required, light outside of this region-of-interest can be blocked by a an iris placed at the Fourier plane of the spatial light modulator. By limiting the field-of-view, one implicitly imposes a low-pass filter at the plane of the spatial light modulator, i.e., blurring the effect of neighboring pixels. Spatial light modulators have on the order of a million pixels, offering far more degrees of freedom than the number of aberration modes that are typically considered. A straightforward way to achieve amplitude modulation is to group 2×2 neighboring pixels into a super-pixel as depicted in Figure 15.5a. The light reflected by these four phase-only pixels will interfere in the sample and enable independent amplitude and phase modulation. Zero amplitude can be simulated by ensuring that two of the sub-pixels impose a phase that differs by half an optical period from the other two sub-pixels. The unwanted light will be diffracted outside the region-of-interest at the highest diffraction orders of the spatial light modulator. Maximum amplitude modulation can be achieved by setting the phase of the 2×2 pixels to an identical value. Phase differences less than half a period yield any in-between amplitude as can be seen from the diagram shown in Figure 15.5b. To simulate a relative amplitude A, the phase difference should be twice the arc-cosine of the amplitude: $2\cos^{-1}(A)$. A super pixel with amplitude A, and phase θ consists of pixels with phases $\phi_\pm$:

$$\phi_\pm = \theta \pm \cos^{-1}(A). \tag{15.6}$$

When the pixels with equal phase are chosen to be diagonally opposing, the excess intensity will be directed to the highest diffraction orders in the four diagonal directions. The pattern formed at these points can be relatively broad, it is thus convenient that such disturbance is as far as possible separated from the region-of-interest.

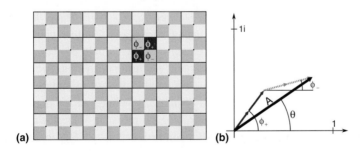

Figure 15.5 Principle of complex field modulation using a single phase-only device. (a) A 14×10 pixel area of a spatial light modulator is divided in 7×5 super pixels. The four phase-only sub-pixels of each super pixel are programmed with phases deviating positively (ϕ_+) or negatively (ϕ_-) from the argument, θ, of the complex function at the super pixel. (b) The four constituent fields of a super pixel are depicted in the complex plane by the two solid arrows (ϕ_+), and the two light dotted arrows (ϕ_-). When $\phi_\pm = \theta \pm \cos(A)$, the field components corresponding to the four components add up to yield the complex modulation $A \exp(i\theta)$.

15.2.3.1 Measures to Improve Accuracy

The modulation scheme approximates the field produced by a complex spatial light modulator, with fewer yet larger pixels, within its field-of-view. Excess light is directed toward four diagonally opposing spots outside the Fourier space of the hypothetical complex modulator, yet within the larger Fourier space of the phase-only modulator. Although this often works well in practice, it is important to understand the limitations of this approach. The complex modulation scheme does not affect the numerical aperture and thus the resolution in image space; however, it does limit the accessible field-of-view in x and y to half of that for phase-only modulation. Modulation outside of this limit requires rapid variations within the super-pixels, and this has a knock-on effect on the field inside the region-of-interest of the complex modulation. In theory such interference can be mitigated by band-limiting the to-be-encoded complex function to half the frequency of the phase-only spatial light modulator. In practice, the band-limitation can be well approximated by smoothing of the complex function before encoding it on the phase-only device.

Instead of using 2×2 super-pixels consisting of four sub-pixels, one could use 2×1 or 1×2 super-pixels consisting of only two sub-pixels. Such an asymmetry avoids the band limit in either the horizontal or the vertical dimension. This may be advantageous when the region-of-interest is centered near the horizontal or vertical axis, respectively. For example, when a vertical or horizontal blazed grating is used to avoid non-modulated light. The excess light will now go to two opposing spots at the extremes of either the vertical or horizontal axis, instead of toward four opposing spots on the diagonals. By displacement of the region-of-interest in the orthogonal direction, one can minimize the possibility of unwanted interference.

15.3 Airy Beams for Light Sheet Microscopy

15.3.1 Introduction

The light sheet microscope achieves high contrast and minimal photo-bleaching by restricting the illumination to a single plane; while wide-field detection, orthogonal to the fluorescence excitation plane allows rapid capturing of the fluorophore density distribution in a large volume. The prodigious combination of minimal irradiation, high contrast, and fast volumetric imaging can perhaps explain the rapid adoption of the light sheet microscope in various branches of biological research. Widespread use drives the demand for further improving resolution, contrast, and speed. Resolution and contrast can be improved by increasing the numerical apertures and confining the light sheet to sub-micron width. However, diffraction limits the distance over which a light sheet can be narrowly confined. A conventional, apertured Gaussian, light sheet is well confined over a distance that is proportional to the square of its width at the waist of the light sheet. Although the light sheet could be confined to approximately 1 micron over a distance of the diameter of a single cell, the rest of the organism will be irradiated by a broad wedge of light, with the consequent negative impact on contrast, resolution, and sample exposure. To overcome this trade-off, propagation-invariant Bessel and Airy beams have been employed to yield uniform light sheets over an extended field-of-view [9–22]. Moreover, such beams are associated with an enhanced tolerance to optical obstacles in the illumination path 23–25 (see also Chapter 14). Perhaps not unrelated, the transverse intensity profile of such propagation-invariant light sheets is significantly broader than that of a truncated Gaussian light sheet at the same numerical aperture. Principles of time-sequenced structured illumination or confocal scanning can be introduced in the light sheet microscopes to filter-out the effect of the transversal structure of the Bessel beam. Such filtering is not required for the asymmetric intensity profile of the Airy beam because it yields images that are well-conditioned for digital deconvolution. In what follows we focus on the Airy beam and show how a light sheet with an Airy function profile can be created and used to convert a conventional light sheet microscope into a high-resolution, wide field-of-view light sheet microscope.

15.3.2   Working Principle

In the same way as digitally scanned light sheet microscopy forms a virtual Gaussian light sheet [8], an Airy beam can be rapidly swiped in the y-dimension to form a virtual Airy light sheet. This process is depicted in Figure 15.6, showing cross sections of the beam and light sheet at various points of the illumination path. As the Bessel beam, the Airy beam has a cross-section with an intensity profile that does not change with propagation. In the theoretical limit, such beams would produce infinitely extended light sheets. Practical implementations, with a finite extent, are achieved using an appropriate mask conjugate to the back aperture of the illuminating objective. The Bessel beam requires a mask that transmits only light through a ring at the edge of the aperture,

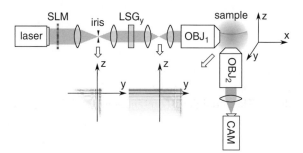

Figure 15.6 Schematic depiction of the Airy beam light sheet microscope. A spatial light modulator (SLM) imposes a cubic phase modulation on the expanded laser beam. An iris filters out undesired refraction orders. The SLM is reimaged onto a light sheet generating element (LSG$_y$) that spreads out the intensity of the Airy beam along the y-axis to form the Airy light sheet. This is often a scanning mirror; however, it can also be a acousto-optic deflector or a cylindrical lens. The Airy light sheet is reimaged into the sample using the excitation objective OBJ$_1$. A z-stack of images is acquired using the fluorescence objective OBJ$_2$ and the camera (CAM). The final image is obtained by deconvolution of the z-stack.

an illumination pattern that is oftentimes achieved using an axicon. The propagation invariance of the Bessel beam is inversely proportional to the annular width used to produce it. The Airy beam is formed by transmitting all light, while imposing a bi-cubic wavefront modulation of the form $\alpha\left(u^3 + v^3\right)$, where u and v are normalized Cartesian coordinates with origin at the center of the aperture, and α is a factor that controls the modulation depth. Typical values of α are between 3 and 15 times the wavelength, with the propagation invariance of the Airy beam is proportional to the α value.

A light sheet microscope can be readily converted into an Airy beam light sheet microscopy by inserting a phase modulating element, or imposing the phase on an existing element such as a mirror or an aberration correcting spatial light modulator. The raw images obtained with the Airy light sheet microscope are slightly blurred in the axial dimension (z), and somewhat warped. However, in comparison to the image of a conventional light sheet image for the same numerical aperture, the Airy light sheet image will already show increased uniformity across the field-of-view. A diffraction-limited image with an order of magnitude larger field-of-view can readily be achieved by digital post-processing. Both the warping and the axial blur can be removed with a single, efficient, one-dimensional linear deconvolution as detailed by Vettenburg et al. [22] (supplementary information).

15.3.3 Imaging Performance of the Airy Beam Light Sheet

As the conventional light sheet microscope, the Airy beam light sheet microscope captures the full field-of-view in a single snapshot without blocking any light. Although the high acquisition efficiency keeps sample exposure to a minimum, it may not be immediately obvious that high contrast and resolution can be attained over an extended field-of-view. In what follows we compare the field-of-view of propagation-invariant light sheets to that of the, conventional, truncated Gaussian light sheet. Next, the performance of those light sheets is compared as a function of the position in the field-of-view.

15.3.3.1 Field-of-View

The truncated Gaussian light sheet, used in a conventional light sheet microscope, has a wedge-like spatial intensity distribution. Optimal axial confinement and illumination efficiency are achieved at its apex, where the light sheet width is limited by diffraction. At other positions, the intensity distribution broadens, thereby limiting the axial resolution and causing unnecessary irradiation of the sample. An important trade-off exists between the width of the light sheet and the field-of-view over which its intensity remains confined. A two-fold reduction in light sheet width causes four-fold reduction in field-of-view.

Such trade-off can be circumvented by the use of propagation-invariant Bessel or Airy beams, which have an intensity cross section that does not change with propagation. Light sheets created with practical implementations of such beams are shown in Figure 15.7a, below the conventional truncated Gaussian light sheet for the same numerical aperture. The beams propagate along the x-axis, are swiped into a light sheet along the y-axis (see Figure 15.6), and the image is taken along the z-axis. Although the

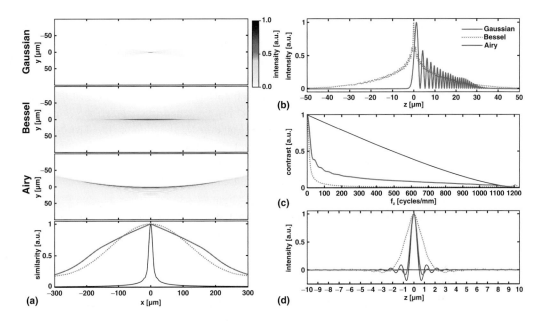

Figure 15.7 Comparison of the intensity profiles of light sheets generated with different beam types. (**a**) The plot shows the similarity of three light sheet intensity profiles with that of their section at $x = 0$, for a truncated Gaussian beam (thin solid line), a Bessel beam (dotted line), and an Airy beam light sheet (thick solid line). It can be seen that the Bessel and the Airy light sheet have intensity cross sections that vary significantly less with propagation distance, x, than that of the truncated Gaussian light sheet. The inverted intensity cross-section image of the respective light sheets are shown above the plot. (**b**) The intensity cross section at $x \equiv 0$ of the respective light sheets, normalized to the maximum value. (**c**) The normalized contrast as a function of spatial frequency. (**d**) The impulse response after axial deconvolution of the image of point-emitter acquired with the three respective light sheets. It can be seen that high contrast values seen in (**c**) yield a high-resolution image reconstruction.

propagation-invariant regions are finite for the practical realizations of the Bessel and Airy beam light sheets, it is clear that their extent is several times wider than the conventional Gaussian light sheet. The propagation-invariance is quantified in the bottom plot of Figure 15.7a. This plots shows the similarity between the intensity profile of the light sheet at x with that at the light sheet waist ($x \equiv 0$). Similarity is defined here as one minus the normalized root-mean-square of the intensity difference. It can be seen that both the Bessel beam light sheet (dotted line) and the Airy beam light sheet (thick solid line) remain uniform for a field-of-view width of hundreds of microns, while the Gaussian light sheet (thin solid line) cannot maintain uniformity for the same high numerical aperture.

15.3.3.2 Resolution

Even though Bessel and Airy beam light sheet are uniform over a larger field-of-view, this does not guarantee high contrast or good axial resolution. When comparing the intensity cross sections at $x \equiv 0$ plotted in Figure 15.7b for the respective light sheets, it becomes clear that only the conventional light sheet (thin solid line near $z = 0$ in Fig. 15.7b) is well confined axially. Without further modifications to the imaging method, propagation invariant light sheets would lead to low axial resolution. Various modifications have been proposed to filter out the fluorescence excited by the transversal structure of the Bessel beam. This avoids the near-zero axial contrast seen in the axial modulation transfer function of the Bessel beam (dotted line) shown in Figure 15.7c. It can also be noted that the axial modulation transfer function for the Airy beam light sheet is significantly higher (thick solid line). This indicates that the fluorescence excited by the transversal structure of the Airy beam light sheet can be recombined in a high-resolution image by deconvolution, without the need to discard any light. Indeed, after deconvolution, the axial intensity cross-section of a point-source imaged by the Airy light sheet is comparable to that of the conventional Gaussian at its best focus (Figure 15.7d, thick solid and thin solid line, respectively). Note that deconvolution cannot restore diffraction-limited axial resolution for the Bessel beam (dotted line). The fluorescence excited by the Bessel-beam must thus be removed by multi-photon microscopy or introducing concepts of structured illumination and confocal scanning as discussed in the previous chapter.

15.3.4   A Compact Implementation Using Common Off-the-Shelf Lenses

The Airy beam is most often produced by the Fourier transform of a two-dimensional cubic phase modulation of the form $\alpha \left(u^3 + v^3\right)$. A spatial light modulator in the illumination path can be used for both aberration correction and cubic phase modulation to accurately produce an Airy beam in the sample. Although such an approach offers maximum flexibility and high accuracy, the cost of a spatial light modulator may be avoided by using static components. Custom phase masks, either diffractive or refractive, are produced commercially; however, for small orders the production cost may still be considerable. An additional obstacle to cost reduction and compactification, now a common requirement in many advanced light sheet microscopy techniques, is the need to digitally scan the shaped beam. In this section we consider an alternative approach that exploits the aberrations introduced by a cylindrical lens.

Unlike the Bessel beam light sheet, the Airy light sheet can be produced without digitally scanning a beam. The cubic modulation that yields the Airy beam is rectangularly separable in a component orthogonal to the light sheet, αu^3, and a component in the light sheet plane, αv^3. For rectangular apertures, and in approximation for circular apertures, the rectangular separability of the Fourier transform leads to the separability of the Airy beam in its focal region. Averaging out the intensity in the plane, either by digitally scanning the beam into a light sheet or by use of a cylindrical lens, does not affect the transverse profile of the light sheet. It can thus be concluded that a cylindrical lens can produce an Airy light sheet from an Airy beam. Moreover, a one-dimensional Airy beam, generated by one-dimensional cubic modulation, αu^3, is sufficient. Such a one-dimensional cubic modulation can be readily approximated by introducing a well-controlled amount of optical aberrations and thus offers a path to a compact, low-cost, low-complexity implementation.

Optical aberrations are deviations of the optical wavefront from the optimal, e.g., planar or spherical geometry. Typically such aberrations are smooth and well-described by the lowest order Zernike polynomials [26]. In the special case of a cylindrical lens, the target wavefront is a cylindrical wavefront and the aberration is straightforward to describe with one-dimensional polynomials, $\sum_{k=0}^{\infty} c_k u^k$. The, typically dominant, lowest orders correspond to a constant phase offset ($\propto u^0$) that doesn't affect the image, a lateral (z) translation of the focal spot ($\propto u^1$), an axial (x) translation of the focal spot ($\propto u^2$), and the cubic modulation ($\propto u^3$).

When used as designed, a cylindrical lens introduces insufficient aberrations to produce the cubic modulation that is required to form an Airy light sheet. However, aberrations can readily be induced by tilting the axis of the cylindrical lens with respect to the optical axis as shown in Figure 15.8. The lowest order aberrations, tip, tilt, and defocus, will rise rapidly with increased angle, θ, of the cylindrical lens. Fortunately these three aberrations correspond to a simple translation of the focal spot. It is thus sufficient to adjust the lens positions in the optical train to compensate for the lower order aberrations, leaving only the cubic modulation that directly produces the Airy light sheet.

As with other forms of light sheet microscopy, a more complex digitally scanned implementation can produce a uniformly illuminated light sheet. Moreover, in the particular case of the Airy beam, it can be expected that the two-dimensional cubic modulation will proffer enhanced tolerance to optical obstacles in the sample [23]. When sample inhomogeneity is less of a concern, the irradiation for the one-dimensional and the two-dimensional cubic modulation can be considered equivalent. In this case the tilted cylindrical lens implementation may be preferable as it offers the important advantage of compactness whilst reducing cost and complexity.

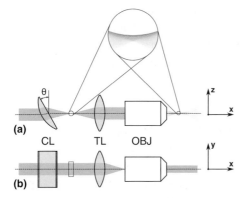

Figure 15.8 The formation of an Airy light sheet using a tilted cylindrical lens as seen from two perspectives. (**a**) View along the y-axis in the light sheet plane (x–z view). (**b**) View along the z-axis orthogonal to the light sheet (x–y view). The cylindrical lens (CL) is tilted by an angle θ to introduce the cubic phase modulation that yields the Airy light sheet profile at its one-dimensional focus. The telescope formed by the tube lens (TL) and objective (OBJ) reimage the Airy light sheet intensity into the sample. A magnified image of the intensity in the focal regions is depicted as an inset above.

15.4   Optical Trapping for Light Sheet Microscopy

If we look at the manner in which we acquire images in light sheet imaging, 3D image stacks are collected by mechanically translating the sample along the detection axis. An alternative is to translate the light sheet and the detection objective along the detection axis at a fixed distance from one other. As might be expected a requirement is to immobilize the specimen during such acquisition, typically achieved using sample mounting gel such as 1.5% agarose gel. However, the use of such gel might be restrictive to the development of the biological sample and can potentially make it difficult to administer appropriate therapeutic or other agents to the sample. If we turn our attention to mobile specimens, such as swimming micro-organisms, the specimen has to be anaesthetized or physically constrained with sufficient force to overcome beating cilia to stop the specimens movement. Again, use of anaesthetics and/or physical force may also compromise the development and normal functioning of the organism, particularly if required for prolonged periods of time.

Recent work has obviated the need for mechanical sample immobilization, rather turning to the use of optical forces [27]. This is a purely light based, non-contact route to hold and translate a sample in its native medium. Optical forces may be used in differing guises: in this work, rather than the more popular optical tweezers, counter-propagating dual-beam trapping was employed for sample confinement and manipulation as shown in Figure 15.9. The counter-propagating beams generate optical gradient and scattering forces that form a much larger potential well than that created by optical tweezers. The potential well of such a "macro-trap" is particularly well suited for the multi-cellular organisms typically imaged with a light sheet microscope.

Since the typical wavelengths used for optical trapping are longer than those used for imaging, optical trapping can be readily integrated into a fluorescence light sheet microscope. It is sufficient to link the optical trapping with the excitation and emission imaging optics using a dichroic mirror before the illumination objective.

The additional optical trapping light path must generate two co-axial beams with an independently controllable focus. A convenient method to generate the beam pair is by separating the two polarizations of a single laser beam using a first polarizing

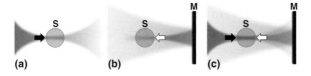

Figure 15.9 Optical immobilization of multi-cellular objects or micro organisms for light sheet imaging. (**a**) A first beam path is focused in front of the sample, causing a right-ward optical scattering force, represented by the black arrow, on the trapped sample (S). (**b**) A second beam, orthogonally polarized to the first, is reflected at a mirror (M) behind the sample before forming a focusing between the mirror and the sample, creating a left-ward scattering force, represented by the white arrow. (**c**) In combination, the two focused beams form a large potential well that serves as a macro-trap for biological samples.

beam splitter. Before recombination using a second polarizing beam splitter, the focus of one polarization is adjusted [27]. The focus adjusting element can be a tunable lens, deformable mirror, or simply a telescope in which a lens is axially translated to adjust the focus. The co-axial beams are introduced through the same objective as the light sheet illumination.

As depicted in Figure 15.9a, the first beam path of the macro-trap is directed toward a point on the optical axis before the to-be-trapped sample (S). The sample will experience a right-ward force that decreases with its distance from the beam's focal point. The second, independently focused, beam is reflected from a mirror behind the sample to create a force gradient toward the left (Figure 15.9b). The macro-trap is formed by emitting both beams simultaneous from the same objective as shown in Figure 15.9c. Since both beams are orthogonally polarized, any potential interference effects are avoided and their independent control enables manipulation of the position and size of the macro trap.

It is important to remark that optical forces may be used in wholly different ways within light sheet imaging. As an example, optical tweezers have been used to study tension forces at cell junctions on the surface of Drosophila embryos [28].

15.5   Outlook and Conclusion

Light sheet microscopy opens the door to live, high-resolution imaging of large samples. Inhomogeneities of the optical properties of thick biological specimen often prevent achieving image resolution and contrast that live up to the imaging potential of the microscope. Not only is the light emanating from the sample affected by a sample-induced blur, also the light sheet entering the sample can spread out to irradiate a far larger volume than intended. In this chapter we discussed various techniques to measure and mitigate the effects of such optical aberrations. While the fluorescence detection path can be corrected using methods similar to those used in epi-fluorescence microscopy; holographic beam shaping techniques proffer advantages for obtaining optimal light sheet confinement in the focal region.

When the light sheet width is pushed below sub-cellular dimensions, its diffraction within the field-of-view becomes an increasingly limiting factor. A conventional light sheet, formed by a truncated Gaussian beam, cannot be confined to a width comparable to the wavelength for much more than a single cell. Light sheets generated with propagation-invariant beams have come to the fore as an elegant solution to maintain a constant light sheet width throughout an extended field-of-view. In this chapter we discussed the propagation-invariant Airy light sheet and how it can be implemented maintaining the key advantages of the light sheet fluorescence microscope, without blocking light and with minimal modifications to the original concept. Although propagation-invariant illumination types can also gain from the adaptive correction techniques discussed in Section 15.2.2, it should be noted that propagation-invariant beams have been shown to have an inherent robustness to optical obstacles in their propagation path.

Light sheet fluorescence microscopy enables time-lapse studies of biological samples mounted in a gel. While this is preferable to mounting specimen on microscope slides or dishes, it may restrict sample development and administration of therapeutic agents. We described a multi-modal approach whereby the specimen under investigation is immobilized by a macro-trap using counter-propagating infrared beams. This obviates the need for a rigid gel and enables controlled studies of the specimen's interaction with external agents.

15.6   References

[1] E. G. Reynaud, U. Kržič , K. Greger, and E. H. K. Stelzer. Light sheet-based fluorescence microscopy: more dimensions, more photons, and less photodamage. *HFSP J.*, 2(5):266–275, 2008. PMID: 19404438.

[2] P. J. Keller, A. D. Schmidt, J. Wittbrodt, and E. H. K. Stelzer. Reconstruction of zebrafish early embryonic development by scanned light sheet microscopy. *Science*, 322(5904):1065–1069, 2008.

[3] P. J. Keller, F. Pampaloni, and E. H. K. Stelzer. Three-dimensional preparation and imaging reveal intrinsic microtubule properties. *Nat. Methods*, 4(10):843–846, October 2007.

[4] C. Bourgenot, C. D. Saunter, J. M. Taylor, J. M. Girkin, and G. D. Love. 3D adaptive optics in a light sheet microscope. *Opt. Express*, 20(12):13252–13261, June 2012.

[5] Z. Zhang, Z. You, and D. Chu. Fundamentals of phase-only liquid crystal on silicon (LCOS) devices. *Light Sci. Appl.*, 3:e213, October 2014.

[6] H. I. C. Dalgarno, T. Čižmár, T. Vettenburg, J. Nylk, F. J. Gunn-Moore, and K. Dholakia. Wavefront corrected light sheet microscopy in turbid media. *Appl. Phys. Lett.*, 100(19):191108, 2012.

[7] Tomáš Čižmár, Michael Mazilu, and Kishan Dholakia. In situ wavefront correction and its application to micromanipulation. *Nat. Photon.*, 4(6):388–394, June 2010.

[8] P. J. Keller and E. H. K. Stelzer. Quantitative *in vivo* imaging of entire embryos with Digital Scanned Laser Light Sheet Fluorescence Microscopy. *Curr. Opinion Neurobiol.*, 18(6):624–632, 2008.

[9] T. A. Planchon, L. Gao, D. E. Milkie, M. W. Davidson, J. A. Galbraith, C. G. Galbraith, and E. Betzig. Rapid three-dimensional isotropic imaging of living cells using Bessel beam plane illumination. *Nat. Methods*, 8:417–423, 2011.

[10] Eugen Baumgart and Ulrich Kubitscheck. Scanned light sheet microscopy with confocal slit detection. *Opt. Express*, 20(19):21805–21814, September 2012.

[11] F. O. Fahrbach, V. Gurchenkov, K. Alessandri, P. Nassoy, and A. Rohrbach. Self-reconstructing sectioned Bessel beams offer submicron optical sectioning for large fields of view in light-sheet microscopy. *Opt. Express*, 21(9):11425–11440, May 2013.

[12] F. O. Fahrbach, V. Gurchenkov, K. Alessandri, P. Nassoy, and A. Rohrbach. Light-sheet microscopy in thick media using scanned Bessel beams and two-photon fluorescence excitation. *Opt. Express*, 21(11):13824–13839, June 2013.

[13] E. G. van Putten, I. M. Vellekoop, and A. P. Mosk. Spatial amplitude and phase modulation using commercial twisted nematic LCDs. *Appl. Opt.*, 47(12):2076–2081, April 2008.

[14] Taro Ando, Yoshiyuki Ohtake, Naoya Matsumoto, Takashi Inoue, and Norihiro Fukuchi. Mode purities of Laguerre-Gaussian beams generated via complex-amplitude modulation using phase-only spatial light modulators. *Opt. Lett.*, 34(1):34–36, January 2009.

[15] A. Dudley, R. Vasilyeu, V. Belyi, N. Khilo, P. Ropot, and A. Forbes. Controlling the evolution of nondiffracting speckle by complex amplitude modulation on a phase-only spatial light modulator. *Opt. Commun.*, 285(1):5–12, 2012.

[16] Stephan Reichelt, Ralf Häussler, Gerald Fütterer, Norbert Leister, Hiromi Kato, Naru Usukura, and Yuuichi Kanbayashi. Full-range, complex spatial light modulator for real-time holography. *Opt. Lett.*, 37(11):1955–1957, June 2012.

[17] R. Liu, D. E. Milkie, A. Kerlin, B. MacLennan, and N. Ji. Direct phase measurement in zonal wavefront reconstruction using multidither coherent optical adaptive technique. *Opt. Express*, 22(2):1619–1628, January 2014.

[18] Long Zhu and Jian Wang. Arbitrary manipulation of spatial amplitude and phase using phase-only spatial light modulators. *Sci. Rep.*, 4:7441, December 2014.

[19] Pierre Gemayel, Bruno Colicchio, Alain Dieterlen, and Pierre Ambs. Cross-talk compensation of a spatial light modulator for iterative phase retrieval applications. *Appl. Opt.*, 55(4):802–810, February 2016.

[20] Liang Gao, Lin Shao, Christopher D. Higgins, John S. Poulton, Mark Peifer, Michael W. Davidson, Xufeng Wu, Bob Goldstein, and Eric Betzig. Noninvasive imaging beyond the diffraction limit of 3D dynamics in thickly fluorescent specimens. *Cell*, 151(6):1370–1385, December 2012.

[21] O. E. Olarte, J. Licea-Rodriguez, J. A. Palero, E. J. Gualda, D. Artigas, J. Mayer, J. Swoger, J. Sharpe, I. Rocha-Mendoza, R. Rangel-Rojo, and P. Loza-Alvarez. Image formation by linear and nonlinear digital scanned light-sheet fluorescence microscopy with Gaussian and Bessel beam profiles. *Biomed. Opt. Express*, 3(7):1492–1505, July 2012.

[22] T. Vettenburg, H. I. C. Dalgarno, J. Nylk, C. Coll Lladó, D. E. K. Ferrier, T. Čižmár, F. J. Gunn-Moore, and K. Dholakia. Light sheet microscopy using an Airy beam. *Nat. Methods*, 11(5):541–544, April 2014.

[23] J. Broky, G. A. Siviloglou, A. Dogariu, and D. N. Christodoulides. Self-healing properties of optical Airy beams. *Opt. Express*, 16(17):12880–12891, August 2008.

[24] F. O. Fahrbach, P. Simon, and A. Rohrbach. Microscopy with self-reconstructing beams. *Nat. Photon.*, 4(11):780–785, November 2010.

[25] F. O. Fahrbach and A. Rohrbach. Propagation stability of self-reconstructing Bessel beams enables contrast-enhanced imaging in thick media. *Nat. Commun.*, 3:632, January 2012.

[26] E. Hecht. *Optics*. Addison Wesley, 2002.

[27] Zhengyi Yang, Peeter Piksarv, David E. K. Ferrier, Frank J. Gunn-Moore, and Kishan Dholakia. Macro-optical trapping for sample confinement in light sheet microscopy. *Biomed. Opt. Express*, 6(8):2778, July 2015.

[28] K. Bambardekar, R. Clément, O. Blanc, C. Chardés, and P.-F. Lenne. Direct laser manipulation reveals the mechanics of cell contacts *in vivo*. *Proc. Natl. Acad. Sci. U.S.A.*, 112(5):1416–1421, 2015.

Part VII

Tomography

16 Incoherent Illumination Tomography and Adaptive Optics

Peng Xiao, Mathias Fink, and A. Claude Boccara

16.1 Introduction

16.1.1 FFOCT Background

Optical imaging usually suffers from scattering due to the heterogeneous structures inside biological samples. In order to obtain images of in-depth structures hindered by scattering, many optical imaging techniques have been developed to be able to select ballistic or more precisely singly backscattered photons, such as confocal microscopy [1], multiphoton microscopy [2] or optical coherence tomography (OCT) [3]. OCT has experienced extraordinary increase of applications in various research and clinical studies since its development in the 1990s, especially in ophthalmology, due to its noninvasiveness, high imaging speed, good sensitivity, and micron-scale resolution. Traditional scanning OCT selects ballistic photons through scattering media based on a broadband light source and coherent cross-correlation detection [3]. Both longitudinal [4, 5] and *en face* scanning [6, 7] OCTs rely on point-by-point scanning with spatially coherent illumination for three-dimensional images acquisition. With specific detectors and methods, parallel OCT systems that take *en face* plane images perpendicular to the optical axis have also been developed. Wide-field OCTs [8–10], which also use spatially coherent illumination, is one kind of these parallel OCT techniques. Higher resolutions are achieved in these systems as *en face* acquisition allows using larger numerical aperture optics. Wide-field OCT systems with powerful laser sources or superluminescent diodes give high sensitivity but the image can be significantly degraded by coherent crosstalks [11].

Our laboratory has developed a specific parallel OCT named Full-Field OCT or FFOCT [12]. The schematic of general FFOCT setup is shown in Figure 16.1. Spatially incoherent broadband light sources are coupled into Linnik interferometer that has both microscope objectives in the two arms to select optical slices that are perpendicular to the optical axis. They do not require the usual large depth of field as standard OCT approaches, thus allowing obtaining microscale resolution in 3D. The use of spatially incoherent illumination also suppresses the cross talk noises significantly. The optical beam from the source is divided into two arms; one goes to the sample and the other goes to the reference mirror. The backscattered beams are recombined and imaged on the camera. Interferometric amplitudes which form the FFOCT images are obtained using the combination of two or four phase shifting images by modulating the path

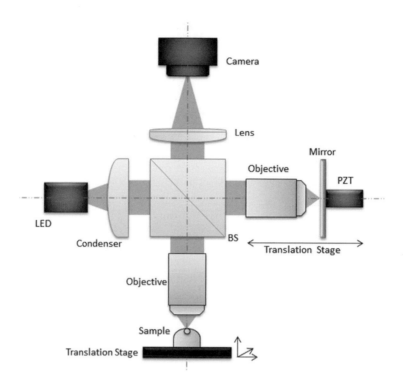

Figure 16.1 Schematic of FFOCT system.

difference with a piezoelectric transducer (PZT) in the reference arm. FFOCT has been working perfectly with ex vivo biological samples and has shown great potential as a noninvasive diagnostic tool for different kind of diseases.

16.1.2   Simplified Adaptive Optics for FFOCT

As well as scattering, there are also multiscale aberrating structures in the eye or bio-logical tissues that would degrade the imaging system performances from achieving high-resolution image quality. Thus adaptive optics (AO) is typically needed to achieve diffraction-limited imaging. Originally proposed and developed for astronomical imag-ing telescopes to correct the atmosphere-induced wavefront perturbations [13, 14], AO has found valuable applications to correct biological tissue-induced wavefront distor-tions in biological and medical imaging [15], especially for retinal imaging to visualize cellular structures [16–18]. AO assisted fundus photography [19], scanning laser oph-thalmoscopy [20–22] and OCT [23–30] systems have achieved reliable images of cones and rods photoreceptors.

In general, the AO part in many systems such as two-photon microscopy [31] or AO-OCT systems [23, 24, 26–30, 32] has strict conjugation of the image focal plane of the microscope objective or of the eye pupil with the wavefront sensors or correction devices. For highorder aberrations, due to the diffractive effects of wave propagation, one cannot rely on simple geometrical optics propagation of the wavefront. In these cases, strict pupil conjugation appears to be mandatory. However, the telescopic systems needed to achieve strict pupil conjugation would increase the system complexity and the optical path length, which would be difficult for FFOCT system since the two arms have to be balanced within less than one micrometer due to the axial sectioning of FFOCT.

The problem would appear differently for low-order aberrations. Many studies on eye aberrations tests have shown that low-order Zernike polynomials are actually dominate (Figure 16.2) [33–35], meaning that the wavefront would look like the same during its propagation. As the ultimate goal of our study is to apply AO-FFOCT for human eye examinations, a transmissive wavefront corrector that could be roughly set in the beam path without strict conjugation would be enough for low-order aberration corrections, analogous to commonly used spectacles for correcting eye's myopia and astigmatism. In this way, we would be able to overcome complex setups realizations for AO-FFOCT. While different wavefront correctors have been developed and applied for eye's diffraction-limited imaging, they all have pros and cons when considering parameters like temporal bandwidth, reflectivity, mirror diameter, number of actuators, etc. [36].

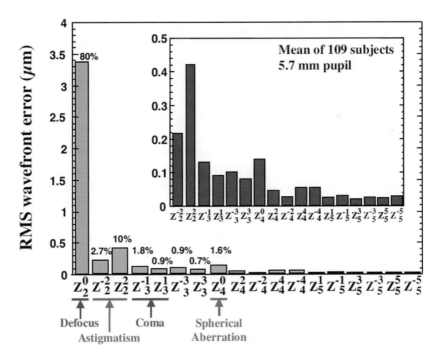

Figure 16.2 Mean absolute rms wavefront error of all 18 Zernike modes for the 109 normal subjects across a 5.7 mm pupil. The percentages listed above the first eight modes represent the percentage of the variance of the wave aberration accounted for by each Zernike mode. The magnitudes of the higher-order aberrations may be seen in the inset figure, which shows all modes except Zernike defocus (Z_2^0) with the ordinate expanded. Reprinted with permission from [33], *JOSA A*.

Transmissive liquid crystal spatial light modulator (LCSLM) [37–41] would fit for our application. With a large number of pixels and a low control voltage, it can alter the wavefront in transmissive way when light passing through. LCSLMs have already been used in some studies to change the refractive state [40] or to correct the aberrations of the eye [41], but in these cases pupil conjugation using telescopes have been used. Of course the confined 2π phase-modulation range of LCSLM might limit the correction of aberrations with large magnitudes. But the adjusting range is doubled as the incoming and outgoing beams both induce optical path difference in our system. Phase wrapping could also be used to extend the dynamic range [40, 42].

In most AO systems, direct wavefront measurements are usually demonstrated with a wavefront sensor [Chapter 2] or coherence-gated wavefront sensing. But due to the lack of generally adaptable wavefront sensors and the inherent complexity of the coherent wavefront sensing, wavefront sensorless methods have also been developed such as hill climbing [43], genetic algorithm [44], simulated annealing [45], pupil segmentation [46 and Chapter 4], etc. Computational AO has also been developed to do postprocessing of experimental data to overcome aberration effects [Chapter 17]. As we are going to discuss in the section 16.3, in FFOCT, aberrations actually do not affect the width of the system PSF but rather the signal level, which is an unexpected property of full-field spatially incoherent interferometry [47, 48]. Therefore, a wavefront sensorless method [Chapter 1] that relies on the improvement of image quality, which is well adapted to the FFOCT detection [49, 50], was used for the optimization process. No wavefront sensor is needed which further simplified the AO-FFOCT system.

16.2   AO-FFOCT Setup

The current AO-FFOCT system schematic is shown in Figure 16.3. Based on a Linnik interferometer, a LED with 660 nm center wavelength and 25 nm bandwidth (Thorlabs) is used as the incoherent light source. The illumination beam is split into the reference arm and the sample arm at a ratio of 50:50 with a nonpolarizing beamsplitter. Two 4X/0.2NA Plan APO objectives (Nikon) are used, one is in the sample arm to simulate the open pupil human eye and the other is in the reference arm. A reference mirror attached to a piezoelectric transducer (PZT) is placed at the focal plan of the objective in the reference arm while the imaging sample would be placed in the sample arm. The back-reflected beams from the reference mirror and the sample are recombined by the beamsplitter and focused with an achromatic doublet lens onto a fast (150 fps) CMOS camera (MV-D1024E-160-CL-12, PhotonFocus). The setup is well aligned to ensure that the focusing of the two arms and their optical paths are matched. The PZT creates a four-phase modulation of the reference mirror and a FFOCT image can be reconstructed with these four corresponding images [12]. Usually several FFOCT images are averaged to improve the signaltonoise ratio (SNR). 5 images were used for the experiments described here, requiring about 150 ms in total. The system has a field of view of 1.7×1.7 mm^2 with the theoretical resolutions of 2 μm (transverse) and 7.7 μm (axial).

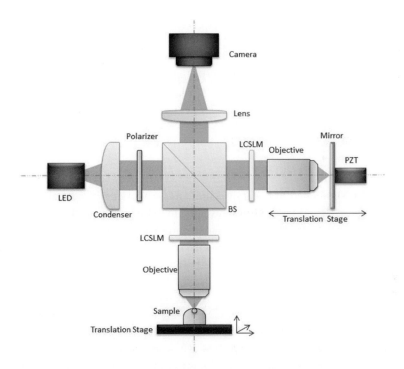

Figure 16.3 Schematic of AO-FFOCT system coupled with LCSLMs. BS: beamsplitter, LCSLM: liquid crystal spatial light modulator, PZT: piezoelectric transducer. Reprinted with permission from [48], *Journal of Biomedical Optics*.

To conduct the wavefront correction, a transmissive LCSLM is inserted into the sample arm beam path at about 2.5 cm after the back aperture of the objective lens. Hence no strict plane conjugations are utilized. Another identical LCSLM is also set in the reference arm beam path for dispersion correction. A polarizer is inserted in the illumination path since the LCSLM works only with polarized light. By electronically varying the orientation of the molecules inside the pixels of the LCSLM, the refractive index of the pixels is altered independently from each other, generating variable phase masks to correct the wavefront of the polarized light passing through them.

16.3   FFOCT Resolution Almost Insensitive to Aberrations

Optical images are obtained by amplitude or intensity convolution of the diffraction-limited images with the optical system PSF. Depending on the nature of the illumination, spatially coherent or incoherent, amplitude or intensity has to be considered [51, 52]. It is commonly known that optical wavefronts would be perturbed when aberrations exist, resulting in aberrated PSF that causes optical image fuzziness. Here, instead of considering the PSF of a classical imaging system such as a microscope, we will focus on the system PSF of interferometric imaging systems in which aberrations only affect the sample arms. Interference would happen between an undistorted wavefront from the reference beam and the distorted wavefront of the sample beam. More precisely we will consider the cases of scanning OCT with spatially coherent illumination, wide-field OCT with spatially coherent illumination and FFOCT with spatially incoherent illumination to show that in FFOCT with incoherent illumination the system PSF width is almost independent of the aberrations that only reduce the signal level.

16.3.1　Theory

To remain committed to the PSF definition, we will consider a point scatterer as our object and will analyze the system response to it. Let's suppose the single point scatterer is at position $(x', y') = (a, b)$, the sample arm PSF of the interferometer is h_s and the reference arm PSF of the interferometer is h_r. For simplification, we will ignore all the constant factors in the following expressions. So in all the three different cases we mentioned above, the sample field at the detection plane would be

$$g_s = h_s\left(x' - a, y' - b\right) \tag{16.1}$$

In the case of traditional scanning OCT, the reference field of each scanning position at the detection plane would be $h_r\left(x - x', y - y'\right)$. Since coherent illumination is used, Interference happens at each scanning position and the final interference would be a sum of the interference term across the scanning filed result in

$$\langle g_s g_r\rangle_s = \iint h_s\left(x' - a, y' - b\right) h_r\left(x - x', y - y'\right) dx' dy' \tag{16.2}$$

Thus, the system PSF of scanning OCT system would be a convolution of the sample arm PSF and the reference arm PSF as shown in Figures 16.4a–16.4c. When aberrations exist, the convolution of the aberrated sample arm PSF with the diffraction-limited reference arm PSF results in an aberrated system PSF for the scanning OCT systems (Figures 16.4d–16.4f).

In the case of wide-field OCT, as coherent sources are used, the optical beams are typically broadened by lenses to form parallel illuminations on both arms of the interferometer [10]. Thus plane waves impinge on both the object and the reference mirror. In the sample arm, the point scatterer will send a spherical wave back that will be focus on the camera plane, which can be described by expression (16.1). For the reference arm, consider it as homogeneous illumination, a plane wave will be reflected back by the reference mirror and form a uniform field at the camera plane. Thus the interference happen between the two arms would be

$$\langle g_s g_r\rangle_w = h_s\left(x' - a, y' - b\right) \tag{16.3}$$

as constant value is ignored. So the system PSF is actually defined by the sample PSF. It is illustrated in Figures 16.4g–16.4i. When aberrations distort the backscattered wavefront of the sample arm, the aberrated sample arm PSF interferes with a uniform reference field results in an aberrated system PSF for the wide-filed OCT systems (Figures 16.4j–16.4l).

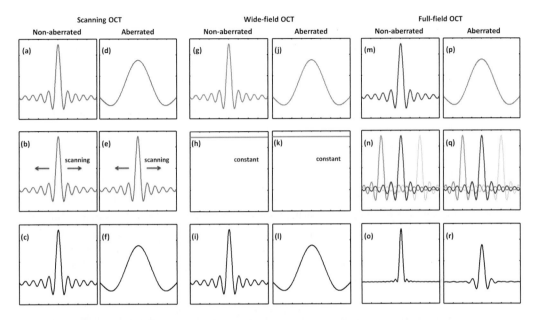

Figure 16.4 Illustration of single point scatterer (PSF) interferences in both nonaberrated and aberrated sample arm PSF situations for scanning OCT and wide-field OCT with spatially coherent illumination and FFOCT with spatially incoherent illumination. (**a, g, m**) Nonaberrated sample arm PSF, (**d, j, p**) Aberrated sample arm PSF, (**b, e**) Scanning reference arm PSF for scanning OCT, (**h, k**) Constant reference field for wide-field OCT, (**n, q**) Reference arm PSFs for FFOCT, (**c, f, i, l, o, r**) The corresponding interference signal (system PSF). Shades in (**n, q**) indicate the spatial incoherence from each other. Reprinted with permission from [47], *Optics Letters*.

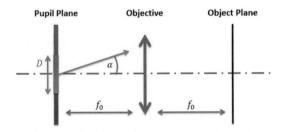

Figure 16.5 Illustration of the Van Cittert–Zernike theorem defining the coherence angle α in the case of an objective with a focal length of f_0 and pupil diameter of D.

When discussing about the case of FFOCT with spatially incoherent illumination, we have to recur to the basic definition of the spatial coherence of both beams that impinge the reference arm as well as the sample arm of the interferometer. As shown in Figure 16.5, consider a circular uniform incoherent source located in the image focal plane of a microscope objective with a focal length of f_0 which could be obtained with a standard Koehler illumination. The very first step is to determine the spatial coherence area in the field of view of the objective. The Van Cittert–Zernike theorem states that the coherence angle is given by the Fourier transform of the source luminance [53]. If the pupil diameter is D, the angle would be defined as $\sin \alpha = 1.22\lambda/D$. At the level of focal plane, it corresponds to a zone of radius $\rho = f_0\lambda/D$ or $\rho = 1.22\lambda/2NA$. We can say that, in absence of aberrations the focal plane is actually "paved" by small coherent areas (CA) of radius ρ. This radius is also the radius of the diffraction spot that limits the resolution of the microscope objective in absence of aberrations. When going from one diffraction spot to the next adjacent diffraction spots the incoherent plane waves impinging the objective are separated by $\pm\lambda$ on the edges of the pupil.

In absence of aberrations for an interferometer like FFOCT, the single point scatterer at the object plane of the sample arm would lies in a single CA (Figure 16.6a) and the backscattered signal will only interfere with signal reflected from the corresponding CA in the reference arm (Figure 16.6c). Note that the size of the CAs is the same as the diffraction spot, the signal from one CA in the reference arm arriving at the camera plane could be expressed as the reference PSF. Thus the interference would be

$$\langle g_s g_r \rangle_f = h_s \left(x' - a, y' - b\right) h_r \left(x' - a, y' - b\right) \qquad (16.4)$$

The system PSF is actually the dot product of the sample PSF and the reference PSF as shown in Figures 16.4m–16.4o. The overall signal reflected from the reference mirror at the camera is still homogenous but we displayed it by combining multiple reference PSFs reflected from different CAs that have different spatial modes.

When aberrations exist in the sample arm, the various CAs in the object plane will be distorted, have larger sizes and overlap with each other (Figure 16.6b). This result in the backscattered signal of the single point scatterer in the sample arm containing not only the spatial mode of the targeted focus CA but also the modes from the overlapped

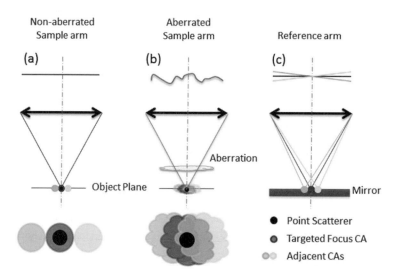

Figure 16.6 Illustration of the sample and reference wavefronts in spatially incoherent interferometer with a single point scatterer in cases of nonaberrated and aberrated sample arm. Different colors in CAs and wavefronts indicate different spatial modes. Reprinted with permission from [47], *Optics Letters*.

adjacent CAs. Thus with aberrations that create a broadened sample PSF, interference will happen not only with the reference beam corresponding to the targeted CA, but also with the beams corresponding to the adjacent CAs.

What we are going to demonstrate and to illustrate is that the interference signal with the targeted focus CA gives a much stronger signal than the ones with the adjacent CAs, resulting in an "interference" PSF that is much thinner than the one of the classical broadened sample PSF. At the level of the image plane, the interference between the sample aberrated beam and the nonaberrated reference beam is only possible in a zone limited by the spatial coherence of the reference beam. In order to be more quantitative, we will use the Strehl ratio approach.

The "best focus" signal intensity damping compared to the diffraction-limited condition is given (for small aberrations) by the Strehl ratio $S = e^{-\sigma^2}$ that is proportional to the peak aberrated image intensity; σ is the root-mean-square deviation over the aperture of the wavefront phase $\sigma^2 = (std\,(\phi))^2$. Suppose $\emptyset$ is the phase of the interference wavefront between the sample signal and the reference signal corresponding to the targeted focus CA, then the phase of the interference wavefront with the reference signals corresponding to an adjacent CAs is $\phi + \phi_1$, where ϕ_1 is a phase that varies linearly from one edge of the pupil to the other in the range of $\pm 2\pi$. A comparison between the signal ratio of the interference signal with the targeted CA and the one with an adjacent CAs

$$s_t = e^{-(std(\phi))^2} \gg s_a = e^{-(std(\phi+\phi_1))^2} \tag{16.5}$$

would show that the influence of off axis CAs is strongly damped.

When consider various aberrations leading to a Strehl ratio of 0.3, numerical calculations results are shown in Figure 16.7. For defocus, the intensity ratio of the interference with adjacent CAs is damped for about 740 times compared with the interference with the targeted focus CA, resulting in a signal damping or an amplitude damping of 27.1 times. The amplitude damping ratio is calculated by

$$Amplitude\ damping\ ratio = \sqrt{s_t/s_a} \tag{16.6}$$

as amplitude instead of intensity is obtained in FFOCT signal. It's easy to prove that this value is fixed for all the axisymmetric aberrations like defocus, astigmatism, spherical aberrations, etc. While for coma with a Strehl ratio of 0.3, the simulated amplitude damping ratio is 13.4–53.0 times depending on the spatial position of the adjacent CAs around the targeted focus CA. In another word, the interference signal was severally damped going from the targeted CA to the adjacent CAs. Thus in the camera plane, as shown in Figures 16.4p–16.4r, the interference signal result in a dot product of the aberrated sample PSF with the reference PSF corresponding to the targeted focus CA since the interference with the reference PSFs corresponding to the adjacent CAs are significantly reduced. This actually matches with equation (16.4) for nonaberrated situation, the system PSF could be calculated by the dot product of the sample PSF and the reference PSF. For distorted sample PSF (mostly broadened), its interference with the reference channel conserves the main feature of an unperturbed PSF with only a reduction in the FFOCT signal level.

Zernike Mode	Targeted Focus wavefront	Adjacent Wavefront	Amplitude Damping Ratio
Defocus	 $s_t = 0.3$	 $s_a = 4.07 \times 10^{-4}$	27.1
Astigmatism	 $s_t = 0.3$	 $s_a = 4.07 \times 10^{-4}$	27.1
Coma	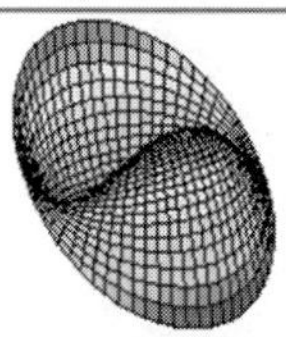 $s_t = 0.3$	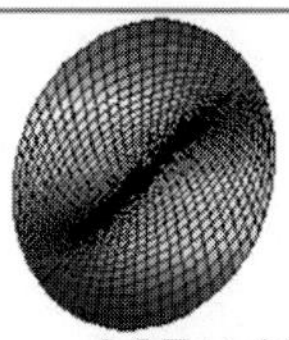 $s_a = 4.07 \times 10^{-4}$	13.4 - 53.0
Spherical Aberration	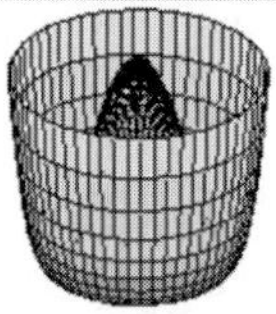 $s_t = 0.3$	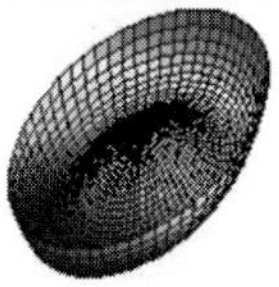 $s_a = 4.07 \times 10^{-4}$	27.1

Figure 16.7 Aberrated interference wavefronts and numerical simulations of the Strehl ratio and amplitude damping for interference with targeted CA and adjacent CAs. Defocus, astigmatism, coma, and spherical aberration are considered. The damping for coma varies depending on the spatial position of the adjacent CAs [47]. A black and white version of this figure appears in some formats. For the color version, please refer to the plate section.

With stronger aberrations, the ratio of the size of the distorted CA in the sample arm and the diffraction-limited PSF would increase. But for axisymmetric aberrations, as we mentioned, the signal damping ratio of the adjacent CAs is the same and it is strong enough to keep the resolution. The shape of the PSF in presence of aberration looks more and more as the amplitude PSF with larger wings. In the cases for nonaxisymmetric aberrations like coma, the position of the best focus is changed, without affecting the resolution, and that only distorts the field of view. We mentioned "almost" for the resolution conservation, because there are also situations in which the product of the reference arm PSF with off-center aberrated sample arm PSF may results in losing some sharpness due to the high side lobes of the Bessel PSF function that are larger than its square.

16.3.2   Experimental Demonstration

16.3.2.1   PSF Determination Using Gold Nanoparticles

Experiments were first conducted using the commercial LLtech FFOCT system Light-CT scanner with 0.3 NA microscope objectives [54] and gold nanoparticles to check how the system PSF would be affected by inducing different level of defocus. In this experiment, 40 nm radius gold nanoparticles solution was diluted and dried on a coverslip so that single particles could be imaged. By moving the sample stage, 10 μm, 20 μm and 30 μm defocus was induced to the targeted particle. The length of the reference arm was shifted for the same value in order to match the coherence plan of the two arms for imaging. Theoretically, the system resolution was 1.5 μm corresponding to about 2.5 pixels on the camera. By adding 10 μm, 20 μm and 30 μm defocus, the sample PSF would be broadened by 2.3 times, 4.6 times and 6.9 times. FFOCT images (Figures 16.8a–16.8d) and the corresponding signal profiles (Figures 16.8e–16.8h) of the same nanoparticle were displayed. With more defocus added the signal level of the gold nanoparticles is reduced, but the normalized signal profiles graph (Figure 16.8i) shows clearly that the size of the particle that corresponds to the system PSF width keeps the same for all the situations. In FFOCT systems with various NA (0.1, 0.2, and 0.8), this phenomenon has also been verified.

16.3.2.2   USAF Resolution Target: Avoiding Blur through Incoherent Illumination Interference Imaging System

Using a negative USAF setting at the best focus position of the sample arm of the customized AO-FFOCT system, the resolution conservation merit of FFOCT system was again confirmed experimentally. Here, a random aberration root-mean-square (RMS) wavefront error $= 0.27\lambda$, corresponding to a Strehl ratio $= 0.06$ (Figure 16.9g) was induced with the LCSLM in the sample arm by generating and applying random voltages within the adjusting range across the LCSLM pixels. The Strehl ratio is calculated by the square of the ratio of mean FFOCT image intensity after and before the aberration was applied. Figure 16.9 shows the sample reflectance images and FFOCT images of the USAF resolution target before and after the random aberration was induced. The reflectance images were recorded by blocking the reference arm in FFOCT thus the system works as a wide-field microscope. The reflectance image is blurred after the aberration is added, while there is no obvious blurring of the line patterns in the FFOCT image but only a reduction of the image intensity. The normalized intensity of the selected line in the reflectance image shows a distortion after the aberration was added, while it shows a conservation of the shape for the FFOCT image. Note that the image contrast of scanning OCT using spatially coherent illumination would be close to the reflectance image from the sample arm.

16.4   Aberration Correction Algorithm

A wavefront sensorless approach is applied for aberration correction based on the FFOCT signal level since aberrations affect only the signal level without reducing the

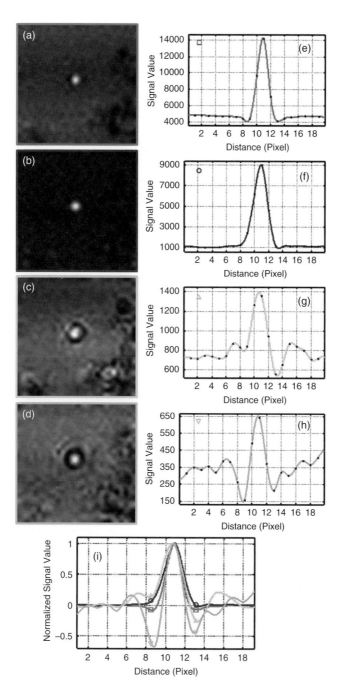

Figure 16.8 FFOCT experiment results of gold nanoparticle by adding different level defocus. FFOCT images (**a–d**) and the corresponding intensity profile (**e–h**) of a targeted nanoparticle are shown for well-focused (**a, e**) and defocused for 10 μm (**b, f**), 20 μm (**c, g**) and 30 μm (**d, h**) situations. Normalized PSF profiles are shown in (**i**) indicating no obvious broadening are observed after inducing different level of defocus. Reprinted with permission from [47], *Optics Letters*.

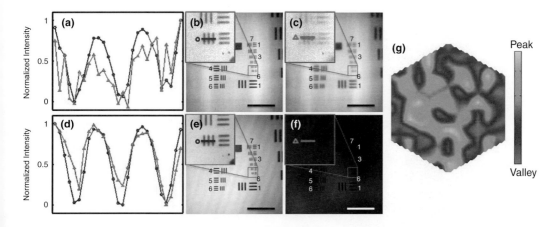

Figure 16.9 Comparison of the reflectance (**b**, **c**) and FFOCT (**e**, **f**) images of a negative USAF resolution target before (**b**, **e**) and after (**c**, **f**) adding a defocus aberration. (**a**, **d**) show the comparison of the normalized reflectance intensity and FFOCT signal of the selected line without (darker gray line with circles) and with (lighter gray line with triangles) aberration added. The plot of the random aberration pattern is shown in (**g**). Scale bar: 100 μm. Reprinted with permission from [48], *Journal of Biomedical Optics*.

image resolution in FFOCT. The wavefront sensorless method consists of the sequential adjustment of the coefficients of low-order orthogonal Zernike polynomial functions applied to the LCSLM to optimize the metric function. For LCSLM-induced aberration correction with USAF resolution target as the sample, the mean intensity of FFOCT image was used as the metric function. For in-depth sample-induced aberration correction, the average intensity of the 300 pixels with maximum intensity values in the FFOCT image was used as the metric function instead because the mean intensity of the overall image would be less sensitive to the AO process since most parts of the FFOCT image has very low or even no signal. Of course the optimization process could also be restricted to specific region of interest. Indeed anisoplanatism does exist as demonstrated in Figure 16.11, but the experiment results show acceptable correction efficiency with this simple AO algorithm. No phase wrapping was used for experiments showed in this paper. Correction coefficients were indeed selected within the adjusting range of the LCSLM. The orthogonality of different Zernike modes ensures that the coefficient of each mode for optimal correction is determined independently [55, 56]. This algorithm has been proposed and used by many groups with different wavefront shaping methods and optimization metrics in specific applications [30, 57, 58]. For the aberration correction experiments mentioned in this chapter, only Zernike modes 3 to 8 were optimized just to demonstrate the feasibility of our system and method. For each mode, FFOCT images were taken for seven different coefficients within the adjusting range. With the extracted metric function values, B-spline interpolations were done and the coefficient that produced the highest metric function was chosen as the correction value. As a result, the entire optimization process could be done in about 6.3 seconds.

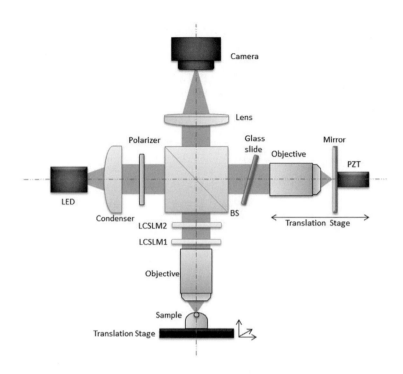

Figure 16.10 Schematic of adaptive optics FFOCT system for LCSLM-induced aberration correction. LCSLM2 was inserted at 50 mm after LCSLM1. LCSLM2 was used for aberration introduction while LCSLM1 was used for aberration correction. Reprinted with permission from [48], *Journal of Biomedical Optics*.

16.5   LCSLM-Induced Aberration Correction

To test the performances of the AO-FFOCT system with no well-defined conjugation and wavefront sensorless algorithm, experiments of LCSLM-induced aberration correction were first conducted by imaging a negative USAF resolution target. As shown in Figure 16.10, in this experiment LCSLM2 were inserted into the sample arm for aberration introduction at about 5 cm after the original LCSLM1 which was used for aberration correction, thus there is no well-defined conjugation between the aberration introduction plane and the correction plane. A glass slide was inserted into the reference arm for dispersion compensation. The USAF target was set at the best focus position in the sample arm and a random aberration mask (RMS = 0.23λ, Strehl ratio = 0.12) was generated and applied to the LCSLM2. Figure 16.11a shows the original FFOCT image with the added aberration. By using the wavefront correction algorithm and applying the correction phase mask onto LCSLM1, defocus, astigmatism, coma, and spherical aberration were corrected successively. Figures 16.11b–16.11g show the images after each correction with a clearly visible improvement of image quality after each optimization process. The black curve in Figure 16.11h shows the increase of the metric function and the red,

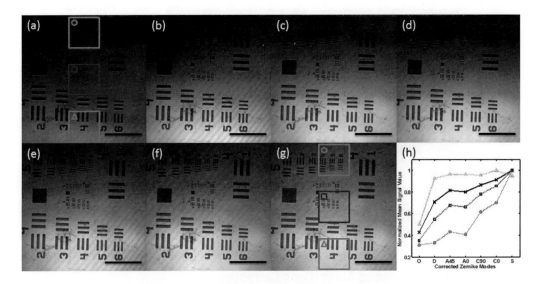

Figure 16.11 FFOCT images of a negative USAF resolution target during the nonconjugate AO correction process of a random aberration. (**a**) original image with a random aberration added, (**b–g**) images after defocus, astigmatism 45, astigmatism 0, coma 90, coma 0, and spherical aberration were corrected respectively, (**h**) graph of the metric function (cross curve) increase after each correction step and mean intensity changes (circle, square and triangle dashed curves) of the corresponding selected regions indicated in (**a, g**). Scale bar: 350 μm. Reprinted with permission from [48], *Journal of Biomedical Optics.*

blue and green dashed curves display the mean intensity changes of the corresponding selected regions indicated with the same colors in Figures 16.11a and 16.11g. The fact that different levels of improvement were achieved for different regions with the same correction phase mask for each Zernike mode implies the existence of anisoplanatism in our experiment. Nevertheless, the mean intensity of the FFOCT image got an increase of 135% after the overall correction, reaching 80% of the nonaberrated FFOCT image, while having diffraction-limited resolution. The experiment was repeated three times with different random aberrations and it results in an average increase of the mean intensity to 78.0% ± 2.2% of the nonaberrated image, corresponding to a Strehl ratio of 0.61 ± 0.035.

For comparison, conjugate AO experiment was conducted by using the same LCSLM for aberration introduction and correction. With the same random aberration induced by LCSLM2, aberration correction was demonstrated also on LCSLM2 itself. With the same algorithm, Zernike modes 3–8 was corrected by applying net voltages of random pattern plus the Zernike modes to LCSLM2. As shown in Figure 16.12, the whole correction result in the mean intensity of the FFOCT image reaching 86% of the nonaberrated FFOCT image. Again, the three repeated experiments result in an average increase of the mean intensity to 84.3% ± 2.1% of the nonaberrated image, corresponding to a Strehl ratio of 0.71 ± 0.036.

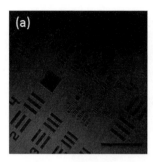 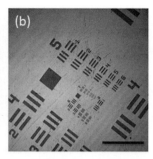

Figure 16.12 FFOCT images of a negative USAF resolution target before and after the conjugate AO correction process of a random aberration. (**a**) original image with a random aberration added. (**b**) Image after defocus, astigmatism 45, astigmatism 0, coma 90, coma 0, and spherical aberration were corrected. Scale bar: 350 μm. Reprinted with permission from [48], *Journal of Biomedical Optics.*

16.6   Sample-Induced Aberrations Correction

16.6.1   Ficus Leaf Experiment: Weak Aberrations Case

Due to the spatial variations of refractive index within biological samples and surface topography, aberration distortion is severe when imaging into the sample volume. In order to further demonstrate the feasibility of our system and method even for weak aberrations correction, experiments of sample-induced aberrations corrections were done with a ficus leaf. The system setup described in Figure 16.3 was used here. By imaging at a depth of 75 μm under the leaf surface only weak aberrations are induced and we can thus check the sensitivity of our correction approach; the low-order contents of the self-induced sample aberrations were corrected step by step with the aforementioned methods. As showed in Figure 16.13, the optimized image (Figure 16.13b) shows an intensity increase compared with the original image (Figure 16.13a) and from the zoomed in images, more structured information appears. This is due to the fact that the correction process increased the SNR and more signals that were buried by the noise before appear after the AO correction. The graph of the metric function while adjusting the coefficients of each Zernike mode is displayed in Figure 16.13c. The highest positions of each curve correspond to the coefficients used for the optimal correction of each mode. Figure 16.13d shows the increase of metric function. The whole correction process results in 13.3% improvement of the metric function. The metric function improvement increases to 35.5% when imaging deeper at 120 μm under the leaf surface in another experiment.

16.6.2   Mouse Brain Tissue Slice Experiment: Strong Aberrations Case

After showing the ability of this AO-FFOCT approach to optimize the signal even with a low level of aberration, we checked another biological tissue of relevance that suffers from strong scattering and stronger aberrations – the brain tissue, where FFOCT signal

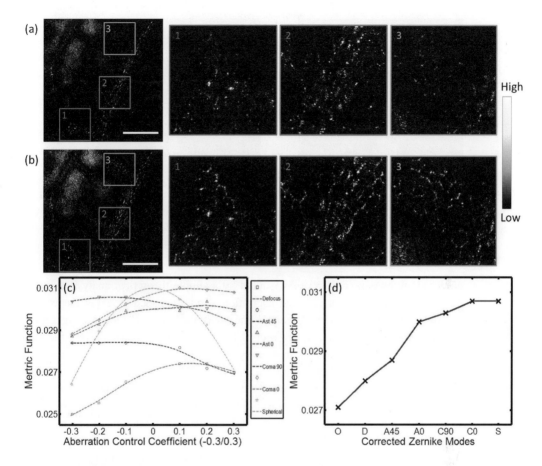

Figure 16.13 Comparison of FFOCT images of a ficus leaf before (**a**) and after (**b**) sample self-induced aberration was corrected when imaging at a depth of 75 μm. (**c**) graph of the metric function during the optimization process, (**d**) graph of the metric function increase after each correction step. Scale bar: 500 μm, Zoomed in area: 425 × 425 μm. Reprinted with permission from [48], *Journal of Biomedical Optics.*

is usually strongly reduced when imaging deep in the sample. Experiments were conducted with a fixed mouse brain tissue slice to correct the wavefront distortion. Imaging was performed at 50 μm under the brain tissue surface without liquid matching fluid and the results are shown in Figure 16.14. The high-signal fiber-like myelin structures appeared much more clearly after the whole correction process because of the increased SNR; indeed the metric function was increased by 121%.

16.7   Discussion and Conclusion

We have demonstrated that in spatially incoherent illumination interferometry like full-field OCT, the system PSF width is almost insensitive to aberrations with only signal

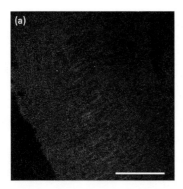 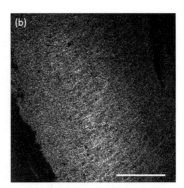

Figure 16.14 Comparison of FFOCT images of fixed mouse brain tissue slice before (**a**) and after (**b**) sample self-induced aberration was corrected when imaging at a depth of 50 μm. Scale bar: 500 μm. Reprinted with permission from [48], *Journal of Biomedical Optics.*

amplitude reduction. This is demonstrated by a simple theoretical analysis as well as numerical simulations for different aberrations, and confirmed by experiments with a full-field OCT system. More precisely the aberration-induced reduction in signal is roughly proportional to the square root of the Strehl ratio. Let us consider the realistic case of a diffraction-limited imaging system with a PSF width of 2 μm that allows for instance resolving the cones in retinal imaging. With a Strehl ratio of 0.1, which is considered to give a low quality image, the PSF would be broadened to about 6 μm that would mask the cell structures. But in full-field OCT system, the same Strehl ratio would only reduce the signal by a factor of 3.1 while keeping the image sharpness.

We also demonstrated that a compact transmissive LCSLM can be directly coupled to an FFOCT system as an AO element for wavefront distortion compensation with a wavefront sensorless algorithm. Our experiments show the potential of this compact AO-FFOCT system for aberrations correction imaging. The conjugation of the LCSLM with the pupil plane was discarded in our AO-FFOCT system. Traditionally, AO devices are usually conjugated with a well-defined plane. For both pupil AO, in which conjugation is done to the pupil plane, and conjugate AO, in which conjugation is done to the plane where the aberrations dominate, a plane is needed for wavefront measurement and the inverse phase mask needs to be applied to the same plane with the conjugated wavefront correctors. The advantages and disadvantages of both conjugations have been recently discussed in [59]. From what we have learned in our experiments, we think that the problem might be easier for applications with metric-based wavefront sensorless adaptive optics because the only criteria are the metric functions of the image. Strict conjugation might be abandoned, especially for loworder aberrations correction cases. The corrected signal level with this nonconjugate AO reaches 78.0% ± 2.2% of the nonaberrated situation. This is slightly inferior but still acceptable compared with a conjugate AO experiment which results in a corrected FFOCT image signal level reaching 84.3% ± 2.1% of the nonaberrated image.

Our approach simulating eye aberration correction in a simple manner opens the path to a straightforward implementation of AO-FFOCT for retinal examinations in the

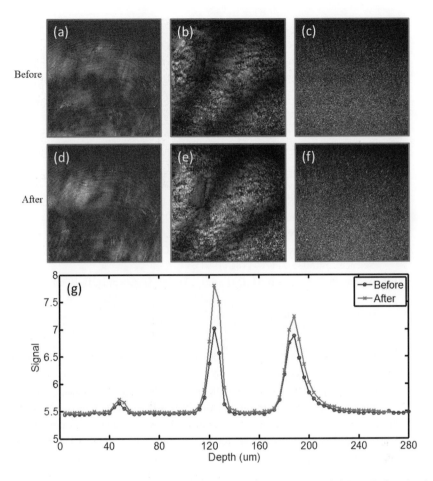

Figure 16.15 AO-FFOCT images of different retinal layers of an artificial eye before (**a–c**) and after (**d–f**) aberration correction and averaged image intensity curves along depth scanning (**g**) showing the signal increase of all the three detected layers.

future research. In the case of eye examinations, we can restrict aberrations correction to the main aberrations (e.g., focus and astigmatism) that will improve the SNR and skip the high-order aberrations. Ultimately, the lens in the eyeball will play the role of the objective used in the sample arm in our experiments, therefore a new reference arm with path and dispersion compensation [60] will need to be designed taking into consideration of the eye characteristics. By using an artificial eye model used to train ophthalmologists [61], we have demonstrated preliminary adaptive optics retinal imaging experiments. By translating the model eye along the optical axis, three retina layers were detected. The corresponding FFOCT images are displayed in Figures 16.15a–16.15c. Based on the second layer, the model eye self-induced aberrations corrections were done to improve the image signal level by using the same algorithm as described before. After the optimization process, the improved FFOCT retinal images are shown in Figures 16.15d–16.15f. The curves in Figure 16.15g shows the average image intensity of the FFOCT images along different depth while the peaks indicating the three detected layers. The signal level for all the layers is increased after aberration correction. Taking the second layer as an example, the signal level has increased by 48% after subtracting the background noise. This work is supported by the HELMOLTZ Synergy funded by the European Research Council (ERC).

16.8   References

[1]   T. Wilson, *Confocal Microscopy*, London: Academic Press (1990).

[2]   W. Denk, J. H. Strickler, and W. W. Webb, "Two-photon laser scanning fluorescence microscopy," Science 248(4951), 73–76 (1990).

[3]   D. Huang, E. A. Swanson, C. P. Lin, J. S. Schuman, W. G. Stinson, W. Chang, M. R. Hee, T. Flotte, K. Gregory, C. A. Puliafito, and J. G. Fujimoto, "Optical coherence tomography," Science 254(5035), 1178–1181 (1991).

[4]   N. Nassif, B. Cense, B. H. Park, S. H. Yun, T. C. Chen, B. E. Bouma, G. J. Tearney, and J. F. de Boer, "*In vivo* human retinal imaging by ultrahigh-speed spectral domain optical coherence tomography," Opt. Lett. 29, 480–482 (2004).

[5]   R. Huber, M. Wojtkowski, and J. G. Fujimoto, "Fourier Domain Mode Locking (FDML): A new laser operating regime and applications for optical coherence tomography," Opt. Express 14, 3225–3237 (2006).

[6]   B. M. Hoeling, A. D. Fernandez, R. C. Haskell, E. Huang, W. R. Myers, D. C. Petersen, S. E. Ungersma, R. Wang, M. E. Williams, and S. E. Fraser, "An optical coherence microscope for 3-dimensional imaging in developmental biology," Opt. Express 6, 136–146 (2000).

[7]   Y. Zhang, J. Rha, R. S. Jonnal, and D. T. Miller, "Adaptive optics parallel spectral domain optical coherence tomography for imaging the living retina," Opt. Express 13, 4792–4811 (2005).

[8]   S. Bourquin, P. Seitz, and R. P. Salathé, "Optical coherence topography based on a two-dimensional smart detector array," Opt. Lett. 26, 512–514 (2001).

[9]   E. Bordenave, E. Abraham, G. Jonusauskas, N. Tsurumachi, J. Oberlé, C. Rullière, P. E. Minot, M. Lassègues, and J. E. Surlève Bazeille, "Wide-field optical coherence tomography: imaging of biological tissues," Appl. Opt. 41, 2059–2064 (2002).

[10]   M. Laubscher, M. Ducros, B. Karamata, T. Lasser, and R. Salathé, "Video-rate three-dimensional optical coherence tomography," Opt. Express 10, 429–435 (2002).

[11]   B. Karamata, P. Lambelet, M. Laubscher, R. P. Salathé, and T. Lasser, "Spatially incoherent illumination as a mechanism for cross-talk suppression in wide-field optical coherence tomography," Opt. Lett. 29, 736–738 (2004).

[12]   L. Vabre, A. Dubois, and A. C. Boccara, "Thermal-light full-field optical coherence tomography," Opt. Lett. 27(7), 530–533 (2002).

[13]   H. Babcock, "The possibility of compensating atmospheric seeing," Pub. Astron. Soc. Pac. 65, 229–236 (2010).

[14]   G. Rousset, J. C. Fontanella, P. Kern, and F. Rigaut, "First diffraction-limited astronomical images with adaptive optics," Astron. Astrophys. 230, L29–L32 (1990).

[15]   J. A. Kubby, *Adaptive Optics for Biological Imaging*, Boca Raton: CRC Press (2013).

[16]   P. Godara, A. M. Dubis, A. Roorda, J. L. Duncan, and J. Carroll, "Adaptive optics retinal imaging: emerging clinical applications," Optom. Vis. Sci. 87 (12), 930–941 (2010).

[17] J. Porter, H. M. Queener, J. E. Lin, K. Thorn, and A. Awwal, eds., *Adaptive Optics for Vision Science*, Hoboken: John Wiley (2006).

[18] D. R. Williams, "Imaging single cells in the living retina," Vision Res. 51(13), 1379–1396 (2011).

[19] J. Liang, D. R. Williams, and D. T. Miller, "Supernormal vision and high-resolution retinal imaging through adaptive optics," J. Opt. Soc. Am. A 14(11), 2884–2892 (1997).

[20] A. Roorda, F. Romero-Borja, W. Donnelly III, H. Queener, T. Hebert, and M. Campbell, "Adaptive optics scanning laser ophthalmoscopy," Opt. Express 10(9), 405–412 (2002).

[21] Y. Zhang and A. Roorda, "Evaluating the lateral resolution of the adaptive optics scanning laser ophthalmoscope," J. Biomed. Opt. 11(1), 14002 (2006).

[22] D. Merino, J. L. Duncan, P. Tirubeedhula, and A. Roorda, "Observation of cone and rod photoreceptors in normal subjects and patients using a new generation adaptive optics scanning laser ophthalmoscope," Biomed. Opt. Express, 2(8), 2189–2201 (2011).

[23] Y. Zhang, B. Cense, J. Rha, R. S. Jonnal, W. Gao, R. J. Zawadzki, J. S. Werner, S. Jones, S. Olivier, and D. T. Miller, "High-speed volumetric imaging of cone photoreceptoes with adaptive optics spectral-domain optical coherence tomography," Opt. Express 14(10), 4380–4394 (2006).

[24] R. J. Zawadzki, S. M. Jones, S. S. Olivier, M. Zhao, B. A. Bower, J. A. Izatt, S. Choi, S. Laut, and J. S. Werner, "Adaptive-optics optical coherence tomography for high-resolution and high-speed 3D retinal *in vivo* imaging," Opt. Express 13(21), 8532–8546 (2005).

[25] O. P. Kocaoglu, S. Lee, R. S. Jonnal, Q. Wang, A. E. Herde, J. C. Derby, W. Gao, and D. T. Miller, "Imaging cone photoreceptors in three dimensions and in time using ultrahigh resolution optical coherence tomography with adaptive optics," Biomed. Opt. Express 2(4), 748–763 (2011).

[26] Y. Zhang, J. Rha, R. Jonnal, and D. Miller, "Adaptive optics parallel spectral domain optical coherence tomography for imaging the living retina," Opt. Express 13(12), 4792–4811 (2005).

[27] E. J. Fernández, B. Hermann, B. Povazay, A. Unterhuber, H. Sattmann, B. Hofer, P. K. Ahnelt, and W. Drexler, "Ultrahigh resolution optical coherence tomography and pancorrection for cellular imaging of the living human retina," Opt. Express 16, 11083–11094 (2008).

[28] E. J. Fernández, B. Povazay, B. Hermann, A. Unterhuber, H. Sattmann, P. M. Prieto, R. Leitgeb, P. Ahnelt, P. Artal, and W. Drexler, "Three-dimensional AO ultrahigh-resolution optical coherence tomography using a liquid crystal spatial light modulator," Vision Res. 45, 3432–3444 (2005).

[29] O. P. Kocaoglu, R. D. Ferguson, R. S. Jonnal, Z. Liu, Q. Wang, D. X. Hammer, and D. T. Miller, "Adaptive optics optical coherence tomography with dynamic retinal tracking," Biomed. Opt. Express, 5 (7), 2262–2284 (2014).

[30] K. S. K. Wong, Y. Jian, M. Cua, S. Bonora, R. J. Zawadzki, and M. V. Sarunic, "*In vivo* imaging of human photoreceptor mosaic with wavefront sensorless adaptive optics optical coherence tomography," Biomed. Opt. Express 6(2), 580–590 (2015).

[31] M. Rueckel, J. A. Mack-Bucher, and W. Denk, "Adaptive wavefront correction in two-photon microscopy using coherence-gated wavefront sensing," PNAS 103(46), 17137–17142 (2006).

[32] S. H. Lee, J. S. Werner, and R. J. Zawadzki, "Improved visualization of outer retinal morphology with aberration cancelling reflective optical design for adaptive optics – optical coherence tomography," Biomed. Opt. Express 4(11), 2508–2517 (2013).

[33] J. Porter, A. Guirao, I. G. Cox, and D. R. Williams, "Monochromatic aberrations of the human eye in a large population," JOSA A, 18(8), 1793–1803 (2001).

[34] J. F. Castejón-Mochón, N. López-Gil, A. Benito, and P. Artal, "Ocular wave-front aberration statistics in a normal young population," Vision Res. 42(13), 1611–1617 (2002).

[35] X. Hong, L. Thibos, A. Bradley, D. Miller, X. Cheng, and N. Himebaugh, "Statistics of aberrations among healthy young eyes," in *Vision Science and Its Applications*, A. Sawchuk, ed., Vol. 53 of OSA Trends in Optics and Photonics, Optical Society of America (2001), paper SuA5.

[36] N. Doble, D. T. Miller, G. Yoon, and D. R. Williams, "Requirements for discrete actuator and segmented wavefront correctors for aberration compensation in two large populations of human eyes," Appl. Opt. 46(20), 4501–4514 (2007).

[37] D3128 Spatial Light Modulator Meadowlark Optics.

[38] G. D. Love, "Wave-front correction and production of Zernike modes with a liquid-crystal spatial light modulator," Appl. Opt. 36(7), 1517–1524 (1997).

[39] A. Vyas, M. B. Roopashree, R. K. Banyal, and B. R. Prasad, "Spatial Light Modulator for wavefront correction," arXiv preprint arXiv: 0909.3413 (2009).

[40] L. N. Thibos and A. Bradley, "Use of liquid-crystal adaptive-optics to alter the refractive state of the eye," Optom. Vis. Sci. 74(7), 581–587 (1997).

[41] F. Vargas-Martin, P. M. Prieto, and P. Artal, "Correction of the aberrations in the human eye with a liquid-crystal spatial light modulator: limits to performance," JOSA A 15(9), 2552–2562 (1998).

[42] D. T. Miller, L. N. Thibos, and X. Hong, "Requirements for segmented correctors for diffraction-limited performance in the human eye," Opt. Express 13(1), 275–289 (2005).

[43] P. N. Marsh, D. Burns, and J. M. Girkin, "Practical implementation of adaptive optics in multiphoton microscopy," Opt. Express 11(10), 1123–1130 (2003).

[44] L. Sherman, J. Y. Ye, O. Albert, and T. B. Norris, "Adaptive correction of depth-induced aberrations in multiphoton scanning microscopy using a deformable mirror," J. Microsc. 206, 65–71 (2002).

[45] S. Zommer, E. N. Ribak, S. G. Lipson, and J. Adler, "Simulated annealing in ocular adaptive optics," Opt. Lett. 31(7), 939–941 (2006).

[46] N. Ji, D. E. Milkie, and E. Betzig, "Adaptive optics via pupil segmentation for high-resolution imaging in biological tissues," Nat. Methods 7(2), 141–147 (2009).

[47] P. Xiao, M. Fink, and A. C. Boccara, "Full-field spatially incoherent illumination interferometry: a spatial resolution almost insensitive to aberrations," Opt. Lett. 41(17), 3920 (2016).

[48] P. Xiao, M. Fink, and A. C. Boccara, "Adapive optics full-field optical coherence tomography," J. Biomed. Opt. 21(12), 121505 (2016).

[49] A. Dubois, G. Moneron, and A. C. Boccara. "Thermal-light full-field optical coherence tomography in the 1.2 μm wavelength region," Opt. Commun. 266(2), 738–743 (2006).

[50] S. Labiau, G. David, S. Gigan, and A. C. Boccara, "Defocus test and defocus correction in full-field optical coherence tomography," Opt. Lett. 34(10), 1576–1578 (2009).

[51] M. Born and E. Wolf, "Principles of optics: electromagnetic theory of propagation, interference and diffraction of light," CUP Archive (2000).

[52] J. W. Goodman, *Introduction to Fourier Optics*, New York: Roberts and Company (2005).

[53] C. W. McCutchen, "Generalized source and the van Cittert–Zernike theorem: A study of the spatial coherence required for interferometry," J. Opt. Soc. Am. 56, 727–733 (1966).

[54] LLTech SAS, France, www.lltechimaging.com/.

[55] R. J. Noll, "Zernike polynomials and atmospheric turbulence," JOsA 66(3), 207–211 (1976).

[56] G. Dai, *Wavefront Optics for Vision Correction*, New York: SPIE Press (2008).

[57] S. Bonora and R. J. Zawadzki, "Wavefront sensorless modal deformable mirror correction in adaptive optics: optical coherence tomography," Opt. Lett. 38(22), 4801–4804 (2013).

[58] D. Debarre, M. J. Booth, and T. Wilson, "Image based adaptive optics through optimization of low spatial frequencies," Opt. Express 15(13), 8176–8190 (2007).

[59] J. Mertz, H. Paudel, and T. G. Bifano, "Field of view advantage of conjugate adaptive optics in microscopy applications," Appl. Opt. 54(11), 3498–3506 (2015)

[60] C. K. Hitzenberger, A. Baumgartner, W. Drexler, and A. F. Fercher, "Dispersion effects in partial coherence interferometry: implications for intraocular ranging," J. Biomed. Opt. 4(1), 144–151 (1999).

[61] OWE Technical Design Inc., www.rowetechnical.com/tpme.html.

17 Computational Adaptive Optics for Broadband Optical Interferometric Tomography of Biological Tissue

Nathan D. Shemonski, Yuan-Zhi Liu, Fredrick A. South, and Stephen A. Boppart

17.1 Introduction

In addition to the many hardware methods for wavefront shaping described in the previous chapters, it is also possible to modify the wavefront computationally. Using interferometric detection, the complex optical wavefront can be measured. The phase of the wavefront can then be adjusted in software in a method analogous to that of a deformable mirror. This method is termed digital adaptive optics or computational adaptive optics (CAO). This is distinct from a previously published method of the same name which used ray tracing to guide amplitude deconvolution of 3-D data sets [1]

Using concepts similar to those initially developed for holography [2–4], interferometric data can be used to reconstruct images of tissue samples without the traditional depth-of-field limitation. Among these solutions are numerical refocusing [5–7], depth-encoded synthetic aperture microscopy [8, 9], holoscopy [10], and interferometric synthetic aperture microscopy [11]. Higher-order aberrations can also be corrected computationally. For thin samples, this was achieved using digital holographic microscopy [12, 13]. Recently, thick tissue samples have been imaged using computational, or digital, adaptive optics [14, 15].

Computational aberration correction methods draw inspiration from hardware adaptive optics (HAO). In HAO, a deformable mirror is used to physically modify the wavefront, while CAO modifies the wavefront by multiplication of the data with a phase filter in the spatial-frequency domain. To determine the appropriate correction hardware methods use image metrics (Chapter 1), a wavefront sensor (Chapter 2), or guide stars (Chapter 3). Methods analogous to each of these can be performed in software using the complex interferometric data [14–16]. This parallelism with existing methods makes CAO applicable in many cases where HAO has been used, without the need for wavefront shaping hardware. This chapter will address key concepts in the theory and implementation of CAO, as well as demonstrate its application for *in vivo* biomedical imaging.

17.2 Theory

This section provides a summary of the theoretical derivation of CAO. This is modeled for a general OCT imaging geometry. Interferometric imaging is briefly discussed,

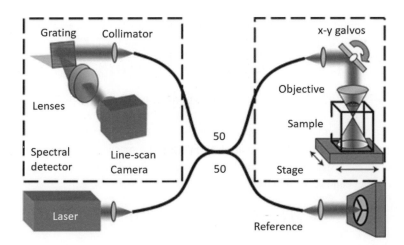

Figure 17.1 Schematic of a spectral domain OCT system. Light in one arm of a fiber-based Michelson interferometer is focused into the sample, while the other arm is used as a reference. The spectral interference is measured with a spectrometer. Adapted by permission from Macmillan Publishers Ltd: Nature Physics [11], copyright 2007.

and then the aberrated forward model and CAO implementation are introduced. A related computational technique, interferometric synthetic aperture microscopy (ISAM), is introduced as a special case for defocus correction. Lastly, we briefly compare CAO and HAO.

17.2.1   Theoretical Model

OCT is an imaging modality based on broadband interferometry which measures the three-dimensional structure of a sample from its scattering radiation. Although OCT can be categorized into different operating mechanisms based on different broadband sources and detection schemes, such as time domain OCT (TD-OCT) and Fourier domain OCT (FD-OCT), the basic principle is the same. Without loss of generality, we begin with the three-dimensional dispersion-corrected intensity signal, which is acquired from FD-OCT (Figure 17.1).

$$
\begin{aligned}
I(x, y; k) &= |E_S(x, y; k) + E_R(k)|^2 \\
&= |E_S(x, y; k)|^2 + |E_R(k)|^2 + 2\mathrm{Re}\left[E_S(x, y; k)\, E_R^*(k)\right],
\end{aligned}
\tag{17.1}
$$

where x and y denote the lateral position, and k is the optical wavenumber. The term $E_S(x, y; k)$ is the field scattered from the imaging sample, $E_R(k)$ is the reference field, and the superscript asterisk (*) indicates complex conjugation. For point-scanning OCT, the lateral information is acquired through two-dimensional transverse scanning of the incident focused beam along the x and y axes. To successfully apply CAO, the lateral scanning step (sampling step) should be smaller than the theoretical lateral resolution of the system and satisfy the Nyquist sampling requirement. For full-field OCT or holoscopy, the transverse spatial information is directly measured by the two-dimensional camera (see Chapter 16).

The complex interferometric cross-correlation signal, which contains the object field information we are interested in, can be extracted from Eq. (17.1) and denoted by

$$S(x, y; k) = E_S(x, y; k) E_R^*(k).$$
(17.2)

Under the first Born approximation, the sample arm signal can be written as a convolution of the (complex) system PSF of the microscope, $h(x, y, z; k)$, with the sample scattering potential $\eta(x, y, z)$ [17, 18], where the focus is fixed at the location $z = 0$. The probing light in the sample arm shares the same spectrum as the reference arm with different intensity ratio. For simplicity, we ignore the relative ratio here, and the sample arm signal can be expressed as

$$E_S(x, y; k) = \iiint E_R(k) h(x - x', y - y', z'; k) \eta(x', y', z') \, dx' dy' dz'.$$
(17.3)

The measured signal can then be derived as

$$S(x, y; k) = |E_R(k)|^2 \iiint h(x - x', y - y', z'; k) \eta(x', y', z') \, dx' dy' dz'.$$
(17.4)

By invoking the convolution theorem, Equation (17.4) can be written in the transverse spatial-frequency domain as

$$\tilde{S}(Q_x, Q_y; k) = \int H(Q_x, Q_y, z'; k) \, \tilde{\eta}(Q_x, Q_y, z') \, dz',$$
(17.5)

where the tilde (˜) represents the 2D transverse Fourier transform, Q denotes spatial frequency, and $H(Q_x, Q_y, z; k)$ is the (depth-dependent) band-pass response of the effective PSF. Note that the source spectrum $|E_R(k)|^2$ has been absorbed into the system transfer function for simplicity.

Under asymptotic approximations for both the near-focus and far-from-focus cases [17], Equation (17.5) can be simplified from a convolution to a multiplication, resulting in the forward model

$$\tilde{S}(Q_x, Q_y; k) = H(Q_x, Q_y; k) \, \tilde{\eta}(Q_x, Q_y, Q_z),$$
(17.6)

where (˜) represents the 3D Fourier transform, and the axial spatial frequency is

$$Q_z = -\sqrt{(2k)^2 - Q_x^2 - Q_y^2}.$$
(17.7)

The factor of 2 in the right side of Eq. (17.7) comes from the double-pass geometry. The transfer function of the system $H(Q_x, Q_y; k)$ is related to the generalized pupil function derived in Fourier optics.

Due to imperfections of the imaging optics or the sample itself, aberrations are present in many scenarios where the optical wavefront is distorted and image quality degraded. In the general case, the effect of aberrations is spatially variant, especially in complex biological tissue samples. This coupling between the space and frequency domains leads to an aberrated forward model that is defined in a piecewise manner over the spatial domain. In the region of space where the wavefront error can be considered to be the same or space-invariant, it would be recognized as an isoplanatic patch, or volume of stationarity, which were introduced in astronomical adaptive optics [19]. Within the isoplanatic volume, the same wavefront correction is valid.

Aberrations can be modeled as an additional phase filter applied to the general pupil function of an ideal imaging system. When restricted to one such isoplanatic volume, Eq. (17.5) can be modified to an aberrated forward model as

$$\tilde{S}_A\left(Q_x, Q_y; k\right)\Big|_{V(x_0, y_0, z_0)} = H_A\left(Q_x, Q_y; k\right)\Big|_{V(x_0, y_0, z_0)} H\left(Q_x, Q_y; k\right)\tilde{\eta}\left(Q_x, Q_y, Q_z\right).$$

(17.8)

Here, within the isoplanatic volume $V(x_0, y_0, z_0)$ centered at position $(x_0 y_0, z_0)$, the aberration phase filter can be written as $H_A(Q_x, Q_y; k)\big|_{V(x_0, y_0, z_0)}$. The aberration effects can be corrected through phase conjugation using the inverse filter

$$H_{AC}\left(Q_x, Q_y; k\right)\Big|_{V(x_0, y_0, z_0)} = H_A^*\left(Q_x, Q_y; k\right)\Big|_{V(x_0, y_0, z_0)}$$

(17.9)

which results in the aberration-free signal

$$\tilde{S}_{AC}\left(Q_x, Q_y; k\right)\Big|_{V(x_0, y_0, z_0)} = H_{AC}\left(Q_x, Q_y; k\right)\Big|_{V(x_0, y_0, z_0)} \tilde{S}_A\left(Q_x, Q_y; k\right)\Big|_{V(x_0, y_0, z_0)}.$$

(17.10)

In hardware-based adaptive optics, aberrations are physically corrected by the deformable mirror during the imaging procedure, which means $H_{AC}(Q_x, Q_y; k)\big|_{V(x_0, y_0, z_0)}$ is physically applied to the wavefront. In CAO, however, $H_{AC}(Q_x, Q_y; k)\big|_{V(x_0, y_0, z_0)}$ is digitally applied to the complex signal $\tilde{S}_A\left(Q_x, Q_y; k\right)\big|_{V(x_0, y_0, z_0)}$ which can be done after the imaging procedure and accomplished in postprocessing. Because of the combined effect of the confocal and coherence gates in OCT/OCM, the aberration correction can be carried out layer-by-layer. If the aberrations can be considered invariant over the transverse field of view for each depth layer, the aberration and correction filters can be simplified to $H_A\left(Q_x, Q_y; k\right)\big|_{V(z_0)}$ and $H_{AC}\left(Q_x, Q_y; k\right)\big|_{V(z_0)}$, respectively. For the region close to the focus, the aberration correction filter could provide an image with diffraction-limited resolution. For other depths, the field can be related to the focal region using angular spectrum propagation. In the transverse spatial-frequency domain, this propagation could be approximated as the defocus aberration, and could be included in the term $H_{AC}(Q_x, Q_y; k)\big|_{V(x_0, y_0, z_0)}$. Consequently, CAO has the ability to achieve 3D diffraction-limited resolution by using the OCT volumetric imaging technique.

17.2.2   Implementation of CAO

The aberration correction filter $H_{AC}(Q_x, Q_y; k)\big|_{V(z_0)}$ is related to both the spatial-frequency and spectral domains, so it can correct both monochromatic as well as chromatic aberrations [16]. In an achromatic system, the k-dependence of the 3D aberration correction filter can be simplified to the monochromatic aberration function, $H_{AC}(Q_x, Q_y)\big|_{V(z_0)}$. Figure 17.2 shows a flowchart of the CAO processing steps. Beginning with the complex OCT data set, the two-dimensional Fourier transform of the

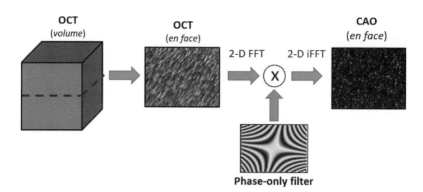

Figure 17.2 A flowchart of 2-D CAO processing from an OCT volume. Beginning with a full tomogram, one *en face* plane is extracted at a time and a 2-D phase filter is applied in the Fourier domain.

en face image field at each depth is calculated. The appropriate aberration correction filter is determined for the image field and then applied in the Fourier domain by complex multiplication at each depth. Afterward, the two-dimensional inverse Fourier transform is performed to retrieve the aberration-free image field. The *en face* image fields are then stacked along depth to obtain the aberration-free volume. It is noted that if the aberrations vary considerably within the 2D field of view at a given depth, one can divide the image field into subregions and implement the processing steps on each subregion. The aberration corrected subregions can then be recombined to retrieve the composite image for each depth [20]. In some special cases where the dominant artifacts are space-invariant along depth, the same aberration correction filter can be applied to the entire 3D volume to speed up the computation.

17.2.3   Determining the Correction Filter

Generally the optical aberrations induced by the imaging sample are unknown. The combination of the sample dependent and system aberrations makes determining the appropriate correction filter a significant challenge. Here we introduce several techniques to estimate the aberrations. One solution is similar to sensorless HAO techniques [21]. In sensorless HAO, a sequence of aberration corrections beginning with an initial guess is imposed on the deformable mirror to modulate the pupil function. The subsequent change of the signal or image quality provides the feedback for optimization. Instead of physically altering the wavefront, CAO digitally applies a sequence of computational aberration correction filters in the spatial-frequency domain of the complex image. The correction filter $H_{\mathrm{AC}}(Q_x, Q_y)\big|_{V(z_0)}$ can be expressed as a modal term as in Zernike polynomial decomposition (Chapter 1), or as a zonal-based wavefront (Chapter 4). Several image metrics such as peak intensity, spatial-frequency content, and image sharpness have been proposed and demonstrated for use in coherent imaging [14, 22, 23]. Figure 17.3 shows an example of a Zernike-polynomial-based correction filter [14]

Amplitude $0.9Z_4\text{-}1.7Z_5\text{-}0.7Z_6\text{+}0.2Z_{11}$ Phase

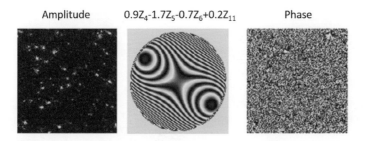

Figure 17.3 Astigmatism, spherical aberration, and defocus correction of TiO_2 phantom data using a 2D aberration correction filter. The filter was applied to the *en face* plane of least confusion. The central plot shows the cumulative pupil phase, resulting from the Zernike polynomial corrections Z5 (astigmatism at $45°$), Z6 (astigmatism at $0°$), Z4 (defocus), and Z11 (spherical aberration). Reproduced with permission from [14], *Proc. Natl. Acad. Sci.*

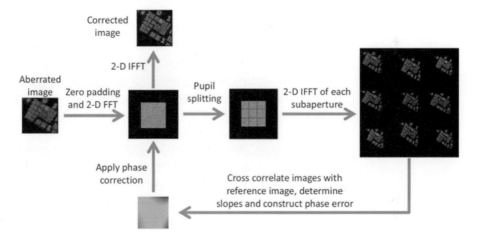

Figure 17.4 Illustration of the subaperture method for estimation of the aberration correction filter. Subapertures in the spatial-frequency domain are compared to the central reference subaperture to determine the slope of the wavefront. Reprinted with permission from [15], *Optics Express.*

A technique termed subaperture correlation, which mimics the operation of a Shack–Hartmann wavefront sensor for measuring extended sources [24], is proposed by A. Kumar [15] to digitally correct the aberrations in scattering OCT data. Just as the Shack-Harman wavefront sensor divides the pupil plane into subregions, the subaperture correlation technique divides the Fourier plane of each *en face* complex image into different subapertures. The wavefront slopes of each subaperture are calculated from the cross-correlation of the reconstructed subaperture images to the center subaperture reference image, as illustrated in Figure 17.4. This method has shown image improvement for scattering samples that have a uniform Fourier spectrum, although only low-order aberrations can be corrected.

Originally developed for astronomy, the so-called guide star (GS) method has also been proposed in CAO [16]. For a sample containing pointlike structures, the guide

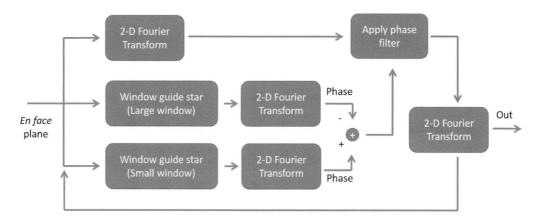

Figure 17.5 A flowchart of the 2D GS-CAO processing steps. Beginning with a single *en face* plane, the full extent of a guide star is windowed in addition to a small window which represents the targeted size and shape. An iterative approach corrects for high-order aberrations.

star method can be used to correct not only the high-order monochromatic but also the chromatic aberrations. The GS-CAO technique can be implemented in 3D coordinates as described in Ref. [16]. For a simplified solution, a flowchart of the GS-CAO processing steps for each 2D *en face* plane is shown in Figure 17.5. One begins with an *en face* plane and windows two regions around the candidate guide star. A large spatial window is used to capture the entire aberrated PSF of the guide star, and a small window is used to crop the guide star to the desired target PSF. By isolating the signal from a guide star, the aberrations can be sensed from the phase difference of the measured and desired wavefront in the Fourier domain. The measured aberration filter can then be digitally conjugated in the Fourier domain and applied to the whole image. Finally, after the inverse Fourier transform, the aberration-free data can be recovered.

17.2.4   ISAM as a Special Case

In scenarios where aberration effects are not significant, a more computationally efficient technique, termed interferometric synthetic aperture microscopy (ISAM), can be used to correct the defocus throughout the whole 3D volume of the OCT data simultaneously. Recall Equations (17.6) and (17.7), reproduced here

$$\tilde{S}\left(Q_x, Q_y; k\right) = H\left(Q_x, Q_y; k\right) \tilde{\tilde{\eta}}'\left(Q_x, Q_y, Q_z\right),$$
$$Q_z = -\sqrt{(2k)^2 - Q_x^2 - Q_y^2}. \tag{17.11}$$

This shows that the OCT signal is a filtered version of the 3D Fourier transform of the object scattering potential. However, note that the signal is acquired as a function of wavenumber, not as a function of axial spatial frequency.

The scattering potential can be recovered by resampling the OCT signal to be a function of axial spatial frequency. The transfer function H is generally smooth and can be ignored without significant degradation of image quality [25, 26]. Therefore, it is sensible to compute the unfiltered solution of the scattering potential in the frequency domain,

$$\tilde{\tilde{\eta}}^{+}(Q_x, Q_y, Q_z) = \tilde{S}(Q_x, Q_y; k)\Big|_{k=\frac{1}{2}\sqrt{Q_x^2+Q_y^2+Q_z^2}}, \tag{17.12}$$

which can be returned to the spatial domain by the 3D inverse Fourier transform

$$\eta^{+}(x, y, z) = F^{-1}\left\{\tilde{\tilde{\eta}}^{+}(Q_x, Q_y, Q_z)\right\}. \tag{17.13}$$

In practice, the range of transverse spatial frequencies Q_x and Q_y is determined by the measurement step size along the transverse dimension according to the Fourier transform relationship. The range of measured axial spatial frequencies is well approximated by the range of measured wavenumbers, $Q_{z,\min} = -2k_{\min}$ and $Q_{z,\max} = -2k_{\max}$. The resampling indices required by Equation (17.12) can then be readily calculated.

It can be seen that ISAM is based on the fact that defocus in the spatial domain manifests as a coordinate warping in the Fourier domain of the signal, which is also known as the Stolt mapping that was originally developed in the field of geophysics [27]. This result, governed by the physics of the data acquisition, relates the 3D Fourier transform of the complex OCT tomogram to the 3D Fourier transform of the object structure through the Stolt mapping. Reconstruction via ISAM resampling corrects defocus by restoring constructive interference across the transverse bandwidth for all depths simultaneously.

17.3   Stability

One of the greatest challenges in computational techniques such as CAO is the requirement of stability. For hardware solutions such as sophisticated optical designs, Bessel beams, or HAO, the benefits of each technique are still achieved even under large amounts of motion. For CAO, the acquired phase is used to reconstruct the amplitude image. Therefore a stable phase relationship along both the fast- and slow-scanning axes is required. Acquiring such data can be challenging and has required special attention.

First, consider why a stable phase relationship can be difficult to achieve. If a scatterer moves up or down by a distance equal to $\lambda/4$, then the phase of the associated sample field will shift by π (note the factor of 2 due to the double-pass nature of the measurement). If this were to occur multiple times while collecting light from that scatterer, the phase relationship between the axial scans would be completely lost. This is depicted in Figure 17.6. Here we show *en face* simulations of a small, pointlike scattering particle which suffers from defocus (Figure 17.6a). Under stable conditions, the phase of this image is as shown in Figure 17.6b. The corresponding CAO reconstruction is shown in Figure 17.6d which successfully recovered the pointlike scatterer. Figure 17.6c shows an unstable configuration where small motion on the order of the wavelength was added to the image. This motion would have been too small to see in an amplitude-only cross section. The same CAO reconstruction was applied to Figure 17.6c resulting in Figure 17.6e. One can see how the pointlike scatterer could not be recovered even under motion on the order of the wavelength of light.

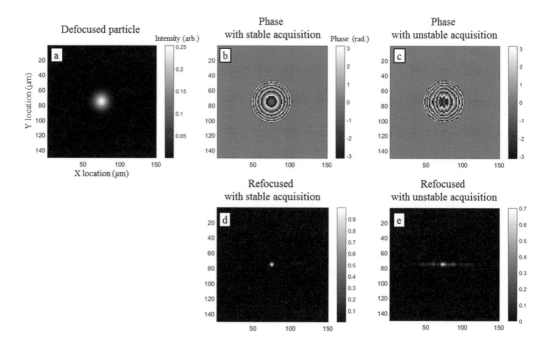

Figure 17.6 Effect of axial motion on digital refocusing. (**a**) An *en face* slice through a simulated defocused particle. (**b**) The associated phase of the defocused particle under stable acquisition conditions. (**c**) The phase of the same defocused particle with small wavelength-scale axial motion. (**d**) *En face* plane after digitally refocusing under stable conditions. (**d**) *En face* plane after digitally refocusing under unstable conditions.

In this section, we will discuss different sources of instability in addition to work which has overcome these instabilities in a variety of configurations. It should become clear that although stability is a large challenge, it does not restrict CAO from a variety of applications.

17.3.1   System Stability: Reference Arm, Galvanometer Scanners, Light Source

When designing a system for CAO, it is important to realize that different components can introduce instabilities and influence the quality of the final processed image, as evident in Figure 17.6. For slowly scanned systems, the reference arm can play a large factor in stability. This is because any variations between the sample and reference paths in an interferometric setup will affect the phase. This can be seen in Figure 17.7 where Figure 17.7a shows a cross-sectional image of pointlike scatterers which are in focus near the bottom and out of focus near the top. Due to the slow acquisition time (5 FPS), an attempted refocusing did not recover the desired pointlike scatterers away from focus (Figure 17.7b). Figure 17.7c shows a successfully refocused image achieved using phase stabilization, to be discussed in a later section.

Another source of instability is the scanning device. Any imperfections such as jitter or nonrepeatable movements can introduce instabilities. In a traditional ophthalmic

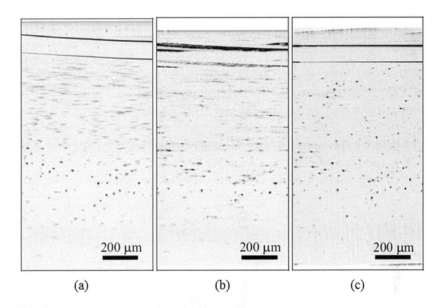

(a)   (b)   (c)

Figure 17.7 Experimental data showing cross-sections of pointlike scatterers. (**a**) Original cross section showing in-focus particles just below the middle and unfocused particles near the top of the image. (**b**) Attempted refocusing under unstable conditions (reference arm fluctuations). (**c**) Refocusing after correcting (stabilizing) the phase. © 2006 IEEE. Reprinted, with permission, from [28].

or microscope-based OCT system, the scanning device is often a pair of galvanometer scanners which provide controlled and repeatable scans. Although jitter and resonance peaks of the galvanometer scanners can cause nonlinear motion and can be modeled as phase instabilities [29], these are often not dominant sources of instability. For catheter-based systems, a rotating fiber is the scanning mechanism which does not allow for repeatable or stable scanning [30]. Although the theory for refocusing in rotational/catheter-based systems has been developed [18], due to these scanning instabilities in addition to a typically low NA, experimental CAO has not yet been performed.

Finally, the source can introduce a certain level of instability for swept-source OCT systems. This occurs when the acquired spectral sweeps vary either from imprecise triggering or other deformations of the wavelength versus time [31, 32].

17.3.2   Phase Stabilization

Although for modern spectral domain and swept-source OCT systems it is often not necessary to correct for instabilities of the system (due to the high scanning speeds and few moving parts), there may still be times that such a situation arises as described above. As such there have been solutions which aim to track and correct the phase disturbances using simple hardware changes.

Ralston *et al.* [28] initially achieved this by including a coverslip on top of the sample of interest. Any fluctuations of the system would then be measured as phase fluctuations along the smooth air-glass or glass-tissue interface. This essentially converted the system into a common-path OCT system [33]. The results of this technique are shown in Figure 17.7 where on the far right, clear pointlike particles are recovered even far from focus. Sun *et al.* [30] also found that it was possible to perform Doppler OCT using a fiber-based endoscope by using the sheath of the probe as a phase reference. By adding additional phase shifters in the sample and reference arms, Moiseev *et al.* [34] found that the phase instabilities could be measured and removed with heterodyne detection. Finally, to correct for swept-source instabilities, the addition of a phase reference [31] or a Mach-Zehnder interferometer (MZI) [32] can achieve phase stability.

17.4   *In Vivo* Imaging

Although much work has been performed on static or *ex vivo* samples [11, 15, 35, 36], these will likely not be the ideal application of CAO. This is because there are often fewer restrictions on the size and price of the system when imaging static prepared samples. The preferred application is likely to be *in vivo* human imaging. To succeed at *in vivo* imaging, one must consider the stability of not only the system itself, but also how it interacts with the sample (due to inevitable sample motion) to ensure a stable data acquisition [25, 37–39]

Initially, it was desired to show that *in vivo* digital refocusing was possible without developing any motion correction techniques. This was first shown by Hillmann *et al.* [40] with a full-field Holoscopy setup (Figure 17.8). Although the various refocusing steps were not explicitly shown in this work, since the data were collected in a nonimaging plane, digital refocusing was required to obtain the final image. Any other digital

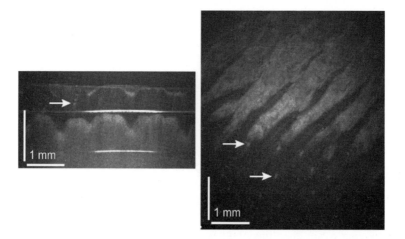

Figure 17.8 Cross section (left) and *en face* (right) images of a human finger acquired with a Holoscopic imaging microscope. Reprinted with permission from [40], *Optics Letters.*

techniques, such as aberration correction, would have also been possible. Later, Ahmad, Shemonski *et al.* [25] showed that with proper mounting practices, stable *in vivo* data could be achieved with a scanned SD-OCT system. It was further found that by viewing the refocused data in real time, repeatable acquisitions were possible.

17.4.1   Motion Correction

Some previous *in vivo* techniques [34] required physical contact with the sample of interest. Although possible for some applications such as skin imaging, other imaging scenarios such as retinal imaging do not easily allow for direct contact. From here, it thus became important to address the challenge of *in vivo* sample motion.

The techniques for motion correction in the context of CAO can be almost directly borrowed from Doppler OCT [41] and SS-OCT [37]. The main philosophy behind these techniques is to look at phase differences between neighboring acquisitions – be it different A-scans or different spectral values. If adequately oversampled, the phase differences should remain closely packed around zero. Thus by applying various filtering techniques, the desired slowly varying phase differences associated with the sample structure can be isolated from the larger higher-frequency phase differences association with motion.

In vivo motion correction to computationally improve transverse resolution was first demonstrated by Shemonski *et al.* [42]. One experimental result is shown in Figure 17.9 (see original publication for colored images). The sample being imaged was a human finger and, due to the technique of mounting, the resulting motion was restricted to the axial dimension. In the top right corner of Figure 17.9, refocused *en face* sections through a single sweat duct are shown with and without motion correction. With motion correction, the crescent shape of the cross section is clearly visible. The dynamic range of the measured motion was approximately 50 radians which at central wavelength of 1.3 μm (and taking into account the double-pass configuration) corresponds to approximately 5 μm of motion along the axial dimension. As the axial resolution of the system was only 6 μm, it was a safe assumption to say that only phase correction was necessary.

The above phase correction technique followed a previous analysis [43] which discussed how stable a data set must be to perform CAO. To ensure the phase correction was robust to different types of samples, a predefined amount of oversampling was necessary. Using the statistical model developed [29], it was shown in [42] that if the spatial sampling is on the same order as the diffraction-limited resolution of the system, then adequate phase motion correction could be performed.

Figure 17.10 shows another example where phase-based motion correction (using the same technique as described in [42]) is applied to data from a handheld OCT probe. After refocusing using ISAM, two locations of localized motion (indicated by white arrows on the left) resulted in horizontal stripes through the image. This is because the fast-scanning axis was left to right in this acquisition. After phase-based motion correction (right), the stripes were not visible. On the far left edge of Figure 17.10, the two instances of localized motion were visible in the motion measured by the phase-based analysis.

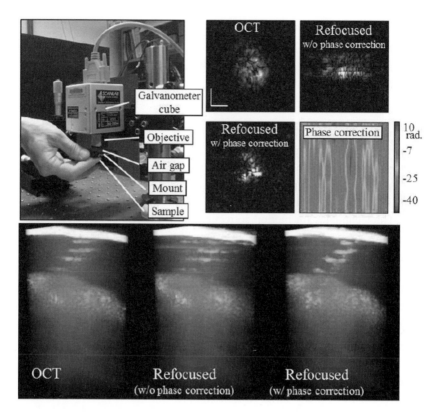

Figure 17.9 *In vivo* digital refocusing with and without motion (phase) correction. In the top left, the imaging configuration is shown. In the top right, sample *en face* slices with and without motion correction are shown. On the bottom, the full 3-D data sets are displayed. Figure adapted from Shemonski *et al.* Adapted with permission from [42], Optics Express.

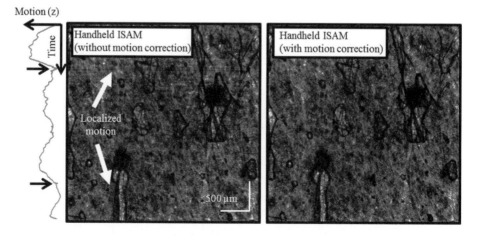

Figure 17.10 Motion correction and refocusing using a handheld probe to image *in vivo* skin. On left, horizontal stripes are results of digital refocusing with small amounts of axial motion. On right, these artifacts are avoided by correcting the phase before digital refocusing.

Finally, we note that some work was focused on correcting both the axial (phase) motion and transverse motion [42, 44]. In particular, the speckle found in OCT [45] or laser-based images can provide high-contrast structures suitable for image registration even when the image is aberrated or out of focus. This approach has been successful for *in vivo* imaging and may be crucial for certain applications.

17.4.2   Retinal Imaging

As often mentioned in the digital refocusing or aberration correction literature [15, 46, 47], a key application of CAO is for retinal imaging. This is because imperfections of the human eye limit the achievable transverse resolution of retinal imaging [48–50]. Although using techniques such as HAO to overcome these imperfections is effective, the additional cost and complexity of HAO has provided a challenging roadblock for widespread commercial adoption. Digital refocusing or aberration correction can supplement the need for complex correction hardware possibly providing a faster route to commercial designs and clinical implementation.

Unlike imaging skin, imaging the human eye poses more challenging motion to overcome. Contact with the eye is uncomfortable for subjects, and other techniques such as using the cornea as a phase reference [36] may prove for difficult alignment procedures. Therefore, motion correction techniques will likely play an important role in the success of digital refocusing and aberration correction. Similar to the progression of *in vivo* skin imaging discussed previously, it is desirable to correct for aberrations of the eye without any significant motion correction.

The approach is thus to image as fast as possible to overcome the natural movement of the eye, but still collect enough photons to provide sufficient SNR. Techniques such as line-field SS-OCT [51], full-field SS-OCT [35, 52], and *en face* OCT [53] all provide high-speed retinal imaging. The first published demonstrations of CAO in the eye was Shemonski *et al.* [39, 54] where an *en face* OCT system was used at 0.4 MHz point-scanning rate (4 kHz *en face* line rate). Other works such as [55–58] previously showed minor improvements using image filters based solely on the amplitude of scattered light. These works, though, were not capable of overcoming the main aberrations of the eye and were thus not fully successful at high-resolution retinal imaging. Also, Fechtig *et al.* [51] showed digital refocusing in the retina using the amplitude and phase information. Unfortunately, these results were only applied along the fast axis. As such, scanning multiple orders of magnitude slower was possible [43] and therefore did not address the main challenge of *in vivo* retinal refocusing.

Figure 17.11 shows a set of images near the fovea of a healthy human volunteer. In Figure 17.11b, the original *en face* OCT image mosaic is shown. Few distinguishable features are visible. After digital aberration correction (Figure 17.11c), cone photoreceptors are visible throughout the full field-of-view. A different aberration filter was used for each acquisition in the mosaic. Although an *en face* line rate of 4 kHz was used for this imaging system, phase correction along the slow-scanning axis was still required.

Finally, for the sake of capturing an accurate historical timeline of this body of work, although a proceeding was not published, Hillmann *et al.* [59] presented work at the

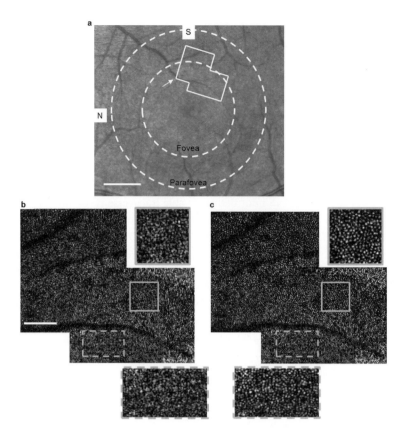

Figure 17.11 Digital refocusing images of the living human retina. (**a**) A fundus image showing the location of the acquired *en face* OCT data. (**b**) Original *en face* OCT data. (**c**) *En face* OCT data after digital refocusing and aberration correction. N, nasal; S, superior. Scale bars represent 2 degrees in (**a**) and 0.5 degrees in (**b, c**). Adapted by permission from Macmillan Publishers Ltd: Nature Photonics [39], copyright 2015.

same time as Shemonski *et al.* [54] demonstrating computational aberration correction in the living retina far from the fovea using a full-field SS-OCT system.

17.5   Opportunities and Challenges in CAO

CAO provides new opportunities for high-resolution imaging in ophthalmology and microscopy. In addition to lower cost and simpler system configurations in comparison with HAO, the postprocessing nature of CAO also allows more flexible aberration sensing and correction.

Sharing the same advantages as sensorless HAO, the magnitude and accuracy of the measured wavefront aberrations in CAO are not limited by the wavefront sensor design. Hence, the usual trade-off of a Shack–Hartmann wavefront sensor between the sampling rate (the accuracy of the wavefront estimation) and the dynamic range (the maximum aberration that can be measured) can be avoided. The aberration estimation is also immune to the noncommon path errors, which are caused by the different paths between the wavefront sensor plane and the imaging plane. As in sensorless HAO, CAO can strongly suppress the measurement errors caused by the specular reflection of optical components or the cornea of the eye. Furthermore, the innate coherence gate of CAO has the ability to reject the out-of-focus scattering light and achieve a depth-resolved aberration estimation, which is particularly important in microscopy.

The separation of image acquisition and aberration correction brings additional advantages. Considering spatially variant aberrations as an example, HAO needs to sense the aberrations in each small isoplanatic volume and then execute each correction while imaging the subject. For sensorless HAO in particular, multiple iterations need to be implemented to obtain the best correction, which further prolongs the imaging procedure. By using CAO, the image acquisition time can be shortened, and varying aberration corrections can simultaneously be applied to each isoplanatic volume afterward. Released from the physical limitations of the deformable mirror, e.g., the number and the stroke range of the actuators, CAO has the potential to correct higher-order and larger-amplitude aberrations. Because CAO manipulates the complex interference signal, in principle, better image quality can be realized by applying not only phase corrections, but also amplitude corrections.

Despite the above advantages, CAO is limited by its nature as an interferometric-imaging-based technique. In contrast, HAO has been demonstrated as a powerful imaging technique for both interferometric imaging (e.g., OCT) and nonlinear optical imaging. Additionally, because HAO has the ability to concentrate the illumination energy on the region of interest, a high signal-to-noise ratio can be achieved in aberrated samples, which is especially important for deep tissue imaging. A different SNR advantage may be achieved with CAO by using an aberrated beam to spread the illumination energy throughout depth. Following removal of the aberrations in postprocessing, high SNR could be achieved throughout the imaging volume in contrast to a single focal plane in standard imaging methods.

The greatest challenge in bringing CAO to widespread use is the phase stability requirement. However, this has recently been largely overcome through high-speed scanning and motion correction techniques. CAO is now available as an alternative wavefront correction tool in scenarios where traditional hardware solutions are not suitable due to prohibitive cost or complexity.

17.6   References

[1] Z. Kam et al., "Computational adaptive optics for live three-dimensional biological imaging," Proc. Natl. Acad. Sci. USA **98**, 3790–3795 (2001).
[2] A. F. Fercher et al., "Image formation by inversion of scattered field data: experiments and computational simulation," Appl. Opt. **18**, 2427–2439 (1979).
[3] P.-C. Ho et al., "Structural measurement by inverse scattering in the first Born approximation," Appl. Opt. **15**, 313–314 (1976).

[4] W. H. Carter, "Computational reconstruction of scattering objects from holograms," J. Opt. Soc. Am. **60**, 306–314 (1970).

[5] Y. Yasuno et al., "Non-iterative numerical method for laterally superresolving Fourier domain optical coherence tomography," Opt. Express **14**, 1006–1020 (2006).

[6] L. Yu et al., "Improved lateral resolution in optical coherence tomography by digital focusing using two-dimensional numerical diffraction method," Opt. Express **15**, 7634–7641 (2007).

[7] A. Kumar et al., "Numerical focusing methods for full field OCT: a comparison based on a common signal model," Opt. Express **22**, 16061–16078 (2014).

[8] J. Mo et al., "Focus-extension by depth-encoded synthetic aperture in optical coherence tomography," Opt. Express **21**, 10048–10061 (2013).

[9] J. Mo et al., "Depth-encoded synthetic aperture optical coherence tomography of biological tissues with extended focal depth," Opt. Express **23**, 4935–4945 (2015).

[10] D. Hillmann et al., "Efficient holoscopy image reconstruction," Opt. Express **20**, 21247–21263 (2012).

[11] T. S. Ralston et al., "Interferometric synthetic aperture microscopy," Nat. Phys. **3**, 129–134 (2007).

[12] T. Colomb et al., "Numerical parametric lens for shifting, magnification, and complete aberration compensation in digital holographic microscopy," J. Opt. Soc. Am. A **23**, 3177–3190 (2006).

[13] L. Miccio et al., "Direct full compensation of the aberrations in quantitative phase microscopy of thin objects by a single digital hologram," Appl. Phys. Lett. **90**, 041104 (2007).

[14] S. G. Adie et al., "Computational adaptive optics for broadband optical interferometric tomography of biological tissue," Proc. Natl. Acad. Sci. USA **109**, 7175–7180 (2012).

[15] A. Kumar et al., "Subaperture correlation based digital adaptive optics for full field optical coherence tomography," Opt. Express **21**, 10850–10866 (2013).

[16] S. G. Adie et al., "Guide-star-based computational adaptive optics for broadband interferometric tomography," Appl. Phys. Lett. **101**, 221117 (2012).

[17] B. J. Davis et al., "Nonparaxial vector-field modeling of optical coherence tomography and interferometric synthetic aperture microscopy," J. Opt. Soc. Am. A **24**, 2527–2542 (2007).

[18] D. L. Marks et al., "Inverse scattering for rotationally scanned optical coherence tomography," J. Opt. Soc. Am. A **23**, 2433–2439 (2006).

[19] E. Marchetti et al., "On-sky testing of the multi-conjugate adaptive optics demonstrator," The Messenger **129**, 8–13 (2007).

[20] A. Kumar et al., "Anisotropic aberration correction using region of interest based digital adaptive optics in Fourier domain OCT," Biomed. Opt. Express **6**, 1124–1134 (2015).

[21] H. Hofer et al., "Wavefront sensorless adaptive optics ophthalmoscopy in the human eye," Opt. Express **19**, 14160–14171 (2011).

[22] Y.-Z. Liu et al., "Computed optical interferometric tomography for high-speed volumetric cellular imaging," Biomed. Opt. Express **5**, 2988–3000 (2014).

[23] J. R. Fienup et al., "Aberration correction by maximizing generalized sharpness metrics," J. Opt. Soc. Am. A **20**, 609–620 (2003).

[24] T. R. Rimmele et al., "Solar adaptive optics at the National Solar Observatory," in Proc. SPIE 3353, **72** (1998).

[25] A. Ahmad et al., "Real-time *in vivo* computed optical interferometric tomography," Nat. Photon. **7**, 444–448 (2013).

[26] T. S. Ralston et al., "Real-time interferometric synthetic aperture microscopy," Opt. Express **16**, 2555–2569 (2008).

[27] B. J. Davis et al., "Interferometric synthetic aperture microscopy: computed imaging for scanned coherent microscopy," Sensors **8**, 3903–3931 (2008).

[28] T. S. Ralston et al., "Phase stability technique for inverse scattering in optical coherence tomography," in *Proc. of the 3rd IEEE Int. Symp. on Biomedical Imaging*: *Nano to Macro* (IEEE, 2006), pp. 578–581.

[29] B. J. Vakoc et al., "Statistical properties of phase-decorrelation in phase-resolved Doppler optical coherence tomography," IEEE Trans. Med. Imag. **28**, 814–821 (2009).

[30] C. Sun et al., "*In vivo* feasibility of endovascular Doppler optical coherence tomography," Biomed. Opt. Express **3**, 2600–2610 (2012).

[31] B. Vakoc et al., "Phase-resolved optical frequency domain imaging," Opt. Express **13**, 5483–5493 (2005).

[32] B. Braaf et al., "Phase-stabilized optical frequency domain imaging at 1-μm for the measurement of blood flow in the human choroid," Opt. Express **19**, 20886–20903 (2011).

[33] A. B. Vakhtin et al., "Common-path interferometer for frequency-domain optical coherence tomography," Appl. Opt. **42**, 6953–6958 (2003).

[34] A. A. Moiseev et al., "Digital refocusing for transverse resolution improvement in optical coherence tomography," Laser Phys. Lett. **9**, 826 (2012).

[35] D. Hillmann et al., "Efficient holoscopy image reconstruction," Opt. Express **20**, 21247–21263 (2012).

[36] A. S. G. Singh et al., "In-line reference-delayed digital holography using a low-coherence light source," Opt. Lett. **37**, 2631–2633 (2012).

[37] D. Hillmann et al., "Common approach for compensation of axial motion artifacts in swept-source OCT and dispersion in Fourier-domain OCT," Opt. Express **20**, 6761–6776 (2012).

[38] N. D. Shemonski et al., "Stability in computed optical interferometric tomography (Part II): *In vivo* stability assessment," Opt. Express **22**, 19314–19326 (2014).

[39] N. D. Shemonski et al., "Computational high-resolution optical imaging of the living human retina," Nat. Photon. **9**, 440–443 (2015).

[40] D. Hillmann et al., "Holoscopy – holographic optical coherence tomography," Opt. Lett. **36**, 2390–2392 (2011).

[41] B. R. White et al., "*In vivo* dynamic human retinal blood flow imaging using ultra-high-speed spectral domain optical coherence tomography," Opt. Express **11**, 3490–3497 (2003).

[42] N. D. Shemonski et al., "Three-dimensional motion correction using speckle and phase for *in vivo* computed optical interferometric tomography," Biomed. Opt. Express **5**, 4131–4143 (2014).

[43] N. D. Shemonski et al., "Stability in computed optical interferometric tomography (Part I): Stability requirements," Opt. Express **22**, 19183–19197 (2014).

[44] J. Lee et al., "Motion correction for phase-resolved dynamic optical coherence tomography imaging of rodent cerebral cortex," Opt. Express **19**, 21258–21270 (2011).

[45] J. M. Schmitt et al., "Speckle in optical coherence tomography," J. Biomed. Opt. **4**, 95–105 (1999).

[46] M. K. Kim, "Adaptive optics by incoherent digital holography," Opt. Lett. **37**, 2694–2696 (2012).

[47] Y.-Z. Liu et al., "Computational optical coherence tomography," Biomed. Opt. Express **8**, 1549–1574 (2017).

[48] W. Drexler and J. G. Fujimoto, eds., *Optical Coherence Tomography*: *Technology and Applications* (Springer, 2015).

[49] L. N. Thibos et al., "Statistical variation of aberration structure and image quality in a normal population of healthy eyes," J. Opt. Soc. Am. A **19**, 2329–2348 (2002).

[50] J. Liang et al., "Aberrations and retinal image quality of the normal human eye," J. Opt. Soc. Am. A **14**, 2873–2883 (1997).

[51] D. J. Fechtig et al., "High-speed, digitally refocused retinal imaging with line-field parallel swept source OCT," in Proc. SPIE **9312**, 931203 (2015).

[52] A. Kumar et al., "Numerical focusing methods for full field OCT: a comparison based on a common signal model," Opt. Express **22**, 16061–16078 (2014).

[53] M. Pircher et al., "Simultaneous SLO/OCT imaging of the human retina with axial eye motion correction," Opt. Express **15**, 16922–16932 (2007).

[54] N. D. Shemonski et al., "A computational approach to high-resolution imaging of the living human retina without hardware adaptive optics," in Proc. SPIE **9307**, 930710 (2015).

[55] I. Iglesias et al., "High-resolution retinal images obtained by deconvolution from wave-front sensing," Opt. Lett. **25**, 1804–1806 (2000).

[56] J. C. Christou et al., "Deconvolution of adaptive optics retinal images," J. Opt. Soc. Am. A **21**, 1393–1401 (2004).

[57] N. Meitav et al., "Improving retinal image resolution with iterative weighted shift-and-add," J. Opt. Soc. Am. A **28**, 1395–1402 (2011).

[58] N. Meitav et al., "Estimation of the ocular point spread function by retina modeling," Opt. Lett. **37**, 1466–1468 (2012).

[59] D. Hillmann et al., "Numerical aberration correction in optical coherence tomography and holoscopy by optimization of image quality," presented at SPIE Photonics West, San Francisco, CA, February 2015.

Index